Evolution

A Developmental Approach

"This book, written as an undergraduate text, is a really most impressive book. Given the burgeoning interest in the role of developmental change in evolution in recent times, this will be a very timely publication. The book is well structured and, like the author's other books, very well written. He communicates with a clear, lucid style and has the ability to explain even the more difficult concepts in an accessible manner."

Dr Kenneth McNamara, University of Cambridge

"There is much more to evolution than mere gene frequency changes in natural populations. Wallace Arthur was among the first to recognize fully the lack of a developmental dimension from the traditional view of evolution and is among the main actors who have been shaping the emerging agenda of evolutionary developmental biology.

From his research experience, the author has distilled in this book an original approach to the study of evolution, written in his uniquely attractive style where immediateness successfully mixes with conceptual clarity."

Professor Alessandro Minelli, University of Padova

COMPANION WEBSITE

This book has a companion website:

www.wiley.com/go/arthur/evolution

with Figures and Tables from the book for downloading

Evolution

A Developmental Approach

Wallace Arthur
National University of Ireland, Galway

WILEY-BLACKWELL
A John Wiley & Sons, Ltd., Publication

Blackwell Publishing was acquired by John Wiley & Sons in February 2007. Blackwell's publishing program has been merged with Wiley's global Scientific, Technical and Medical business to form Wiley-Blackwell.

Registered Office
John Wiley & Sons Ltd, The Atrium, Southern Gate, Chichester, West Sussex, PO19 8SQ, UK

Editorial Offices
9600 Garsington Road, Oxford, OX4 2DQ, UK

For details of our global editorial offices, for customer services and for information about how to apply for permission to reuse the copyright material in this book, please see our website at www.wiley.com/wiley-blackwell

Library of Congress Cataloguing-in-Publication Data

Arthur, Wallace.
Evolution : a developmental approach / Wallace Arthur.
p. cm.
Includes bibliographical references and index.
ISBN 978-1-4051-8658-2 (pbk.) – ISBN 978-1-4443-3720-4 (hardcover)
1. Developmental biology. 2. Evolution (Biology). 3. Comparative embryology. I. Title.
QH491.A769 2011
571.8–dc22

2010040517

A catalogue record for this book is available from the British Library.

This book is published in the following formats: eBook [ISBN]; ePub [ISBN]

Set in 10/12pt Photina MT by SPi Publisher Services, Pondicherry, India
Printed and bound in Malaysia by Vivar Printing Sdn Bhd

1 2011

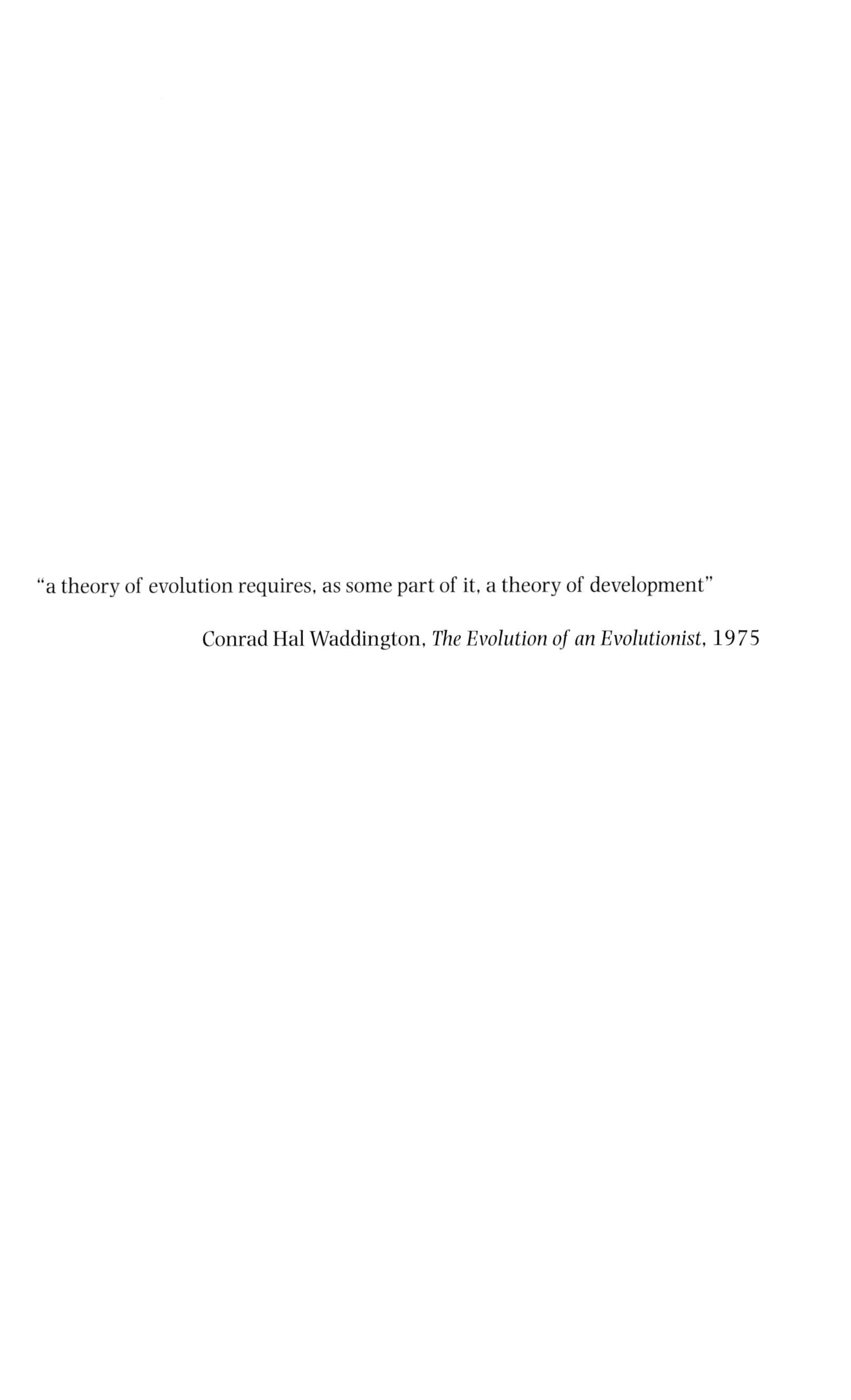

"a theory of evolution requires, as some part of it, a theory of development"

Conrad Hal Waddington, *The Evolution of an Evolutionist*, 1975

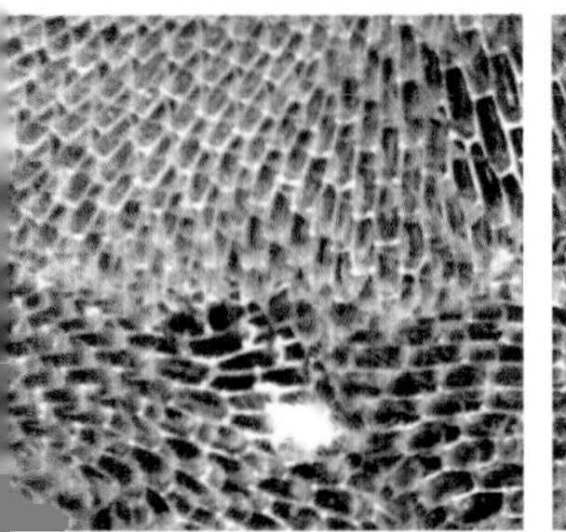
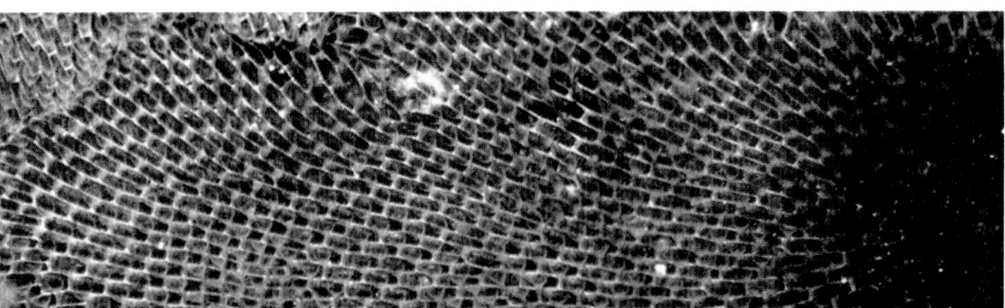

Contents

This book has a companion website: www.wiley.com/go/arthur/evolution

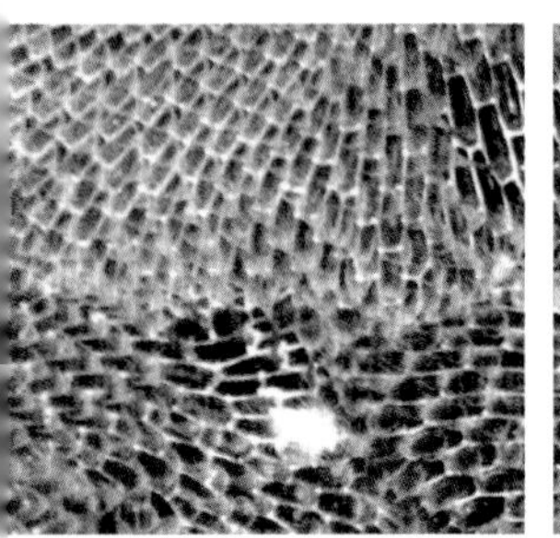
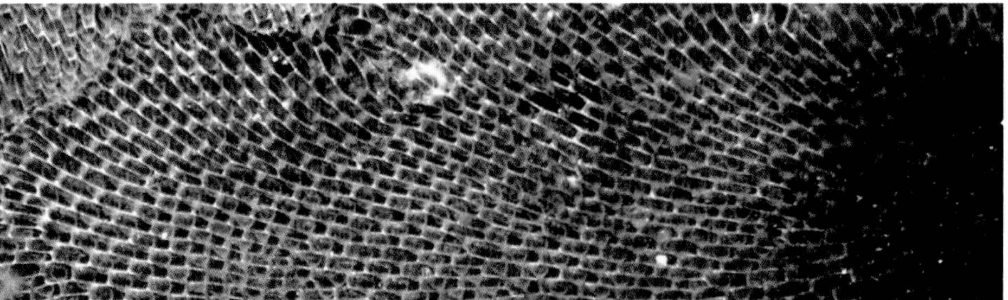

Preface

I have written this book first and foremost for students who are taking a course in evolution at a university or college. Although there are many evolution texts 'out there', there are none that cover the ground in the same way as this one. This book adopts a very specific approach to the evolution of animals and plants – an approach in which the central theme is how evolution works by altering the course of egg-to-adult development.

When I was a student, I bought a book that defined evolution as 'a change in the gene frequency of a population'. While evolution does indeed involve change at the population level, it involves changes at other levels too – most importantly at the level of the individual organism, like you and me. What makes us different from our closest living non-human relatives, chimpanzees, lies in our different structures and behavioural capabilities, as well as in the genetics of our populations.

The overly population-based approach to evolution is now giving way to a more integrative approach – one in which the process of development that turns fertilised eggs into adults is seen as being important to the evolutionary process in a variety of ways. However, it is seen as being important *as well as*, not instead of, changes in gene frequency caused by Darwinian natural selection. This is a crucial point, because some previous developmental approaches to evolution advocated a dismissal of population genetics and a denial that micro-evolutionary changes within species form the basis of most long-term evolution; this denial is now seen to be mistaken.

The recent resurgence of a developmental approach to evolution has several sources, all of them clustered into a short period of a few years around 1980. The most important of these was the discovery that the genetic basis of development in widely different animals is much more similar than was previously thought – a conclusion that was later extended to plants.

This discovery gave birth to the science of evolutionary developmental biology, or 'evo-devo' for short, in which comparisons of the developmental roles of homologous genes among different taxa became the focus of attention. But this is not a book devoted solely to evo-devo. There are some good such books already, and thus no need for another one just now. Rather, this is a book about how evo-devo can be integrated with other approaches to evolutionary biology, giving us a more complete view of evolution than has ever been available before.

I have tried to keep the book short and its level in the early chapters introductory, so that it will be useful to undergraduates taking their first university courses in evolution. However, as is appropriate for educational books in general, the level of discussion rises as the book progresses. Some of the later chapters would thus work well as the subject of seminars and tutorials in more advanced courses. This is particularly true from Chapter 10 onwards.

I should now say a few words about the book's style and structure and, associated with those things, how best to use it.

This book, like other textbooks in science, represents a journey of exploration of a particular field, with the author leading the reader through a sequence of topics. Because of this, I have adopted the 'we' style of speech traditionally used by mathematicians but increasingly favoured in science too. So, instead of 'thus it can be seen that evolution is caused by ... ,' the text reads 'thus we can see that evolution is caused by...' I hope that my chosen style will be perceived as it is intended: friendlier and less formal.

With regard to structure, the book's 20 chapters are grouped into four Parts. At the beginning of each of these, there is a cover-page giving a brief synopsis of what each chapter in the relevant Part deals with. So, while one way to read the book is from start to finish (and of course I would recommend that!), another is to read those four cover-pages first, and to choose a different route. This might be especially useful for readers with enough background in the subject to omit a few chapters, particularly some of the early ones.

At the end of each chapter are a few suggestions for further reading; and at the end of the book there is a list of references. In the text, readers are directed to this reference list (if they wish to be) by superscripts. The rationale behind this split in the pointers to additional reading is that the chapter-end lists are short and usually include only books and review articles, rather than the harder-to-penetrate 'primary literature'. The latter is grouped together at the end. This way, beginning students can ignore the superscripts in the text and the back-of-the-book reference list, and just look at the short chapter-end lists of more reader-friendly sources. More advanced students can do the opposite, or adopt an intermediate approach.

A word of caution on the references: because this is a student text, I have not attempted to give an exhaustive coverage of the literature. To do so, given the breadth of the field, would result in a reference list of perhaps 5000 publications, which would be inappropriate here. Instead, you will find about 250 publications, the aim being to provide just a *sample* of original research papers (and some historical texts) spread over the various topics that are covered.

At the back of the book, you will also find additional sources of information. There is a Glossary, which provides brief definitions of many scientific words that are used in the book. The first time a Glossary word is used in each chapter it is highlighted in bold. There are also four appendices: one that gives a little bit of history; one that deals with gene names, which can be inscrutable to newcomers; one that describes the geological time-scale, which is often used in the book to place major evolutionary events in the appropriate temporal context; and one that gives a very introductory account of how to infer patterns of relationship (evolutionary trees) from comparative data on different species.

Finally, a few words of thanks. I am grateful to all of the following for their help. Alec Panchen, Alessandro Minelli, Ken McNamara, and Patricia Moore kindly read and constructively criticised the entire draft manuscript. My sons Michael and Stephen produced much of the original artwork. I am also very grateful to the many colleagues, authors and publishers who granted permission for the use of artwork that was reproduced (in some cases in modified form) from elsewhere. The editorial staff at Wiley-Blackwell have helped in many ways. I would especially like to thank Ward Cooper for his support throughout the project; also Rosie Hayden, Emily Tye, Delia Sandford, Revathy Kaliyamoorthy and Carla Hodge. Much of the writing was done while on sabbatical leave from my University; I am grateful to the University for granting this leave. I am also grateful to Darwin College Cambridge for giving me a Visiting Associate position during my sabbatical. Finally, I would like to thank the various friends, colleagues and students with whom I have discussed the subjects presented herein over many years: you are too numerous to name, but you know who you are! Discussion, including argument between the proponents of different points of view, is a key element of the progress of science.

Wallace Arthur
September 2010

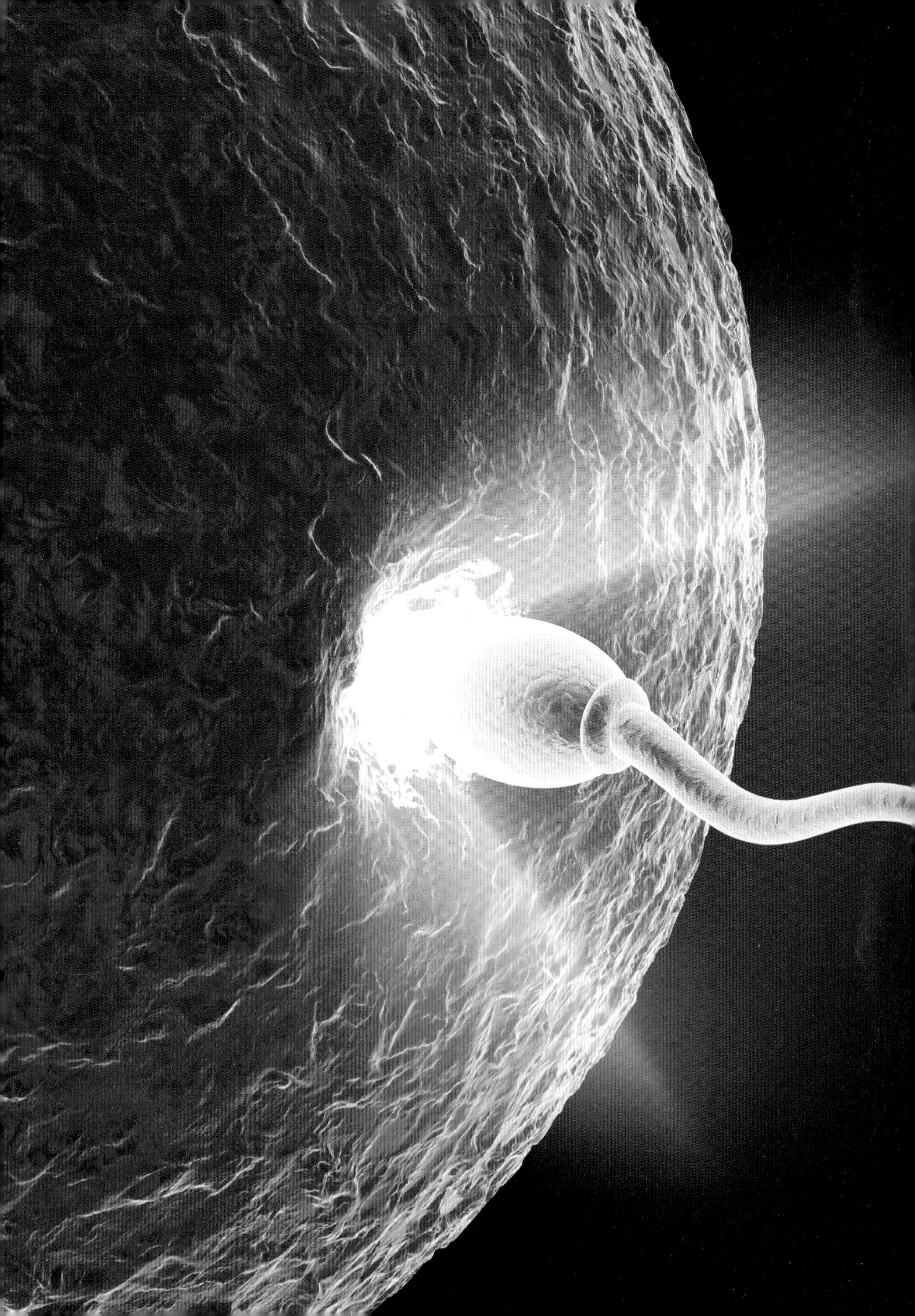

PART I

Foundations

'Science is, I believe, nothing but trained and organised common sense'.

Thomas Henry Huxley ('Darwin's bulldog') *Collected Essays*, vol. 3, 1893

The exciting challenge that this book sets itself is to explain to you what is known about the ways in which the development of animals and plants evolves; and also, in the final chapter, to explain what is not yet known, and thus what are the key questions for future research. We begin to approach this challenge on the foundations provided by the first four chapters, as follows:

Chapter 1: We start by examining the reason why understanding egg-to-adult development is absolutely crucial for a complete picture of the evolution of multicellular creatures.

Chapter 2: This chapter deals with the forerunners of present-day evolutionary developmental biology (or evo-devo), the way in which evo-devo has arisen, and the nature of evo-devo today, including the range of approaches that can be included within it.

Chapter 3: Here, the focus is on the nature of the developmental process, in terms of the things cells do, which are the results of an interplay between genes, their products, an array of other molecules and, often, environmental effects.

Chapter 4: Our focus now shifts to the structure of natural populations, since it is in these that all evolution happens. We examine their spatial, age and genetic structure. We then look at ways in which the genetic structure of populations can be altered by natural selection.

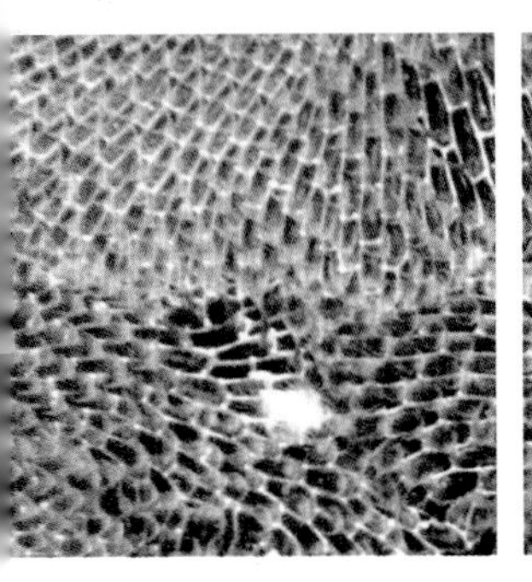

CHAPTER 1
Introduction

1.1 From Darwin to Development

The theory of evolution, established by Charles Darwin more than 150 years ago, and still itself evolving, is one of the most impressive products of science. As Darwin said[1] in the closing paragraph of *The Origin of Species*: 'There is grandeur in this view of life' – a view in which many diverse creatures, both past and present (Fig. 1.1), have been brought into existence by natural processes, and in particular by the interplay between two such processes – heritable variation and natural selection.

Darwin marshalled a wide range of evidence in support of his theory. He drew on information from animal and plant breeding, fossils, behaviour, morphology, embryology and geography, among others. And he used all of these to build a sound basis for his key contribution to evolutionary theory: natural selection.

But evolution is, as noted above, the result of an *interplay* between two things – heritable variation and natural selection – it is not explicable by either of these on its own. The fact that Darwin was unable to enlighten us as much about the former as about the latter was hardly due to an oversight on his part; rather, it was due to limitations on what was generally known at the time in this area of 'heritable variation'.

It is worth dissecting this phrase, because it includes both the inheritance of genetic variation and the process of development through which **phenotypic** variation is produced. Darwin was aware of the problem that there was not, in the 1850s, a clear understanding of how inheritance worked, but proceeded as best he could regardless. He later tried to supply a theory of inheritance – 'pangenesis' – but got it wrong. He was doubtless also aware that there was not a clear understanding of how egg-to-adult development worked, in terms of causal mechanisms, but he proceeded to use

Figure 1.1 A sample of creatures, present and past. All these forms and countless others have been produced from earlier ancestral forms by evolutionary modification of the course of development. This process can in each case be traced back to one of the several origins of multicellularity in the distant evolutionary past. The organisms included here are all discussed as examples later in the book.

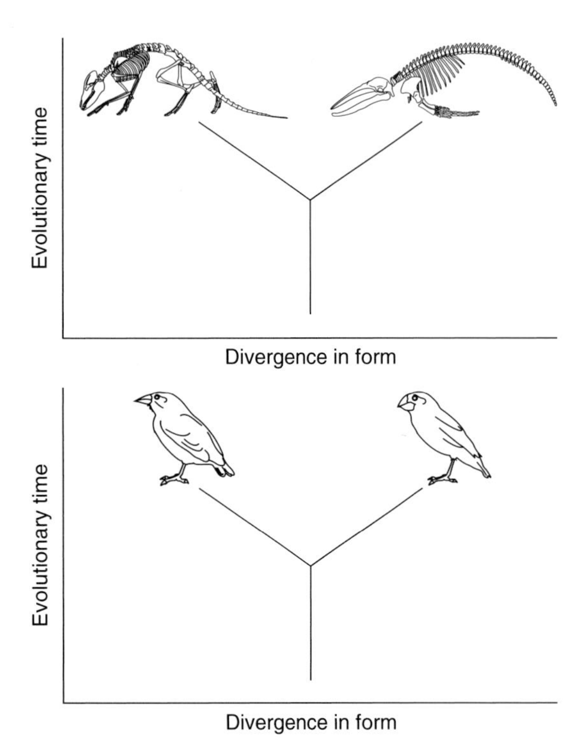

Figure 1.2 Two typical evolutionary trees, as often found in books, articles and posters on the subject. Top: a tree depicting divergence in general form between whales and their sister-group (the extinct mammal *Indohyus*). Bottom: a tree depicting divergence in a particular character – the depth of a bird's beak. Note that in the former case the top of the time axis cannot be the present because *Indohyus* is extinct. (*Indohyus* redrawn from Thewissen *et al.* 2007, *Nature*, **450**: 1190–1194.)

the information on descriptive and comparative embryology that did exist in the 1850s to good effect. As he remarked (Chapter 13), 'community in embryonic structure reveals community of descent.'

The two most important things that have happened since Darwin's synthesis of the evidence for evolution in 1859 have been the incorporation of genetics and developmental biology into the 'big picture', with the result that it has even more grandeur than before. The incorporation of genetics, which came first (Appendix 1), had both positive and negative effects on the incorporation of developmental biology that followed, is still in full swing, and is the subject of this book.

But why, actually, *is* development so important for evolutionary theory? There is a very specific and compelling answer to this question. It relates to the ways in which evolution *can* and *cannot* produce one type of animal or plant from another. This point is best made in relation to the type of evolutionary trees typically found in papers and books on the subject, two of which are shown in Fig. 1.2. Notice that in both trees the vertical axis is some measure of time, while the horizontal axis is some measure of difference in the morphology of the animal concerned, in one case 'generalised' and hard to quantify, in the other case a very specific measure of a particular structure (the depth of a bird's beak).

In a very important sense, both of these trees represent impossible evolutionary transitions. They both employ the familiar shorthand method of representing an animal by one particular stage of its life-cycle – the adult. But evolution cannot make one kind of adult directly from another. Rather, it can only make a new kind of adult by altering, over a period of generations, the egg-to-adult **developmental trajectory** (Fig. 1.3).

So, an adequate theory of evolution must include not only an account of how **fitness** differences cause changes at the population level, but also an account of how the developmental differences that natural selection acts upon arise in the first place. And these latter differences cannot simply be written off as **mutations**, because a mutation is merely a change in the DNA sequence of a gene. If the gene that mutates causes the developmental trajectory to alter, then we need to know how this happens. Furthermore, developmental trajectories can in most cases be influenced by environmental factors as well as by genes. This is true not just of extreme cases, such as the production of male or female forms in turtles by egg incubation temperature, but also of more subtle cases, such as slight differences in the amount of left-right body asymmetry (often referred to as **fluctuating asymmetry**) that can be the result of variation in temperature and other environmental factors.

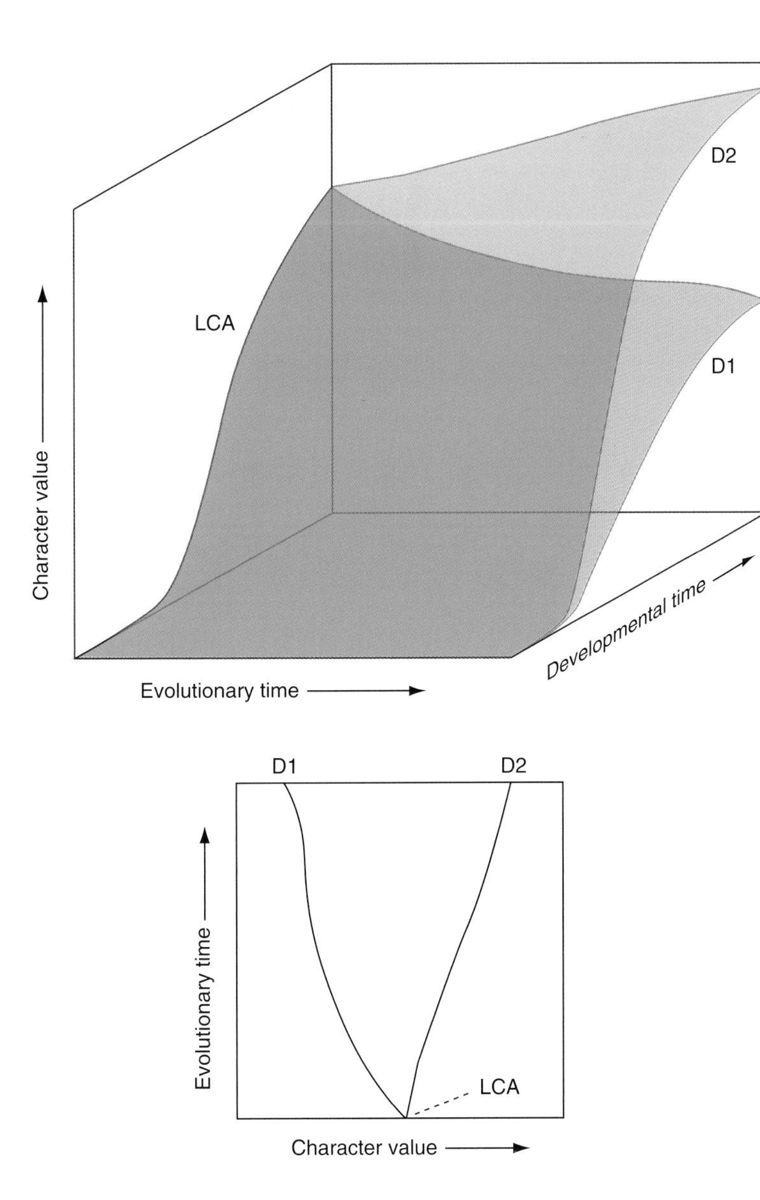

Figure 1.3 A three-dimensional evolutionary tree (top) including the extra dimension of developmental time. This shows how the value of a character (such as the bird's beak depth shown in Fig. 1.2) changes during development, and how this pattern of change itself changes in evolution. Note that the beak depth starts at zero because early bird embryos do not have beaks; and that if the developmental time axis is ignored, the tree reduces to two dimensions (bottom) and resembles that shown in Fig. 1.2 (bottom). LCA – last common ancestor; D1 – daughter species 1; D2 – daughter species 2.

Such considerations give the environment not just one role in evolution – that of selective 'sieve' – but rather two, with the other being a role in the production of the variation in the first place. Of course, non-heritable variation, or phenotypic **plasticity**, cannot itself contribute to evolution, precisely because of its non-heritable nature. But if different **genotypes** differ in their pattern of developmental response (or their developmental **reaction norm**) to environmental variation, as is now widely known to occur, then this provides material for evolutionary change. Indeed, all evolutionary theory that deals with phenotypes that are completely genetically determined can be regarded as a subset of more general evolutionary theory in which the determination of developmental trajectories, and hence of phenotypes, is more complex.

What evolve, therefore, are not just adult animals or plants, but rather complete life-cycles. Furthermore, we should not think of pre-adult stages as evolving 'in order

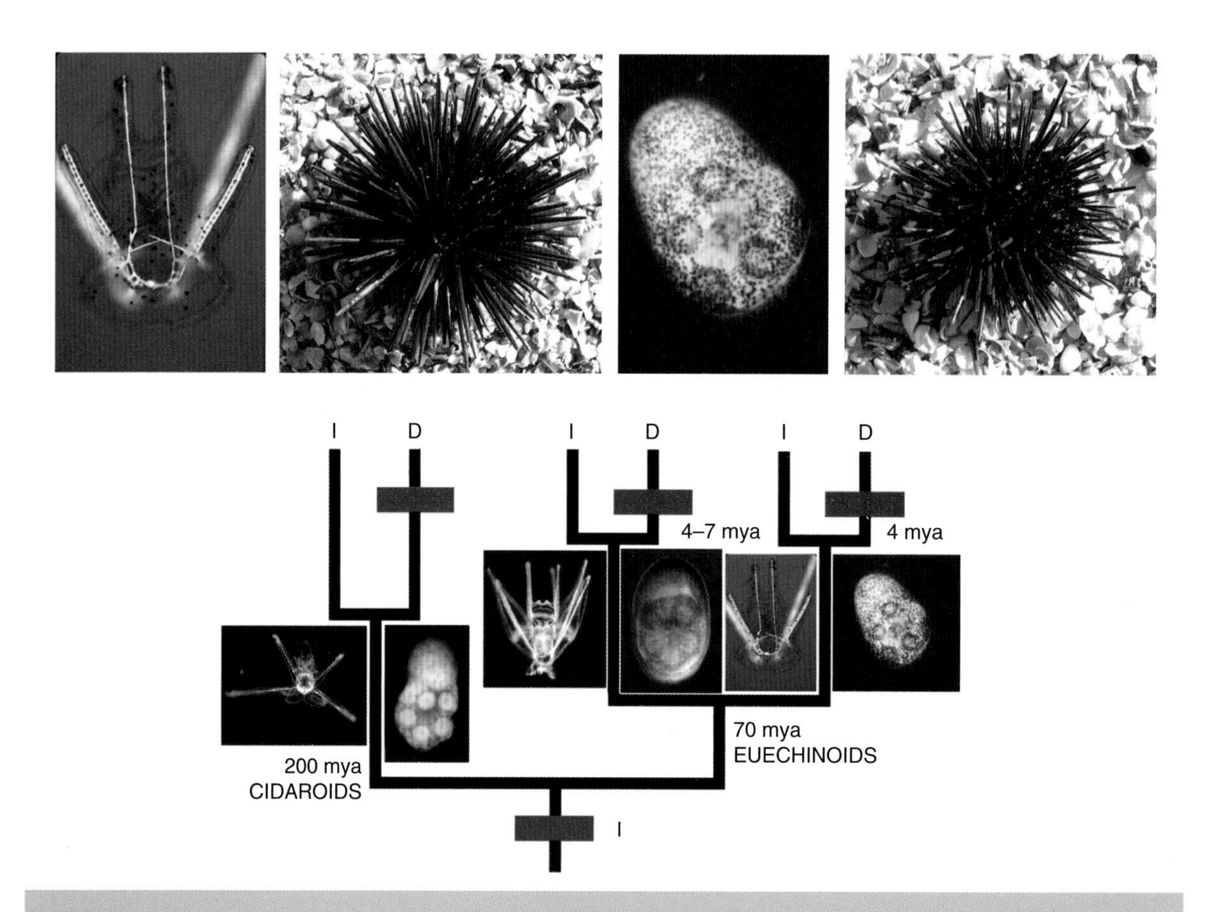

Figure 1.4 Evolution of early developmental stages, as exemplified by the evolutionary divergence of yolk-feeding and plankton-feeding echinoderm larvae. Top: Comparison of the larvae and adults of the congeneric species *Heliocidaris tuberculata* and *H. erythrogramma* (photographs courtesy of E.C. Raff and R.A. Raff). Bottom: Phylogeny showing multiple origins of yolk-feeders (red bars) from ancestral plankton-feeders. I – indirect development; D – direct development. (See Appendix 4 for an introduction to the business of inferring phylogenetic trees from comparative data.) (Reproduced with permission from *Int. J. Dev. Biol.*, **47**: 623–632.)

to enable adult forms to evolve'. This overly 'adultocentric' view of things (as recently criticised by the Italian biologist Alessandro Minelli[2]) is misleading. Instead, what happens is that there are variations at *all* developmental stages. In each case, some variants may be fitter than others, *either* because of the advantages they possess at that stage *or* because of advantages that accrue further **downstream** in the developmental pathway or, of course, for both reasons. In some cases, particularly in animals with complex life-cycles (or 'indirect development' – i.e. development to adult via a **larval** form), evolution of larval stages may occur quasi-independently of evolution of the adult. This is true, for example, in the case of evolutionary switches between plankton feeding and yolk feeding in **echinoderm** larvae[3], where the plankton-feeders have 'arms' that the yolk-feeders do not (Fig. 1.4), but this does not lead to a corresponding difference in the adults.

Having now seen that the case for the centrality of development in the evolutionary process is unassailable (but with a caveat to be discussed in Section 1.3), we need to examine development itself, and also to ask about the ways in which it can evolve. This approach (Section 1.2) will reveal several problems, some of which can be easily remedied at our current stage of knowledge, but some of which cannot. These problems include: the absence of some key terms; the previous over-emphasising of some processes (e.g. **heterochrony**); the need to connect organism-level observations with both molecular and populational ones; and the crucial issue of whether development in some sense guides evolution. This last issue is perhaps the most fascinating of all but is also the hardest to deal with and the most controversial. At stake here is the question of whether the structure of the developmental variation available to natural selection can influence the direction that evolution takes, rather than this direction being entirely set by selection alone; and if the former, then whether the role of development is merely negative ('constraint') or both positive and negative ('bias'). This will be discussed in detail in Chapter 13.

1.2 Development; and Evolutionary Changes in Development

The development of any animal or plant can be thought of as a time-sequence of more or less well-defined stages. The simplest kind of development is 'direct development', as in mammals. Indirect development, whether in echinoderms, insects or amphibians, is more complex in that the route to the adult takes what might be thought of as a 'detour' via immature stages that are radically different from the adult as opposed to miniature versions of it. Some plants, notably trees, and also some colonial animals, such as **bryozoans** ('moss animals'), are of a **modular** nature, which means that within one individual tree or one bryozoan colony, a major phase of development repeats itself multiple times. This is readily apparent to the casual observer of deciduous trees in successive springs, as the development of leaf modules occurs on a massive scale. It is also apparent, though only readily through a microscope, to observers of the growth of bryozoan colonies (Fig. 1.5), by the development of additional **zooid** modules around a colony's periphery.

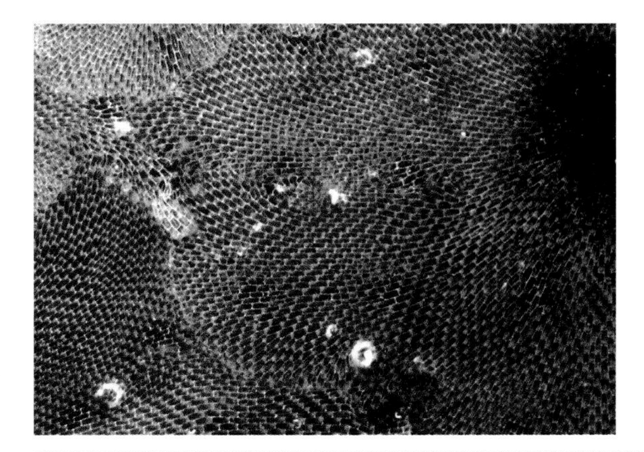

Figure 1.5 Modular development, as exemplified by a bryozoan colony. Each zooid is genetically identical to every other one in the colony, which is formed by the repeated asexual reproduction and development of zooids from the first one, which is referred to as the 'ancestrula'. A large colony may consist of many hundreds of zooids (photograph courtesy of Peter Wirtz).

In any animal or plant, both the overall developmental process, and any particular component of it, such as the developmental pathways leading to the appearance of segments, limbs or leaves, can be thought of as a trajectory. Each such trajectory represents a very specific route for a cell population that is different from other possible routes.

But equally, each trajectory can vary, both within an individual (the leaves on a single maple tree are not all identical) and, more importantly from an evolutionary perspective, among the various individuals that make up an ecological population. It is here that we move from thinking about development itself to how altered development can arise in evolution.

This is an absolutely crucial moment in our approach to the whole subject of this book. We can now see the nature of what we are dealing with: a change in something that is itself a process of change. This is very different to the old 'ecological genetics' approach to evolution, the practitioners of which were usually observing evolutionary change in something that was the fixed outcome of development. An example of the latter is industrial **melanism** in moths, where the phenotype studied was the external pigmentation of the *adult* (Chapter 4). Development itself was ignored in most such studies.

Given the more recent 'change in a process of change' approach, we need to be very careful that we have a suitable language to use to deal with such a complex situation. The complexity was nicely described by the American developmental biologist Scott Gilbert[4] in 2007, as follows: 'For evolutionary developmental biology, the current challenge is producing a 5-dimensional representation: the four standard dimensions of space and time placed into the context of the paleontological temporal dimension.'

So, now to the issue of a suitable language. Biologists call a change in a gene 'mutation'. Although this can be extended to the phenotypic level by talking of mutant phenotypes, this usage is not helpful. It connects better with the old ecological genetics approach of ignoring development and concentrating on the *adult* phenotype. What we need instead is a different term that indicates clearly that we are referring to a change in development. We can't just call it 'developmental change', as that would be ambiguous and more likely to be interpreted just as going along one particular trajectory (in developmental time) rather than switching from one trajectory to another (in evolutionary time). But if 'developmental change' won't do, what will?

One term that definitely will not do as an overall term for evolutionary change in **ontogeny** (i.e. development) is 'heterochrony', despite the book title by McKinney and McNamara[5], *Heterochrony: The Evolution of Ontogeny*, with its implication that the two are synonymous. Rather, heterochrony (evolutionary change in developmental timing) is merely a subset of the overall evolution of development, as indicated by the chapter title 'It's not all heterochrony' in Rudolf Raff's book[6] *The Shape of Life*.

Two phrases that have already been used in the context of evolutionary changes in development may be considered as candidate 'umbrella-terms' to cover *all* such changes, whether heterochronic or other. These are **developmental repatterning**[7] (sometimes in the form ontogenetic repatterning[8], which is synonymous) and developmental reprogramming[9]. There are two reasons why the former phrase is preferable. First, 'developmental program(me)', 'programming' and 'reprogramming' are too philosophically loaded, and are interpreted by some biologists as smacking of 'genetic imperialism' (Appendix 1). Second, 'developmental reprogramming' has become used in a different, and narrower, way[10] in the last few years. So, developmental repatterning now seems the obvious choice as the umbrella-term for all evolutionary changes in development, and we will use it throughout this book.

Any developmental process can be thought of as a pattern in time, space or (usually) both. Already 'pattern' is part of the language at different levels of developmental study. For example, at the molecular level we speak of the **expression pattern** of a gene in an embryo (Fig. 1.6); and at the tissue level we speak of **pattern formation**, to indicate, for example, the different developing patterns of the five digits of our hand.

If the development of one individual can be thought of as patterning (in a multitude of senses), then the evolution of development can be thought of as repatterning. Logically, considering any aspect of development – for example, the expression pattern of a gene – there are four types of repatterning that can occur: changes in time, place, amount or type. There are well-established terms for the former two – **heterochrony** and **heterotopy**; and there are more recently-introduced terms for the latter two – **heterometry** and **heterotypy**. This series of terms provides a broad categorisation of the types of evolutionary change that can occur in the developmental process. It will be useful to keep them in mind later (and so they are used as chapter titles in Part II) as the intricacies of particular **case studies** emerge.

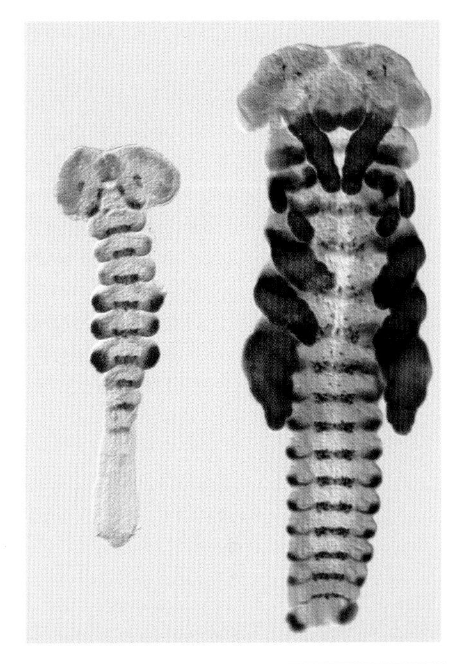

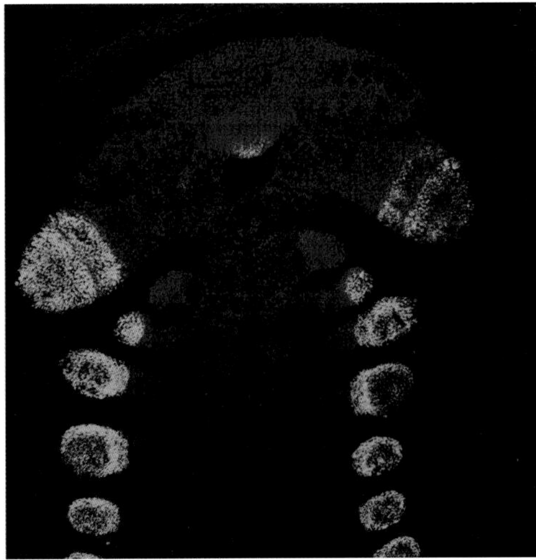

Figure 1.6 The expression pattern of a gene in embryos. The example shown is the expression of a gene called *distal-less* (see Appendix 2 for information on gene names) in crustacean (top) and centipede (bottom) embryos. In the crustacean (*Parhyale*) the red staining reveals expression of the gene in developing limbs in early-stage (left) and later-stage (right) embryos (photograph courtesy of Nipam Patel); in the centipede (*Strigamia*) the expression of the gene is shown in green (photograph courtesy of Cornelius Eibner).

1.3 Development and the Realm of Multicellularity

There is a restriction to the importance of development in evolution that does not apply to the importance of genes; specifically, development is only a necessary part of evolutionary theory when the creatures that are evolving are multicellular. This is because creatures that are unicellular throughout their life-cycle lack 'development', at least in the sense in which the word is normally used. Thus evo-devo deals mainly or wholly with what we will call here the realm of multicellularity.

Note that this term – realm – is not part of the taxonomic hierarchy (in the way that **domain**, kingdom and **phylum** are). The reason for adopting such a term is that if we take as a group all multicellular creatures, they do not form a single **clade**. Indeed, far from it: multicellularity has arisen at least five times in evolution, and probably much more often.

The five major origins of multicellularity – in animals, plants, fungi, brown algae and cellular slime moulds – are shown in Fig. 1.7. More minor origins, in the sense that they have led to more restricted invasions of multicellular **morphospace**, have occurred (*inter alia*) as follows: in the cyanobacteria (strings and mats of cells); in the diatoms (which have some multicellular forms, despite being very largely a clade of unicells); and in other groups of 'slime moulds' unrelated to the group shown in Fig. 1.7. It should be noted that the deep divisions of the living world shown in the figure are different from those that can be seen in comparable trees produced a mere decade ago; and our picture of deep **phylogeny** may well yet change further. Despite this, the conclusion that multicellularity has arisen several times in evolution is likely to be robust.

How did multicellularity originate? A recent clue has come from the study of **choanoflagellates** – a clade of unicells that appears to be the **sister-group** of the animal kingdom. The first choanoflagellate genome project[11] has revealed that these creatures possess many genes previously known only from animals and associated with multicellularity, such as genes that make cell adhesion and cell-cell signalling molecules. These proteins may have been used in the unicellular ancestors of animals (and in present-day choanoflagellates) to interact with the environment, including **conspecifics** (potential mates) and other unicells (potential prey). If so, this would represent an example of **exaptation**, in which something initially evolved for one selective reason later becomes useful (and hence selected) for another. This fascinating topic will be discussed further in Chapter 12.

As ever, there is a caveat. One main strand of the evidence for a sister-group relationship between choanoflagellates and animals is the similarity in the form of

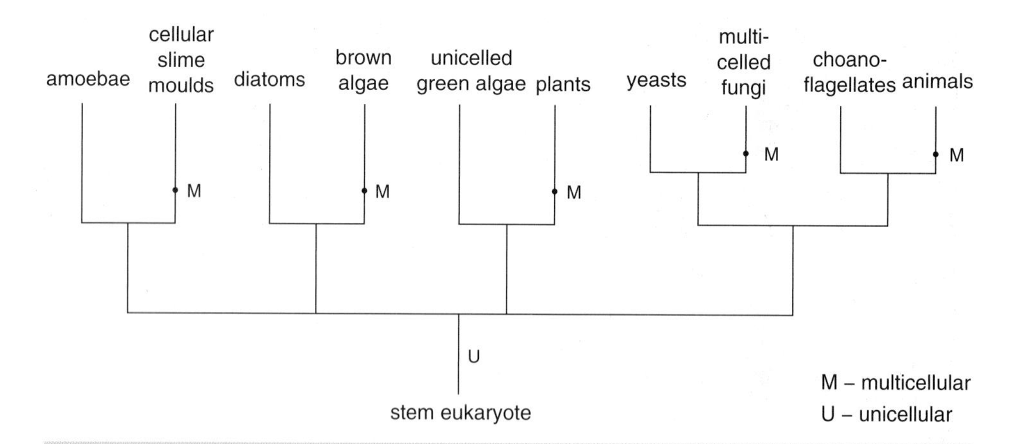

Figure 1.7 The five 'main' evolutionary origins of multicellularity, each marked with an M on the lineage concerned. In each case the probable sister-group is also indicated. Note that in a typical multicellular creature there is also a *developmental* origin of its multicellular state in each generation. In some branches, what is shown is 'exact' – for example, there was probably only a single origin of multicellularity in animals; but in other cases, for example fungi, what is shown may be a simplification.

the typical choanoflagellate cell (possessing a collar) and the collar-cells of sponges (Porifera), which have long been regarded as the most primitive animals. But if a recent molecular phylogeny of the animal kingdom[12] is to be believed, the **ctenophores** (comb jellies) are more basal than sponges. However, the authors concerned admit that their placing of the ctenophores as basal should only be regarded as a hypothesis for now. Figure 1.8 shows this hypothesis and its main alternative, along with pictures of collar cells. A complication to both hypotheses is that the sponge group Porifera may be **paraphyletic** or even **polyphyletic**.

An important general point emerges from this recent dispute about the relationships among the most **basal** animals. In general, it is best to have an agreed phylogeny of

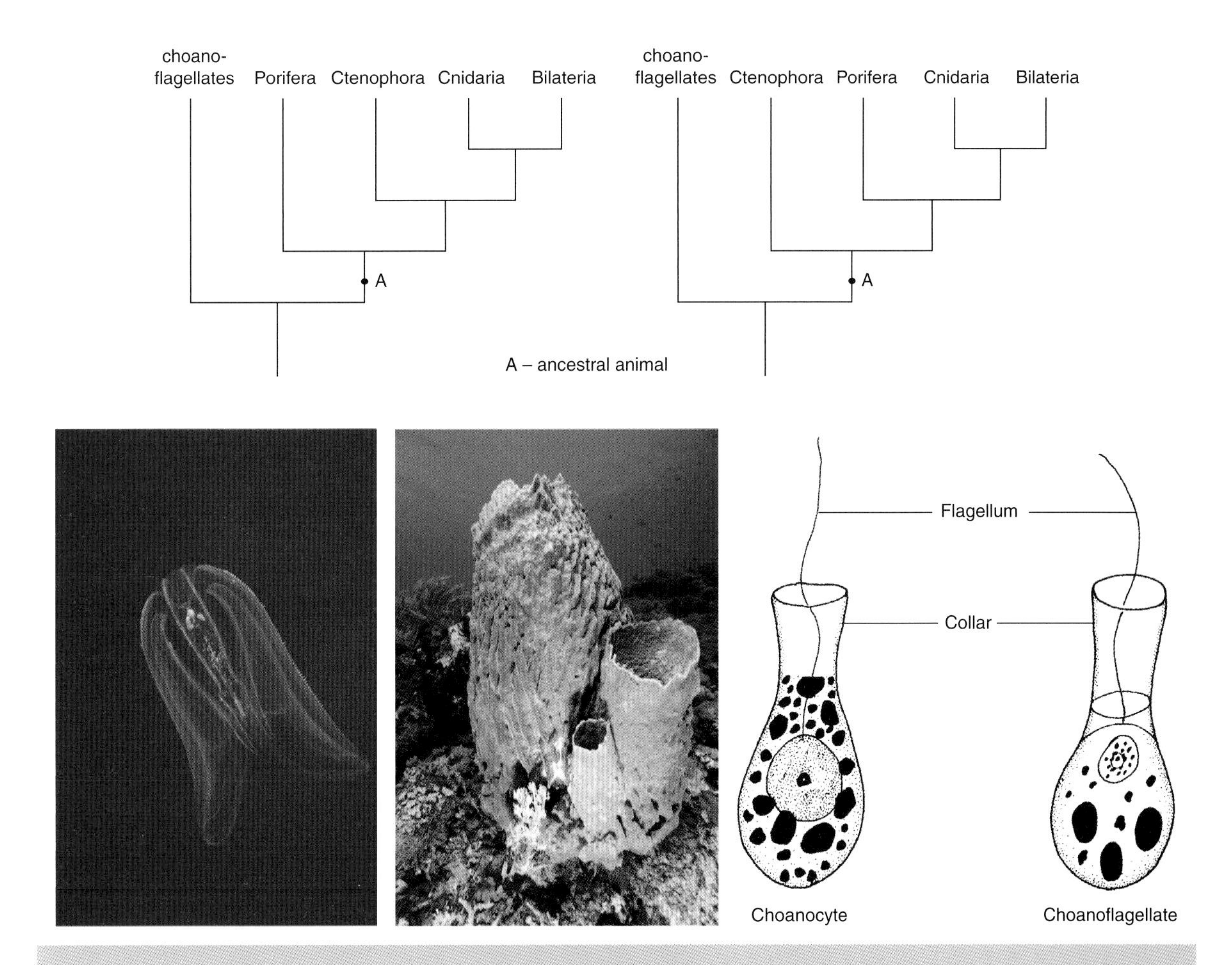

Figure 1.8 Our understanding of the pattern of relationships among the most basal animal groups is still very imperfect. Top: Conflicting hypotheses as represented by alternative trees. Bottom: Pictures of a ctenophore, a sponge, a collar-cell (choanocyte) of a sponge and a choanoflagellate.

any group of animals or plants before mapping onto it (as in Chapter 11) evolutionary changes in development. For some groups, this is not a problem; but for others, our current view of phylogeny may alter radically in the future. If it does indeed alter, then so too must our views on the nature and temporal sequence of evolutionary changes in development – that is, of developmental repatterning.

SUGGESTIONS FOR FURTHER READING

Two good, but very different, introductory books on the 'new science of evo-devo' and on developmental approaches to evolution more generally are:

Minelli, A. 2009 *Forms of Becoming: The Evolutionary Biology of Development.* Princeton University Press, Princeton, NJ.

Carroll, S. 2005 *Endless Forms Most Beautiful: The New Science of Evo-Devo and the Making of the Animal Kingdom.* Norton, New York.

CHAPTER 2
What is Evo-Devo?

2.1 Forerunners of Evo-Devo

As Winston Churchill said: 'The farther backward you can look, the farther forward you are likely to see.' This was intended as a comment on history in general, but it can be applied just as well to the history of science. Any branch of science can only be thoroughly understood in the context of how it has progressed over time.

A good example of how an understanding of present-day evo-devo is informed by a study of its forerunners is the proposed **recapitulation** by embryos of their evolutionary past. Initially regarded (wrongly) as a law of how development evolves, it was later completely rejected (wrongly) and then eventually regarded (correctly) as one of several possible patterns, all of which we need to be able to explain.

So, looking at the historical background of present-day evo-devo is not optional – it is essential. If you are seriously history-averse, you can skip to the skeletal version in Section 2.4 – but I would advise against doing so.

2.2 Nineteenth-Century Comparative Embryology

In the decades following publication of Darwin's classic text in 1859, the leading evolutionary embryologist was Ernst Haeckel, whose publications included three major books. While English translations of the first two have never been undertaken (a project for the future, perhaps), the last – a more popular account of Haeckel's views – was translated[13] into English in 1896 as *The Evolution of Man*. This translation is important because it allows the many English-speaking biologists who cannot read German (alas a large number, and including myself) to understand Haeckel's key points.

Haeckel's work has frequently been misunderstood, as has the relationship between his 'law' and the 'laws' of the pre-Darwinian comparative embryologist Karl Ernst von Baer, whose principle of embryonic divergence is shown in Fig. 2.1. Some leading figures have been guilty of such misunderstanding, including the English biologist Gavin de Beer (see below and Section 2.3). It is important to clarify this matter here, because the evo-devo of today must build on the comparative embryology of the past – but only on those parts of it that we believe to be correct.

Von Baerian divergence, as can be seen from Fig. 2.1, is a pattern of early similarity giving way to later differences. Von Baer[14] gave four laws to describe this pattern; but it can adequately be described by the first of these: 'The general features of a large group of animals appear earlier in the embryo than the special features.' (Source: Alec Panchen's[15] *Classification, Evolution and the Nature of Biology*).

Haeckel famously produced the 'biogenetic law' or 'law of recapitulation'. 'This fundamental law...', writes Haeckel (1896: p. 7) states:

> that Ontogeny is a recapitulation of Phylogeny; or, somewhat more explicitly: that the series of forms through which the Individual Organism passes during its progress from the egg cell to its fully developed state, is a brief, compressed reproduction of the long series of forms through which the animal ancestors of that organism (or the ancestral forms of its species) have passed from the earliest periods of so-called organic creation down to the present time.

Now, two very important questions arise. First, did Haeckel mean that the **ontogenies** of descendant species went through stages similar to the *adult* forms of their ancestors? Second, and related to that, is Haeckel's law compatible or incompatible with von Baer's laws?

Regarding the first question, here is the answer provided by de Beer[16], who interprets Haeckel's law as follows (p. 5): 'The *adult* stages of the ancestors are repeated during the development of the descendants' [my italics]. De Beer's next section is entitled 'The rejection of the theory of recapitulation'. But should we accept that a thorough student of embryology, such as Haeckel, really believed this? Probably not. Here, again, is Haeckel (1896: p. 18):

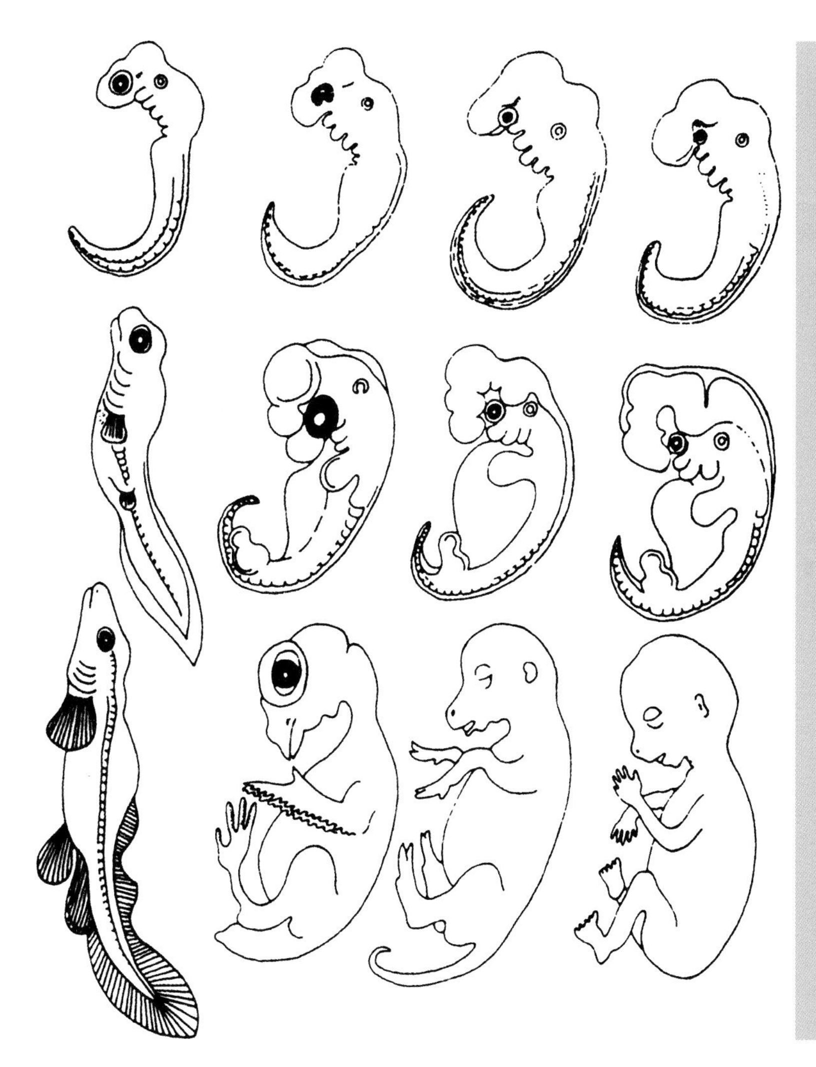

Figure 2.1 The pattern known as von Baerian divergence, as illustrated by the embryos of four vertebrates, fish, hen, cow and human, shown at three different stages – early, middle and late. Note the pattern of early similarity giving way to later differences. (Redrawn from *A Theory of the Evolution of Development*, John Wiley & Sons, Ltd., 1988.)

> The fact is that an examination of the human embryo in the third or fourth week of its evolution [= development] shows it to be altogether different from the fully developed Man, and that it exactly corresponds to *the undeveloped embryo-form* presented by the Ape, the Dog, the Rabbit, and other Mammals, at the same stage of their Ontogeny [again, the italics are mine].

Clearly, the picture in Haeckel's mind was like that of von Baer, not like the 'caricature recapitulation' involving ancestral adult forms in descendant embryos.

Why has Haeckel frequently been misinterpreted as being both so stupid as to see *adult* ancestors in descendant embryos and also in some sense anti-von Baer? The answer must lie at least in part in his choice of phrase: recapitulation. At first sight, this would not seem to be compatible with embryonic divergence.

But this is where we must recall that von Baer's[14] *magnum opus* (published in 1828) was pre-Darwinian, and in later life von Baer was anti-Darwinian; in contrast, Haeckel was pro-Darwinian. To a non-evolutionist such as von Baer, the comparisons of vertebrate embryos shown in Fig. 2.1 were not comparisons of animals that were related in an ancestor-descendant way – rather, they were comparisons among a series of independently-created forms. But to an evolutionist such as Haeckel, they were comparisons that related to shared descent – indeed to what Darwin frequently called 'descent with modification'. Of course 'the Dog' is not a human ancestor, and Haeckel could hardly have believed so, but the two embryos show features that derive from their common descent – such as the rudimentary gill-clefts that both dog and human embryos transiently possess as a result of their shared aquatic ancestry. So they both recapitulate (non-adult) features of their ancestors, while also progressively diverging from each other over developmental time. There is simply no conflict between these views.

There may not be a conflict between them, but there is definitely a problem with both views – specifically their description, by their respective authors, as 'laws'. Generally, in science, 'law' is used for something that is either universally true or at least very nearly so. It also tends to be used for something that can be stated quantitatively/mathematically, rather than merely verbally or in pictures. In physical science, Newton's laws of motion provide a good example. They are very generally true (except at speeds approaching that of light, where Einstein's relativity takes over); and they can be stated quantitatively. In biological science, Mendel's laws of inheritance[17] provide a good example. They are generally true of **diploid** organisms, though the second law, of independent assortment, requires that the genes under consideration are on separate chromosomes. And they make quantitative predictions, such as the well-known Mendelian ratios that result from various sorts of cross – for example, a 3:1 **phenotypic** ratio in the F_2 generation from a cross involving **homozygous** 'wild-type' and mutant parents.

It is clear that neither von Baer's nor Haeckel's 'laws' can be expressed in quantitative form. So they both fail to attract law status in this respect. But it is also the case that they both fail according to the other criterion, that of being universally, or even generally, true.

This second point requires some elaboration. The particular comparisons of developmental trajectories shown in Fig. 2.1 are few in number, and highly non-random in that all involve vertebrate embryos. What of other comparisons? Inter-taxon comparisons could be made within some other major **clade** than vertebrates (e.g. insects). Also, comparisons could be made between representatives of different major clades (e.g. between a vertebrate and an **echinoderm**).

Most insects go through a common early developmental stage called the **germ band**. They are recapitulating this stage, which we believe was present in the development of the 'stem insect'. But since they go on to be very different adults (e.g. dragonfly and wasp), they are also undergoing von Baerian divergence.

Vertebrates and echinoderms both show radial **cleavage** as their earliest embryonic process, in contrast to most other animal groups, which show spiral cleavage (Fig. 2.2). We believe this to be the case because they are both descended from a common ancestor (the ancestral **deuterostome**) that also possessed this feature. The retention of this feature of their ancestor's development can be thought of as recapitulation. But, since their adult forms are incredibly different (think of a mouse and a sea-urchin, for example), the fact that they are similar in their earliest developmental process, cleavage (albeit starting from very different eggs) means that we can think of the pattern also as being von Baerian divergence.

Note, however, that the use of these terms, recapitulation and divergence, is getting somewhat strained. Very few features of the deuterostome ancestor's development (inasmuch as we can infer that development) are recapitulated. Furthermore, the ontogenies of, say, a mouse and a sea urchin, become so different at such an early developmental stage that 'divergence' hardly seems a sensible term for differences in later development.

We must always remember that evolution is a messy process. This means that no single pattern will prevail in all cross-**taxon** comparisons of developmental trajectories. Thus there are no patterns whose generality of occurrence is sufficient to warrant the invention of a 'law'. All developmental stages change in evolution. Some stages, in some taxa, are more resistant to change than others, and this can give rise to patterns, both von Baerian/Haeckelian and others. Although these patterns are statistical rather than absolute or law-like, they are nevertheless of considerable interest. One of them, where the very earliest stages in development are *less* similar than the stages that follow them, giving a specific departure from the pattern suggested by von Baer, will be discussed in Chapter 19.

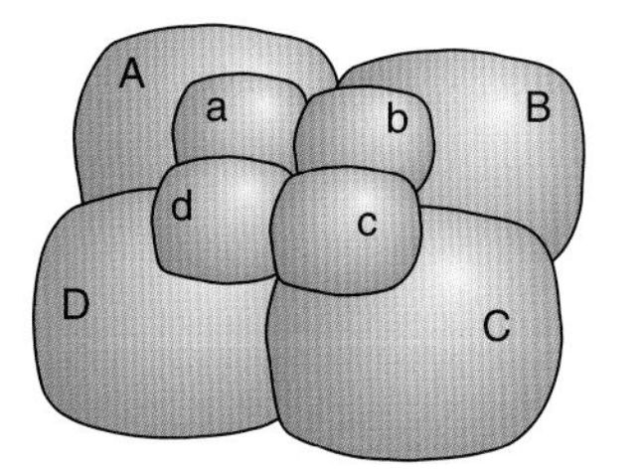

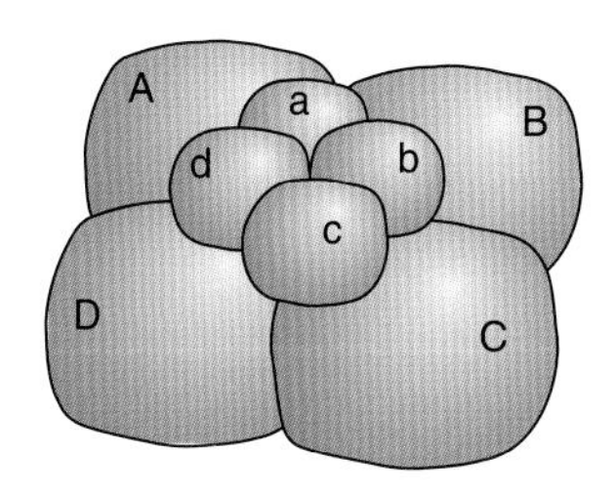

Figure 2.2 Two types of cleavage – the earliest developmental process in animals. Left – radial cleavage, as found in vertebrates, echinoderms and other deuterostomes. Right – spiral cleavage, as typically found in protostomes, including molluscs. Note that the degree of size difference between micromeres and macromeres varies considerably among different animal groups; in particular this difference is typically smaller (even non-existent in some cases) in radial than spiral cleavage. Also note that these are not the only two types of cleavage found; for example, insects have a type of cleavage referred to as syncytial.

2.3 Diverse Antecedents—1900–1980

The leading figures working on the relationship between evolution and development during the period from 1900 to the advent of evo-devo in the 1980s were remarkably heterogeneous in their approaches. They certainly do not constitute a school of thought, in the way, for example, that population genetics did over this same period. We will briefly examine the work of six prominent people here, in historical sequence: D'Arcy Thompson; Gavin de Beer; Richard Goldschmidt; C.H. Waddington; Lancelot Law Whyte and Stephen Jay Gould.

Before doing so, it is interesting to note a statement made by J.B.S. Haldane[18] (a population geneticist, not a student of development) in 1932. This statement presages an important paper published in 1979 by American evolutionary biologists Stephen J. Gould and Richard Lewontin – we will look at this issue in Chapter 13. For some historical background to the Haldane quote, see Appendix 1. Here is the quote:

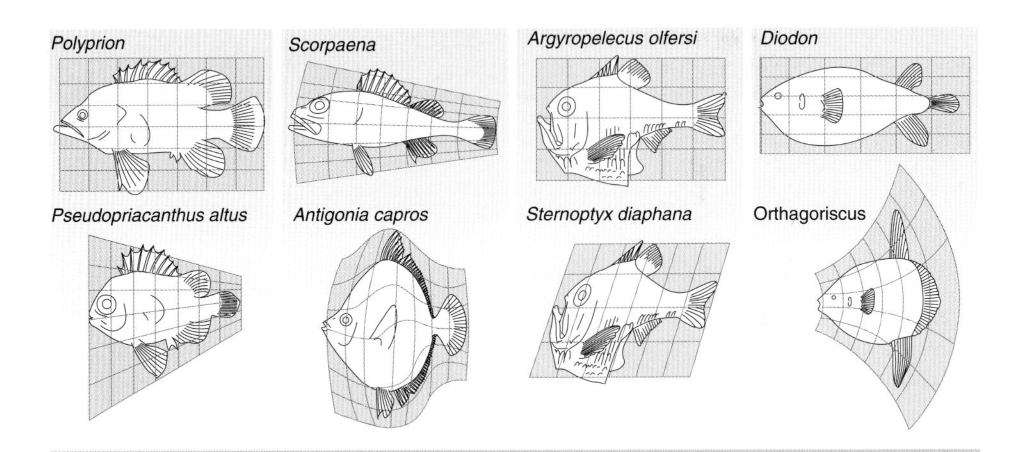

Figure 2.3 The transformations that can be applied to the form of one species within a group (in this case fish) to produce the forms of others. These transformations were devised by D'Arcy Thompson as early as 1917 (see text). The fact that they are possible indicates that evolution often modifies animal (and plant) forms in a co-ordinated, rather than piecemeal, manner. (From Arthur, 2006, *Nature Reviews Genetics*, **7**: 401–406.)

To sum up, it would seem that natural selection is the main cause of evolutionary change in species as a whole. But the actual steps by which individuals come to differ from their parents are due to causes other than selection, and in consequence evolution can only follow certain paths.

D'Arcy Thompson

The most enduring legacy of Thompson's[19] *On Growth and Form* is his 'theory of **transformations**'. Thompson pointed out that changes in development over the course of evolutionary time were not piecemeal, in the sense that each of an animal's characters evolved independently, but rather were *co-ordinated* changes in body form. These could be represented visually by plotting the morphological outline of one animal on a Cartesian grid and then subjecting that grid to some systematic distortion (Fig. 2.3). Ironically, despite Thompson's interest in development as well as in evolution, his pictures of evolutionary transformations generally were of adults. Nevertheless, we can easily extend his idea of a transformation to cover the later phases of ontogeny (when **allometric growth** prevails) rather than just the adults that ultimately result. The take-home message in either case is the same: evolution often involves co-ordinated changes in many aspects of body form.

Thompson's transformations can be regarded as an attempt to quantify what Darwin referred to as 'correlation of growth'. However, since there is always a limit to the range of forms that can be connected up by any one series of transformations, such limits can be taken as being against the pan-gradualist spirit of Darwinism, as Thompson pointed out:

Our geometric analogies weigh heavily against Darwin's concept of endless small continuous variations; they help to show that discontinuous variations are a natural thing, that 'mutations' – or sudden changes, greater or less – are bound to have taken place, and new 'types' to have arisen, now and then.

Gavin de Beer

De Beer[16] published *Development and Evolution* in 1930; and a revised version entitled *Embryos and Ancestors* in 1940, with further revised editions (but no further changes of title) in 1951 and 1958. His central theme was evolutionary change in the timing of developmental events, i.e. **heterochrony**. He emphasised the multitude of types of heterochronic process, highlighting in particular **neoteny**, the retardation of somatic development relative to the development of the reproductive system, with the result that in a neotenous **lineage** what was previously a juvenile (or even **larval**) form becomes reproductively mature. However, he also dealt with many other kinds of heterochrony, and so was essentially pluralist in his approach to the evolution of development. But his pluralism did not extend to recapitulation. Instead of acknowledging that it is one of many patterns that can be found in the evolution of development, he states in his concluding chapter (1958: pp. 170–171) that 'Recapitulation...does not take place.' He reached this conclusion due to taking a rather narrow view of what recapitulation means (Fig. 2.4).

Richard Goldschmidt

Richard Goldschmidt is remembered most for his **saltationist** views[20]. He was much influenced in his overall interpretation of the evolutionary process by his studies of **homeotic** mutations in *Drosophila* fruit-flies[21] – mutations causing the appearance of 'the right structure in the wrong place' – such as the one that causes an extra pair of wings to grow out of the third thoracic segment in place of the small balancing organs that are normally found there (the 'bithorax' phenotype: see Fig. 2.5). He argued (i) that since some of these mutations caused a change that was of sufficient magnitude to bridge the gulf (in some characters anyhow) between higher taxa, such as insect orders; and thus (ii) that these mutations might have formed the basis for the *origins* of the higher taxa (such as orders) concerned.

This theory did not end up being accepted, largely because it became clear that all known homeotic mutants of *Drosophila* are unfit compared to the **wild type**, probably in all relevant environments. Nevertheless, the fact that less drastic mutations in the same genes have a major role in the evolution of development has now become apparent.

Conrad Hal Waddington

Waddington[22] gave us three linked concepts, and new terms to go with them. He pictured developmental trajectories in the form of a ball running down grooves in an **epigenetic landscape** (Fig. 2.6). As can be seen, the landscape is like a system of river

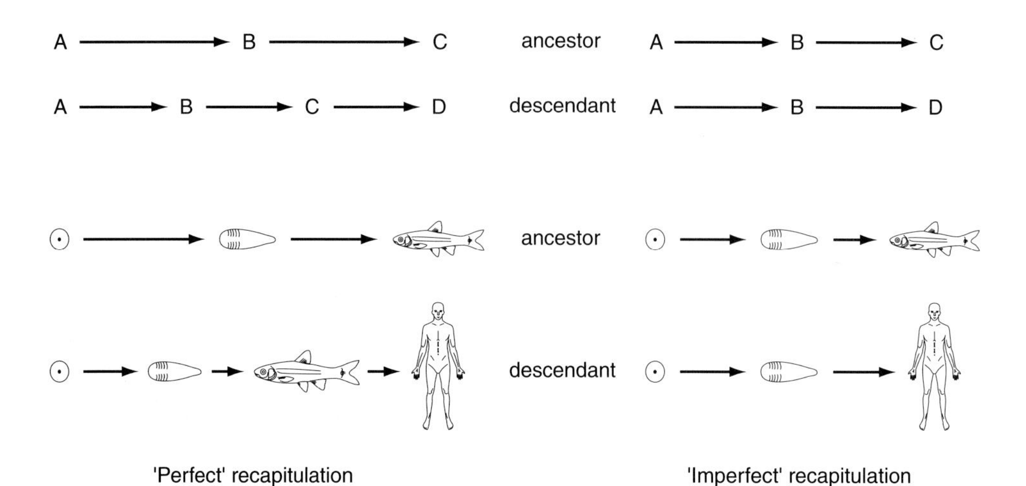

Figure 2.4 Two views of recapitulation. Left: 'perfect' recapitulation, in which evolution adds new developmental stages at the end of the ancestral ontogeny. This does not occur. Right: 'imperfect' recapitulation, in which some features of ancestral ontogenies are repeated (or recapitulated) in the development of descendants, even though they may not lead to functional adult structures in the descendants. This form of recapitulation does occur, and is due to the fact that it is often hard for natural selection to alter early stages. The two forms of recapitulation are shown in terms of abstract developmental stages (A to D) in the top panel; and in terms of a particular example (bottom panel) where A is a vertebrate zygote; B is a vertebrate embryo with gill clefts; C is an adult fish; and D is an adult human.

Figure 2.5 Normal (left) and homeotic mutant (right) forms of the fruit-fly *Drosophila melanogaster*. The particular mutant shown is 'bithorax', in which there is a second pair of wings. This is due to the transformation of the third thoracic segment, which normally produces small flight-balancing structures called halteres, into a repeat of the second thoracic segment, which produces wings. (Reproduced with permission from Carroll, S. *et al.* 2005, *From DNA to Diversity*, Blackwell.)

tributaries, but with the overall topography being the reverse of that which characterises real geography: as we go downhill (= later in developmental time), the rivers diverge rather than converge. Such a system involves a degree of **canalisation**, in that there is local stability of the trajectory – with the amount of stability depending on the depth of the groove. This was Waddington's abstract way of picturing the fact that real developmental systems are to some extent robust in the face of both genetic (mutational) and environmental perturbations.

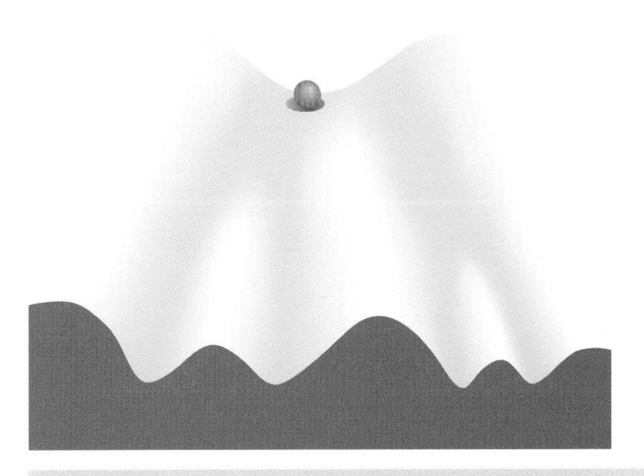

Figure 2.6 The 'epigenetic landscape'. Here, development is pictured in terms of several possible courses (valleys), with the actual course taken in any individual organism of the species concerned being represented by a ball running through particular valleys of the landscape from top (start) to bottom (finish – the adult). (Redrawn from Waddington, 1957, *The Strategy of the Genes*, Allen & Unwin.)

Some perturbations, however, are too extreme for canalisation to be retained, and so development takes a different course – that is, it becomes repatterned. Waddington showed[23] that one particular environmental perturbation – exposure of eggs to ether vapour – could cause *Drosophila* development to switch from production of a normal fly to production of the bithorax phenotype that can also be produced by genetic perturbations (the mutations in the Bithorax gene-complex whose effects Goldschmidt had studied). Furthermore, Waddington also showed that some **genotypes** were more sensitive than others to switching their developmental trajectory in response to ether vapour. So, when he performed an artificial selection experiment and bred only those flies that most readily underwent the switch, he could eventually produce bithorax phenotypes without ether treatment. This process, which he called **genetic assimilation**, looks, but is not, **Lamarckian**. It is a special case of the evolution of phenotypic **plasticity**. This subject (Chapter 16) was being studied in parallel by the Russian biologist Schmalhausen, though the two men had little contact.

Lancelot Law Whyte

Whyte produced two papers and a book[24] in the 1960s, dealing with what he called first 'developmental selection' and later 'internal selection'. The point he was emphasising is that developmental (and more generally organismic) integration – the compatibility of interacting parts – can act as a selective agent just as can an external, ecological agent such as a predator. Thus he was championing internal **coadaptation** rather than external adaptation to particular environmental conditions. We will look further at Whyte's ideas in Chapter 12.

Stephen Jay Gould

We will look in particular at Gould's famous 1977 book[25] *Ontogeny and Phylogeny*. Whether something that is only 40 to 50 years old should be regarded as historical is a moot point. However, we will take the approach here that 'evo-devo' originated around 1980, and that everything before then thus constitutes the historical basis for it rather than being part of it. The rationale for this approach is that before 1980 most studies of the evolution of development were at the organismic level. None of them were at the level of the molecular structure of key groups of developmental genes, as this structure did not begin to be revealed until about 1980 (Section 2.5).

Ontogeny and Phylogeny is very much a book of two parts: the first is historical (and entitled 'recapitulation'); the second is focused on heterochrony. Gould concludes his historical study of recapitulation in a very different way than de Beer did several decades earlier. Instead of concluding that recapitulation does not occur, Gould states:

> But recapitulation was not 'disproved'; it could not be, for too many well-established cases fit its expectations. It was, instead, abandoned as a universal proposition and displayed as but one possible result of a more general process.

The second part of Gould's book, however, has much in common with de Beer. Both concentrate on, and in fact overemphasise, changes in developmental timing (heterochrony). Gould developed a particular way of looking at these, a 'clock model', which is perhaps useful but hardly revolutionary. The historical part of *Ontogeny and Phylogeny* is by far the more valuable of the two. It goes into much more detail than I have been able to go into in my very condensed version of history here; and is therefore a good source for those who want a longer version.

2.4 Conclusions from History; Messages for the Present

By 1980, the following conclusions could be drawn from the largely organismic studies that had been conducted over the previous 150 or so years:

1. All developmental stages in the life-cycles of all organisms can and do change over the course of evolution.
2. There is no universal way in which development evolves. Therefore it is not sensible to try to elevate any particular pattern to the status of a law.
3. This does not mean that the patterns are not interesting. On the contrary, establishing what patterns occur, and what factors favour the occurrence of one pattern over another, is interesting and important.
4. This endeavour must include elucidating the ways in which natural selection can interact with the developmental system – building on the bases provided by Waddington, Schmalhausen, Whyte, and many others.
5. Haldane was right when he said that selection was not the cause of the variation in development upon which it acted; thus the molecular causality of any one

developmental trajectory, and of the differences between different trajectories (at several levels), must become a major part of our attempt to explain, in a comprehensive way, how development evolves.

The last point leads neatly into the key focus of the evo-devo studies of the last 30 years. These studies now become our focus of attention.

2.5 The Advent of Evo-Devo in the 1980s

Arguably, the most important advance in biology, since Watson and Crick's[26] working out of the structure of DNA in 1953 and the subsequent elucidation of the genetic code in the 1960s, was the discovery of the **homeobox** in 1984. This 'box' (Fig. 2.7) is a sequence of some 180 DNA base-pairs found in many genes within the **genome** of any animal, whether it is a primitive **basal** form, such as a sponge, or an advanced and more complex animal, such as an insect or a bird. This discovery is important for two reasons:

First, because the homeobox sequence, when transcribed and translated, gives a 60-amino-acid section of the corresponding protein, and this protein section, called the **homeodomain**, is known to have DNA-binding properties. This relates to the protein's job, which is to move into the cell's nucleus, to attach to the DNA of various genes, to turn them on or off, or to regulate their activity more subtly. Thus homeodomain-containing proteins help to control the process of development through which the fertilised egg becomes an adult.

Second, their distribution throughout the entire animal kingdom means that homeobox-containing genes provide a common basis for how development occurs, regardless of the eventual adult form reached. **MADS-box genes** provide a similar common basis for development in plants. Thus it is possible to generalise about how development works, just as the theories of Darwin and Mendel enabled us to generalise about how the processes of evolution and inheritance work. Before 1984, many biologists thought that development was too complex within any one organism, and too varied between different taxa, to allow any kind of generalisation about how it takes place. Now, in the early 21st century, it is hard for many biologists, especially the younger ones, to recall this pessimistic, pre-1984 view. So here, as a reminder, is a quotation from the opening page of a leading 1960s developmental biology text[27]: 'no general theory of development has emerged, in spite of the mounting mass of observational and experimental information.'

Of course, the discovery of the homeobox does not quite amount to a general 'theory of development' in the same way that Darwin's natural selection does amount to a general 'theory of evolution' (though we will question that second assertion later). And even when other important recent discoveries in developmental genetics (see Section 2.6) are added to the picture, we still do not have a general 'theory of development'. But at least a collection of disconnected facts has been replaced by a few robust generalisations, such as 'development works through a cascade of interacting genes and their direct and indirect products'.

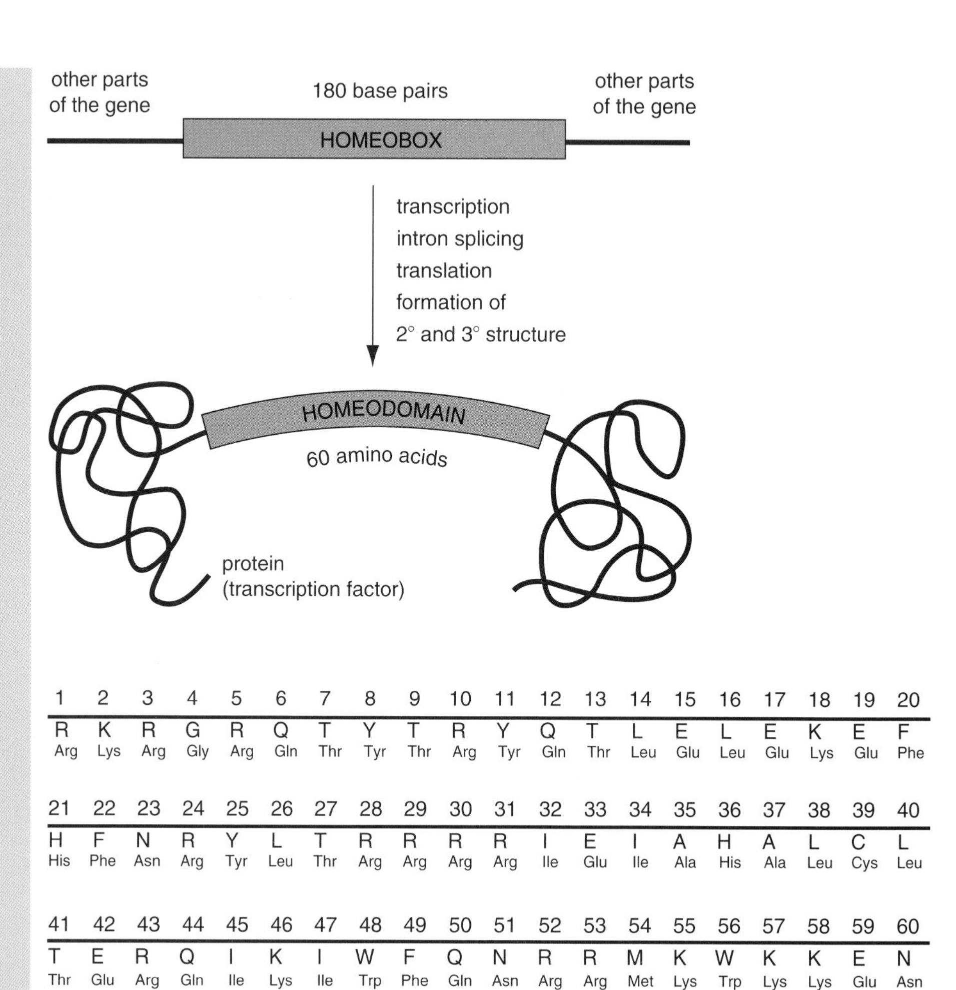

other parts of the gene
180 base pairs
HOMEOBOX
other parts of the gene
transcription
intron splicing
translation
formation of
2° and 3° structure
HOMEODOMAIN
60 amino acids
protein
(transcription factor)
1 2 3 4 5 6 7 8 9 10 11 12 13 14 15 16 17 18 19 20
R K R G R Q T Y T R Y Q T L E L E K E F
Arg Lys Arg Gly Arg Gln Thr Tyr Thr Arg Tyr Gln Thr Leu Glu Leu Glu Lys Glu Phe
21 22 23 24 25 26 27 28 29 30 31 32 33 34 35 36 37 38 39 40
H F N R Y L T R R R R I E I A H A L C L
His Phe Asn Arg Tyr Leu Thr Arg Arg Arg Arg Ile Glu Ile Ala His Ala Leu Cys Leu
41 42 43 44 45 46 47 48 49 50 51 52 53 54 55 56 57 58 59 60
T E R Q I K I W F Q N R R M K W K K E N
Thr Glu Arg Gln Ile Lys Ile Trp Phe Gln Asn Arg Arg Met Lys Trp Lys Lys Glu Asn

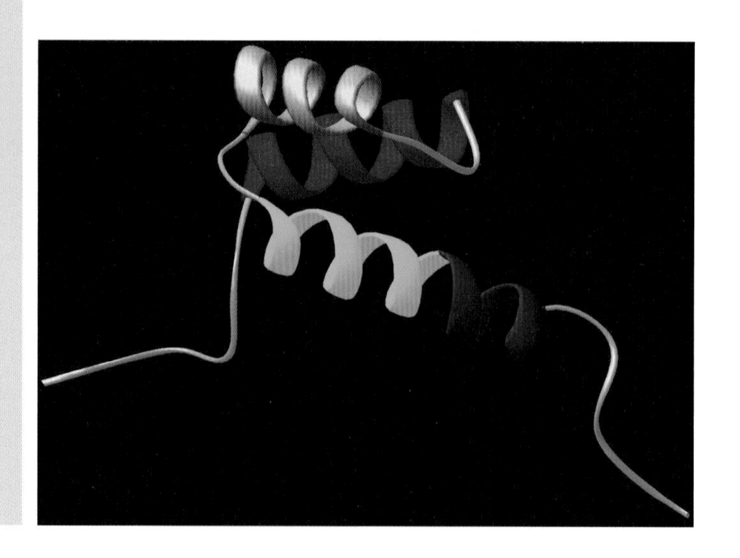

Figure 2.7 The homeobox and homeodomain. Top: the relationship between the two. Centre: the primary structure of one particular homeodomain – that of the *Drosophila* gene *antennapedia*. The 60 amino-acid homeodomain is highly conserved and functions to bind to the DNA of a target gene. The rest of the protein (often several hundred amino acids) is less conserved. Bottom: the three-dimensional structure of a homeodomain (referred to as helix-turn-helix). (Reproduced with permission from Rutgers University website.)

What are these other 'important recent discoveries'? In 1980, before the discovery of the homeobox, a large-scale search for mutations of genes that affect the developmental process was conducted on that classic **model animal**, the fruit-fly *Drosophila melanogaster*. Some of the resulting mutant **phenotypes**, such as the prickly larva known as 'hedgehog' (Fig. 2.8; more on hedgehog's developmental role in Chapter 3), have become quite famous. It can be argued that this study, more than any other, took developmental biology from its earlier descriptive, comparative and experimental phases to a new genetic phase.

Another important discovery was that a certain sub-group of homeobox-containing genes, the **Hox genes**, are responsible for determining the pattern of the antero-posterior axis of the body in most animals; and that, amazingly, the linear order of these genes, in the gene-complex in which they are found on a particular chromosome, usually maps to the antero-posterior order of their zones of expression in the animal's body – a phenomenon known as **colinearity**. We still await a satisfactory explanation of this fascinating finding.

So, the 1980s saw the beginnings of evo-devo in the discovery of homeobox-containing (and other) developmental genes; description of the mutant phenotypes that can arise when these genes are defective in the *Drosophila* model system; and a gradual spreading out of comparative studies across both model and non-model systems, covering gene sequences, expression patterns, and the interactions that make up developmental cascades or pathways. But where has evo-devo gone since then? And what, indeed, is evo-devo? How all-encompassing has it become?

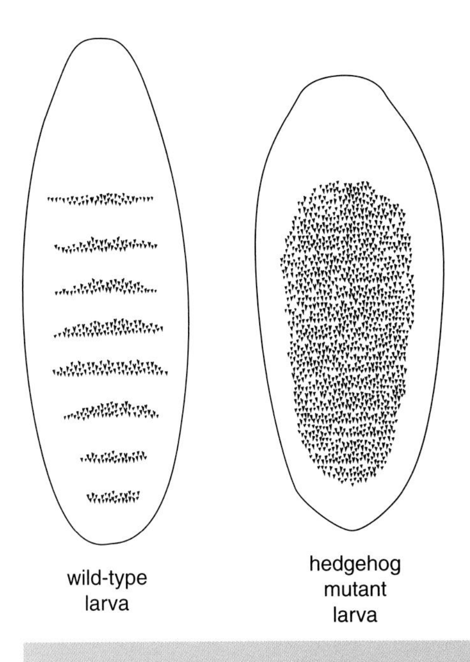

Figure 2.8 Normal (left) and *hedgehog* mutant (right) larval phenotypes of the fruit-fly *Drosophila melanogaster*. In the latter, the bands of denticles have been replaced by a complete coverage, giving a prickly appearance, hence the gene's name. (Ventral view; anterior at the top.)

2.6 Broad and Narrow Views of Evo-Devo

In fact, the new approach to evolution that is now called evo-devo can be thought of in a spectrum of ways. This spectrum can best be envisaged by picturing its extremities.

At one end of the spectrum we have the view that evo-devo is simply comparative developmental genetics – such as the discovery of the homeobox 'writ large' and extended to lots of other developmental genes over and above that group whose protein products contain homeodomains. This view is implicit in a 'conversation' between authors Pigliucci and Kaplan[28] and an imaginary sceptic, when they describe evo-devo as having become 'a molecular study of master switches'. They go on to criticise it as being able to 'provide us with only a (very) partial answer' to key questions in the evolution of development, such as how **body plans** and, more generally, evolutionary **novelties**, originate.

At the other end of the spectrum is the view that 'evo-devo' covers a wide range of types of study – effectively all recent studies that have sought to elucidate the relationship between evolution and development. As should already be clear, I take this broader view herein.

Each **developmental trajectory** or route is produced, in evolution, from a different one. Looked at another way, each route is the 'raw material' with which evolution works. Thus development is both a source for, and a result of, evolutionary change. In this book, we will consider evo-devo to include all recent studies (essentially post-1980) that try to address this two-way relationship, from whatever perspective. Comparative studies of the expression patterns and functions of developmental genes in different taxa are clearly central to this endeavour, and will be expanded on later (starting in Chapter 3).

But there are several other approaches that should also be included: *'eco-devo'*, including the evolution of developmental **reaction norms**, which encompasses genetic assimilation; parts of *quantitative genetics*, especially those that deal with the co-variation of phenotypic characters that are results of the developmental process; *palaeontological evo-devo*, including such classic **case studies** as the fin-to-limb transition; and in general, many branches of *comparative biology*, especially those in which the focus is on some aspect of comparative development. These will all be covered in later chapters.

As well as including different practical approaches to understanding how development evolves, a broad view of 'evo-devo' should include what have emerged as its key conceptual themes. These include gene **co-option** (or recruitment), **developmental bias** (or constraint) and many others (e.g. **modularity**, **evolvability** and the origin of evolutionary novelties). Again, all these will be covered in later chapters.

Taking a broad view of evo-devo, one way to chart its progress is to look at the diverse array of books on various aspects of the subject that have appeared over the last 30 years. These are listed in Table 2.1. Despite the large number of books shown, this list is far from comprehensive in that it excludes most multi-author collections and most 'popular' books on the subject. Clearly, the book-length literature of evo-devo reveals an explosion of activity compared to the rate of production of books on the interface between evolution and development prior to 1980. The recent appearance of four new evo-devo journals indicates a similar explosion of primary literature in this field. Some readers will know these journals already, others not. For the benefit of the latter, they are: *Evolution & Development*; *Development, Genes and Evolution*; *Molecular and Developmental Evolution* and *EvoDevo* (online only) (Fig. 2.9).

2.7 Too Few Laws, Too Many Facts?

A problem that will become very apparent as we enter the following chapter is that we will find it hard to see the wood for the trees, as the old saying goes. The earlier apparent security of having 'laws' of the evolution of development had the advantage of providing a view of a grand-scale wood. But the disadvantage, as we now know, is that this wood was false: a claimed monoculture obscuring the reality of something that is more akin to the diversity of a tropical forest, given the number of ways in which development can evolve. Having abandoned the laws, and witnessed the

Year	Author(s)	Title	Publisher
1983	Raff & Kaufman	*Embryos, Genes and Evolution*	Macmillan
1984	Arthur	*Mechanisms of Morphological Evolution*	Wiley
1987	Buss	*The Evolution of Individuality*	Princeton
1988	Arthur	*A Theory of the Evolution of Development*	Wiley
1988	Thomson	*Morphogenesis and Evolution*	Oxford
1991	McKinney & McNamara	*Heterochrony: The Evolution of Ontogeny*	Plenum
1992	Hall	*Evolutionary Developmental Biology*	Chapman & Hall
1993	Salthe	*Development and Evolution*	MIT
1994	Rollo	*Phenotypes: Their Epigenetics, Ecology and Evolution*	Chapman & Hall
1996	Raff	*The Shape of Life*	Chicago
1997	Arthur	*The Origin of Animal Body Plans*	Cambridge
1997	Gerhart & Kirschner	*Cells, Embryos, and Evolution*	Blackwell
1998	Schlichting & Pigliucci	*Phenotypic Evolution: A Reaction Norm Perspective*	Sinauer
1999	Hall	*Evolutionary Developmental Biology (2nd edn)*	Kluwer
2001	Carroll, Grenier & Weatherbee	*From DNA to Diversity*	Blackwell
2002	Wilkins	*The Evolution of Developmental Pathways*	Sinauer
2003	West-Eberhard	*Developmental Plasticity and Evolution*	Oxford

Table 2.1 A list of books in the general area of evolutionary developmental biology, from the early 1980s to the present. Subtitles are not included for all. The list is not exhaustive, partly because it is hard to define the exact limits of the subject, and partly because many multi-author collections and most popular science books in the area have been excluded. A few books on the history and philosophy of evo-devo are included. Two of the books listed have run to a second edition; in these cases the two editions are listed separately because they are quite different (and in one case they have different publishers). (*Continued on p. 30*)

Table 2.1 Contd

Year	Author(s)	Title	Publisher
2003	Minelli	*The Development of Animal Form*	Cambridge
2003	Hall & Olson (eds)	*Keywords and Concepts in Evolutionary Developmental Biology*	Harvard
2004	Robert	*Embryology, Epigenesis and Evolution*	Cambridge
2004	Arthur	*Biased Embryos and Evolution*	Cambridge
2005	Carroll *et al.*	*From DNA to Diversity (2nd edn)*	Blackwell
2005	Amundson	*The Changing Role of the Embryo in Evolutionary Thought*	Cambridge
2007	Laubichler & Maienschein (eds)	*From Embryology to Evo-Devo*	MIT
2008	Minelli & Fusco (eds)	*Evolving Pathways: Key Themes in Evolutionary Developmental Biology*	Cambridge
2009	Minelli	*Forms of Becoming: The Evolutionary Biology of Development*	Princeton
2009	Gilbert & Epel	*Ecological Developmental Biology*	Sinauer

generation of a wealth of comparative molecular data to add to the earlier comparative organismic information, the pros and cons have been reversed: the wood has disappeared and we are confronted with too many different trees. What is the best way to deal with this situation?

The answer, as often when information seems to have become overwhelming, is to organise that information into categories. Indeed, we can go further than that, and organise it into multi-level hierarchies of categories – Darwin's 'groups within groups'. Taxonomy can be applied not just to organisms but to any other entities – for example genes. Indeed, a hierarchical classification of homeobox genes is available in book form[29], and a simplified classification of some *Drosophila* genes is given in Fig. 2.10.

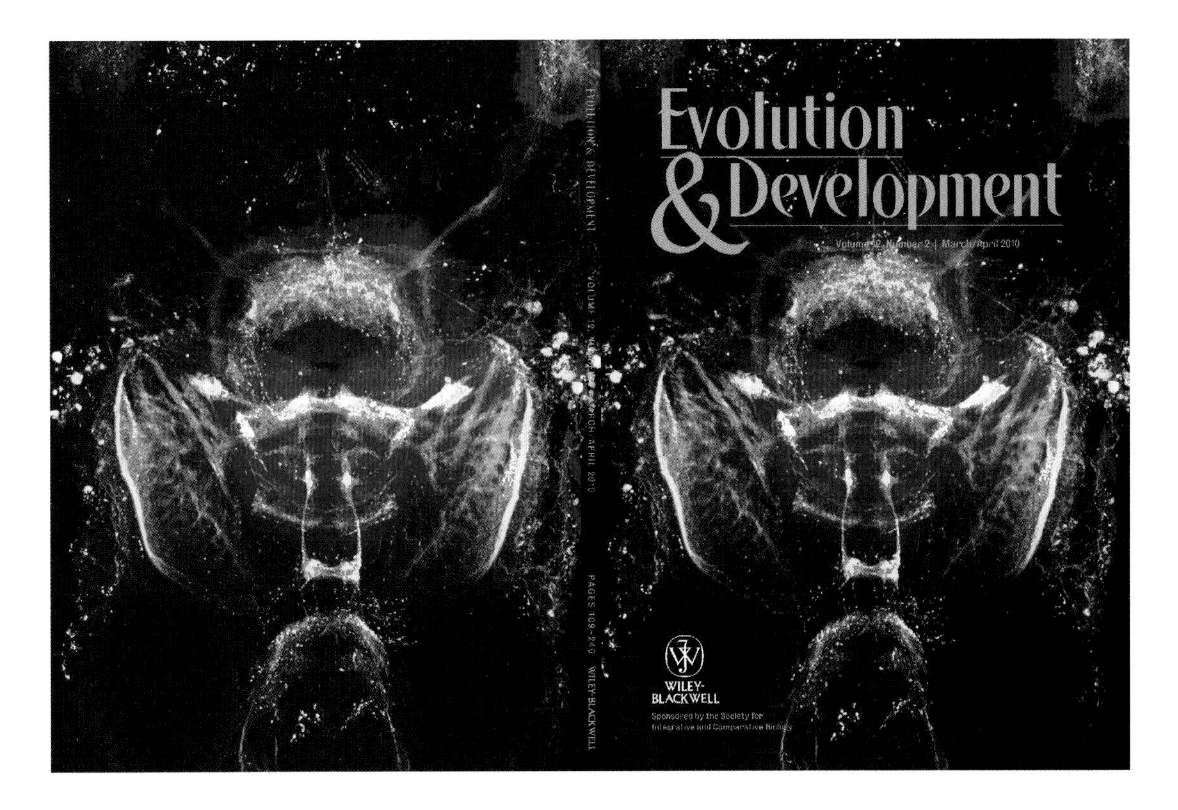

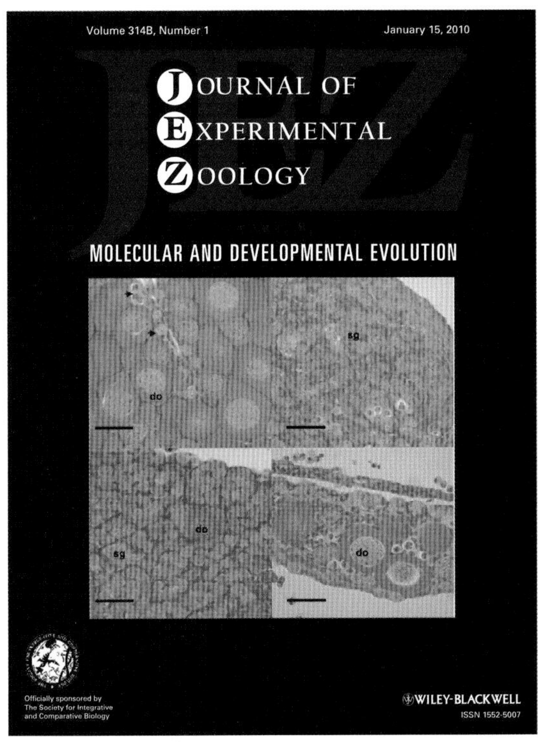

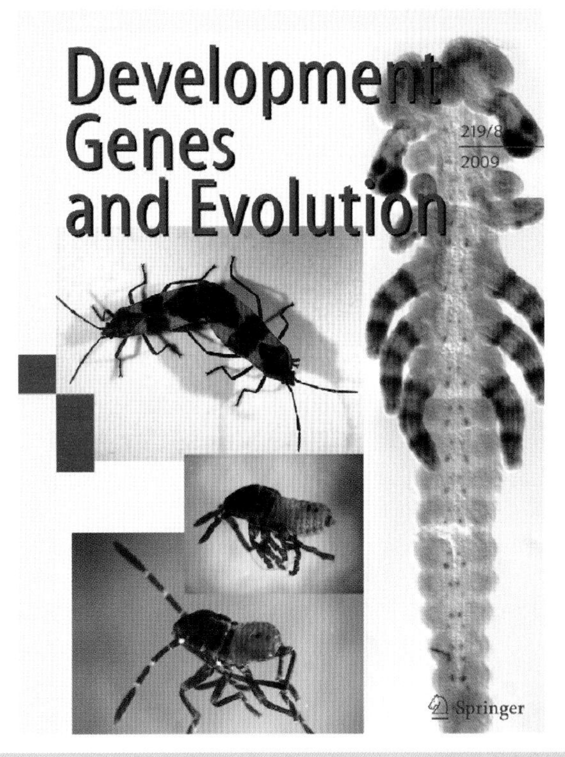

Figure 2.9 Four new or re-branded journals that are vehicles for the publication of research findings in evo-devo and related fields: *Evolution & Development*; *Development, Genes and Evolution* (reproduced with permission from Springer); *Molecular and Developmental Evolution*; and *EvoDevo* (online only, so no cover shown).

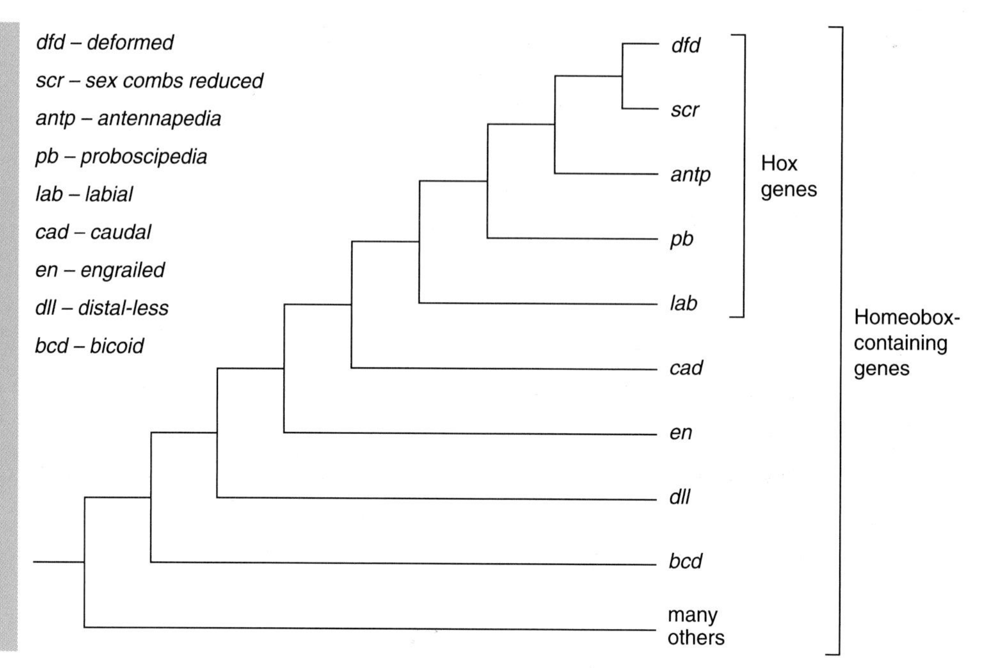

Figure 2.10 A simplified hierarchical classification of the homeoboxes of some *Drosophila* developmental genes. More complete classifications can be found in the literature, both for homeobox genes in general and for human, *Drosophila* and other species' homeobox genes in particular[30].

Types of evolutionary change in development – in other words, types of developmental repatterning – can also be ordered into categories: **heterochrony**, **heterotopy**, **heterometry** and **heterotypy**, as noted earlier. These will form the subjects of Chapters 6 to 9. Adding subcategories, such as **neoteny** and **progenesis** within heterochrony, again gives us 'groups within groups'.

Another categorisation that can be attempted involves the **signalling pathways** that are key components of any developmental system. For example, there are Wnt signalling and Hedgehog signalling pathways, both of which are explained in Chapter 3. Within each of these categories of signalling there are variations. This has led, among other things, to recognition of the '**canonical** Wnt pathway' and variations upon it: again, groups within groups, giving us a hierarchical classification.

Perhaps the biggest challenge that lies ahead is two-fold: (i) to extend and refine such classification systems as new data continue to accumulate; and (ii) to make connections between them. These connections will in some cases be easier, as in relating gene classes to classes of signalling, in other cases harder, as in relating gene classes to classes of change at the organismic level – for example, changes in developmental reaction norms.

In this book, I will try to give as much order as possible to all the individual pieces of information that derive from individual case studies on the evolution of development, conducted at various levels and on various taxa. This approach should ensure that the proverbial wood is not being obscured by its constituent trees. Towards the end of the book (especially in Chapter 19), we will examine what general conclusions seem possible at this stage in the progress of the overall evo-devo endeavour.

SUGGESTIONS FOR FURTHER READING

For anyone interested in more detail on the history of efforts to understand the relationship between evolution and development prior to the advent of molecular evo-devo in the 1980s, with a particular focus on 19th century comparative embryology, a good source is:

Gould, S.J. 1977 *Ontogeny and Phylogeny*. Harvard University Press, Cambridge, MA.

For reviews of the key concepts, and the overall conceptual structure, of evo-devo, broadly defined, see:

Arthur, W. 2002 The emerging conceptual framework of evolutionary developmental biology. *Nature*, **415**: 757–764.

Hall, B.K. and Olson, W. M. (eds) 2003 *Keywords and Concepts in Evolutionary Developmental Biology*. Harvard University Press, Cambridge, MA.

Minelli, A. and Fusco, G. (eds) 2008 *Evolving Pathways: Key Themes in Evolutionary Developmental Biology*. Cambridge University Press, Cambridge, UK.

See also Table 2.1 for a list of books on evo-devo over the last 30 years.

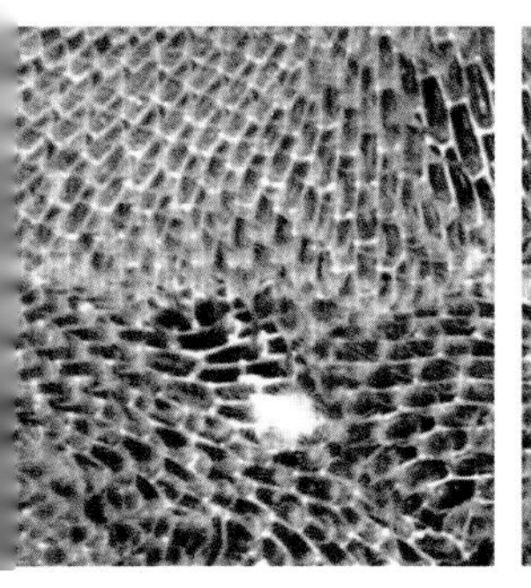
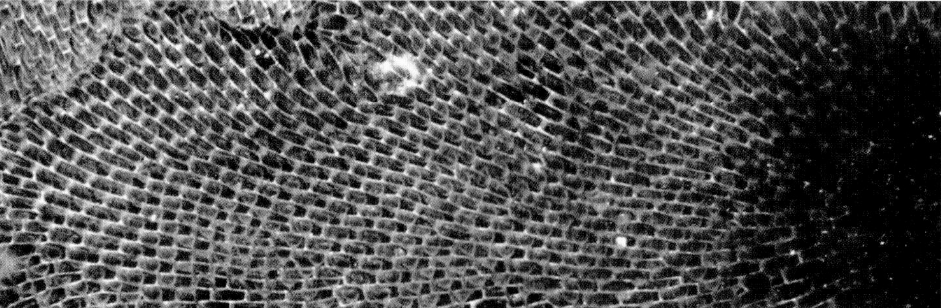

CHAPTER 3
Development, Cells and Molecules

3.1 Analysing the Developing Organism

The early students of comparative embryology, such as von Baer and Haeckel, were restricted in what they could do because of the limited range of techniques available at the time. Obtaining the picture of embryonic divergence that was shown in Fig. 1.6 involved little more than preservation and microscopy. In order to become less descriptive and more analytical, additional techniques were needed. These began to appear in the early-to-mid 20th century with surgical manipulation of embryos and, later, the alteration of developmental and adult **phenotypes** by **mutation**, followed by classical genetic analysis[31]. However, analyses at the molecular level have only become possible relatively recently. Since 1980, the range of molecular approaches available has expanded rapidly[32].

This chapter provides an overview of developmental processes at the cellular and molecular levels, to indicate what is now known of their causality. Before going into the details, however, it is worth emphasising the fact that none of these will remove the classic 'chicken and egg' problem. A late developmental process has as its inputs the results of an earlier one; and the earlier one has as its inputs the results of an earlier one again. Ultimately, this process of regressing backwards through developmental time takes us to the egg. But it does not stop there either.

No egg is homogeneous. Rather, the typical egg is a large cell that is heterogeneous: its cytoplasm contains localisations of many developmentally-active substances, which derive from the mother. For example, in a *Drosophila* egg (and in the eggs of many other flies), there is a localisation of maternally-derived *bicoid* mRNA at the anterior end. Indeed it is this, together with other such localisations, which causes this end to be anterior. (The gene name *bicoid* derives from the mutant phenotype – two-tailed or bi-caudal – that arises in the offspring of mothers in which the gene is absent or defective: see Appendix 2 for explanation of gene names in general.) So, the **developmental patterning** of any one multicellular organism can be traced back, not just to the egg from which it arose, but beyond that to the previous generation – and so on indefinitely. An evolutionary **lineage** can be thought of as a series of developmental patternings and repatternings that extends back to a particular origin of multicellularity several hundred million years ago.

Of course, we cannot study the molecular basis of the development of an ancient ancestor, though we may be able to infer it to some degree by comparisons among its various surviving descendants. All of what follows in this chapter relates to the development of **extant** rather than extinct creatures. But even with this restriction, the variety is overwhelming. Some aspects of development are indeed shared among all the members of very large **clades** – such as the use of the gene *engrailed* (Section 3.3) to determine the polarity of segments in all (as far as we know) arthropods; and 'all' means more than a million species. But other aspects are more specific: the *bicoid* gene, mentioned above, has its anterior-determining role only in one comparatively small group of insects – the 'higher' **Diptera**.

Much molecular developmental biology has been, and still is, carried out on a small number of **model organisms**. The main vertebrate models are the mouse *Mus musculus*, the chick *Gallus gallus*, the frog *Xenopus laevis* and the zebrafish *Danio rerio*. In the invertebrates – a much larger number of species with a far greater range of **body plans** – there are just two main models, the fruit-fly *Drosophila melanogaster* and the round-worm *Caenorhabditis elegans*. The most advanced botanical model is the small flowering plant *Arabidopsis thaliana* belonging to the family **Brassicaceae**. All these model organisms are shown in Fig. 3.1.

It is always worth bearing in mind, when dealing with developmental information derived from studies on any one of these model organisms, the possible breadth of taxonomic group to which it applies. Some information on mouse development will be specific to mice, while other information will be more general – but will it be general to rodents, mammals, vertebrates, or to the whole animal kingdom? As molecular developmental studies spread out to non-model organisms, a process that has begun but still has a long way to go, we will be better able to answer such questions.

It is not just at the molecular level that questions of generality should be borne in mind; such questions apply at the gross phenotypic level too. Because of the use of *Xenopus* as a developmental model organism, biologists often talk of the development of 'the frog'. But the clade containing all frogs (and toads) – the **Anura** – has over 4000 species. While most of these have indirect development, via a tadpole **larva**,

Figure 3.1 The seven main 'model organisms' of developmental biology (for names, see text). As research extends out across an ever-wider spectrum of taxa, the gap between these seven and the many thousands of 'non-model' organisms whose development has been little studied has become less clear. For example, some would now regard the amphioxus *Branchiostoma* as a 'model non-vertebrate chordate'; and the amphipod *Parhyale* might be considered as a 'model crustacean'.

some tropical frog species have evolved direct development, so that what hatches out of the egg is not a tadpole but rather a miniature frog, as we will see in Chapter 18.

For the present, however, we will leave the comparative aspect of development behind; it will re-emerge later in due course. The comparative developmental biology of the 21st century will only be better than the comparative embryology of the 19th century, if it is based on a thorough understanding of the causality of development at the cellular and molecular levels, in at least some animals and plants. Presenting the current state of such understanding is the task now at hand. And the best way to undertake it is to focus in on particular systems – often the model ones – and to show how, in those systems, the visible course of development that has been apparent to biologists for a long time is brought about through cellular and molecular interactions. Throughout the following account, we will focus on interactions that are known or suspected to be important in the development of a wide range of organisms.

3.2 Cells and Development: The Basics

Development is the process in which, from a single-celled beginning (usually), a continuous sequence of events produces, ultimately, an adult consisting of many cells – nearly a thousand in a round-worm, more than a million in a fly, and trillions in a human. At all stages, this process must be very tightly controlled, because the number of possible cellular arrangements is vast, and only a very few of these arrangements represent viable developmental stages that will, through further development, go on to produce a viable adult.

The early stages of development of *Xenopus*, such as the **blastula**, are shown at the cellular level in Fig. 3.2. Later developmental stages are not shown at the same (i.e. cellular) level, simply because the number of cells is too large. Nevertheless, it is easy to imagine, at least in broad terms, the massive-scale cell proliferation and differentiation that must be going on in the development of a later stage, such as a leg.

How are these exceedingly complex sequences of events controlled? The answer can be given at the level of the cell itself, but ultimately the behaviour of cells must be explained in molecular terms. So, in the present section, we start with cells and end up with molecules. In the next section, we focus on molecules: largely on the DNA of the genes, and their RNA and protein products. Then, having looked at how gene expression contributes to development, we climb back up again in terms of levels of organisation, because no gene acts in isolation – rather genes and their products form interacting signalling systems that govern cell behaviour. Finally, we ascend to the level of the organism itself – i.e. back to the level at which we started.

Cells do four main things in the course of development, over and above the 'ordinary' things they do in the adult, such as carrying out the basic metabolic processes that keep them alive. These four things (Fig. 3.3) are: division/proliferation; differentiation; movement; and programmed death, or **apoptosis**. We will now look at these in turn.

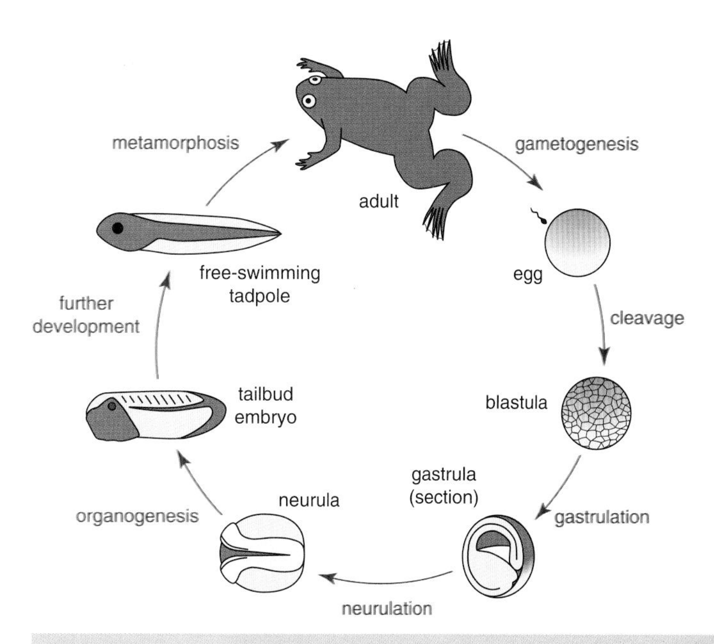

Figure 3.2 The main stages of development of one of the seven 'model organisms' – the frog *Xenopus*. Names of stages are in black text; names of developmental processes are in blue. (Redrawn with permission from Wolpert *et al. Principles of Development*, 3rd edn, Oxford University Press, Oxford, p. 4.)

Cell Proliferation

Whether the adult concerned has hundreds of cells or trillions of them, its production from a zygote (or other unicellular starting point) necessarily involves large-scale cell proliferation by division of 'parental cells' at each developmental stage into a greater number of daughter cells at the next. Aspects of this process that need to be regulated if the correct adult form is to

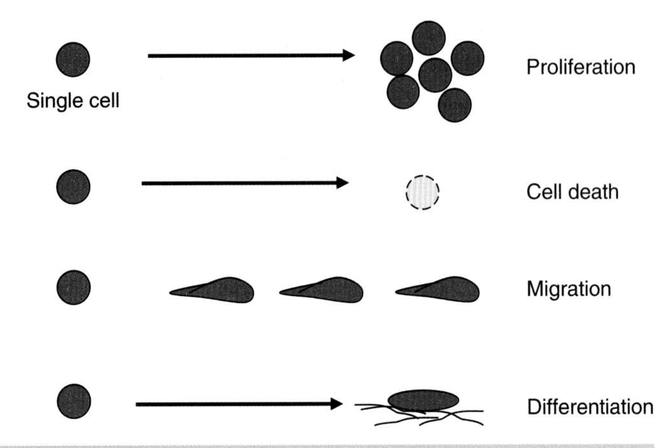

Figure 3.3 Diagrammatic illustration of the four main things that cells do in the course of the developmental process in all multicellular organisms: proliferation (i.e. increasing in number through repeated cell divisions); programmed cell death (apoptosis); migration from one part of the developing organism to another; and differentiating into one of a range of specific cell types (such as those shown in Fig. 3.4 in the case of a developing vertebrate).

ensue include: rate of cell division; plane of division; and duration of period of proliferation. With respect to the last of these, division must eventually either stop completely (in some tissues) or reduce (in others) to the low level that is appropriate to balance the death of cells, so that their number remains constant. Failure to achieve this reduced level of division can result in the pathological condition of tumour formation.

Cell Differentiation

The cells of an early embryo, such as those that make up the *Xenopus* blastula, are rather similar to each other. In contrast, those of the adult belong to many different types – over 200 in most vertebrates (for examples, see Fig. 3.4). It is the process of differentiation through which, in a lineage of cells, one particular differentiated type results, ultimately, from a non-differentiated starting-point, namely a **stem cell**.

It has been clear for many years that virtually all cells, in the body of any particular animal or plant, contain the complete **genome** that is appropriate for the species concerned. Thus almost all human cells contain about 25,000 genes (far fewer than the 100,000 that had been speculated about before the results of the human genome project were published). By coincidence, all cells in an *Arabidopsis* plant also contain about 25,000 genes; and all fruit-fly cells contain somewhat fewer: about 15,000 genes. This means that the process of differentiation does not involve the loss of genes that are not needed in the relevant cell-type; rather, those genes are still present but are switched off.

When a gene is switched on, i.e. making its RNA product, it is said to be *expressed*. Conversely, when it is switched off and inactive, it is sometimes said to be *silent*. Any particular differentiated cell, whether nerve, muscle, skin or whatever, is characterised by a particular pattern of expressed and silent genes. So our central question with regard to differentiation is: how do these patterns come about? As we will see, the answer lies partly in the genes themselves, and partly in the signals they receive from their surrounding environment, itself made up of other cells and the extracellular matrix.

Cell Movement

Some developmental stages are characterised by much cell movement. This is particularly true, in most animals, of the process of **gastrulation** (Fig. 3.2), in which there is large-scale 'streaming' of cells, and indeed whole populations of cells, from

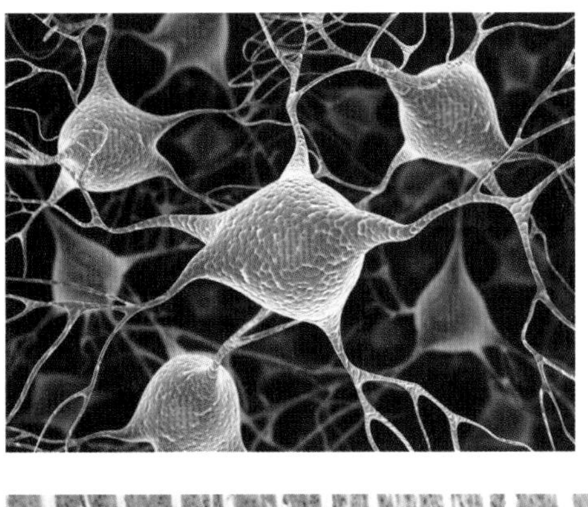
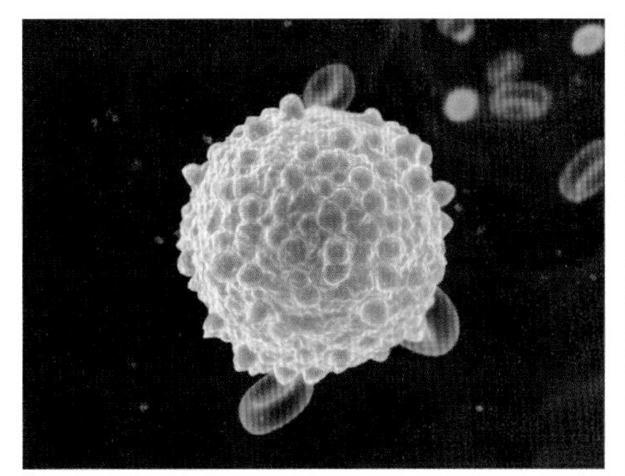
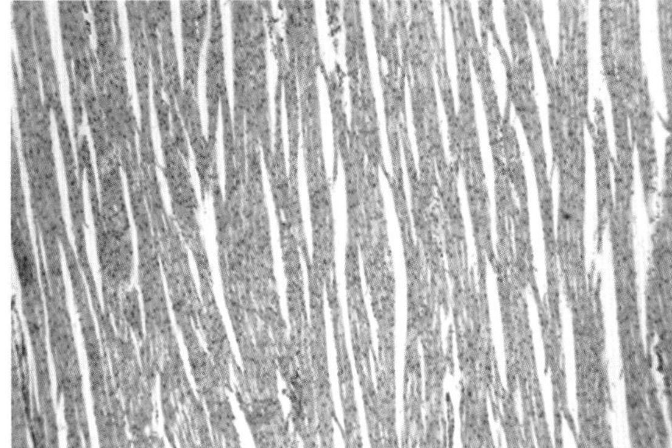
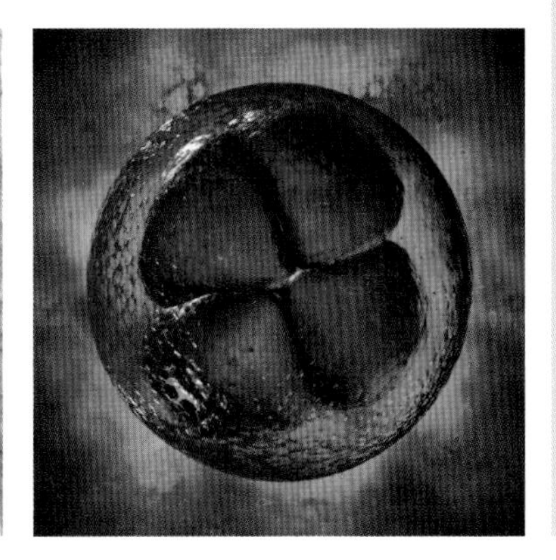

Figure 3.4 An adult vertebrate, such as a human, has more than 200 different cell types. Here, just a few of these are shown to indicate how very different cells can be in the same organism; specifically: nerve cells, a white blood cell, muscle cells and the 'blastomere' cells into which the fertilized egg cell divides.

one part of the embryo to another. As with division and differentiation, the causality of the cellular process must be found at the molecular level.

Most non-moving cells are fixed to their neighbours by various transmembrane cell-cell adhesion molecules, including those of the group called cadherins (calcium-dependent adhesive proteins). For an initially static cell to begin to move, this must be facilitated by the switching off of the genes making such proteins and the degradation of any such proteins already made. But the ability to move, thus generated, is not the same as actually moving. The movement itself can be brought about in a number of ways. One of these involves a different group of transmembrane proteins called **integrins**. These can cause a dynamic pattern of binding and unbinding to the cell's surroundings that can be likened to a cell moving over its neighbours on molecular 'caterpillar tracks'.

Apoptosis

Like all organisms, all cells ultimately die. This can happen for three reasons. First, and most obviously, on the death of an organism, all its constituent cells die with it. Second, during the normal functioning of an organism, different cell-types have different life-spans, with those exposed to particularly harsh conditions dying most

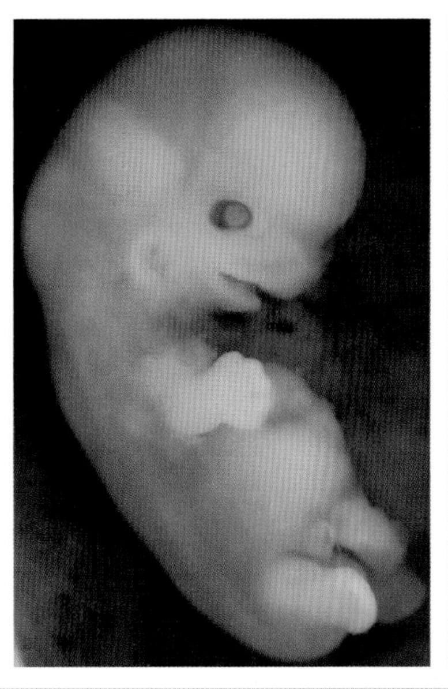

Figure 3.5 Human embryos (left) have webbed hands and feet. The reason our digits become separate is that there is programmed cell death (apoptosis) in the interdigital regions as development progresses. In other kinds of vertebrate where the adult has webbed feet, such as ducks (right), apoptosis is reduced.

readily. A good example of this is an **epithelial** cell in the stomach of a human or other mammal, where exposure to hydrochloric acid ensures a short cell life. When such cells die, they are replaced by the progeny of nearby stem-cells. Third, cells can die because they have been 'programmed' to do so. This category of cell death is called apoptosis, and it is important in a variety of developmental processes. One of these is the production of the digits of the foot Human embryos have webbed feet (Fig. 3.5), but adults, of course, do not. This is because the cells of the inter-digital regions are programmed to die at a certain stage of **embryogenesis**. This apoptosis is reduced or absent in those vertebrates that have webbed feet as adults (e.g. frogs, ducks).

3.3 Genes: Structure, Expression and Developmental Function

We have seen that development can be explained in terms of the things cells do; and that these things can be grouped into four categories: division, differentiation, movement and apoptosis. In turn, these cellular activities can be explained in terms of molecular ones, including patterns of gene expression. So now we turn our attention to genes and how they work.

Much of the information that follows is true of genes throughout the **Eukaryota** – the group that includes animals, plants and fungi. However, a good deal of it is not true of **prokaryote** cells – those of bacteria for example – where gene structure and function are somewhat different.

The Location, Physical Context and Types of Genes

In a eukaryote cell, the vast majority of genes are located in the nucleus. There are also small numbers of genes in the mitochondria and, in plant cells, in the chloroplasts. These cellular organelles have their own genomes due to their origin as prokaryote

endo-symbionts; the gene numbers involved are generally in the range from 10s to 100s, compared with the 10,000-plus in the nucleus.

In the nucleus, genes exist within the physical fabric of the chromosomes (Fig. 3.6). Each eukaryote chromosome consists of a long DNA molecule which is coiled and supercoiled; and is associated with protein molecules, in particular a group of proteins called **histones**. The physical structure of the DNA-histone complex is very dynamic, changing both in the short term as genes are turned on and off, and on the longer time-scale of the cell-cycle (i.e. from one cell division to the next).

The DNA of a typical eukaryote chromosome is not simply a string of genes, as used to be thought; far from it. It is now clear that much of the DNA is non-coding. The proportion varies considerably among different eukaryotes, but often exceeds 90 per cent. It is not yet clear whether most of this is of relevance to development, but *some* non-coding sequences are absolutely crucial for development, notably the regulatory regions that play a major role in gene expression. Whether these are considered part of 'the gene' or not is a semantic issue. Here, we will use 'genic DNA' to include them and 'coding DNA' to exclude them (Fig. 3.7).

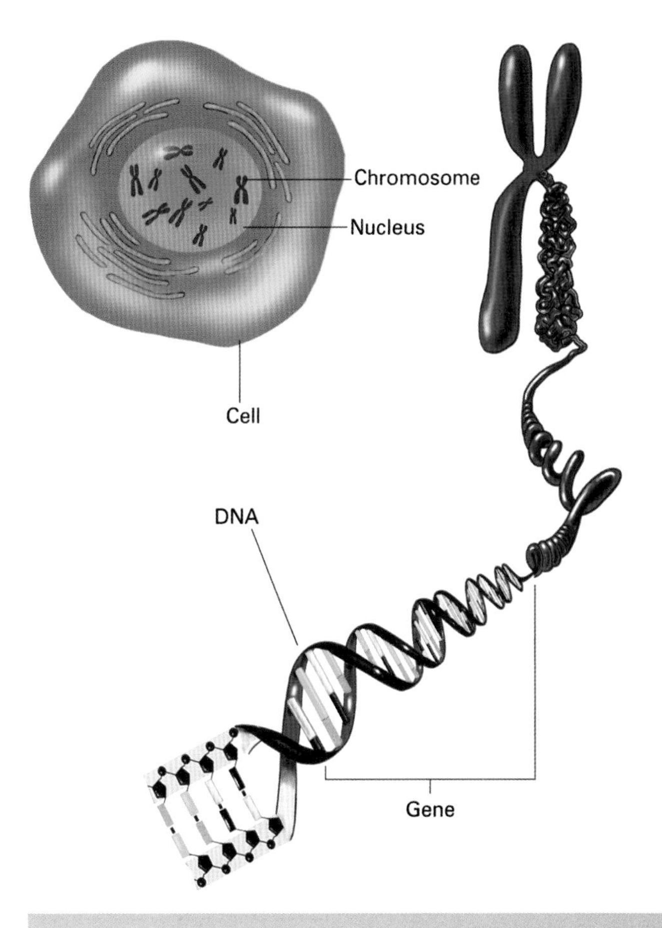

Figure 3.6 The relationship between a gene and a chromosome. A gene is a stretch of DNA double-helix, as shown towards the bottom of the picture in highly magnified form. The double-helix is coiled, 'supercoiled', and associated with proteins, to form the chromosomes (top of picture) that can be seen down an ordinary light microscope in the cell's nucleus (photographs courtesy of National Institutes of Health).

As well as distinguishing genes from other sorts of DNA, it is important to distinguish between different types of genes. From a developmental viewpoint, it is useful to recognise the following three categories of genes, while at the same time recognising that they are not mutually exclusive: some genes will fall into just one category, but others may fall into two.

First, there are the *housekeeping genes* – those that are expressed in most or all cells, because they produce enzymes or other proteins that contribute to the basic metabolism that keeps the cell alive. These include the genes that make the enzymes involved in glycolysis and the Krebs cycle. Second, there are what can be called *terminal target genes* – those whose products are found only in certain cell types, and indeed help to characterise those cell types – such as haemoglobin in red blood cells. The proteins made by these genes are the result of differentiation, rather than a cause of it. Third, there are *regulatory genes*, whose primary role is to control the activities of other genes – both qualitatively (on/off) and quantitatively (e.g. varying the rate of expression). Clearly, the genes of this third category are especially important in development.

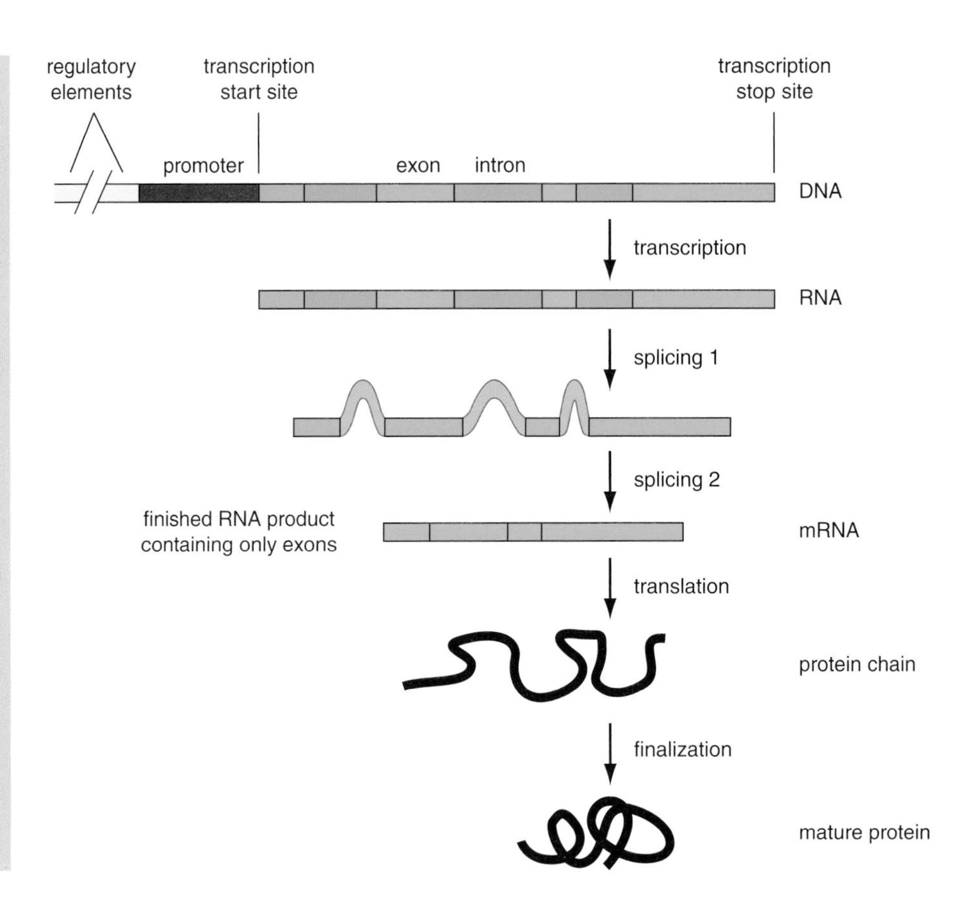

Figure 3.7 A typical eukaryote gene is composed of many different parts. The coding region is interrupted by introns; these are transcribed but then removed from the RNA before it is translated. The mature mRNA corresponds to the coding regions of the gene (in the sense of coding for the resultant protein). Regulatory regions are often upstream of the coding regions; but they may also be found downstream and even within introns.

Gene Expression and Function

Here is a brief and simplified account of what happens when an initially silent gene is switched on (Fig. 3.8). One or more activating molecules, called **transcription factors**, bind to one or more of the regulatory regions of the gene. Often, different regulatory regions are used in different tissues, or in different regions of a developing embryo. This activation ultimately causes the enzyme **RNA polymerase** to bind to the DNA of the gene at the start-site of transcription. Again, other molecules may also bind here too. The polymerase then moves along the gene, producing an RNA sequence that is complementary to the sequence of the coding strand of this gene. Recall that DNA is a double-helix: in any particular gene, usually only one helix is used as the basis for transcription; the other is not transcribed. However, in some cases it appears that the complementary strand can also be transcribed, giving a different product.

The link between the initial binding of one or more activating factors at regulatory sites and the initiation of transcription at the start site is complex and not entirely understood. It seems that in most cases there is a conformational change in the DNA, causing a looping effect, whereby the regulatory site is brought into close proximity with the start site; though this may not always be a necessary part of the process.

When the transcription of the RNA molecule corresponding to the gene is complete, the RNA polymerase and the fully-formed RNA transcript detach from the DNA. At this point, the transcript is still in the nucleus; and it undergoes a process of 'splicing' there, in which the sections of RNA corresponding to the **introns** of the gene are removed. The remaining, **exonic**, bits of RNA are joined together to give the final messenger RNA (or mRNA) that will leave the nucleus through one of the nuclear pores[33] (Fig. 3.9). In some genes, alternative splicing patterns are possible – removal of some versus other intronic sequences – thus enabling the production of several different mRNAs, and hence proteins, from the same gene.

Once in the cytoplasm, the mRNA gravitates to ribosomes, which are the sites of translation – the production of a protein from the 'message'. The protein has a particular primary structure, namely the sequence of amino acids, which is dictated by (i) the DNA sequence of the gene concerned, (ii) the splicing pattern, and (iii) the genetic code through which particular triplets of mRNA bases give rise to the incorporation of particular amino acids into the gradually elongating protein molecule. The forming protein will also fold in various ways, giving it a three-dimensional shape that is crucial to its function (this folding constitutes secondary and tertiary structure). In some cases, such as haemoglobin, the functional

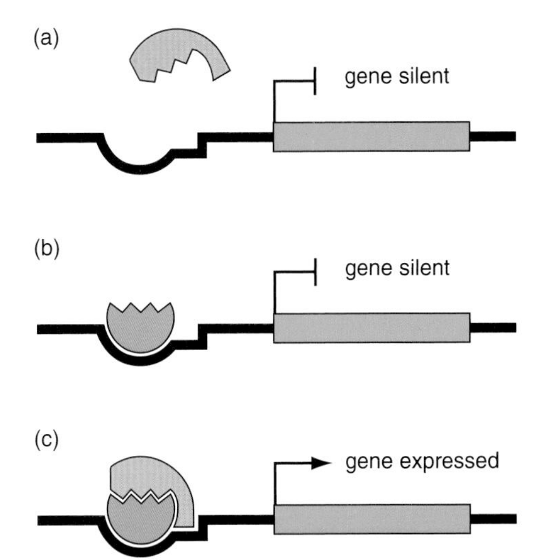

Figure 3.8 Genes are switched on by the binding of transcription factors to regulatory regions. This picture shows a diagrammatic view of how a combination of different transcription factors (c) may be necessary to initiate transcription, whereas one factor on its own cannot do so (a, b).

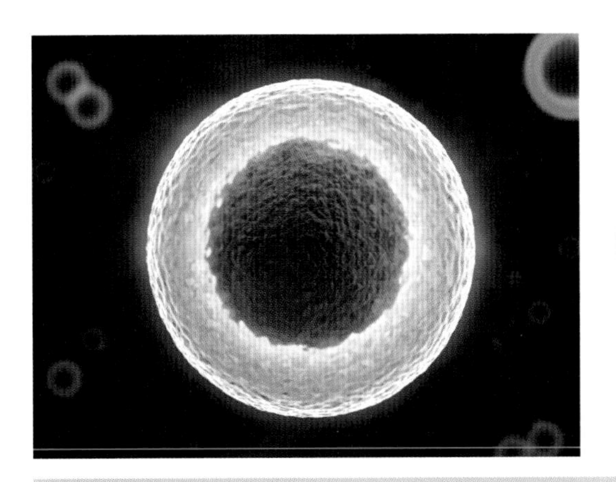

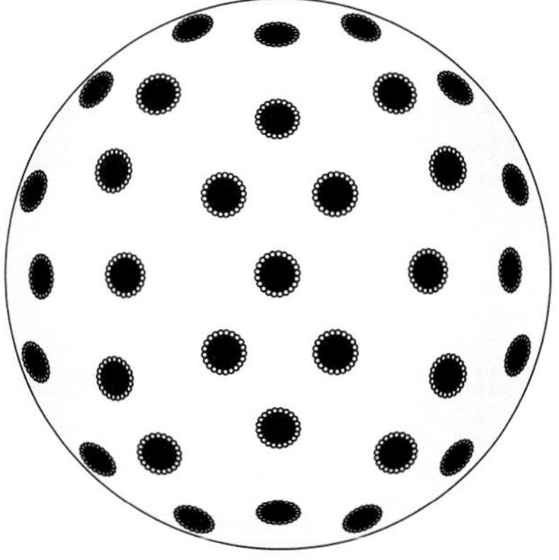

Figure 3.9 In some cells viewed microscopically, like the one shown (left), the nucleus can be seen, often because a nuclear stain has been used. It is approximately spherical, and few features can be seen. However, molecular studies have revealed in considerable detail the structure of pores in the nuclear membrane (right) that permit the ingress and egress of large molecules such as RNAs and proteins. (Right: redrawn from Alber *et al.* 2007, *Nature*, **450**: 695–701 and from the cover-picture of that issue of *Nature* – 29 November 2007.)

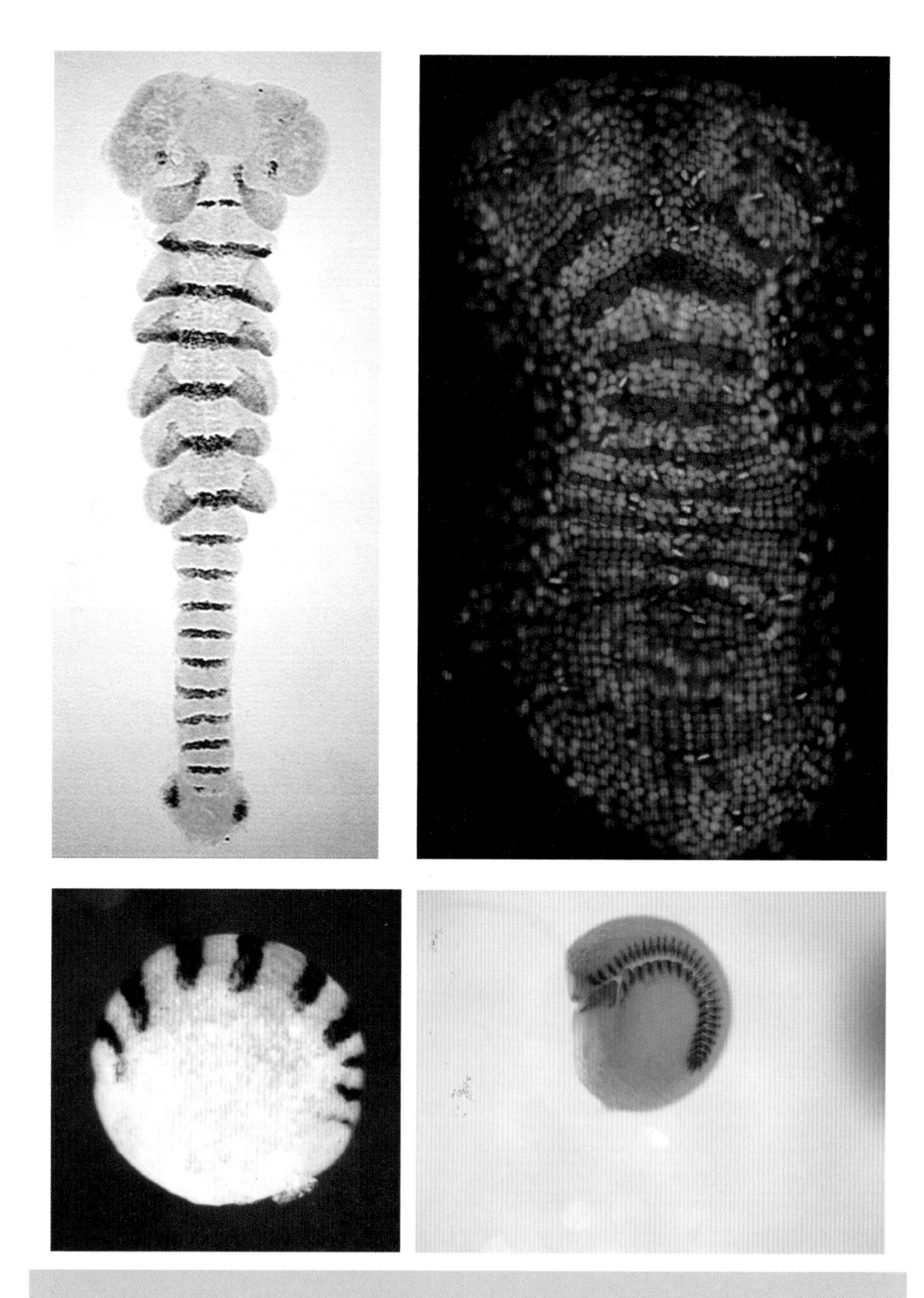

Figure 3.10 Segmental stripes of expression of the *engrailed* gene in arthropod embryos from all four subphyla: an insect (top left); a crustacean (top right); a spider (bottom left) and a centipede (bottom right). (Photographs courtesy of William Browne, Nipam H. Patel, Nicole Pirkl, Wim Damen and Vincent Vedel.)

protein consists not just of one polypeptide chain but of two or more, in which case the association between these chains is called the quaternary structure.

What happens to the protein next depends on its functional role. Many proteins that are involved in controlling the developmental process are transcription factors – so they will go into the nucleus (again via the nuclear pores), bind to the regulatory region of one or more genes, and thus start the cycle of activity off again. Others will be secreted out of the cell, to be picked up by receptors on the outside of other cells, where they will initiate a signalling pathway (Section 3.4).

An example of a transcription factor is the product of the *engrailed* gene, which is involved in arthropod segmentation (see Appendix 2 for an explanation of gene and protein names). Like any other developmental gene, *engrailed* has a particular **expression pattern** in an embryo: that is, it is switched on only in certain places (and at certain times). *Engrailed* is in fact switched on in the cells of the posterior part of each developing segment, both in the model organism *Drosophila* and in other arthropods (Fig. 3.10). This is because it is a 'segment polarity gene', conferring posterior identity – a sort of **positional information**[34] – on the cells in which it is expressed.

3.4 Signalling Pathways Within and Between Cells

Development involves a very complex network of interacting signals. We now recognise several particularly important classes of signalling pathway. In each of these, as in development as a whole, we are confronted by the 'chicken and egg problem' – i.e. the question of deciding which point is the start of the process. In what follows, we will consider the receipt of a signal, or **ligand**, at the exterior of the cell in focus as the start – while acknowledging that this is to some extent an arbitrary choice.

Given the complexities of signalling pathways, it makes sense first to outline a kind of generalised and simplified pathway, and then to look at examples of particular pathways in more detail. The simple picture is shown in Fig. 3.11.

The ligand is attracted to a particular type of receptor. Once bound to that receptor, a signal is transmitted through the cell membrane and relayed into the nucleus via one or more intermediaries. The result, in the nucleus, is that certain genes (depending on the signal concerned) are switched on or off. What happens next depends on what the target genes do. One possibility is that they make proteins that are secreted from the cell and become ligands for a different signalling pathway.

Now for examples of two specific signalling pathways in more detail. We will focus first on the Wnt pathway and then on the Hedgehog pathway. For details of other major pathways, see the latest edition of Scott Gilbert's *Developmental Biology*, full details of which are given in the Suggestions for Further Reading at the end of this chapter.

The Wnt Pathway

The name 'Wnt' derives from the w of *wingless*, a gene affecting wing development (among other things) in *Drosophila* and the nt of *int*, which is short for 'integration site genes' in mice. These latter genes are sites where mouse mammary tumour virus can insert – i.e. they are **oncogenes**. Molecular sequence analysis shows that *wingless* and

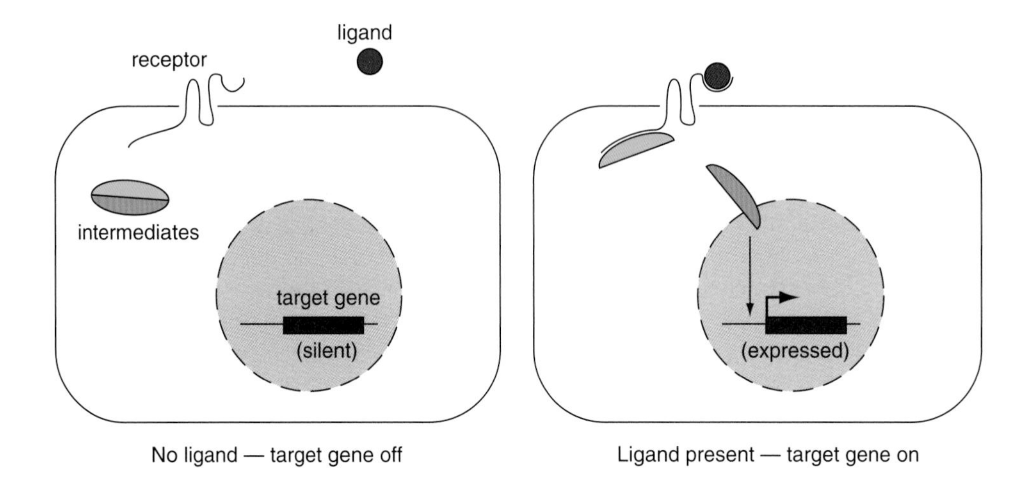

Figure 3.11 Diagram of a generalised (and simplified) signalling pathway. Left: ligand not yet arrived at receptor; target gene switched off. Right: ligand binds to receptor and the target gene is switched on.

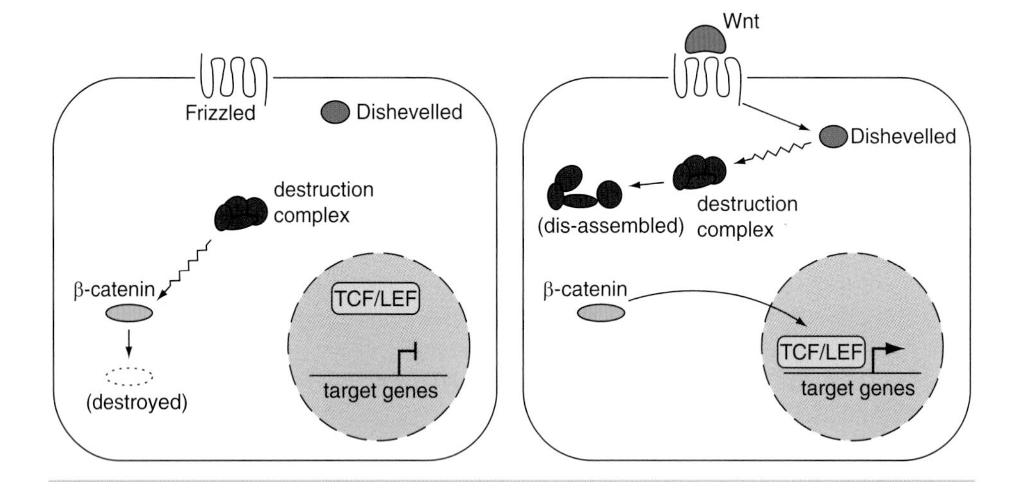

Figure 3.12 The canonical Wnt signalling pathway. Left: target genes silent due to lack of the Wnt ligand. Right: target genes expressed when Wnt is received by the Frizzled receptor in the cell membrane.

int genes are **homologous**. This is a good example of how two different lines of research carried out on very distantly related animals have converged to reveal common signalling pathways.

It is now clear that, for any named signalling pathway, there are different versions of it. Often, the first-discovered one is referred to as the **canonical** one. What is shown in Fig. 3.12 is the canonical Wnt signalling pathway. This is known to operate in all animals, from **basal** ones such as cnidarians to the most 'advanced' ones, for example, vertebrates.

Figure 3.12 shows two situations: one (LHS) when no Wnt signal is reaching the cell concerned and the other (RHS) when the Wnt signal is received. It is the difference between these two that is crucial. Note that the end result, in the nucleus, is the difference between the silence and the transcription of Wnt target genes. The initiation of this transcription by receipt of Wnt is a highly complex process, and what follows is a very simplified account. Wnt binds to a receptor: this is the transmembrane protein Frizzled. This in turn signals to a cytoplasmic protein called Dishevelled, which partially dis-assembles a 'destruction complex' that normally destroys a further protein called beta-catenin. When beta-catenin is not destroyed its concentration rises substantially and it is able to migrate into the nucleus where it binds together with the proteins TCF and LEF to initiate transcription of the target genes. Some of these are genes that initiate cell proliferation; hence the importance of Wnt signalling both in normal development and in pathological development, especially tumour formation.

The Hedgehog Pathway

The *hedgehog* gene was first discovered, as noted in Chapter 2, because one of the developmental effects of mutations in it in fruit-flies is to render the mutant larvae prickly – like hedgehogs. But what are the molecular mechanisms going on behind such observable morphological features?

The Hedgehog protein, like the Wnt protein, is a secreted signalling molecule that is received by receptors on the outside of target cells. The main receptor in this case is a transmembrane protein called Patched. The signalling pathway works as shown in Fig. 3.13 and as described below.

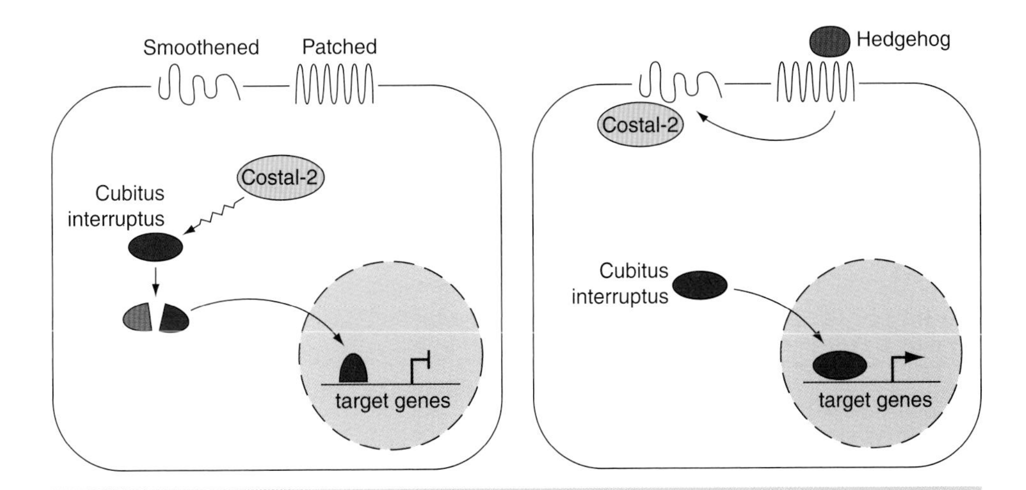

Figure 3.13 Hedgehog signalling. As in Figs 3.11 and 3.12, the situation without ligand is shown on the left (target genes off), while the situation after receipt of ligand, which in this case is the Hedgehog protein, is shown on the right (target genes on).

When a target cell receives no Hedgehog signal, a protein in the cytoplasm called Cubitus interruptus is attacked and shortened. The shortened protein (Ci-S) moves into the nucleus and acts as a transcriptional repressor. A key player in the shortening of Ci into Ci-S is another protein called Costal-2. However, when Hedgehog signal is received by Patched, the latter protein signals to another transmembrane protein called Smoothened, which binds Costal-2 and retains it at the membrane, thus preventing it from diffusing throughout the cytoplasm and attacking Ci. The full-length Ci, unlike its shortened version, is a transcriptional activator, and switches on target genes. Exactly which genes these are depends on the developmental context. We will look at one such context below: segment formation in *Drosophila*.

3.5 Signalling: From Cell to Embryo

Signalling pathways transcend the 'just a few cells' level of organisation, and relate to blocks of tissue containing many cells, often hundreds of them. As an example of this, a vertebrate homologue of the *hedgehog* gene, *sonic hedgehog*, produces a corresponding protein, usually abbreviated to Shh, which has many signalling roles in the development of all vertebrates. One such role involves the secretion of Shh by cells in the **notochord**, the movement of Shh through the extracellular matrix, and its receipt by cells of the developing **neural tube** and flanking blocks of **mesoderm**. The effects of Shh receipt depend on two things: the nature of the cells in the receiving tissue and the concentration of Shh received (Fig. 3.14).

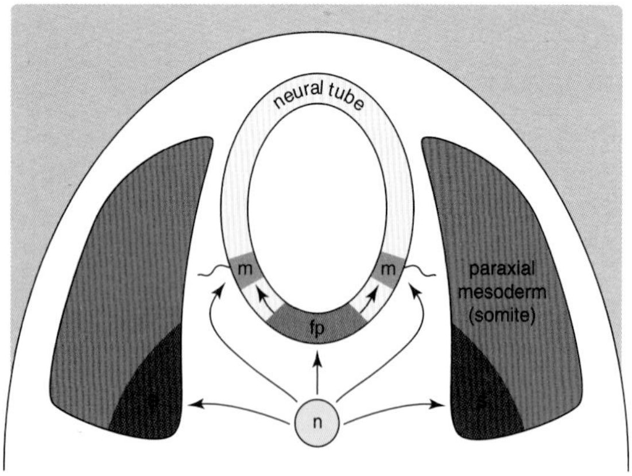

Figure 3.14 Signalling by the Sonic Hedgehog protein in the mid-dorsal region of a vertebrate embryo. The response to receipt of the signal depends on the concentration of Shh, which decreases with distance from its source (the notochord), and on the nature of the responding tissue. n – notochord, m – motor neuron, fp – floorplate, s – sclerotome.

The first-discovered role for *hedgehog* in *Drosophila* was as a segment polarity gene. That is, its product helps to establish and maintain the antero-posterior polarity of the body segments. In other words, it conveys positional information, so that cells 'know' whether they are in the anterior or posterior part of a segment. This segment polarity role also depends on the interaction between Hedgehog and Wingless signalling, as follows (see also Fig. 3.15).

Hedgehog protein is produced and secreted by cells at the posterior end of each segment (or the anterior end of each **parasegment** – the embryonic units that precede, and are out-of-phase with, segments[35]). It is received by Patched receptors in adjacent cells, which respond by expressing the *wingless* gene. Wingless protein is also secreted and is received by the Hedgehog-producing cell, in which it results in the expression of, among others, the *engrailed* gene, whose segmental expression we saw previously (Fig. 3.10). One of the targets of the

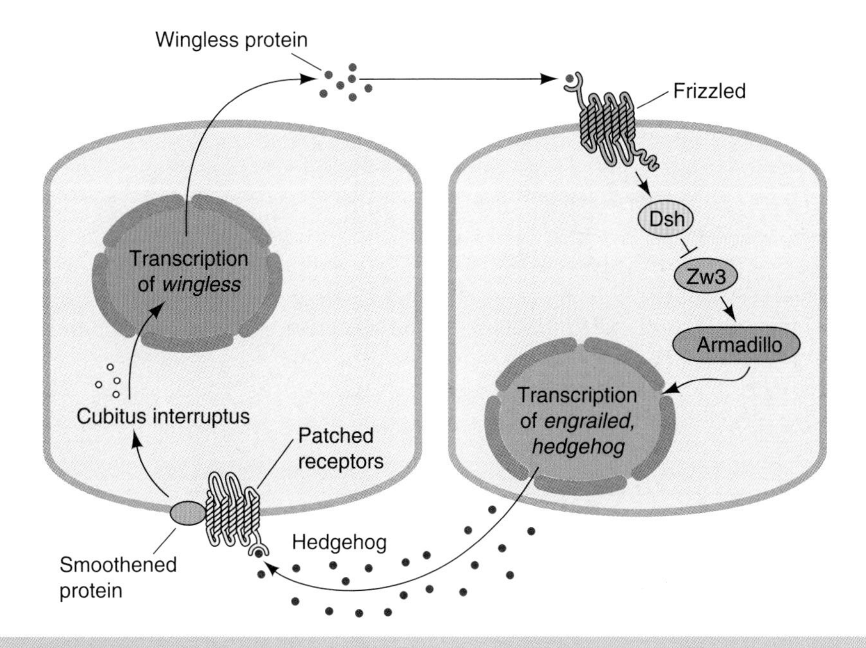

Figure 3.15 The interaction between Hedgehog and Wingless signalling that occurs in the maintenance of segment boundaries in *Drosophila* and other arthropods. As can be seen, the receptor protein for Hedgehog is Patched, while that for Wingless is Frizzled. Several other proteins are also involved. Dsh – Dishevelled; Zw3 – Zest white 3. (Courtesy of Scott Gilbert and Sinauer Associates.)

Engrailed transcription factor is the *hedgehog* gene, so this form of signalling is essentially circular – it acts to maintain segment boundaries after they have been established.

The link with the prickly appearance of hedgehog-mutant larvae is that normally only anterior parts of segments have bands of little projections called **denticles** on their cuticle. However, when *hedgehog* is mutant, the posterior parts make these denticles too – hence the overall prickly appearance rather than an alternation of prickly and naked bands of cuticle.

The interacting roles of Hedgehog and Wingless signalling to maintain segment polarity and segment boundaries is a conserved system across most if not all arthropods. So it applies to a million species, not just one. However, the series of interactions that goes on prior to (**upstream** of) the activation of the segment polarity module is much more variable among different arthropod groups.

The interactions that occur in *Drosophila* were the first to be worked out. The whole series of interactions between genes and proteins that is involved in establishing segments in *Drosophila*, from its beginnings in the egg, is often referred to as the segmentation cascade. This has become a classic story in developmental genetics. Most of the genes involved had been determined by the late 1980s.

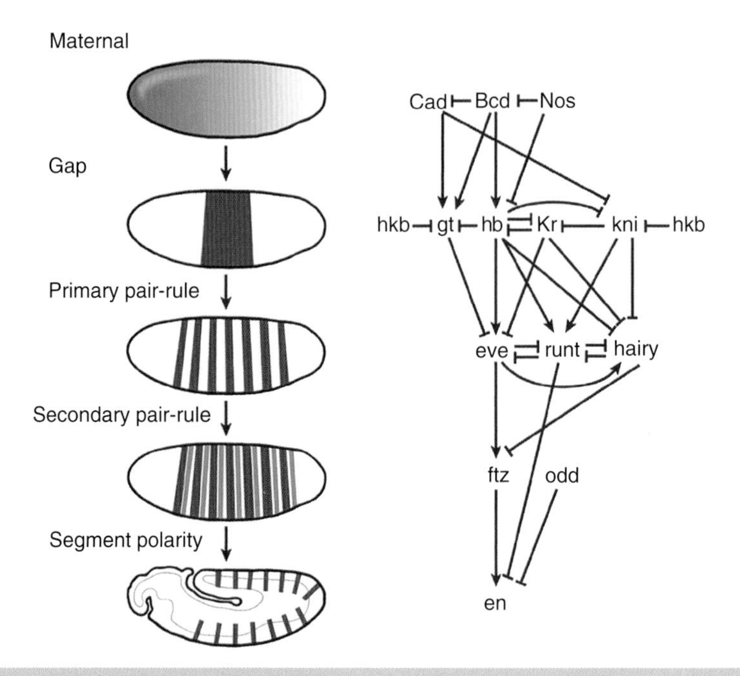

Figure 3.16 The cascade of interactions among genes and proteins that patterns the antero-posterior axis of a *Drosophila* embryo, starting with definition of the ends of the axis and ending with the specification of individual segments. The proteins involved in the earliest interactions are Caudal, Bicoid and Nanos. The overall pattern of interactions is very complex, as can be seen. The names of the intermediate-stage proteins are given just in abbreviated form. The most downstream protein is Engrailed – this provides a connection with Fig. 3.15. (Reproduced with permission from Carroll, S. *et al.* 2005, *From DNA to Diversity*, Blackwell.)

An outline of the process is shown in Fig. 3.16, in which it can be seen that the genes involved fall into four classes – maternal, gap, pair-rule and segment polarity. However, it has recently become clear that the nature of the upstream interactions varies much more between different arthropod groups than does the segment polarity module.

The *Drosophila* cascade is appropriate to a fast-developing insect with a **syncytial blastoderm** (no cell boundaries) and near-simultaneous formation of segments. All the transcription factor products of upstream genes, such as *bicoid* (maternal), *giant* (gap) and *even-skipped* (pair-rule), can freely diffuse through the undivided cytoplasm. However, it is now clear that in slower-developing arthropods, such as centipedes and spiders, the details of the upstream part of the cascade are very different. This finding re-emphasises the general point that we always need to ask questions about the phylogenetic scope of any developmental process – that is, the level of **taxon** within which it applies.

3.6 Long-Range Signalling and Developmental Processes

Many quasi-independent parts of a developing organism are referred to as **modules**. This is true of the leaves of a tree or the **zooids** of a colonial animal, as noted earlier. However, 'module' is now increasingly used to cover a wider range of quasi-autonomous developmental units, such as limb buds, segment **primordia** and developing eyes. While this usage seems helpful, it does not work so well for the development of structures which are less autonomous and more embedded in the whole developing organism, for example parts of the trunk of an unsegmented bilaterian animal such as a flatworm or **gastropod**. So, what alternative term should be used in such cases?

There has been a too loose and inconsistent usage of terms to refer to large-scale developmental processes, which has generated confusion. In the literature the following are all used: developmental pathway, developmental cascade, developmental hierarchy, developmental network, developmental system and **developmental trajectory** (and doubtless less common others). Are these synonymous? If not, how do they differ?

Developmental Pathways

This term is widely used but is problematic for two reasons. The first is confusion with 'signalling pathways', as used in Section 3.5, especially given that the bigger-scale 'pathways' also involve signalling. The second is that 'pathway' could be taken to imply a single route, which is rarely the case.

Developmental Cascades

This term, which was used above in relation to the 'segmentation cascade', is less problematic. It connects well with the usage of upstream and downstream in developmental processes in general, and it carries across from real cascades (waterfalls) the idea that both splittings and convergences of constituent streams of flow are possible. However, it captures less well the possibility of feedback from later stages to earlier ones, as water does not flow uphill.

Developmental Hierarchies

This term (and the related one of 'morphogenetic trees'[36]) suffers from a variant form of one of the problems that besets 'pathway'. That is, instead of implying a single route, it implies a series of branching routes, and thus does not cope well with either convergences or feedback.

Developmental Networks

A network implies a complex structure of interactions – not just linear or hierarchical but potentially including both of those within smaller parts of it. From the perspective of the desirability of starting out with an open mind to explore development and its evolution, 'network' is thus a good choice.

Developmental Systems

This phrase is used in both a rather vague, general sense (as is the broader 'biological systems') and a very specific one: in developmental systems theory or DST[37]. It is better not to use it in intermediate ways, as a synonym for a cascade or network.

Developmental Trajectories

When an entire developing organism is observed over an extended period of time, like one of von Baer's vertebrate embryos (Fig. 1.6), and there is a clear directionality to its progress, then 'trajectory' seems a useful term.

Developmental Processes

This is a good general 'catch-all' term, which implies nothing of the causal structure of the particular process concerned in any case.

Readers of the evo-devo literature in general should bear in mind that these usages have not yet settled down, and are likely to remain author-dependent for a while.

Now, putting aside the issue of which terms are used to describe long-range developmental processes, we should briefly consider an example of one of these processes to see how it differs from the shorter-range signalling already discussed.

In 'higher' insects there is holometabolous development. That is, the larva and adult are quite distinct morphologically, and the latter is produced from the former by a radical developmental change that we call **metamorphosis**. This applies to Diptera (flies), Lepidoptera (butterflies and moths), Coleoptera (beetles) and Hymenoptera (ants, bees and wasps). In contrast, in 'lower' insects such as Orthoptera (grasshoppers and crickets), various stages of nymphs lead gradually to the adult form through a series of moults.

In both this latter case, which is called hemi-metabolous development, and the former one, involving metamorphosis, an important player in long-range signalling is the **hormone** ecdysone. In the holometabolous insects, this acts to switch off many larva-specific genes and switch on others that are specific to the process of metamorphosis, which ultimately leads to the switching on of many adult-specific genes. Ecdysone is produced by the prothoracic glands towards the anterior end of the larva. It uses the larva's already-developed transport system – the haemolymph – to spread throughout the entire larval body. This long-range spread of a hormone is very different from the much more limited spread of a locally-acting **morphogen** such as Sonic hedgehog, discussed in Section 3.5. Effectively, the production of a transport system by local agents when the embryo is small enables that system to be used later by very different signalling agents.

From an evolutionary perspective, metamorphosis provides several interesting questions. Most obvious is, how did it evolve in the first place? But also, once a metamorphic system is in place, how does this affect the way in which natural selection acts on the overall life-cycle? Does holometabolous development mean that the larval and adult stages of insects can evolve quasi-independently of each other? The answer to this last question would appear to be 'yes', given the great

Figure 3.17 Caterpillars and adults of two species of butterflies of the genus *Colias*. This is an example of the relatively unusual situation where the larvae (caterpillars) are more different from each other than are the adults. Left: The pale clouded yellow, *C. hyale*. Right: Berger's clouded yellow, *C. alfacariensis*. Top: male adults; centre: female adults; bottom: caterpillars. (Courtesy of Richard Lewington, 'Butterflies of Britain and Ireland'.)

diversity of life-cycles seen in various holometabolous insect groups, including the divergence of larval form in some cases where the adults remain very similar (Fig. 3.17) and *vice versa*.

SUGGESTIONS FOR FURTHER READING

This chapter has provided just an overview of key entities and processes involved in developmental patterning. The idea has been to provide enough so that, later, when dealing with the evolutionary process of developmental *repatterning*, we have an insight into the complex nature of the processes that are being altered. More detailed information on development is readily available in some excellent texts. The following two books are particularly good. The first is at an introductory/intermediate level; the second is at a more advanced level (and is about twice as long):

Slack, J.M.W. 2006 *Essential Developmental Biology*, second edition. Blackwell, Malden, MA.
Gilbert, S.F. 2010 *Developmental Biology*, ninth edition. Sinauer, Sunderland, MA.

Also, the classic story of the working out of the genetic basis of development in the fruit-fly *Drosophila* is told in the following book:

Lawrence, P.A. 1994 *The Making of a Fly: The Genetics of Animal Design*. Blackwell, Oxford.

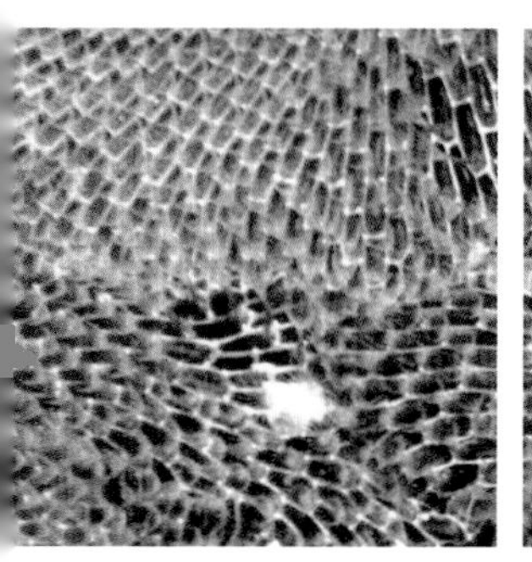

CHAPTER 4
Natural Populations

4.1 The Ecological Theatre and the Evolutionary Play

A book of this title was written in the 1960s by the ecologist G.E. Hutchinson[38]. It was an outstanding choice of title – both memorable and accurate. It reminds us of a very important fact: that all evolutionary changes happen against the backcloth of populations interacting with each other in ecosystems that exist in particular places at particular times.

This reminder is especially important in relation to 'big' evolutionary events, whether at the genetic or developmental/phenotypic levels. For example, the duplication of a **homeobox** gene or the addition of a body segment do not happen on the imaginary lines that we connect up on a page to produce an evolutionary tree diagram. They can indeed be *represented* that way, but they *happen* 'out there' in the real world, in some particular geographical location. If favoured by natural selection they spread, either throughout the species' entire **geographical range** or, because of barriers to their spread, to a certain subset of that range. In the latter case they may, together with other accumulated changes that are stopped by those same barriers, give rise to the splitting, or **cladogenesis**, of the original species into two **daughter species**.

Many evo-devo publications have omitted this real-world backcloth against which development evolves. Perhaps this is justified on the basis that other disciplines within evolutionary biology, notably population/ecological genetics and evolutionary ecology, provide this part of the overall picture. But, since the aim herein is to try to integrate all aspects of the evolution of development, we will include a brief look at natural populations; that is the role of the current chapter.

4.2 Types of Creature; Types of Population

Much of the early work of field-based population genetics and evolutionary ecology was focused on a small number of animals and plants. Among the best known of these are the following (Fig. 4.1): Dobzhansky's work on chromosomal inversion **polymorphisms** in *Drosophila* species[39]; a variety of studies of lepidopteran wing pigmentation patterns[40]; the many studies on shell pigmentation in land-snails of the genus *Cepaea*, initiated by Cain and Sheppard[41]; and the classic case study of beak shape variation in Darwin's finches[42] on the Galapagos islands (Chapter 12).

Since the early, pioneering studies, work on genetic and **phenotypic** variation has spread out across many other taxa. To some extent, the findings of all the work (both early and late) can be generalised; but, before attempting to generalise, it is necessary to look at differences in the types of natural populations, which arise from biological differences in the types of creatures concerned. The American ecologist Robert MacArthur noted the generalising nature of science, stating[43] that science is a 'search for repeated patterns' – and so indeed it is, as well as being a search for their causes. But we who are doing the searching must always be careful not to *over*-generalise, beyond what is consistent with the facts.

We will take human populations as the starting point for looking at different types of populations generally, simply because these are the populations with which we are most familiar. Human populations have the following five features, connecting with the biology of the human organism:

1. Individuals are physically discrete entities, clearly separated from each other, so that, at least in principle, it is possible to count their number in any particular human population, and thus to state the population size, N. (Exception: pregnancies, especially early, unconfirmed ones.)
2. Reproduction is sexual and each individual within a population is genetically unique. (Exception: identical, i.e. monozygotic, twins.)
3. The sexes are separate, so each individual is either male or female, with sex being determined by an XX-XY chromosomal system, and with females being the homogametic (XX) sex. (Exception: rare chromosomal abnormalities such as XXY – Klinefelter's syndrome, or 'intersex'.)
4. Since the genetic system is **diploid**, with each individual possessing two copies – a maternal and a paternal one – of each gene, variation in the population at the gene level includes both **homozygous** and **heterozygous** individuals for the gene

Figure 4.1 Some of the main study animals used in the early days of ecological genetics. The fruit-fly *Drosophila pseudoobscura*, studied by Theodosius Dobzhansky (photograph courtesy of Nicolas Gompel); the butterfly *Maniola jurtina*, studied by E.B. Ford; the land-snail *Cepaea nemoralis*, studied by Arthur Cain and Philip Sheppard; and one of the many species of *Geospiza* (Darwin's finches) studied by David Lack (photograph courtesy of Jeff Podas).

concerned. (Complication: most of our DNA is non-genic, though of course we are also diploid for our non-genic sequences.)

5 There is no larval stage (in contrast to many animal species); so the morphology of individuals in a 'natural' human population is similar in broad terms regardless of age. (Complication: few current human populations are natural, in the way that populations of early humans, or indeed of other species of *Homo*, were.)

It is now appropriate to look at the extent to which populations of other species differ in relation to these five features, because this information will reveal the extent to which we can, and cannot, generalise about population structure. Since populations represent the realm in which natural selection operates, the ability to recognise the limits of generalisations about them is essential.

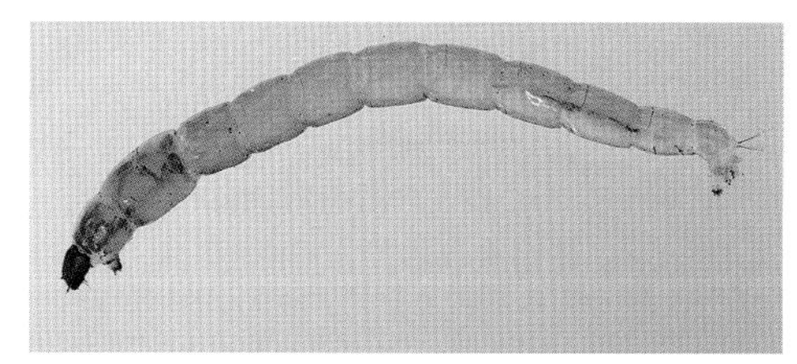

Figure 4.2 Aquatic larvae of two groups of dipterans – midges (top) and mosquitoes (bottom). The ecology of the larvae, and hence the kind of natural selection they will be subject to, is very different from the corresponding adults with a terrestrial/aerial existence (photographs courtesy of Steve Marshall).

The Species of the Early Case Studies Mentioned Above Differ as Follows:

First, *Drosophila* populations: Despite the huge phylogenetic separation between humans and flies, *Drosophila* populations differ from human ones in only the last of the five features mentioned above: they have **larvae**. And the ecology of the larvae is qualitatively distinct from that of the adults. The details of this difference vary from species to species – and there are more than 2000 species in the genus *Drosophila*. A few *Drosophila* species, and many species of insect more generally, have aquatic (freshwater) larvae (Fig. 4.2); so the larva and the adult inhabit different major habitat types; they will thus experience very different kinds of natural selection.

Second, lepidopteran populations: As these are also insects, and indeed quite closely related to the Diptera (flies), it is not surprising that in general their populations differ from those of humans in the same way as *Drosophila*. This is certainly true of the most celebrated (from an ecological genetics perspective) lepidopteran genera, including the peppered moth *Biston* (Section 4.6), and the various butterflies exhibiting **mimicry** (e.g. *Papilio*). It is also true of the butterfly *Bicyclus* used in more recent combined population-based and evo-devo studies[44], which we will discuss in subsequent chapters. However, lepidopteran populations show a difference from human populations that, strangely, is in common not with dipterans but with birds – see below – the female is the heterogametic sex.

Third, populations of the land-snail *Cepaea*: These do not have a larval stage (though marine snails usually do). But they differ from human populations in relation to the third of the list of five features given above: separate sexes. Like many gastropods, *Cepaea* is hermaphrodite: each individual is both male and female. However, most reproduction is by cross-fertilisation rather than self-fertilisation. This is clear to observers of *Cepaea* populations in the mating season (May) when, in dense populations, many copulating pairs of snails can be found. The same is true of the common garden snail, familiar to most biologists as *Helix aspersa* (though its official name is now *Cornu aspersum*), which is a close relative of *Cepaea*; and also true of the related edible snail *Helix pomatia*.

Fourth, populations of Darwin's finches: There are several species of these birds, most belonging to the genus *Geospiza*, and all found in the Galapagos archipelago of many islands and islets. Given that we have now returned to our vertebrate starting point, it is not surprising that these populations are similar to those of humans in relation to most of the five features listed. However, there is one difference: like birds in general, the chromosomal system of sex determination is the opposite of that of mammals in the sense that the male is the homogametic sex. (Students of bird genetics normally use Z and W instead of X and Y for the sex chromosomes, with the male being ZZ, the female ZW.)

So far, using only examples of birds, flies, lepidopterans and snails, we have already seen departures from features 3 and 5 of human populations. To see departures from the other features (1, 2 and 4), we now turn our attention to other taxa.

Figure 4.3 The grass *Agrostis tenuis*. Top: a field population of *A. tenuis* and other species, indicating the difficulty in distinguishing different grass species and individual plants. Bottom: an individual 'plant' of *A. tenuis*; this is a ramet, but if it reproduces asexually, an interconnected series of such plants will form a genet.

Other Species Differ as Follows:

Lack of clear physical separation of individuals is common in the plant kingdom. Indeed, precisely what constitutes an 'individual' plant is often unclear, which has led to the use of **ramet** and **genet** for defining individuals that are physically distinct and genetically distinct, respectively. Take, for example, the grass *Agrostis tenuis* (Fig. 4.3), which was the species used in a well-known study of evolution of heavy-metal tolerance[45] in populations inhabiting areas contaminated by mining waste. This species can reproduce vegetatively as well as sexually. So if we consider a field of grass that is a monoculture of this species (a very idealised situation, as virtually all real fields of grass contain several species), it is impossible to count individuals and thus to state, or even estimate, the population size, N. The problem of individuals not being physically separate is not restricted to plants. It is also found in animals that have a colonial growth form, such as the **bryozoans** mentioned earlier, and others, for example some **tunicates** (Fig. 4.4).

These plant and animal examples of the lack of physical separation of individuals also provide examples of asexual reproduction. However, the latter can also occur in some cases where individuals are indeed separate. A good example is the animal group **Rotifera** (Fig. 4.5). In particular, the members of one large group of rotifers, the Bdelloidea, appear to reproduce exclusively by asexual means: each female produces a diploid egg by mitosis[46] (as opposed to a haploid egg by meiosis). Each of these eggs develops into a new diploid female that is genetically identical to her mother. This group of rotifers is thought to have evolved, many millions of years ago, from another group in which asexual reproduction was the norm but occasional sexual reproduction also occurred.

Figure 4.4 An example of a colonial animal group: tunicates. Although some species of this group of protochordates are solitary, many, including the one shown, are colonial in their growth form. Note that the tunicate colony is yellow – the green animals are sea anemones growing up through the tunicate colony.

Finally, there are also departures from diploidy, even in the world of multicellular creatures (i.e. the realm of evo-devo: see Chapter 1). Some primitive land plants, for example ferns (pteridophytes), have a life-cycle in which there is an alternation between haploid (gametophyte) and diploid (sporophyte) generations. The two types of plant are very different: what we normally think of as 'a fern' is the sporophyte, with the gametophyte being smaller and of a completely different growth form (Fig. 4.6). Higher plants, notably the **angiosperms**, are more similar to the typical animal, in that the haploid phase of the life-cycle is restricted to a handful of cells, including the germ cells.

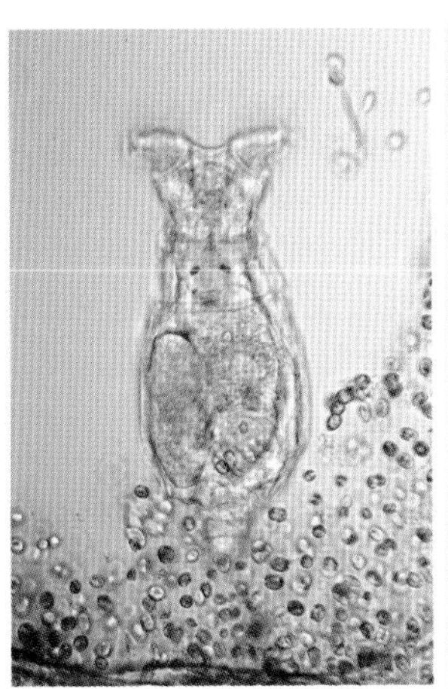

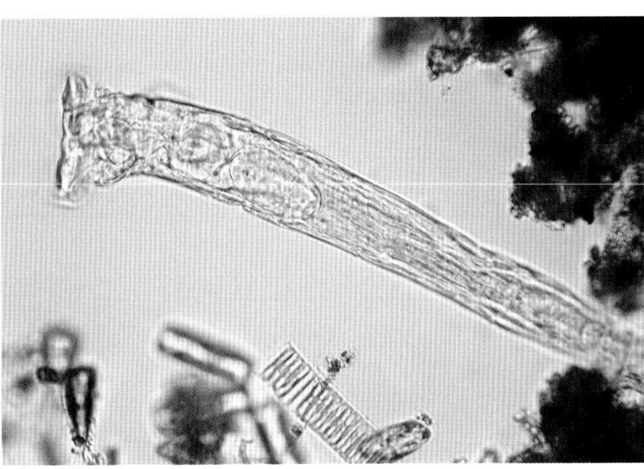

Figure 4.5 Two different species of rotifers. These are tiny animals. Typically rotifers range in size from less than a millimetre to just a few millimetres.

Figure 4.6 What we think of as a 'fern' is the sporophyte stage of the life-cycle (left). However, the spores that can sometimes be seen on the undersides of the leaves develop not into more sporophytes but rather into the gametophyte stage of the life-cycle, which has a very different growth form (right).

Instances of haploidy can also be found in the animal kingdom. In particular, in the insect group Hymenoptera (ants, wasps and bees), reproduction often involves a system known as haplodiploidy. In its simplest form, males are haploid, females diploid. A fertilised female with stored sperm can either fertilise her eggs or not, thus producing, respectively, female or male offspring (Fig. 4.7).

We will now look at spatial (Section 4.3), age-related (Section 4.4) and genetic (Section 4.5) aspects of population structure. In each case it is important to keep in mind, when focusing on particular examples, that inferences may or may not be generalisable to taxa other than the one under consideration, depending on variation in the five key features of populations discussed in this section. As a preview of one example of this: the widespread polymorphism of enzyme-producing genes first revealed in the 1960s in human and fruit-fly populations turned out to be generalisable to most species – but not to those that only reproduce asexually.

4.3 Spatial Structure

The spatial distribution of a population, and indeed of the whole network of populations that make up a species' overall geographical range, is of much importance in relation to the possible patterns and rates of spread of new variants that arise by mutation initially in one or a few individuals living in one particular place. Spatial distributions can be considered at several levels, as follows.

We can start with the context of a single population. But even this apparently simple context begs the question: how is a population delimited? The answer to this depends on both the creature (including its mobility) and the physical nature of the habitat (including the existence of barriers to movement). With regard to the latter point, islands have often been used to study populations because they have clear boundaries. For example, intensive studies have been carried out on the red deer population of the Scottish island of Rum, which is approximately 10 km across, in both N-S and E-W directions. From the combined perspective of the animal's size and mobility, and the size of the island, as well as its distance from the mainland, the red deer of Rum can be considered to be at least potentially a single population. (Human effects, including culling and the construction of fences or other barriers[47] may change that). In contrast, for a patchily-distributed small invertebrate, this same small island may be home to many populations, perhaps hundreds of them.

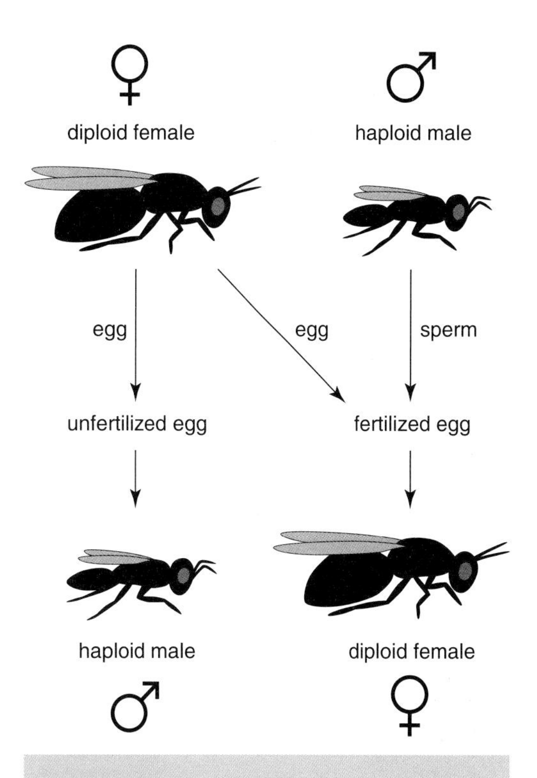

Figure 4.7 The basis of a typical haplodiploid life-cycle, in which males are haploid and females diploid.

The same principle applies to what are called 'ecological islands'. For example, a coniferous forest of about the same size as Rum, so measuring about 10 × 10 km, and separated from other such patches of forest by vast swathes of open farmland, may be home to a single population of long-eared owls but to many separate populations of lichens that live only on dead wood.

Within a population, individuals can, at any moment in time, be aggregated (clumped) in their distribution or may be distributed randomly or regularly (Fig. 4.8). Aggregations are commonest; causes include affiliation to a particular microhabitat and various types of herding behaviour. Completely random distributions are rarest. Quasi-regular distributions arise in particular circumstances: for example due to territoriality (e.g. in many bird species) or to water shortage (e.g. the distribution of the individuals of a tree species in an African savannah).

Often, there is migration among an array of populations in neighbouring areas. Such a system has been described as a **metapopulation**, and much work has been conducted on the dynamics of these systems over the last few decades, notably by the Finnish ecologist Ilkka Hanski[48]. One of the most important points about metapopulations (Fig. 4.9) is that they can act as a buffer to **stochastic** effects taking place within any of the smaller populations of which

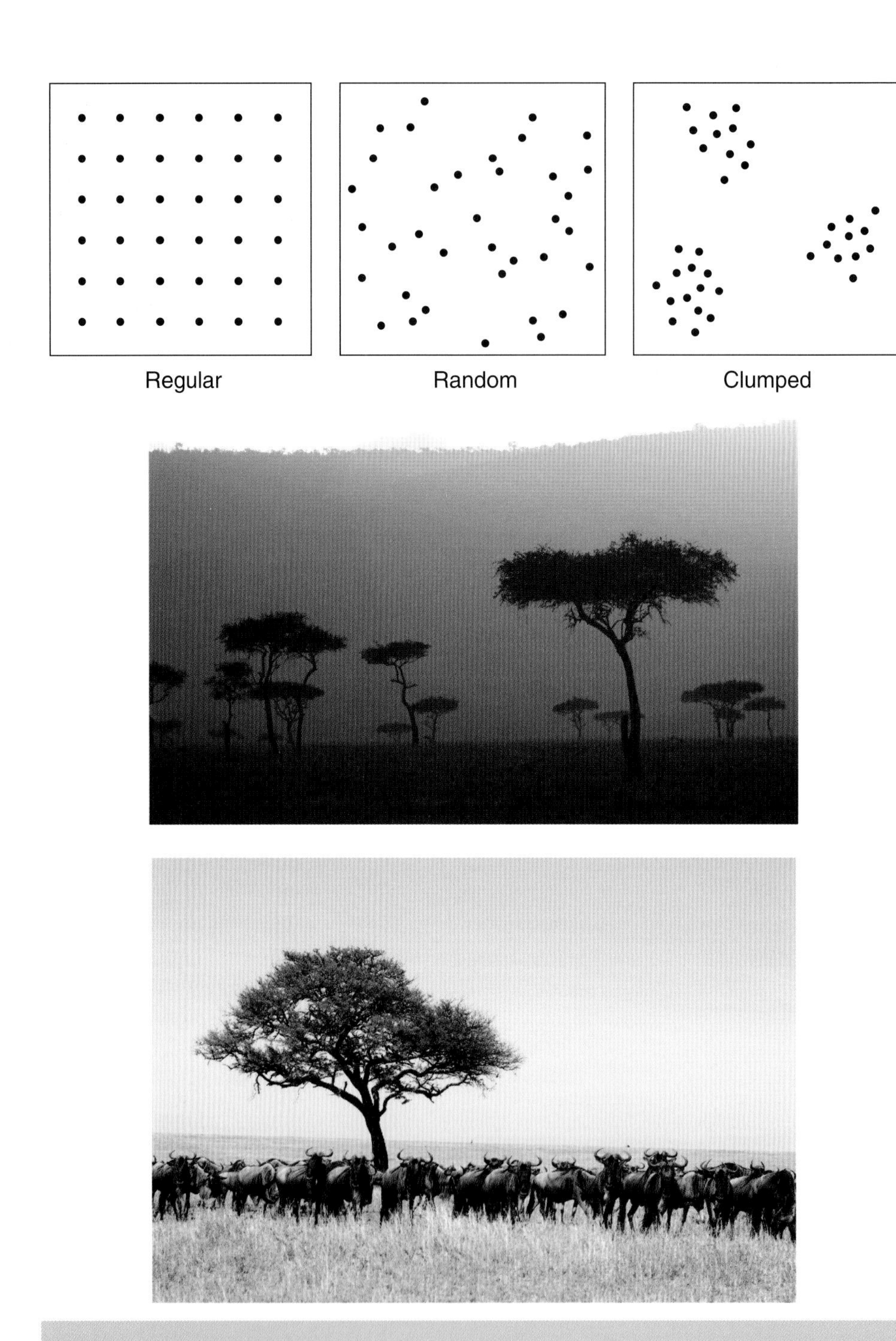

Figure 4.8 Types of spatial distribution of individuals within a population, shown in abstract form (top). Also shown are one example of a quasi-regular distribution (savannah trees; centre) and one example of a clumped distribution (wildebeest; bottom). Note that distributions can change over time, as in the switch from random/clumped to quasi-regular in a bird species that is territorial only in the breeding season.

they are composed. For example, a local extinction of one constituent population can be 'made good' by re-colonisation from another part of the regional metapopulation.

Just as local populations can be linked up into regional metapopulations, the latter can be considered to be linked up at a higher level again, ultimately extending to the whole species' geographical range. Species' ranges vary enormously in their geographical extent. Some species are 'endemic' (restricted) to a single, small island – such as the snail species of the genus *Partula*, many of which are endemic to the tiny Pacific island of Moorea[49] (part of the Tahiti group). At the other end of the spectrum is the distribution of bracken (the fern species *Pteridium aquilinum*), which extends across all continents except Antarctica. Three intermediates between those two extremes are shown in Fig. 4.10.

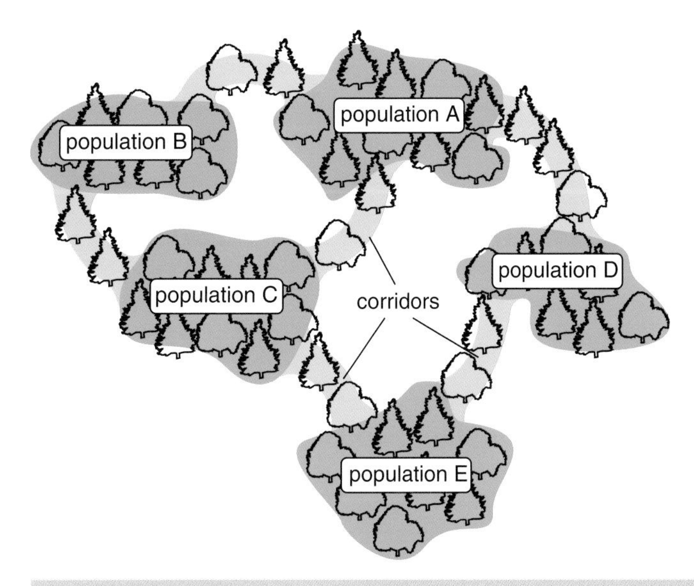

Figure 4.9 The idea of a metapopulation. Almost all species of animal and plant are distributed this way in nature. That is, separation between local populations is incomplete and migration occurs between them, sometimes via 'corridors' of suitable habitat.

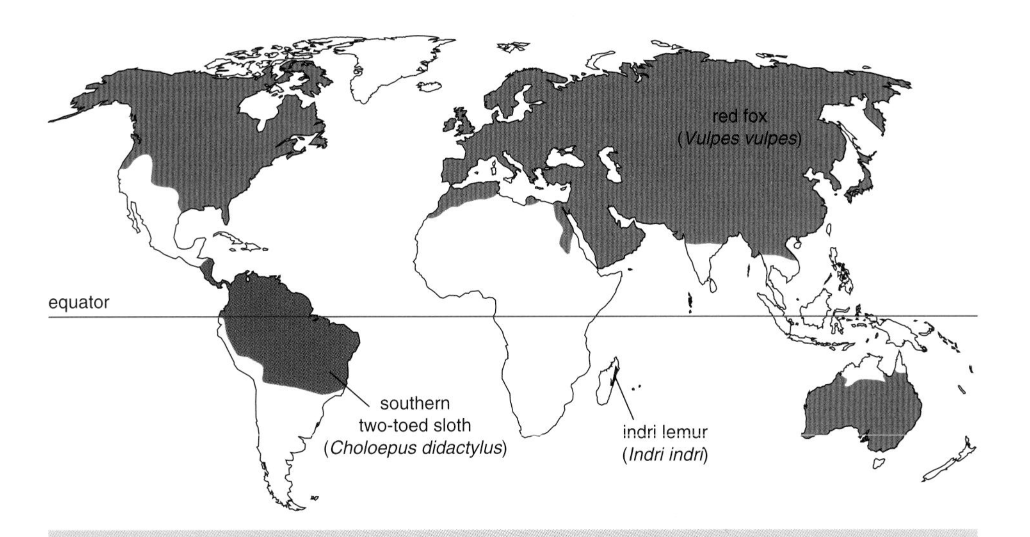

Figure 4.10 Examples of species ranges chosen to show how variable in extent these can be – from a small area of Madagascar in the case of the indri (a lemur species), through the sizeable area of northern South America occupied by a sloth species, to the vast area of the world inhabited by the red fox (whose occupancy of Australia is due to human introduction in the 19th century).

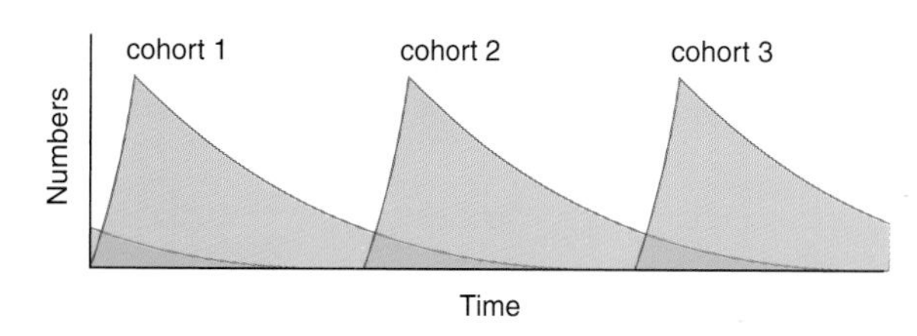

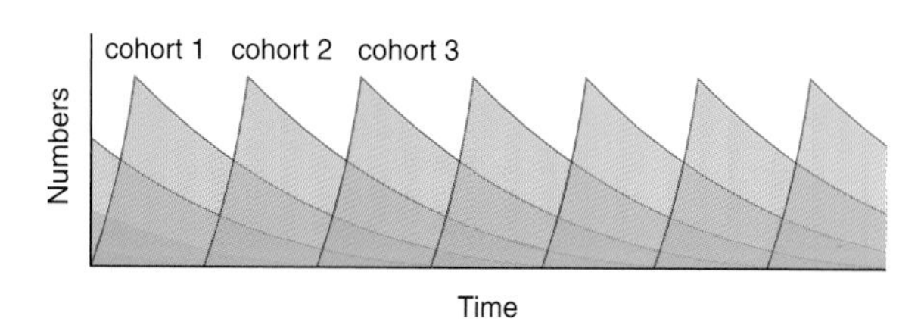

Figure 4.11 Diagrammatic representation of populations with semelparous (top) and iteroparous (bottom) breeding patterns. In the former, at any moment in time there is normally either a single cohort or two cohorts. In the latter case, at any point in time there are multiple cohorts.

4.4 Age Structure

The age structure of a population, like its other features, depends on the type of organism concerned. Most species have breeding seasons (humans are an exception to this rule). Thus each population is composed of a series of **cohorts**, often annual ones. However, the extent of this structuring depends on the reproductive biology of the species (Fig. 4.11). The simplest sort of age structure is found in populations of 'semelparous' species. This refers to a breeding pattern in which, towards the end of their lifespan, individuals reproduce 'all at once', and then die. An example is the pond-snail *Lymnaea stagnalis*, which lays its eggs, in gelatinous egg-masses, on the undersides of the leaves of aquatic plants. For most of the year, the population consists of a single cohort – growing juveniles. But in the summer it consists of two – the adult and egg/hatchling cohorts. A complication here is that this pattern seems not to hold true for all populations of *L. stagnalis*. Rather, in some climatic conditions adults live on for another year, giving a 2/3-pattern of cohorts instead of a 1/2-pattern.

Species in which adults normally reproduce over several breeding seasons are referred to as iteroparous. Here, each population consists of several cohorts (Fig. 4.11). In those species of subterranean geophilomorph centipedes that have been studied, the number appears to be about five. But in a typical tree species, such as maple, the number of annual cohorts in a population might reach over 100.

From an evolutionary perspective, one of the most important things about age structure is that natural selection may act in different ways – i.e. favour different characteristics – at different ages. The original body of population genetics theory largely ignored this possibility. But since then, population genetics has come to deal explicitly with it – a notable work in this area being that of Brian Charlesworth[50].

Notice that all the examples of age structure we have examined so far have involved species with direct development: humans, freshwater snails, centipedes and maple trees. Where there is indirect development – via one or more larval stages – one aspect of age structure (perhaps the most important one) is *stage* structure. Thus any natural population of *Drosophila melanogaster* will have a stage structure that can be measured as the number of individuals currently at egg, larval, pupal and adult stages; or in more detail by also splitting the numbers of larvae between the three larval **instars**.

The subject of a later chapter (Chapter 6) is the group of evolutionary processes known as heterochrony – the evolution of differences in developmental timing. One

Figure 4.12 The paedomorphic salamander known as the axolotl (left) in which the aquatic larval form, breathing through gills, has become reproductively mature and the previous adult stage has been lost. The picture on the right is of a non-paedomorphic salamander species, showing the kind of morphology that would be expected of a terrestrial, air-breathing adult salamander.

particular such process is the speeding up of the development of reproductive maturity so that what was previously a larval stage becomes adult (i.e. reproductively mature), and the original adult stage is lost – a process known as **paedomorphosis**. This has happened in the evolution of salamanders; a well-known example is the Mexican axolotl, *Ambystoma mexicanum* (Fig. 4.12). Interestingly, some other salamander species have remained facultatively paedomorphic, rendering both 'adult larvae' and metamorphosis to 'normal adults' possible.

4.5 Genetic Structure

Little was known of the genetic structure of natural populations prior to the 1960s. Before then, the studies that had been carried out were almost all at the phenotypic level, and were of two sorts: those that focused on continuous variation in characters such as body size; and those that focused on visible polymorphisms, such as the studies on pigmentation in land-snails and lepidopterans already mentioned. The former were hard to interpret because of the only partially **heritable** nature of the variation; while the latter were unrepresentative of polymorphisms in general, most of which do not affect the visible phenotype.

At that early stage of study, there were two opposing views of the 'typical' gene in the 'typical' natural population (in so far as there are such things). Under one view, most genes were thought to be invariant in a population, each being fixed for the fittest, **wild-type** allele. Under the other view, there was thought to be lots of hidden variation, with the typical gene being **polymorphic**, such as those underlying pigmentation patterns. A good account of these conflicting views is given by Richard Lewontin[51].

The vast amount of work that has been carried out since the mid-1960s to investigate the genetic structure of populations has overwhelmingly corroborated the 'widespread polymorphism' view. The presence of large amounts of genetic variation

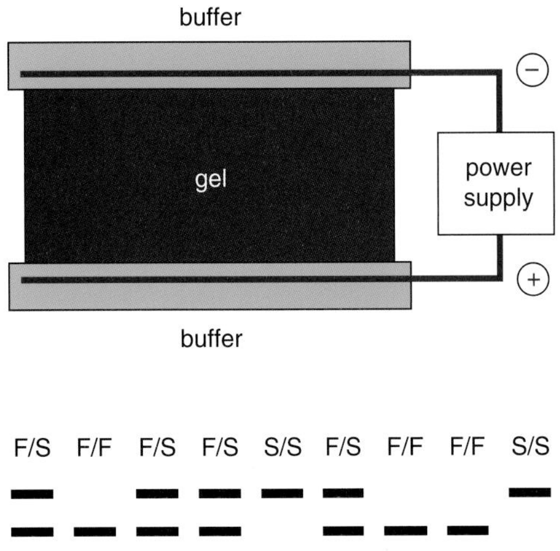

buffer

gel

power supply

buffer

F/S F/F F/S F/S S/S F/S F/F F/F S/S

Figure 4.13 Gel electrophoresis was first used to reveal widespread polymorphism at the molecular level in natural populations in the 1960s. Top: a diagram of the apparatus, showing that the gel is exposed to an electric field, which causes proteins to migrate at different rates if they have different overall charges. Bottom: a banding pattern on a gel, showing the simplest kind of polymorphism, in which each of the single-banded 'electro-phenotypes' is a homozygote (F/F and S/S), the double-banded one a heterozygote (F/S). F and S simply stand for 'fast' and 'slow', referring to comparative speed of movement along the gel. More recently, electrophoretic methods have been used to examine variation at the levels of DNA and RNA also.

was revealed first at the level of gene products (proteins) and then at the level of the gene itself (DNA), including also variation in non-genic regions of DNA, which, in eukaryotes, are often collectively longer than the genic regions, as noted in Chapter 3.

The proteins examined in the population-wide studies of the 1960 to 1980s were generally routine metabolic enzymes, the products of 'housekeeping' genes. The fact that there is widespread polymorphism in these genes (Fig. 4.13) was (and is) not easy to interpret. The selectionist view was that some form of balancing selection (Chapter 12) was actively maintaining the variation. In contrast, the neutralist view (pioneered by the Japanese geneticist Motoo Kimura[52]) was that the variation in the proteins concerned did not affect their function and thus was 'invisible' to natural selection; in other words, it was selectively **neutral**. In this case, the gene frequency of each polymorphism was effectively on a 'random walk', caused by the vagaries of **genetic drift**. Although the action of genetic drift had previously been thought to be restricted to very small populations, Kimura showed that it could also affect large ones. And it now looks as if his neutral theory is largely correct, with most polymorphisms being neutral and only a small proportion of them being actively maintained by balancing selection – though there is still not unanimity in relation to this issue.

More recently, studies of genetic variation in natural populations have moved to the level of the genetic material itself – DNA. These studies have revealed yet more variation, and of a variety of sorts, as indicated below.

At this stage, it is necessary to ask: what kinds of DNA sequence make up the **genome** of a eukaryote? We will focus in particular on the human genome for the moment, since so much is known about it. However, the types of DNA sequence found in the human genome are also found in most (perhaps all) eukaryote genomes, though their relative contributions to the overall genome may vary considerably among eukaryote taxa.

The human genome contains the following types of sequence:

1. *Genic DNA*: This may be divided in two ways: first, within a gene, into coding (for amino acids in the protein product) and non-coding, with some of the latter being regulatory regions. Second, between genes, into the DNA of housekeeping, terminal-target and developmental genes, as discussed in Chapter 3 (though some genes fall into more than one of these categories).
2. *Pseudogenic DNA*: Pseudogenes are related to actual genes (e.g. globin pseudogenes). However, they differ in seeming to be non-functional, and they often contain frame-shift mutations or stop codons that would appear to make them so. But recent work urges caution in assuming that these stretches of DNA have *no*

function. Some pseudogenes are now known to produce small RNAs that are important in gene regulation.

3 *DNA that is transcribed but not translated*: This DNA is responsible for the transcription of various types of RNA other than mRNA. These include tRNA, rRNA; also various types of small RNAs (generally ~20–30 bases long), so there is overlap with **2** above. (There is also overlap with **1** above: intron DNA is transcribed but not translated).

4 *Various classes of repetitive DNA*: It is tempting to label these as non-genic DNA but, while that is true of some classes, it is not true of others. Tandemly repetitive DNA (satellite DNA plus mini- and micro-satellite DNA) consists of short sequences, from a few bases long to hundreds of bases long, each repeated in tandem many times, often in several different parts of the genome. In particular, there is much tandemly repeated DNA in the **centromere** and **telomere** regions of chromosomes. Other repetitive DNA is repeated not in tandem but rather in an interspersed manner. Much of this DNA belongs to transposons – mobile genetic elements that can replicate and insert themselves at various points in the genome. The DNA of full-length transposons *does* include genic DNA, including the genes for transposase enzymes that make transposition possible.

We now ask four questions, with the fourth being the most important in the context of this book. First, how much of the human genome is composed of the different types of DNA? Second, to what extent do these proportions vary across other eukaryote taxa? Third, are all types of DNA variable within populations? Fourth, which types are most relevant to the evolution of development?

Rough answers to the first three questions are as follows. The human genome contains about 2 per cent of the first three types of sequence together, and about 98 per cent of repetitive DNA. The exact proportions vary across different eukaryote groups. One recent study suggests that humans are unique in their abundance of pseudogenes[53]. All types of DNA sequence exhibit intraspecific (and indeed intrapopulation) variation. Looking at the finest scale possible, that of just one DNA base-pair, single-nucleotide polymorphisms or SNPs (pronounced 'snips') have been found to characterise both genic and non-genic DNA. Also, in tandemly repeated DNA, the number of repeats is polymorphic within populations (this being referred to as variable number tandem repeats or VNTRs).

With regard to the fourth, most important, question – about relevance to the evolution of development – the most obvious answer is that the DNA, both coding and regulatory, of *developmental genes* is crucial. However, not a single other category of DNA sequence can yet be conclusively labelled as unimportant to developmental evolution. Mutations in developmental genes can be caused by transposon insertion; some pseudogenes have a function after all; and, related to that, many types of small RNA molecule have recently been implicated in gene regulation.

4.6 Natural Selection

In this section we will look at how natural selection acts to change the genetic and phenotypic structure of populations. The discussion will be brief; fuller accounts can be found in many other sources, two of which are given in the Suggestions for

Further Reading at the end of this chapter. We will look at the classic case study of industrial **melanism** in the moth *Biston betularia*, in which darkly-pigmented (melanic) phenotypes were selectively favoured in soot-polluted environments, and then broaden out from that starting point.

Selection on Pigmentation Polymorphism in *Biston betularia*

To gain an insight into the context in which natural selection has been acting in this species, we should first note the main features of its populations, in relation to the various criteria of population structure discussed in the previous four sections.

To summarise in this respect:

1 The species is semelparous and has a single generation a year, which is important in relation to the speed with which selection acts.

2 Resting adults are aggregated in particular microhabitats, something that is important for the nature of the selection.

3 There is migration between local populations (and hence a metapopulation structure), which may slow down the effect of selection. This migration may include wind-blown larvae[54] as well as flying adults.

4 The genetic basis of the phenotypic difference observed between pale and melanic individuals is a polymorphism at a single locus with two main alleles. The melanic allele that arose by mutation in a population of pale moths is **dominant**. Thus selection was able to begin to act as soon as the first mutations occurred in an environment that favoured them.

5 The molecular and developmental bases of the melanic phenotype are still poorly understood, though molecular studies on melanic forms in general are in progress[55].

The basic story is as follows. By now this is probably familiar to most biology undergraduates, but a brief summary here will be useful nevertheless, both as a reminder of the key points and to rebut some recent criticisms of this case study.

In a rural environment in which trees have pale-coloured bark, often partially covered with pale green lichens, the original form of *B. betularia* is well-camouflaged (Fig. 4.14). That is, it is **cryptic** rather than readily visible to predators hunting by sight – mostly birds. In contrast, melanic mutants that appear by mutation in this sort of environment are conspicuous.

However, after the onset of the industrial revolution in Britain (where the first studies were done) and in many other countries, the tree trunks and other surfaces on which moths rested by day in industrial areas became blackened with soot. In this different kind of environment, the selective differential was reversed and the melanic mutant was favoured. The original studies were done in Manchester, and showed that melanics, first discovered as rarities around 1800, rose in frequency to around 99 per cent by 1900. However, with the introduction of pollution control legislation, and indeed the changing nature

Figure 4.14 The different degrees of crypsis of pale and melanic forms of the peppered moth against various backgrounds: a soot-blackened tree trunk (top left); 'clean' birch bark (top right); a wall (centre and bottom left); and a lichen-covered tree-trunk (centre right). (Photographs courtesy of Laurence Cook, Michael Dockery and Ilik Saccheri).

of pollution (less soot, more 'invisibles'), the frequency of melanics began to decrease and, by the year 2000, had fallen back to less than 50 per cent in the relevant populations (Fig. 4.15).

This case study, as an example of natural selection taking place in field populations, has many strengths: the simple genetic basis of the phenotypic change; the contrast with rural populations; the parallel increases in the frequency of melanic forms both in other industrial areas and in other species; the speed at which it took place; and the

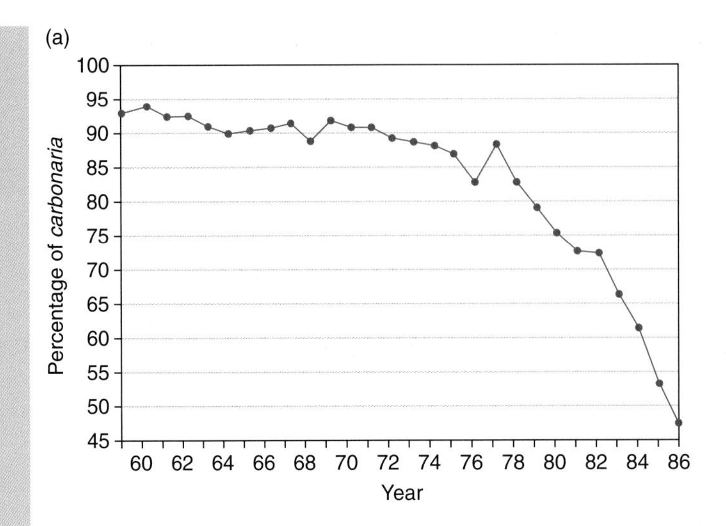

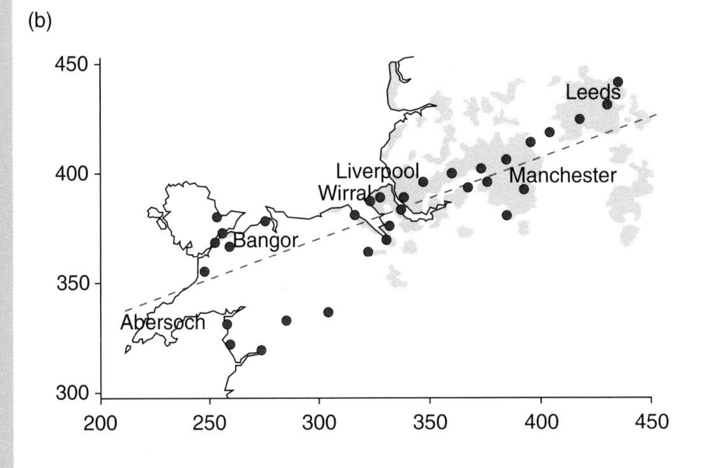

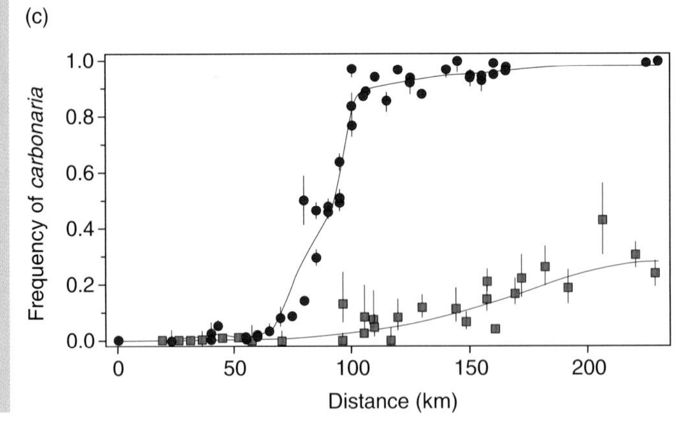

Figure 4.15 Top: Decline in the frequency of melanic moths over time in a particular urban population of *B. betularia*, after the end of the soot-polluted period and the return of pale lichens to the tree-trunks and branches where the moths rest. Centre: Series of sampling points arranged approximately in a line, stretching from rural North Wales to the Liverpool-Manchester-Leeds conurbation. Bottom: Cline in the frequency of melanics along this series of sampling points in the 1960s/1970s (top graph) and in the 2000s (lower graph). (Redrawn with permission: (a) Arthur, 1987, *The Niche in Competition and Evolution*, Wiley; (b) Saccheri *et al.* 2008, *PNAS*, **105**: 16212–16217. Copyright National Academy of Sciences, USA.)

apparently clear nature of the selection process. It is for these reasons that it has become a classic study.

However, it has also been criticised. Some of the criticisms have come from the creationist camp and are not worth spending time on. Others have come from a popular book[56] on the subject, *Of Moths and Men*, which accuses some of the scientists involved in demonstrating selective predation of fraud – an accusation that does not stand up to scrutiny. But others have come from scientists themselves, and include questioning of whether the nature of the selection is as simple as was first thought.

In all probability, the selection going on in the case of melanism in *Biston betularia* has many complexities. However, recent work has demonstrated:

- that moths of this species do indeed frequently rest on tree trunks and branches during daylight hours;
- that many species of birds do predate them, often to a significant extent;
- that this predation is selective, favouring the cryptic form; and
- that although moths are eaten by various other predators, such as bats, this predation appears *not* to be selective.

So, the main conclusions of the early work have been found to be true. Thus our interpretation must be that this is indeed a classic case study of natural selection in action, albeit one that still awaits elucidation of its developmental dimension.

Broadening Out to Selection on Polymorphisms in General

Only a small proportion of genes affect the external pigmentation of an animal or plant, and even fewer have such a major effect as the gene converting a pale speckled moth into a melanic one. To paint a broader picture of the nature of selection acting on polymorphic variation, it is necessary *either* to take a more abstract general approach *or* to look at a more 'typical' gene, if there is such a thing. In this section, we will do the former.

In general, in a population of any animal or plant, many genes are polymorphic. If there is complete dominance of one allele over another, the three genotypes (assuming two alleles) are often represented as AA, Aa and aa. Alternatively, to deal with cases where there is no dominance, the representation is often A_1A_1, A_1A_2 and A_2A_2. The way in which selection works depends on the relative fitnesses, in the Darwinian sense, of the three genotypes. Usually the fittest genotype is (arbitrarily) assigned a fitness value of 1, whereas the others are given fitnesses between 0 and 1 to indicate their relative lack of fitness.

What we looked at in *Biston* populations in polluted environments was a situation in which the fitnesses were as follows: AA and Aa both 1 (due to dominance of the melanic allele); aa fitness 1-s, where s is the selective coefficient measuring the strength of the selection against the pale speckled moths. Selection of this kind is referred to as directional, because its effect is to drive the genetic (and phenotypic) constitution of the

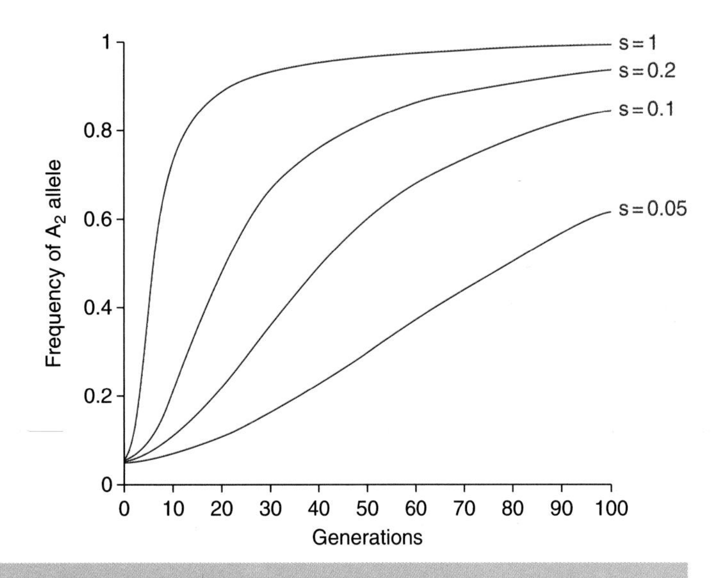

Figure 4.16 The relationship between the strength of selection a population experiences (measured by the 'selective coefficient', *s*) and the speed of its response in terms of altered genetic structure.

population in a particular direction, with the speed at which it does so depending on the strength of the selection (Fig. 4.16). The logical end result of this process is the *fixation* of the allele conferring increased fitness and the *loss* of the alternative allele.

Note that fixation is a process affecting just a single *gene*, though it may have consequences for other genes that are closely chromosomally linked to the one under investigation. This is quite distinct from *speciation*, the origin of a new species (Chapter 17), which normally involves many genes. In the former case the variants can freely interbreed (e.g. pale and melanic *B. betularia*), while in the latter they cannot. When a new species arises, it is given a new Latin name. The use of Latin names for polymorphic variants within a species is thus confusing and, although it is a widespread practice in the *Biston* case study, I have avoided such intraspecific Latin names here.

The main alternative to directional selection is balancing selection, which, as its name suggests, acts to maintain genetic variation within a population rather than to cause its loss. There are two main causes of this: (i) where the most favoured genotype is the heterozygote (heterozygous advantage); and (ii) where the fitness of a genotype is not fixed but instead is negatively related to its frequency, so that increased commonness leads to decreased fitness. Both of these situations can be demonstrated mathematically to lead to an equilibrium gene frequency in the population, which will persist until the fitness relationships change, perhaps as a result of environmental change. Balancing selection has been a subject of much interest to population geneticists, especially those who focus on the maintenance of variation in populations. However, it is of less interest to students of evo-devo, because we are interested in the ways in which development can *change* in evolution. **Developmental repatterning** connects more obviously with directional selection than with balancing selection.

Standing Variation versus 'Waiting for Mutation'

We now need to note an important distinction. As we saw in Section 4.5, most populations are characterised by many polymorphic **loci**. Hence, if a new form of selection is applied, it will often be possible for the population to begin responding to that selection immediately, making use of this **standing variation**. However, in other cases, the available variation may be neutral in relation to the new form of selection, and the population may thus not be able to respond to the selection until a non-neutral form arises by mutation. Unfortunately, we do not know whether a typical pre-industrial revolution population of *Biston betularia* contained a few melanic individuals or none. If it was the latter, then it would have been a case of 'waiting for mutation'.

While the difference between the ability for immediate evolutionary change based on standing variation and the need to 'wait' for a new variant form to appear is very real, ultimately *all* evolutionary change in development is based on mutation and developmental repatterning, because the standing variation (genetic and phenotypic) also came about in this way in the past. The next part of the book explores these crucial processes, with the emphasis on repatterning, which, in contrast to mutation, has generally not been treated in a coherent way.

SUGGESTIONS FOR FURTHER READING

The books that relate to the subject matter of this chapter are those in the general field of population biology. This field includes population ecology and population genetics, though these are often treated separately. One good introduction to population genetics is:

Gillespie, J.H. 2004 *Population Genetics: A Concise Guide*, second edition. Johns Hopkins University Press, Baltimore, MD.

For readers interested in pursuing the story of industrial melanism further, see:

Majerus, M.E.N. 1998 *Melanism: Evolution in Action.* Oxford University Press, Oxford.

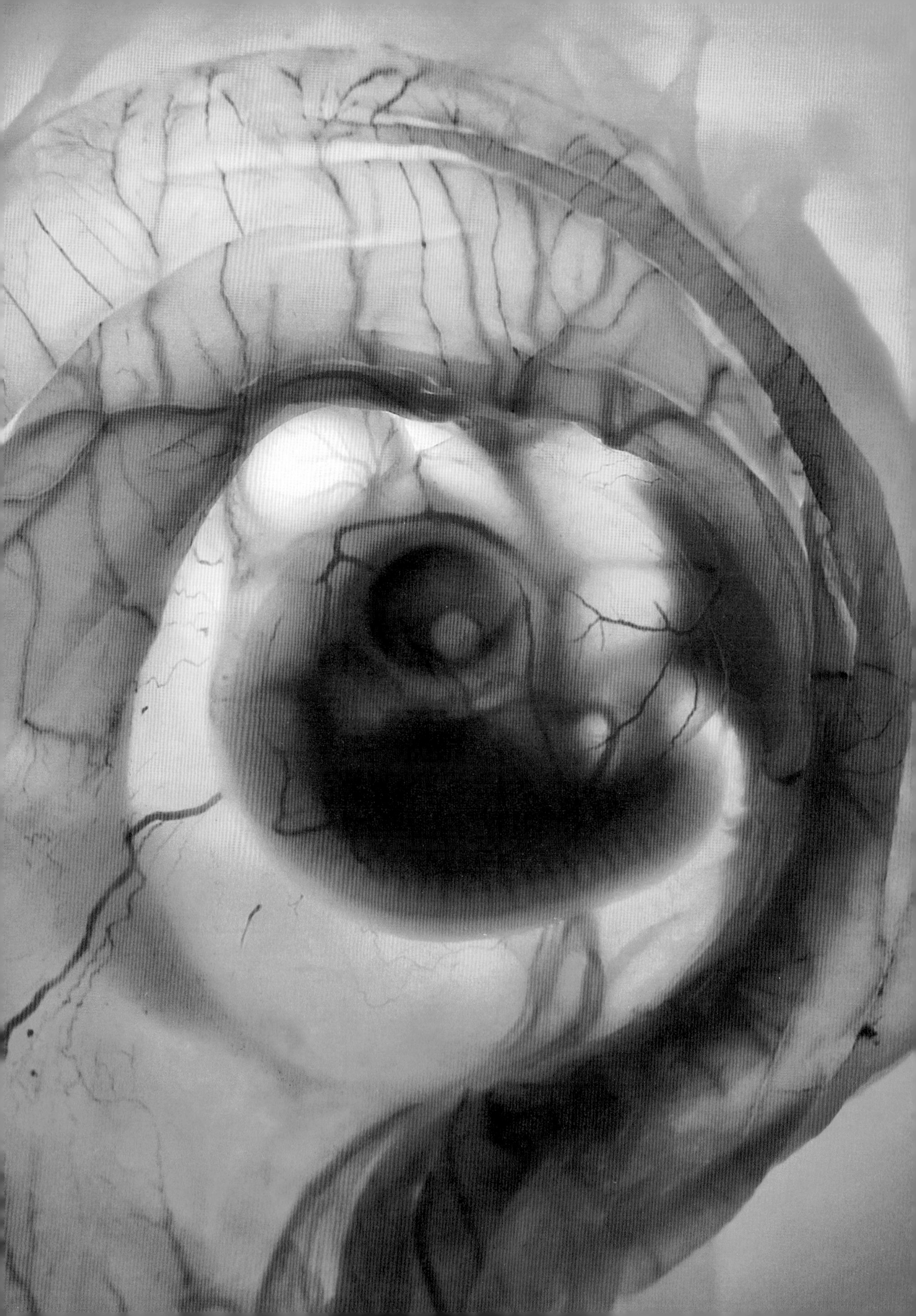

PART II

Developmental Repatterning

'evolution is not just a statistical genetical problem but also one of the developmental potentialities of the organism'

Richard Goldschmidt, *The Material Basis of Evolution*, 1940

The aim now is to examine all the different ways in which development can evolve. It is important to do this because in the past there has been a tendency for many biologists to focus on, and to exaggerate the importance of, one particular kind of developmental evolution at the expense of others. This has been true, for example, of recapitulation (in the 19th century) and heterochrony (in the 20th century). So, to begin with, we need an overall word that covers all possible kinds of evolutionary change in development, equivalent to 'mutation' as a word that covers all possible kinds of change in the DNA sequence in any part of the genome. This word, as we saw in Part I, is 'repatterning'. If we start from the perspective of developmental repatterning in general, and try to see how particular kinds of repatterning fit into the overall picture, then we are starting with open minds rather than a desire to stress the importance of one kind of repatterning over another. This balanced approach will now proceed in the following way:

Chapter 5: Mutation is the ultimate cause of developmental repatterning. Here, we look at the consequences of mutations in developmental genes for embryogenesis, post-embryonic development, the resultant adult forms, and fitness.

Chapter 6: Some evolution involves changes in the *relative timing* of developmental processes of descendants compared with their ancestors (usually compared indirectly via related extant forms). Such 'heterochrony' is widespread in evolution.

Chapter 7: Other evolutionary changes result in alteration of the *relative spatial positioning* of developmental processes. Like heterochrony, this 'heterotopy' is a widespread evolutionary phenomenon.

Chapter 8: Here we examine evolutionary changes in the *relative sizes* of parts of an organism, or in the *relative concentrations* of different gene products; that is to say, evolutionary changes in which something differs primarily in *amount* rather than in time or place. This 'heterometry' is also widespread in evolution.

Chapter 9: Now the focus shifts to evolutionary changes in development that produce outcomes which we regard, however subjectively, as *different types* of structure. Such 'heterotypy' is harder to measure than changes in time, place or amount. It is also harder to define, involving, as it does, a subjective judgement of the limits of any one 'type'.

Chapter 10: Here we examine the typically compound nature of developmental repatterning, which involves a mixture of various kinds of change in development happening together, rather than 'pure' forms of heterochrony (etc.) happening one after the other.

Chapter 11: Having looked at all the different types of developmental repatterning, we now consider mapping various instances of repatterning to phylogenetic trees. This is important in relation to being able to distinguish homology from convergence.

Chapter 5
Mutation and Developmental Repatterning

5.1 Mutation in Terms of Altered DNA Sequence

It is an undisputed fact that evolution depends upon inheritance and, in particular, upon **heritable** variation. Even in cases where evolution involves developmental **plasticity**, such as winged and wingless forms of an insect species, the ability to respond in a plastic way to environmental variation is inherited, despite the fact that offspring do not inherit the form taken by a parent as a result of a plastic response.

The main source of heritable variation is **mutation**, broadly defined to include all changes in the DNA sequence anywhere in the nuclear or organelle genomes, whether large or small, and whether in coding, regulatory or other regions. Heritable mutation can also be called germ-line mutation, to distinguish it from the somatic mutation that arises in other tissues, such as that which causes the formation of a tumour.

Until recently, it was often said that mutation is the *sole*, rather than the 'main' (as it was referred to above) cause of heritable variation. But we now know that, in addition to the DNA sequence itself being heritable from parents to offspring, so too may be modifications of the 'state' of stretches of DNA, where the sequence itself is unaffected. An example is that offspring may inherit the methylation pattern of a stretch of DNA from their parents. DNA methylation involves the addition of methyl groups to nitrogenous bases, often with the result of suppressing gene expression. The study of such processes is referred to as **epigenetics**, because the methylation is something that is 'outside of', but superimposed on, the genetic realm of the DNA sequence itself. This field is very new; it is likely that there will be many interesting discoveries from it in the near future. See Suggestions for Further Reading, at the end of this chapter, for a reference to a recent book on the subject.

Many geneticists have studied mutation from about 1900 onwards. The result is a massive body of knowledge. Here, we will be selective in examining this knowledge and will in particular look at how mutation in the germ line of a parent affects the development of their offspring – i.e. **developmental repatterning**. But before linking mutation to development, we should briefly consider mutation itself.

As noted above, mutations can be very small or very large, or anywhere in between. So can their consequences, but the two do not map together in a simple way, because the tiniest of mutations – a single base change – can have massive developmental effects. As a broad approximation, the size of a mutation, in terms of DNA sequence change, can be considered as belonging to one of four classes, as follows, connected with the mechanism of mutation and the type of DNA sequence involved:

Point Mutations

Each nitrogenous base of the millions that make up the sequence of the entire **genome** can be regarded as a 'point', in which case a 'point mutation' involves just a single base. The possible types of point mutation are the switching, loss, or gain of one base. These are referred to as substitutions, deletions and insertions. In a coding region, these have predictable consequences (Section 5.2).

Although the frequency of occurrence of point mutations can be increased by many factors (such as X-rays), there is a baseline level of such mutations that is regarded as 'spontaneous' – that is, it has no obvious cause. Spontaneous point mutations probably reflect the inherent vulnerability of DNA-replication to errors. The surprising thing is not that such errors occur, but rather that their rate is so low. The high degree of accuracy of DNA replication is doubtless a result of the accumulated action of natural selection in the past.

Mutations involving Several Consecutive Bases

This seems to be a common type of mutation in regions of tandemly repeated DNA, that is, sequences that are repeated many times, one immediately after the other, along a particular stretch of a chromosome. For example, if we have a 5-base sequence, say

AATTG, repeated hundreds of times in tandem, there is often variation between individuals within a population in the precise number of such repeats. The **polymorphism** in the population, which takes the form of a variable number of tandem repeats – or VNTRs – ultimately arises from 'slippage' of the DNA, as these long regions are duplicated. VNTRs are the basis for the technique of DNA fingerprinting, devised by British geneticist Alec Jeffreys[1], in which every individual in a human population (except monozygotic twins) is seen to be different in terms of the pattern of bands on a gel; these differences can be used to determine paternity and to identify suspects in forensic science (Fig. 5.1). This individual uniqueness applies in most other species too; fingerprinting has been used for determining parentage in (for example) promiscuous bird species[2], as well as in humans.

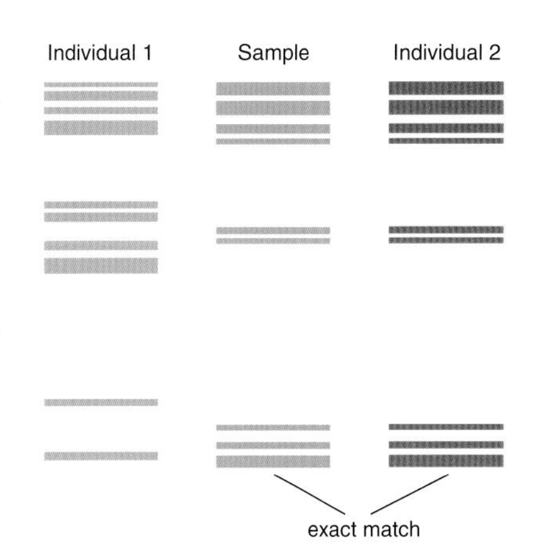

Figure 5.1 An example of the use of DNA fingerprinting to determine which of two individuals was responsible for a crime. Left: suspect individual 1; centre: sample of blood from perpetrator left at crime scene; right: suspect individual 2. It is clear that individual 2 is the perpetrator. The technique for visualising the unique banding patterns of individuals is a modified version of the gel electrophoresis technique described in Chapter 4.

Mutations involving Several Hundreds or Thousands of Bases

One important cause of mutations on this scale is the movement of transposable elements, or **transposons**, through the genome. These elements are generally thought to be 'selfish DNA' that is effectively parasitic in the genome of its host. There are two main types: those elements that are transcribed into RNA and then converted back to DNA by reverse transcriptase enzymes; and those that can transpose directly from one site to another in the genome using transposase enzymes, the genes for which are often contained within the overall DNA sequence of the element concerned. Their capability for replicative transposition means that they can come to constitute a substantial component of the host genome. Their relevance to development lies partly in the fact that they can 'jump' to a position within a developmental gene, thus mutating it and often rendering it functionless (Fig. 5.2).

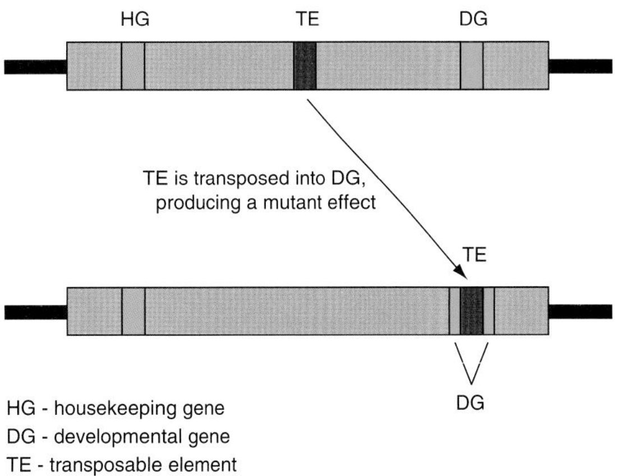

Figure 5.2 Diagrammatic version of the jumping of a transposable element from an inter-genic region to the middle of a developmental gene, thus causing altered development and a mutant phenotype.

Mutations involving Whole, or Large Fractions of, Chromosomes

It has long been known that some mutations (chromosomal aberrations) are large enough to be seen by ordinary light microscopy, especially in certain systems – notably the giant **polytene chromosomes** found in the salivary glands, and some other tissues, of the larvae of flies such as *Drosophila*. In such cases, the different versions of the chromosome concerned can be seen by their different banding patterns. Types of chromosomal mutation include inversions (Fig. 5.3),

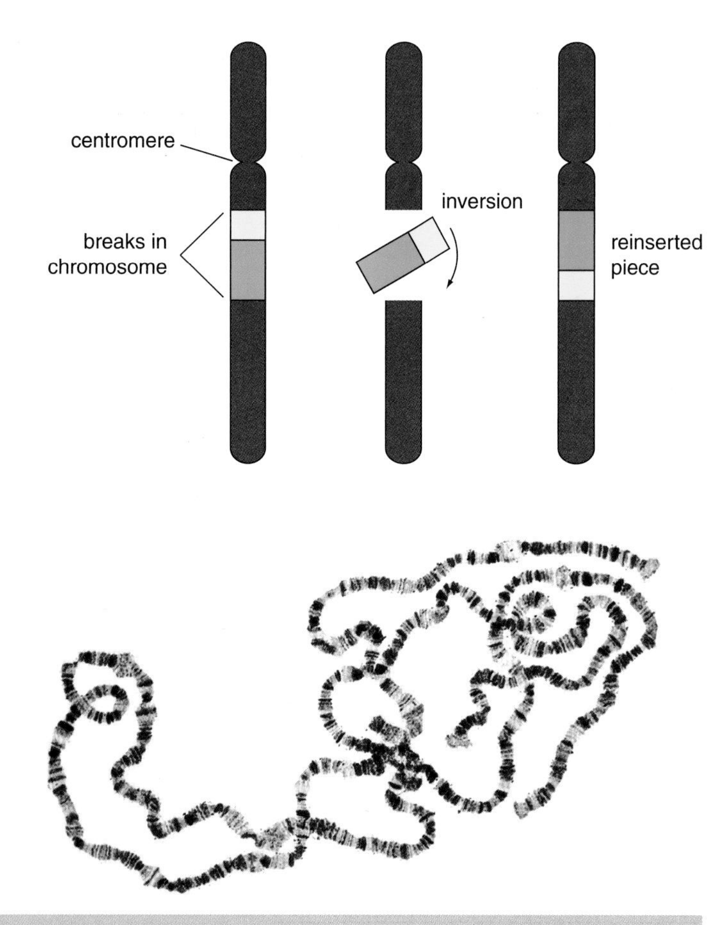

Figure 5.3 Top: Diagram of a chromosomal inversion. Bottom: Polytene chromosomes as seen under the microscope, showing the characteristic banding pattern, part of which is in reverse sequence after an inversion (photograph courtesy of Victor Corces).

translocations and deletions. Many natural populations are characterised by inversion polymorphisms, which typically have less detrimental effects on fitness than the other types of chromosome-level mutations. As we noted earlier, inversion polymorphisms were extensively studied in the early days of ecological genetics by Theodosius Dobzhansky.

It is clear that flies carrying different inversion types can be acted upon by natural selection: populations of some *Drosophila* species show repeatable seasonal cycles in the frequency of different inversion types, which is indicative of selection rather than **genetic drift**. However, the nature of the selection is not obvious, partly because the inversion incorporates many different effects. These include damage to the DNA sequence at the break-points; also 'position effects', in which a gene's activity is affected by its location on the chromosome. Some large inversions, such as those known in *Drosophila*, and also the seaweed fly *Coelopa*[3], can contain hundreds, or even thousands, of genes.

Polymorphism involving chromosome number rather than gross structure seems to be rare in natural populations. The reason for this is that having a mutant chromosome number is usually seriously detrimental to fitness, as in the case of Down's syndrome in humans, caused by an extra copy of chromosome 21, and characterised by particular developmental effects. However, some animals, and a larger number of plants, have small accessory chromosomes called B-chromosomes, and these often do vary in number in natural populations. Also, whole-genome duplication, or **polyploidy**, occurs in nature, again more commonly in plants than in animals, and has been implicated in the formation of new species, for example the marsh-grass *Spartina townsendii*.

5.2 Mutation in Terms of Proximate Functional Consequences

Ultimately, mutations can affect all aspects of development, and also the **fitness** of the organisms carrying them. But before looking at these 'ultimate' effects, we will examine the 'proximate' ones – those that are just one or two steps away from the altered DNA sequence itself. Both proximate and ultimate effects depend very much on where in the genome a mutation occurs, as follows:

Coding Regions of Genes

The proximate effects of mutations in these regions have been known since the genetic code was worked out in the 1960s. The code is shown in Fig. 5.4. A single base substitution can have three kinds of effect: it can alter the mRNA but not the protein, due to the 'redundancy' of the code (e.g. UUU to UUC); it can alter both the mRNA and a single amino acid within the protein (e.g. AGC to AGA); or it can have a more drastic effect on the protein (e.g. premature termination as a result of mutation that leads to a stop codon (e.g. UCG to UAG). A single insertion or deletion will cause a frame-shift mutation, the result of which is often the same as in the last of the three possible effects of a base substitution – that is, premature termination of the protein.

second base in mRNA triplet code

first base in mRNA triplet code	U	C	A	G	third base in mRNA triplet code
U	UUU Phe	UCU Ser	UAU Tyr	UGU Cys	U
	UUC	UCC	UAC	UGC	C
	UUA Leu	UCA	UAA **stop**	UGA **stop**	A
	UUG	UCG	UAG **stop**	UGG Trp	G
C	CUU Leu	CCU Pro	CAU His	CGU Arg	U
	CUC	CCC	CAC	CGC	C
	CUA	CCA	CAA Gln	CGA	A
	CUG	CCG	CAG	CGG	G
A	AUU Ile	ACU Thr	AAU Asn	AGU Ser	U
	AUC	ACC	AAC	AGC	C
	AUA	ACA	AAA Lys	AGA Arg	A
	AUG Met	ACG	AAG	AGG	G
G	GUU Val	GCU Ala	GAU Asp	GGU Gly	U
	GUC	GCC	GAC	GGC	C
	GUA	GCA	GAA Glu	GGA	A
	GUG	GCG	GAG	GGG	G

Figure 5.4 The genetic code, showing the links between triplets of bases in mRNA and amino acids in the corresponding protein (or in a few cases, 'stop' signals).

Regulatory Regions of Genes

It is now known that regulatory regions can occur in many different places. They can be both upstream and downstream of the coding sequence, and in either case can be very variable distances from it. There are also some regulatory regions within the gene – in the transcribed but untranslated **introns**. The effects of mutations in these regions are harder to specify than those in coding regions. However, in general, any mutation in a regulatory region can affect the binding of transcription factors, both qualitatively (Fig. 5.5) and quantitatively. The changes shown do not affect the gene product, but they may well affect where and when it is produced, and in what quantities. Since most genes have multiple regulatory regions, mutation in just one of them may affect the expression of the gene in just one part of the embryo or at just one developmental stage.

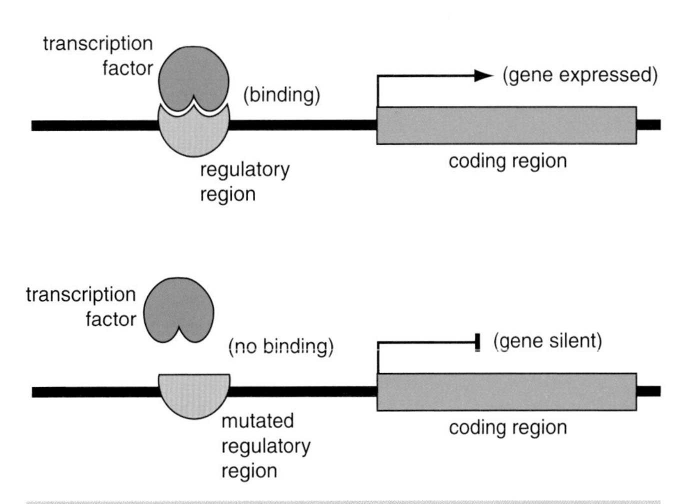

Figure 5.5 Depiction of a mutation in the regulatory region of a developmental gene and its possible effect in terms of altered binding of transcription factors and, hence, altered gene expression.

Other Regions of the Genome

In some regions, there may be very little in the way of proximate or ultimate effects of a mutation: indeed, if the sequence that a mutation occurs in is neither coding nor regulatory, the DNA sequence itself may change without any consequence at all, either for the development of the organism or its fitness. This may be true, for example, of the VNTRs mentioned above and in Chapter 4. Although the altered DNA can be visualised by a biologist in terms of the type of DNA fingerprinting gel that was shown in Fig. 5.1, the organism itself may be unaffected. However, it is important to be cautious about such a conclusion, because some DNA sequence changes in non-genic regions, although not directly regulatory, may cause conformational changes in their region of the chromosome that make regulatory sequences more or less accessible to a variety of agents that can bind to them.

5.3 Developmental Repatterning at Molecular and Higher Levels

We will now focus on mutations occurring in developmental genes, and the effects these may have on the developmental processes controlled or affected by such genes. Thus we will ignore purely **housekeeping genes** and also those non-genic regions of the genome in which mutations have no developmental effect.

At this point, it is worth recalling the definition of **developmental repatterning**, since it is perhaps the single most important term in the book. Developmental repatterning refers to a change from a previous pattern of development (e.g. in a parent organism or parental species) to an altered pattern of development (e.g. in the parent's offspring or in the **daughter species** that arise from the parental one). Here 'pattern' is used in its most general sense. Suppose that developmental trajectories 'B' and 'C' characterise daughter species that have both descended from a common ancestor with **developmental trajectory** 'A'. Then, assuming that neither B = A nor C = A, both lineages have undergone repatterning.

Although an altered developmental pattern can arise for environmental reasons (e.g. winged offspring from wingless parents in some insect species due to increased population density: see Chapter 16), we will concentrate here on repatterning that has a genetic cause. Moreover, we will initially concentrate on a case in which the repatterning has a single mutational origin. This is a good starting point because of its simplicity. But shortly we will also focus on: (i) inter-**taxon** repatterning that is the result of the accumulation of many mutational changes over long periods of time; and (ii) the fact that any two members of a natural population of a sexually reproducing species are likely to have subtly different patterns of development, due to the effects of **allelic** differences between them at multiple gene **loci**.

Repatterning of Arthropod Segments

At the molecular level, repatterning can most readily be seen in terms of the altered spatial and/or temporal **expression pattern** of a developmental gene.

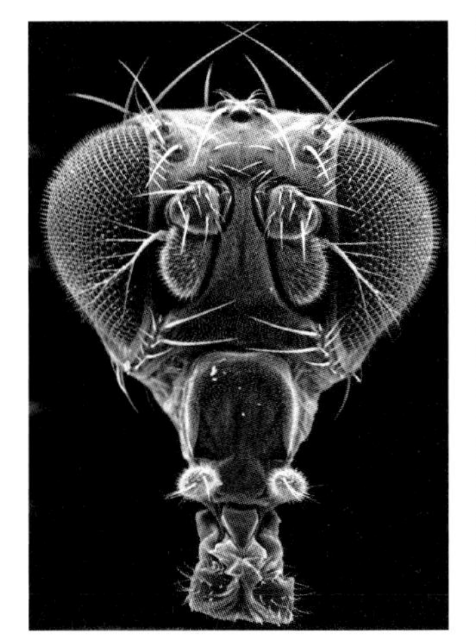

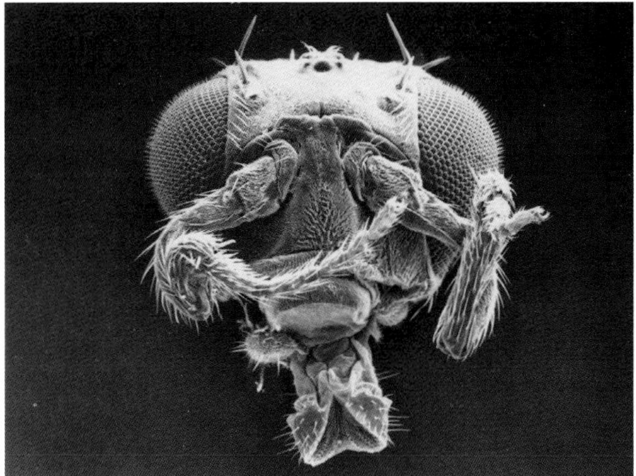

Figure 5.6 Top: Normal (left) and antennapedia mutant (right) forms of *Drosophila*, seen from the anterior end, using scanning electron microscopy (SEM) (photograph courtesy of Flybase). Bottom: Antennapedia fly, dorsal view (photograph courtesy of Carroll, S. *et al.* 2005, *From DNA to Diversity*, Blackwell.)

Take, for example, the Hox gene *antennapedia*, or *antp*, in *Drosophila*, and its homologues in other taxa. This, like other Hox genes, influences antero-posterior axis patterning in most bilaterian animals, and is a **segment identity** gene in those that are segmented.

In *Drosophila*, *antp* is normally expressed in the thorax, and its expression is associated with thoracic segment identity. However, a mutation in the regulatory region of this gene can result in its **ectopic expression** in the developing head – specifically in the 'eye-antennal disc' – one of the many **imaginal discs** in the larva from which adult structures are made during metamorphosis. In such mutants, an extra pair of legs develops from the head, in the position where the antennae would normally be (Fig. 5.6).

This is a very drastic example of repatterning, a type called **homeosis** (the right structure in the wrong place; a sub-category of **heterotopy**), which helps us to understand the developmental role of the gene concerned but not, at least in any detail, its role in evolution. There is no doubt that mutations in the *antp* gene have contributed to the evolutionary divergence of arthropod (and other) taxa (see below), but not mutations that have such a major effect. As we saw in Chapter 2, the geneticist Richard Goldschmidt proposed (in the 1940s and 1950s) that homeotic mutations such as that involving *antp* could contribute to evolutionary change[4]. However, this

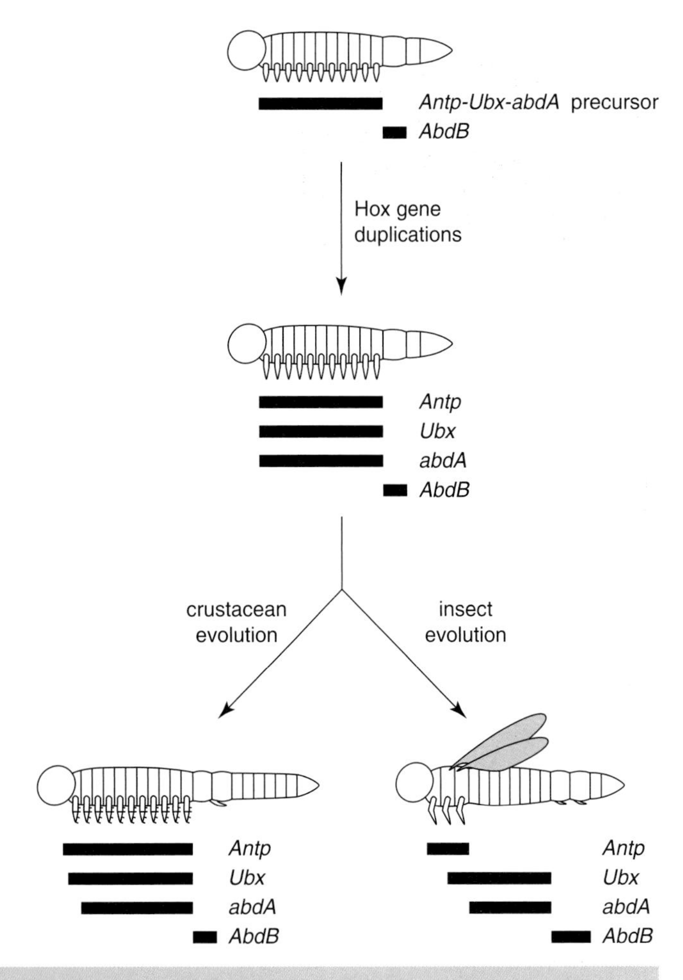

Figure 5.7 The antero-posterior expression zone of the gene *antennapedia*, in terms of spans of segments, in an insect and a crustacean. Note the correspondence, in both cases, between *antp* expression and leg development. (Expression patterns of three other genes are also shown.) (Redrawn from Averof and Akam, 1995, *Nature*, **376**: 420–423.)

hypothesis has generally not been accepted by biologists, because these mutations are characterised by severely reduced fitness, as opposed to the enhanced fitness that would be required for them to be spread through populations by natural selection.

When comparisons of *antp* expression are made, not between wild-type and mutant *Drosophila*, but between insects and branchiopod crustaceans[5], it is clear that the evolutionary divergence of these groups has involved changes in the expression pattern of this gene (Fig. 5.7). It appears that from an ancestor with many leg-pairs, the insect body plan of just three leg-pairs (on the three thoracic segments) has been derived by a shortening of the zone of expression of *antp* along the antero-posterior axis. What we do not know, however, is how this shortening came about, in terms of the number of changes, their magnitudes and their distribution through evolutionary time.

Another example of repatterning involves the genes *engrailed*, *wingless* and *hedgehog*, all of which were discussed, in relation to expression patterns and signalling pathways, in Chapter 3. Although these genes have many functions, one that they share is that of the determination of segment polarity in arthropods. In relation to this, all are expressed in a number of segmental stripes. Changes in the number of these expression stripes give rise to an altered number of morphological segments. Such changes can be seen in Fig. 5.8. The number of stripes can be altered by two kinds of mutation: (i) mutations in the coding regions of genes upstream of the segment polarity ones; and (ii) mutations in the regulatory regions of the segment polarity genes themselves. Also, as in the evolution of segment identity via Hox genes, we do not know how many such changes have occurred in any particular divergence.

Repatterning of Brassicacean Flowers

Although it is clear that homeotic mutants in *Drosophila* are never found in large numbers (if at all) in natural populations, this should not lead us to conclude that the same is true of homeotic mutants in all other taxa. An interesting example of whole

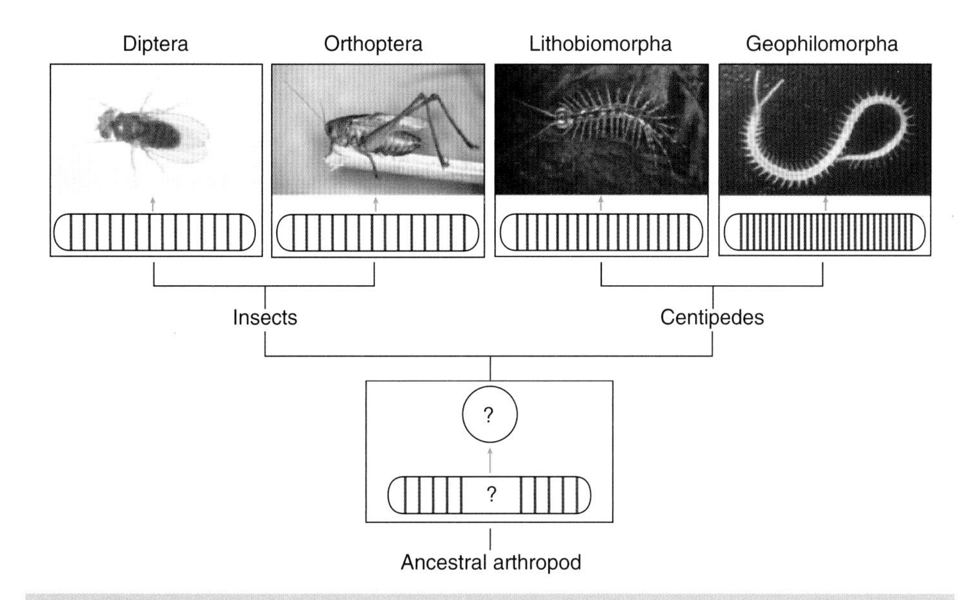

Figure 5.8 Phylogenetic relationships among four arthropods – two insects and two centipedes. The *engrailed* expression stripes that form part of the genetic basis for the segments are shown diagrammatically. (Reprinted with permission from Arthur, 2002, *Nature*, **415**: 757–764.)

populations of plants with a homeotic flower-type in the wild has recently been documented[6]. This involves the **brassicacean** species *Capsella bursa-pastoris* (shepherd's purse), a close relative of the 'model plant' *Arabidopsis*.

The mutant form (Fig. 5.9) has its petals transformed into extra stamens, and so has been called 'stamenoid petals' (spe). The underlying molecular mechanisms are not yet clear, but they probably cause a change from ABE to CBE gene expression in the transformed petals (Fig. 5.10). Again, as in *Drosophila*, the alteration of development that occurs in the spe mutant is a homeotic one, and thus falls into a sub-category of heterotopy. Although mutant plants in the wild are visited less often by pollinators – an inevitable consequence of the conversion of insect-attracting petals to stamens – this does not seem to significantly reduce fitness, as the species is largely a self-pollinator.

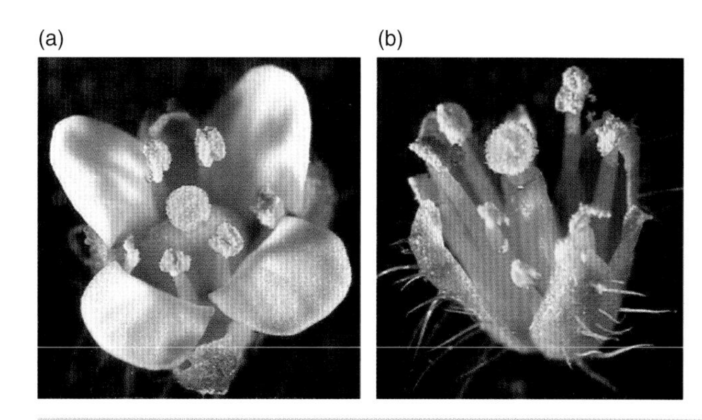

Figure 5.9 (a) Normal and (b) Mutant forms of the brassicacean plant *Capsella bursa-pastoris.* This particular mutant phenotype has 'stamenoid petals' due to a homeotic transformation of the petals to a stamen-like state. (Reproduced from Hintz *et al.* 2006, *J. Exp. Botany*, **13**: 3531–3542.)

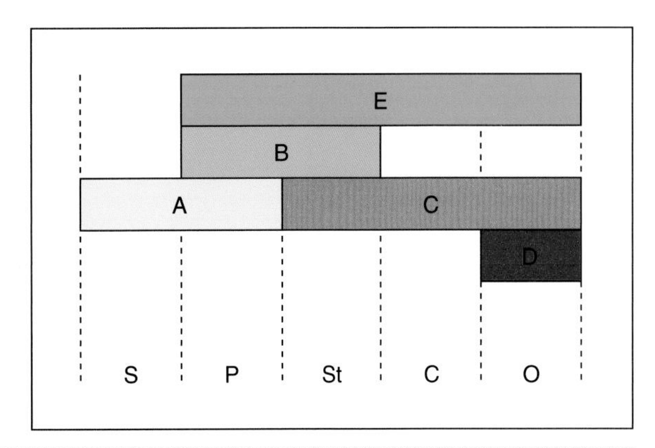

Figure 5.10 The ABCDE model of the genetic basis of flower development. According to this model, each whorl of structures in the flower is determined by a particular combined expression pattern of genes belonging to the five classes A to E. The structures are: S – sepals; P – petals; St – stamens; C – carpels; O – ovary.

Body Size in Peas and People

The homeotic mutants of *Drosophila* and *Capsella* discussed above are **macromutants** in the sense that they stand out clearly from the normal range of variation in segmental, or floral, structure in the species concerned. However, most developmental and phenotypic variation in body form in natural populations is not like this – rather, it consists of many intergrading forms that collectively give rise to a normal distribution of values for the character being studied – whether it is measured in the adult or at some earlier stage of development. It is interesting to contrast these two types of variation, both of which represent developmental repatterning.

It seems appropriate to start with Mendel's work[7] on the height of pea plants (genus *Pisum*). One of his famous experiments, which led to the now-familiar 3:1 phenotypic ratio in the F_2 generation, involved crossing normal and dwarf peas. Like stamenoid petals in *Capsella*, dwarf pea plants can be regarded loosely as 'macromutants' because they stand out clearly from the normal distribution of height that characterises **wild-type** pea plants (and indeed dwarf ones too). The situation is that of two non-overlapping normal distributions (Fig. 5.11).

Such a pair of distributions would also result from measuring the heights of two groups of humans from the same (e.g. Irish) population: people of 'normal' height and achondroplastic dwarves. The latter typically constitute a very small fraction of any

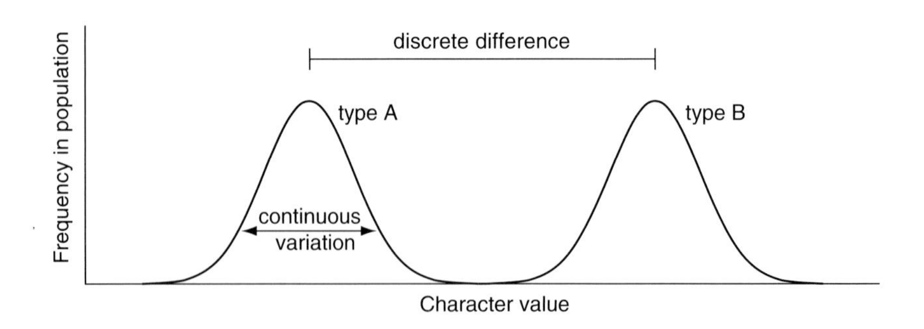

Figure 5.11 Two non-overlapping normal distributions. This pattern of variation is found where there are two 'types' in a population (A and B), but neither is fixed – rather each is characterised by continuous variation among individuals of that 'type'.

human population (<1 in 20,000). They have a mean height approximately 50 cm less than other humans. Their development has been repatterned by a dominant mutation of the gene FGFR3[8]. (FGFR stands for fibroblast growth factor receptor.)

The Nature of Developmental Repatterning

Consideration of this phenomenon of non-overlapping normal distributions, whether in peas, people or any other species, leads to some important conclusions about the nature of developmental repatterning – specifically about its directionality and its frequency of occurrence.

When large-scale repatterning is observed, the direction in which it has occurred is usually clear. This is true both of macromutational repatterning within a species and of accumulated evolutionary repatterning between species. Consider first the case of homeotic mutants. It is clear, both in *Drosophila* and *Capsella*, that the homeotic forms have originated from wild-type ancestors. So, while a homeotic transformation can potentially happen in either direction, the actual direction that it took (as a result of mutation) is known. This is also true in many (but not all) cases of accumulated evolutionary large-scale repatterning. If we knew nothing about arthropod phylogeny, the direction of the repatterning that led to three pairs of thoracic legs in insects but many more in branchiopod crustaceans could have occurred in either direction. However, since it now seems that insects are a **clade** within the **Pancrustacea**, and have evolved from a more 'typically' crustacean ancestor, the direction of the repatterning seems clear, as noted above: a shortening of the A-P zone of *antp* expression in the derived insect **lineage** (as opposed to a lengthening of this zone in crustaceans that derived from an insect-like ancestor).

When we turn our attention to normally-distributed variation in developmental patterns and in the resulting adult characters, often thought of as being underlain by **micromutational** changes at many gene loci, the issue of directionality becomes more complex. Also, its frequency is seen to be very high, more similar to that of genetic recombination (which occurs in each generation) than to that of mutation (which occurs, in any given gene, perhaps only once in a million generations).

Although genes persist through many generations until altered by mutation, genotypes and phenotypes are created anew in each generation through the reshuffling of gene combinations by the two main processes of sexual reproduction – meiosis and fertilisation. Since developmental patterning is simply a four-dimensional way of looking at phenotypes, it follows that developmental patterns are also reshuffled in each generation. So repatterning is going on all the time, but not in any given direction. Within the normal distribution of height in a population of wild-type humans or pea plants, some families will show mean offspring height to be taller than the mid-parental value, while in other families the opposite will be seen.

Analysis of the genetic causality of phenotypic normal distributions has shown that, rather than being underlain by hundreds of individually tiny micromutational variants, they are often underlain by the segregation of alleles at a smaller number, perhaps 50 or so, of gene loci, of which just a few (maybe 4 or 5; **mesomutations**, if you like) contribute much more than the others. These 50 or so loci are referred to as the **quantitative trait loci** (QTLs)[9] underlying the character concerned. We will return to QTLs in Chapter 12.

For now, the important point is that developmental repatterning within a species can be seen to be large and unidirectional or small and multidirectional. An important question is the extent to which evolutionary change is based just on the latter (often said to be the neo-Darwinian view) or on a mixture of the two. A related question is whether the two are distinct or are connected by a variety of intermediates. We will return to such questions in later chapters.

5.4 Developmental Repatterning at the Level of the Whole Organism

Some of the examples looked at in the previous section involved particular parts of the organisms concerned – segments and flowers. However, characters such as height involve the development of the whole organism. Whether changes in height (or body size in general) are really repatternings is a moot point. It can be argued that the pattern is the same and it has simply been scaled up. But conversely it can be argued that *all* changes in development are repatternings, including changes in size. One reason to take this latter view is that the growth of organisms is usually **allometric**, meaning that shape changes as size increases, rather than **isometric**, where there is no change of shape with increasing size.

Allometric developmental changes have been studied for many decades, with some of the early work being carried out by Julian Huxley[10], grandson of 'Darwin's Bulldog', Thomas Henry Huxley. A famous example is the relative size of the head, which diminishes relative to the size of the trunk, in the growth of humans (Fig. 5.12). In evolution, changes in size may be accompanied by changes in shape because of an *unchanged* allometric growth pattern. However, often patterns of allometric growth themselves change in evolution, as can be seen, though indirectly, in D'Arcy Thompson's pictures of what he called **transformations**[11], further examples of which are given in Fig. 5.13 (see also Fig. 2.3).

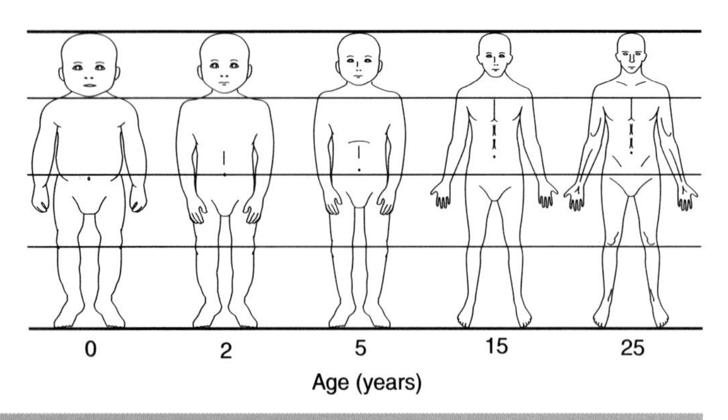

Figure 5.12 Decreasing relative head size with growth in humans. This is one of the classic examples of allometric growth in which organisms change shape as they grow, because some parts grow faster than others.

The reason that these pictures show evolutionary change in the pattern of allometric growth only indirectly is partly that D'Arcy Thompson drew transformations that were idealised rather than ancestor-descendant ones; and partly that the comparisons he made were generally of *adults*. Ideally, we would like to see the whole growth trajectory in both actual ancestors and their descendants. Of course, this is rarely if ever possible.

D'Arcy Thompson was fascinated by the mathematical description of the transformations he observed, but did not delve into their causality, partly because his interests did not lie in that direction, but partly also because the relevant tools were unavailable at the time. Much has

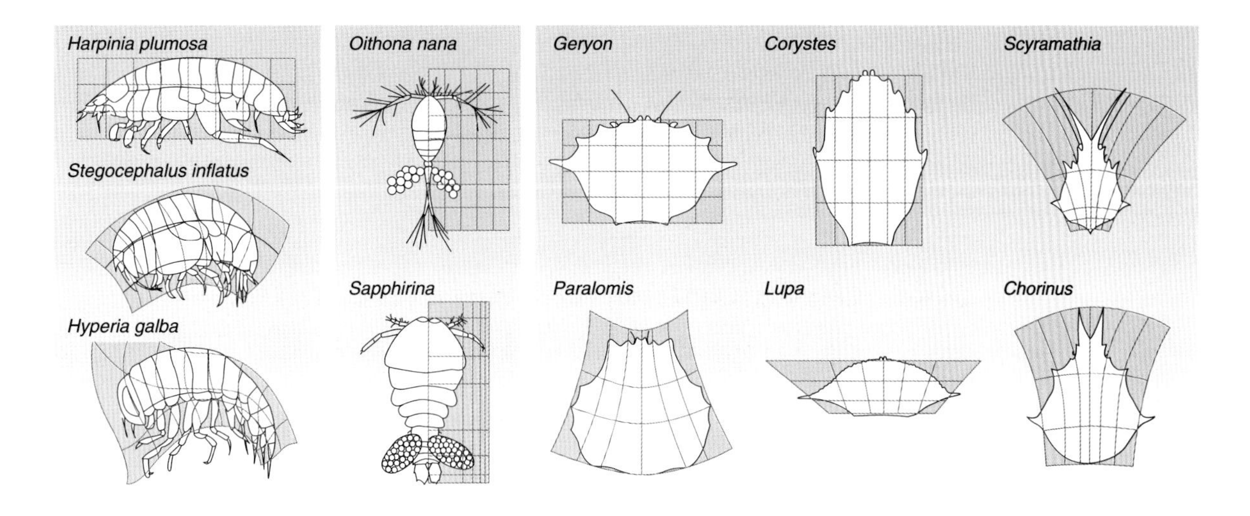

Figure 5.13 Thompsonian transformations in three groups of crustaceans: amphipods, copepods and crabs. In each case (as in the fish pictured in Chapter 2), systematic distortion of the outline body form of one species can produce the body form of another. (Redrawn from Arthur, 2006, *Nature Reviews Genetics*, **7**: 401–406.)

changed since he published his magnum opus[11] *On Growth and Form* (first edition 1917, second edition 1942). We can now consider more readily at least the general types of morphogenetic agents that must underlie Thompsonian transformations.

Especially in those cases of transformations involving the geometry of the whole body, the cause must lie not in **morphogens** whose field of action is a few tens or hundreds of cells across, but rather in long-range agents such as **hormones**, an example of which, ecdysone, was discussed in Chapter 3. These are carried through the circulatory system and have effects that pervade the whole developing organism, particularly through its later developmental stages.

Thompsonian transformations, like homeotic ones, fall within the category of developmental repatterning called heterotopy, i.e. changes in the spatial pattern of events. However, it is likely that most evolutionary changes of form are a *combination* of two or more types of repatterning. Even when the most obvious thing happening is a change in spatial pattern, a change in temporal pattern may well be involved too. There are few generalisations that can be made with confidence about evolution, but an exception to this rule is that it is a complex and often messy process.

5.5 Developmental Repatterning and Fitness

The relationship between an organism's genetic constitution and its suite of phenotypic characteristics is often referred to as the **genotype-phenotype map**[12]. However, from an evo-devo perspective, wherein we are taking a four-dimensional view of phenotypes, a better name is the *genotype-developmental pattern map*. In fact, from a general perspective also, this second term is preferable. After all, an organism retains its genotype throughout life while, for much of that life, the phenotype undergoes considerable change, this being especially true of indirect developers such as frogs, butterflies and sea urchins. So, in the genetic case, each organism is of a specific 'type'; but in the phenotypic one, 'type' is a singularly inappropriate word.

There is general agreement that the genotype-developmental pattern map is very complex. The same is true of another kind of map – that between developmental pattern and Darwinian fitness, which we can refer to as the pattern-fitness map.

Although we could imagine a third type of map – the genotype-fitness map – this one makes little sense, because it jumps over a logically connecting phenomenon (development). As is often said, natural selection acts on phenotypes (or in fact on developmental patterns) rather than on genotypes. So we will not discuss this third kind of map further.

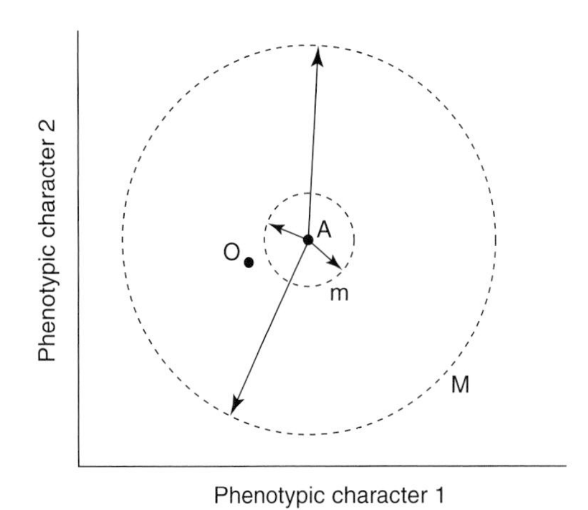

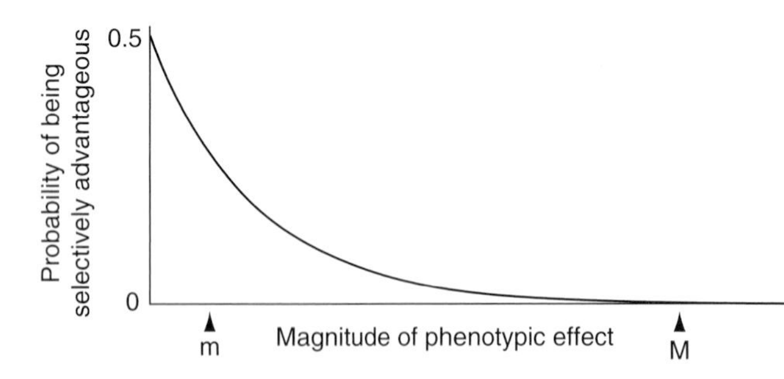

Figure 5.14 Fisher's geometrical argument against the evolutionary importance of macromutations. Top: From any given starting point (A) a small-effect mutation (m) may move the phenotype closer to the optimum (O); while a large-effect mutation (M) will necessarily take it further away. Bottom: Proposed negative relationship between selective advantage and magnitude of mutational effect that results from generalising this argument to mutations of all possible magnitudes of effect. (Redrawn from Arthur, W. 1997, 2000, *The Origin of Animal Body Plans*, Cambridge University Press, Cambridge.)

The pattern-fitness map is an absolutely crucial determinant of the direction of evolutionary change – a subject that will be expanded on in Part III. This map is at the heart of understanding what is possible and what is not in the evolution of organismic form. Much of the debate about the relative importance of different agencies in 'steering' evolution can be directly linked to how the advocates of different views see this map – either explicitly or implicitly.

A particularly informative conflict in this area occurred in the mid-20th century between Fisherian and Goldschmidtian views of the nature of the evolutionary process – with these terms being named after the **gradualist** Ronald Fisher and the **saltationist** Richard Goldschmidt. In the Fisherian view, no mutations with large phenotypic (i.e. developmental) effect would contribute to evolution, because they would inevitably lead to decreased rather than increased fitness. This argument was made geometrically[13], and can be seen in Fig. 5.14. Goldschmidt, on the other hand, argued that large-effect mutations, including those leading to homeotic repatterning, were especially important in evolution, and underlay some of the major divergences, such as those leading to the origins of new **body plans**.

Both Fisherian and Goldschmidtian views can now be seen as seriously flawed, though this is more generally accepted of Goldschmidt's view than of Fisher's. In both cases we can see the flaws by thinking in terms of the pattern-fitness map.

In fact, Fisher's argument, although originally cast in terms of mutations and without any reference to development, is an argument specifically about the relationship between the magnitude of developmental repatterning and the direction of the change in fitness. Fisher concludes his argument[13] by saying that while small changes (in developmental pattern, though of course he did not use that term) could readily lead to increases in fitness, 'for greater changes the chance of improvement diminishes progressively, becoming zero, or at least negligible, for changes of a sufficiently pronounced character.'

There are at least three reasons why this argument is flawed. First, it assumes a single phenotypic optimum. In reality there may be two or several, in which case some major repatternings may accidentally hit upon one of the others. Second, it is far too abstract. How is the 'magnitude' of a change in development measured in reality? There are many *types* of repatterning, and a certain 'magnitude' of one type may give a very different result from a comparable 'magnitude' of another. Third, there is a great difference between 'zero' and 'negligible', especially in the vastness of geological time. Indeed, if something happens with negligible frequency in the short-term (e.g. mutation in 'gene A' in 'generation *i*' in a small population of *Drosophila melanogaster*), that 'something' becomes a near-certainty in, say, a single dipteran evolutionary lineage over 100 million years.

The Goldschmidtian view is equally flawed but for different reasons. Goldschmidt argued that homeotic repatterning, such as that producing 'bithorax' *Drosophila* individuals, could form the basis for the origin of a new insect order. He suggested that there might be some ecological conditions in which bithorax (and other homeotic) mutants might be favoured by selection. However: (i) no such conditions have ever been found; bithorax flies have not spread through *any* natural populations; and (ii) the direction of change is 'backwards' in relation to the direction of actual evolution. Regarding this latter point, insects with a single pair of wings have evolved from ancestors with two pairs of wings independently in several groups, the best-known of these being the **Diptera** and **Strepsiptera**[14], which are not sister-groups[15]; whereas, as far as we know, the reverse change has never happened.

To conclude this chapter we will look at two other schools of thought, one old and one new, in relation to how they connect with pattern-fitness maps. First, the 'old' one: orthogenetics. This view of evolution, which was prominent in the late 19th and early 20th centuries, saw the evolutionary process as something driven internally rather than externally. It envisaged lineages having a kind of innate tendency to go in 'direction X' – for example, larger body size – and saw this as unrelated to changes in fitness. In this view, therefore, natural selection was relegated to the role of a minor player in evolution, doing some fine-tuning here and there, but not responsible for major trends. Because no satisfactory mechanism was proposed for the claimed innate tendencies of lineages, this school of thought eventually fell into disrepute.

Finally, the comparatively new school of thought called *eco*-devo. Although this school can trace its history back to C.H. Waddington (in the 1950s) and beyond, it has blossomed recently, with much work on adaptive phenotypic plasticity. When development is plastic, and thus its pattern is determined partly by the genotype but partly also by the environment, the pattern-fitness map becomes more complex. If developmental pattern 'A' is the fittest only in populations living in particular environmental conditions, genes producing that pattern in an obligate manner might not necessarily be spread by natural selection; rather, genes allowing that pattern to be produced in one environment, and another pattern to be produced in a different environment, may be spread by selection. We will look further at the importance of plasticity for the evolution of development in Chapter 16.

SUGGESTIONS FOR FURTHER READING

This chapter has looked at the relationship between gene mutation and developmental repatterning. This approach is a developmentally-explicit way of considering the more written-about relationship between genotype and 'phenotype', though the latter term is often unhelpful because the 'phenotype' changes through development, and so is not a fixed 'type' at all.

For further information about mutation at the level of the gene, and indeed at the level of DNA more generally, see the relevant chapters of any recent Genetics text, such as:

Pierce, B. 2008 *Genetics: A Conceptual Approach*, third edition. Freeman, San Francisco.

For a review of evolutionary developmental biology in which developmental repatterning is central (albeit therein called 'developmental reprogramming') see:

Arthur, W. 2002 The emerging conceptual framework of evolutionary developmental biology. *Nature*, **415**: 757–764.

For reviews of aspects of the new field of epigenetics, see the edited collection of works in:

Allis, C.D., Jenuwein, T., Reinberg, D. and Camparros, C.L. 2007 *Epigenetics*. Cold Spring Harbor Laboratory Press, CSH, NY.

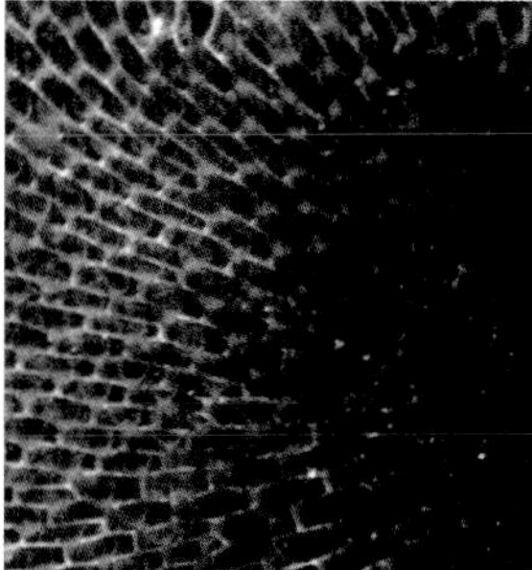

CHAPTER 6
Heterochrony

6.1 What is Heterochrony?

The simple answer to this question, given already in Chapter 1, is: 'evolutionary change in developmental timing'. Yet this is one of those apparently simple answers that are not simple at all. We should therefore dissect what is meant by a change in developmental timing. We will do this first in an abstract and general way; later in the chapter, we will look at the details of specific examples.

Consider a developmental gene, simply called *g*, which effectively represents all such genes. Now consider its period of expression in the last common ancestor (LCA) of two descendant **daughter species**. This gene will have had a certain period of expression in a developing individual organism belonging to the LCA. For example, it might have been expressed for a short period during **gastrulation** or a longer period during subsequent **allometric growth**. Although different individuals of this ancestral species probably had slight variations in the period of expression of *g*, we will ignore this complication for the moment.

The period of expression can be characterised by its start-point and its end-point. For now, we will assume that these two 'markers' are all that we require to examine **heterochrony** of gene expression. So, we can consider heterochrony by looking at how these markers of gene expression period could change in evolution. Specifically,

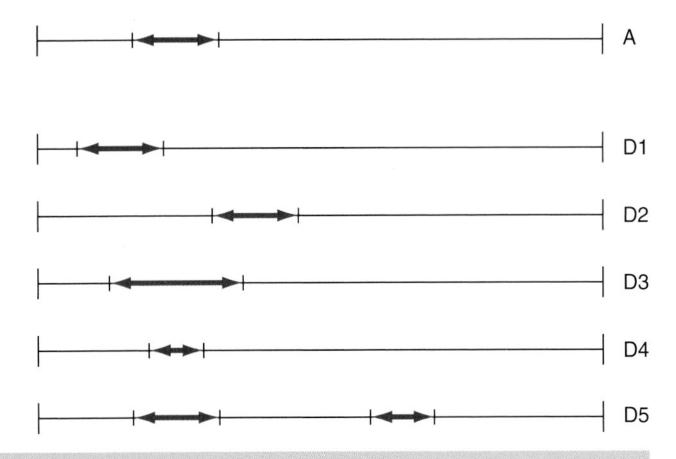

Figure 6.1 Types of molecular-level heterochrony illustrated in terms of evolutionary modification of the temporal expression scope of a developmental gene in ancestral (A) and descendant (D) ontogenies. The temporal expression has been made earlier in D1, later in D2; it has been extended in D3, shortened in D4; and an additional expression period has been added in D5.

if we compare the LCA and one of its daughter species, various forms of heterochrony can be envisaged (Fig. 6.1).

There are, however, two further aspects of the temporal pattern of gene expression to be considered. First, during the period when a gene is switched on, in other words being expressed, the rate of production of transcripts, and ultimately proteins, per unit time, may vary for many reasons. Second, it is possible that, in the LCA, gene *g* is switched back on again during another phase of development, perhaps in a different **germ layer** or tissue, giving it a double (or treble, etc.) period of expression. Taking account of possible evolutionary changes in these additional parameters yields a vast range of possible types of heterochronic repatterning. So the deceptiveness of the 'simple' definition of heterochrony should now be very clear.

6.2 Types and Levels of Heterochrony

The account given above was clearly at the molecular level, whereas the example of paedomorphic salamanders mentioned in Chapter 4 was at the organismic level. At both levels (and indeed at intermediate ones) there are many different types of heterochronic process. While many types of molecular heterochrony are illustrated in Fig. 6.1, they are not accompanied by an elaborate terminology. Unfortunately, at the morphological level, there already *is* an elaborate terminology, much of it due to Gavin de Beer[16] (Chapter 2); it can also be found in subsequent books that are largely or wholly on heterochrony, notably those by Gould[17] and by McKinney and McNamara[18]. We are not going to use most of this terminology, but here, for reference, are the terms concerned: **paedomorphosis**, peramorphosis, **progenesis, neoteny**, post-displacement, hypermorphosis, acceleration and pre-displacement.

There are at least two problems with this suite of terms: first, many of the words themselves are off-putting; and second, they generally relate to one particular comparison – that between the timing of **somatic** growth versus reproductive maturity. So we have the worst possible combination: an elaborate suite of off-putting terms for a very restricted realm of heterochronic processes and their results.

Having said that, let us look at just one example to illustrate the rationale behind three of the terms, as follows. Recall again the phenomenon of paedomorphic salamanders, in which the late-stage larva has become reproductively mature and the adult form is lost. This is a very real, and interesting, example of heterochrony. Furthermore, the single term paedomorphosis is, despite the adverse comments above, a reasonable one for describing the outcome, as it roughly translates to 'juvenile form', reflecting the fact that the stage which is now reproductively mature (and hence is 'the adult' in that sense) does indeed have what is in other species only the form of a juvenile.

Now, if we forget about absolute time and concentrate on the *relative* timing of somatic change and reproductive maturity, we can see that what has happened in the paedomorphic salamanders is that the latter has occurred sooner, in relation to the former, than used to be the case in a non-paedomorphic ancestor. So far so good. But for most purposes we do not need to distinguish between neoteny, where paedomorphosis is supposedly produced by retarded somatic development, and progenesis, where paedomorphosis is instead supposedly produced by accelerated reproductive development. In the first place, not all authors use the terms in exactly the same way; and second, everything in development is relative. There are no absolutes, since all developmental processes and characters, including the length of the overall developmental period, can and do change over evolutionary time (and sometimes on the much shorter ecological timescale too, due, for example, to changing ambient temperature).

The third reason why we will avoid most of the existing terminology becomes apparent by considering the following question: What happens if, instead of comparing reproductive versus somatic development, we decide to compare cardiovascular versus neurological development, or to make some other comparison? Surely then we would need a whole new range of terms? A much better solution, and the one adopted here, is to use as few terms as possible, but to make as clear as possible exactly what is happening in each specific example. I hope you will find this strategy more acceptable than the alternative one.

With regard to the connection between different *levels* of heterochrony, this may be complex. Altered timing of gene expression, that is, molecular-level heterochrony, may give rise to (say) spatial changes, i.e. heterotopy, at the morphological level. Combinations of kinds of developmental repatterning like this will be the subject of Chapter 10.

6.3 Heterochrony at the Organismic Level

Here is an example of heterochrony involving a change in the relative timing of two important events in arthropod development: egg hatching and segment formation. The group we will use to illustrate this change is the class Chilopoda (centipedes), consisting of over 3000 species. In particular, we will compare centipedes of the 'common garden' type (genus *Lithobius*, order Lithobiomorpha), which represents a fairly **basal** centipede design, with another order (Geophilomorpha), which is more **derived** (Fig. 6.2).

The smallest adult centipede is less than 1 cm long, while the largest is around 30 cm. Notwithstanding this almost 50-fold size variation, and the range of climates experienced by various centipede groups (from tropical to subarctic), all centipedes have long periods of **embryogenesis** by more general arthropod standards. There are no rapid developers like *Drosophila melanogaster*, where embryogenesis can take only about a day (and the total developmental period to adulthood can take as little as 10 days). In contrast, all centipedes take some weeks of embryogenesis before the hatchling emerges from the egg; and a year or more to reach adulthood.

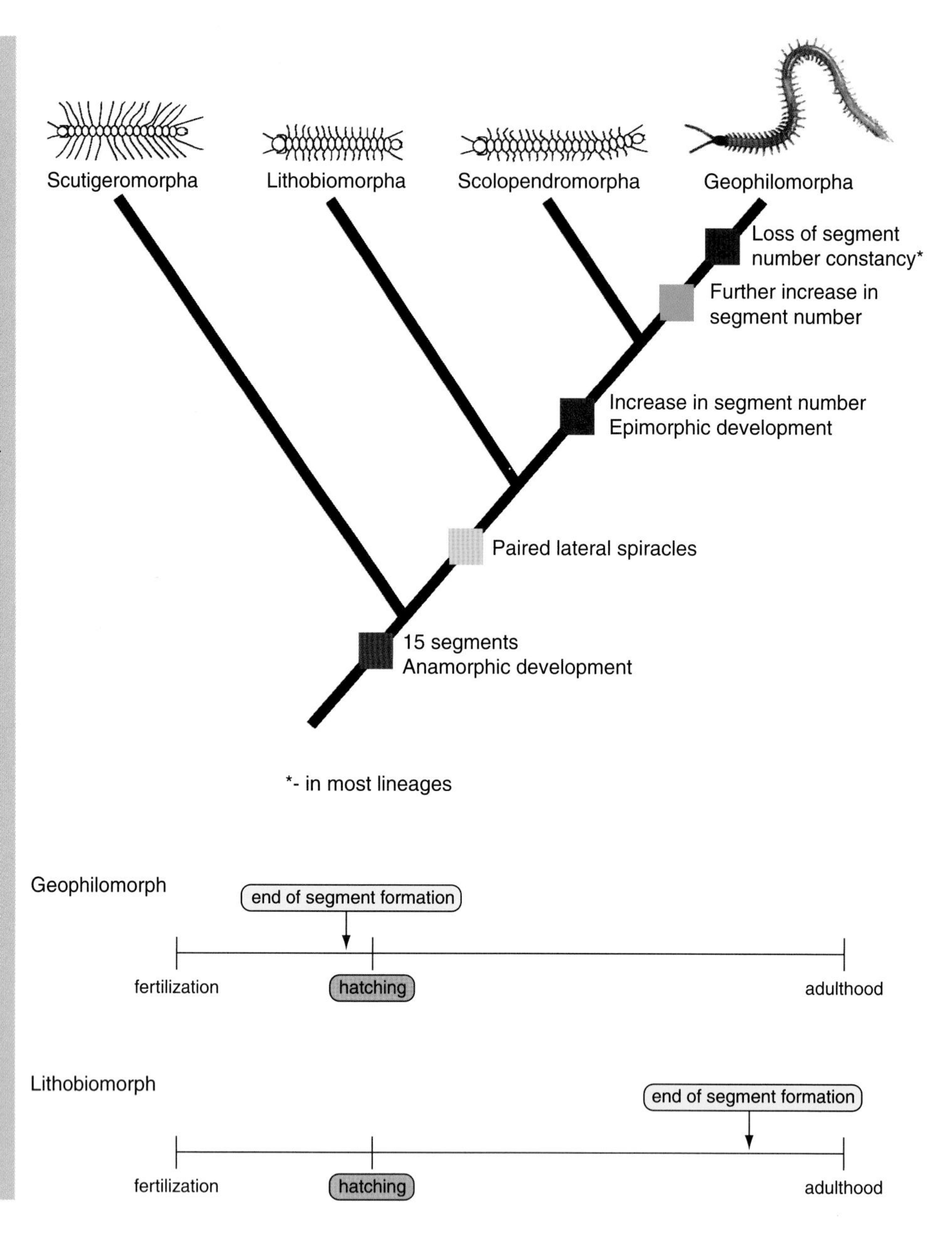

Figure 6.2 Top – Relationships between the four main centipede orders. Diagrams of the typical body forms of each are included. The two orders Scolopendromorpha and Geophilomorpha form the group called the Epimorpha. Bottom – Heterochronic change in the relative timing of egg hatching and segment formation between the orders Lithobiomorpha and Geophilomorpha. (Top diagram from Arthur & Chipman, 2005, *BioEssays*, **27**: 653–660.)

The development of several species of *Lithobius* has been studied[19], and the general pattern is that the hatchling has 7 fully developed leg-bearing segments, with the adult number (15) being reached through posterior segment addition during a series of moults over the course of about a year. So the temporal relationship between hatching and segment formation is as shown in Fig. 6.2 (bottom). All other lithobiomorph genera are thought to have a broadly similar pattern.

Turning now to the Epimorpha (the group made up of the two orders Scolopendromorpha and Geophilomorpha; see Fig. 6.2), we find a remarkably different pattern. The period spent in the egg is broadly similar (a few weeks rather than days or months), but the hatchling has the *full adult complement* of segments. And this is despite the fact that the adult segment number is higher (between 21 and 191). The period of post-embryonic growth is still long (variable but often a few years). This period is still characterised by several moults but, while the size of the segments (and hence of the animal) increases in these moults, the *number* of segments does not. Thus in this case the relationship between the timing of hatching and segment formation has been significantly altered (Fig. 6.2).

We can now ask three important questions: *when* did this change occur in evolution; *what* was its molecular basis in terms of altered patterns of gene expression; and *why* did it occur, in terms of possible ways in which it might lead to enhanced **fitness**? We will address these questions in turn.

Centipede fossils are few and far between (in both space and time). Nevertheless, there are well-preserved specimens from the late Carboniferous (Pennsylvanian) Mazon Creek formation in Illinois, USA, the age of which is approximately 310 MYA. These include a genus called *Mazoscolopendra*, specimens of which are very similar in structure to present-day scolopendromorphs. So at least we can say that the heterochronic change that is shown in Fig. 6.2, assuming that it took place in the stem-group of the Epimorpha, happened earlier than this; though how much earlier is not clear.

Turning to the molecular basis of this organism-level heterochronic change, the central question is how a cascade of interacting genes and their products can be speeded up to such an extent that the final outcome, in terms of the adult segment number, is produced in approximately two weeks rather than about a year. This question cannot yet be answered completely, but we are a lot closer to being able to do so than a decade ago, given what has been learned about the developmental genetics of the production of centipede segments.

The main features of the developmental genetics of segmentation in the arthropod model system *Drosophila* had been established by the late 1980s[20], as was noted in Chapter 3. Since then, evo-devo workers have examined the developmental genetics of segmentation in wide range of other arthropods, including other insects, but also representatives of the other arthropod sub-phyla – crustaceans, chelicerates and myriapods. Particularly relevant here is the work done in this area on centipedes – both lithobiomorphs[21] and geophilomorphs[22].

It is now clear that some of the findings from the *Drosophila* system can be generalised – such as the deployment of *engrailed* as a **segment polarity** gene. However, it is equally clear that some other findings cannot be generalised. For example, segmentation in spiders[23] and centipedes[24] involves pulses of activity of the Notch-Delta signalling system (Fig. 6.3), as does segmentation in vertebrates (Chapter 10). Perhaps *Drosophila* is unusual because of the near-simultaneous formation of all its segments, in contrast to the antero-posterior sequential segment formation that is found in most segmented animals.

As a very rough categorisation, we can divide centipede segment formation into 'early' and 'late' phases, the former including Notch-Delta signalling, the second including the expression of segment polarity genes. It seems likely that the timing of segment formation is determined primarily in the earlier phase. There is some experimental evidence for this in geophilomorphs. However, even if this is true, and the rate of segment formation is determined by the speed of the Notch-Delta 'clock', we still do not know what alters the clock's timing.

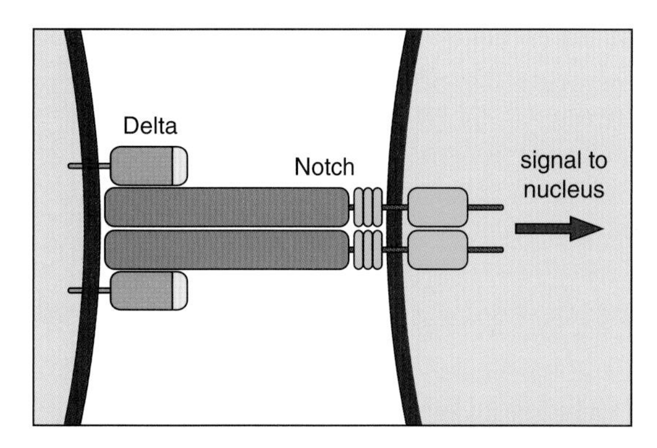

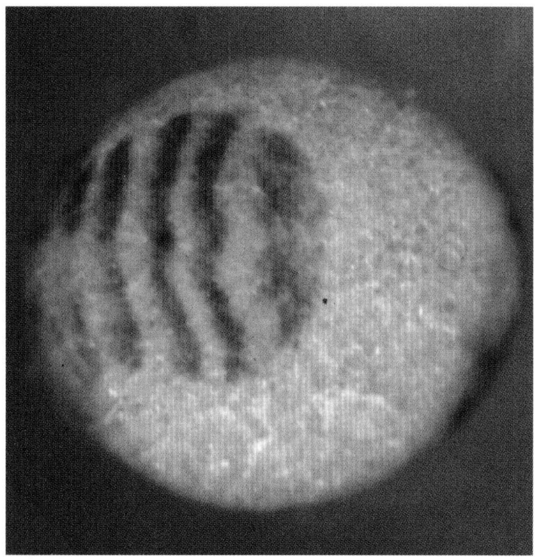

Figure 6.3 Top – Diagram of Notch-Delta signalling; Bottom – Waves of expression of *delta* in the early stages of segmentation in a spider embryo (photograph courtesy of Wim Damen).

Now to the third question about the acceleration of segment formation in epimorphic centipedes. What might be the advantage of this? Or, in other words, why might it have been favoured by natural selection? Here we meet one of the big problems faced by students of evo-devo. Given that many of the most fascinating forms of developmental repatterning happened a very long time ago – in this case more than 300 MYA as we saw earlier – it is impossible to be certain about the ecological conditions prevailing at the time, and hence to know which sorts of developmental repatterning would have been advantageous and which not.

Students of evolution in general – not just of evo-devo – have often been accused of 'adaptive story-telling'. The basis of this criticism is as noted above: namely, lack of firm knowledge about the relevant environmental conditions, in which case the claimed adaptiveness of some trait is simply speculation – an imaginary **adaptive scenario** that may never have taken place.

This is a very apt criticism and one that should always be borne in mind. However, there are two possible responses to it, of which one is infinitely preferable to the other. Basically, we can either decide never to propose adaptive scenarios, or propose them sparingly, and make very clear indeed that they are *hypotheses*. Taking the latter, preferable, approach in the case of the evolution of accelerated segment formation in epimorphic centipedes, here is one adaptive hypothesis based on the manoeuvrability of hatchlings.

Lithobiomorph centipedes are surface-dwellers. They inhabit the leaf-litter layer of terrestrial ecosystems. Although they can often be found under stones or rotting logs, they are rarely found beneath the soil surface. Geophilomorphs, in contrast (and as their name implies) are subterranean. Again they can be found under stones or logs, but at any one time the vast majority of any population is within the upper layers of the soil. This was demonstrated very forcibly by a field trial at an agricultural experimental station, where large-scale turning over of earth yielded a density of geophilomorphs more than 100 times that which results from myriapodologists sampling an area of habitat by hand – turning over stones and logs.

As soon as geophilomorph hatchlings are fully capable of movement, they burrow into the soil. To remain on the surface would be to invite an early death through desiccation or predation. But if they hatched with only seven segments like a typical lithobiomorph, their ability to burrow would be much impaired. Imagine an earthworm composed of just seven segments, and then add the extra difficulty of a hard exoskeleton. Thus we can see how selection might have favoured production of all segments prior to hatching.

But there is a problem with this hypothesis: we do not know the exact morphology or ecology of the original epimorphic centipede. In particular, since this creature gave rise to both the mostly non-subterranean scolopendromorphs and the soil-dwelling geophilomorphs, a possibility is that the LCA of these two groups was still surface-dwelling. If so, the adaptive hypothesis fails. This consideration reinforces the point made above that all such hypotheses should be treated with caution.

6.4 Heterochrony at the Molecular Level

We will look at two examples here: first, an invertebrate example that connects, albeit tangentially, with the centipede example discussed above; and second, a vertebrate example involving opsin genes.

Heterochronic Shift in the Expression of a Pair-rule Gene in *Drosophila* Species

As we saw in Chapter 3, the stage immediately preceding the expression of the segment polarity genes in the segmentation cascade of *Drosophila melanogaster* was the expression of pair-rule genes. One of the primary pair-rule genes, *hairy*, has been the subject of a comparative study of three species of *Drosophila*, which showed that the expression of this gene has undergone heterochronic shifts during the evolution of the genus[25].

The three species studied were *D. melanogaster*, *D. simulans* and *D. pseudoobscura*. The first two are quite closely related, the last much less so (Fig. 6.4). (Recall that *Drosophila* is a very **speciose** genus: there are more than 2000 named species.)

Given this pattern of relationship, it should come as no surprise that *D. melanogaster* and *D. simulans* are much more similar, both morphologically and ecologically, than either of them is to *D. pseudoobscura*. One important aspect of this pattern of differences involves body size and the duration of the developmental process. *D. pseudoobscura* individuals are much larger than individuals belonging to either of the other two species; and they take longer to develop, at any particular temperature. This means that it does not make sense to compare the temporal **expression patterns** of *hairy* among the three species in terms of absolute time. A better approach, as noted earlier, is to compare such patterns with other developmental events. In this case, those events included the initiation of the cellularisation of the **blastoderm**. (Recall from Chapter 3 that *Drosophila* initially has a **syncytial** blastoderm lacking cell membranes.)

Taking *D. melanogaster* as a reference point, the time to full (7-stripe) expression of *hairy* (Fig. 6.4) is shifted as follows in the other two species:

- *D. simulans*: 14 minutes earlier
- *D. pseudoobscura*: 24 minutes later

We can then compare these figures with those relating to the initiation of cellularisation, which are as follows:

- *D. simulans*: 16 minutes earlier
- *D. pseudoobscura*: 5 minutes later

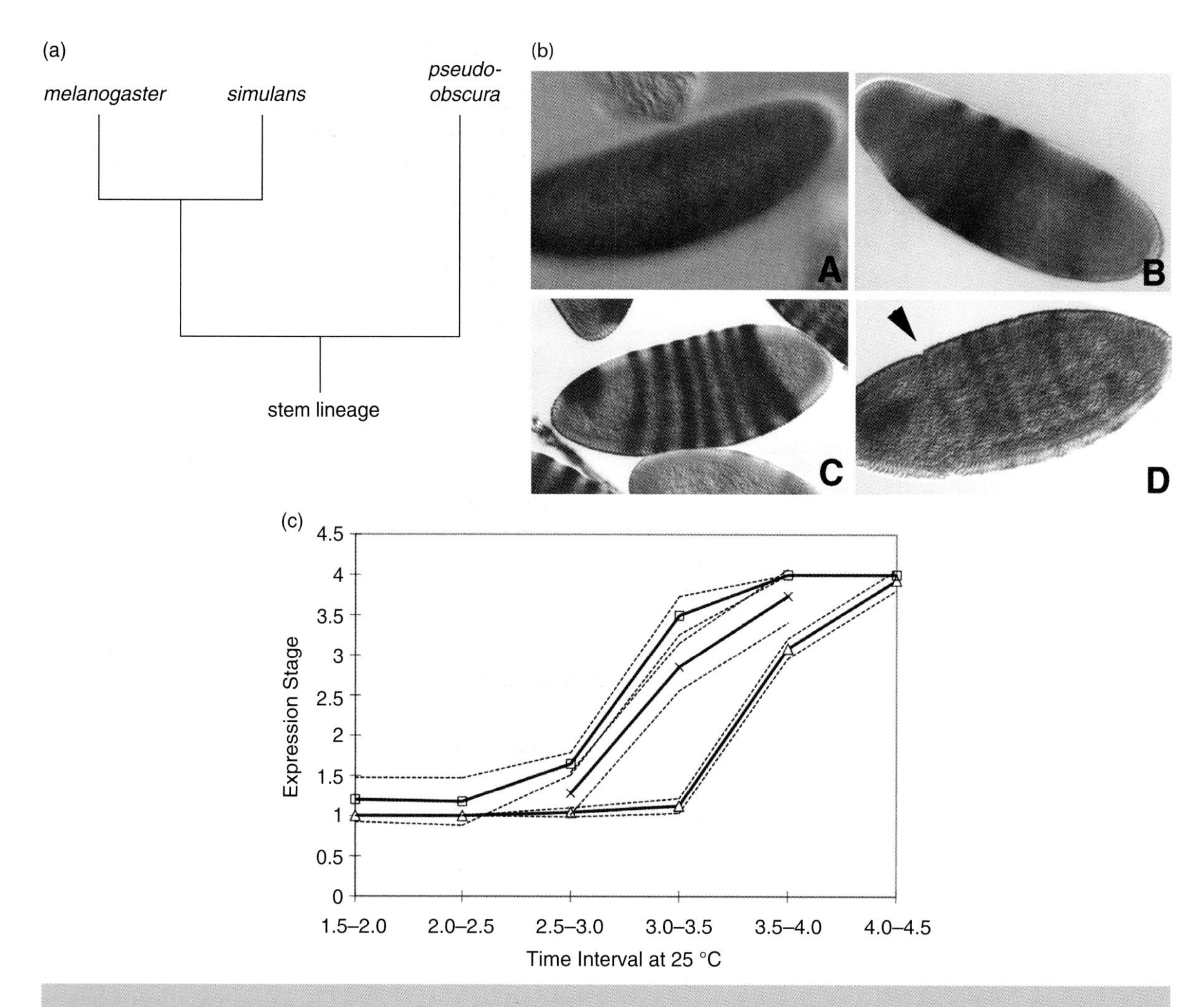

Figure 6.4 (a): Phylogenetic relationships among three species of *Drosophila*. (b): Stages of expression of the pair-rule segmentation gene *hairy*. At stage A, *hairy* expression is just beginning; at stage B, at least three stripes of expression can be seen; at stage C, the full 7-stripe pattern is clear; and at stage D, the cephalic furrow (arrowhead) appears. (c): Differences in the timing of *hairy* expression among the three species. Expression stage 2 corresponds to picture B. Time is measured in hours from egg deposition. Squares – *D. melanogaster*; crosses – *D. simulans*; triangles – *D. pseudoobscura*. Solid lines – averages; dotted lines – confidence limits. (From Kim *et al.* 2000, *PNAS*, **97**: 212–216.)

What these comparisons show is that the faster onset of full *hairy* expression in *D. simulans* is roughly equivalent to the accelerated initiation of cellularisation (and of later developmental events, such as cephalic furrow formation). It therefore requires no special explanation, and does not constitute a heterochronic shift occurring in the evolutionary divergence of these species. In contrast, the delayed onset of full expression of *hairy* in *D. pseudoobscura* is disproportionately great compared to the delays in other aspects of development. Therefore, this does indeed represent a heterochronic change, albeit the adaptive reason for it (if there is one; see Section 6.5) is unclear.

Note that absolute times (in minutes) were used above, despite the earlier criticism of their use. At first sight this looks inconsistent, yet it is not. The absolute times are being used here, not as the primary frame of reference, but rather as a way of enabling the relative times of two different developmental events to be compared.

Heterochronic Shifts in the Expression of Opsin Genes in Cichlid Fish

The family of teleost fish called Cichlidae contains about 1700 species, many of them in Africa, including the substantial radiations of species that have taken place in the great lakes Victoria and Malawi. As in other vertebrates, their eyes function by directing light onto the cells of the retina, from which signals go to the brain via the optic nerves. One of the major cell types in the retina is the cone cell. Cells of this type are particularly important for colour vision. Cone cells contain light-sensitive proteins called opsins.

In the Nile tilapia, *Oreochromis niloticus* (Fig. 6.5), there is a change in exactly which opsin genes are expressed as development proceeds. Using this species as an **outgroup** for comparative purposes, a recent study[26] examined several species from Lake Malawi – some that

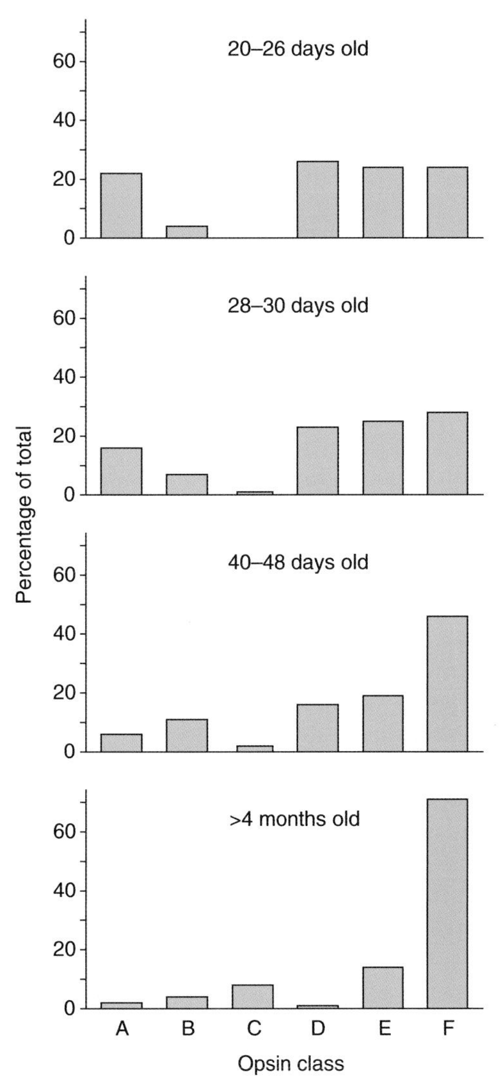

Figure 6.5 Top: The Nile tilapia, *Oreochromis niloticus*. This species of cichlid fish was used as the 'outgroup' in a study of the opsin genes/proteins in several other cichlid fish from the African Great Lakes. Bottom: Altered opsin gene expression over developmental time in the tilapia. As can be seen, some classes of opsins decrease in expression with age (e.g. A) while others increase (e.g. F). (Bar charts modified from Carleton *et al.* 2008, *BMC Biology*.)

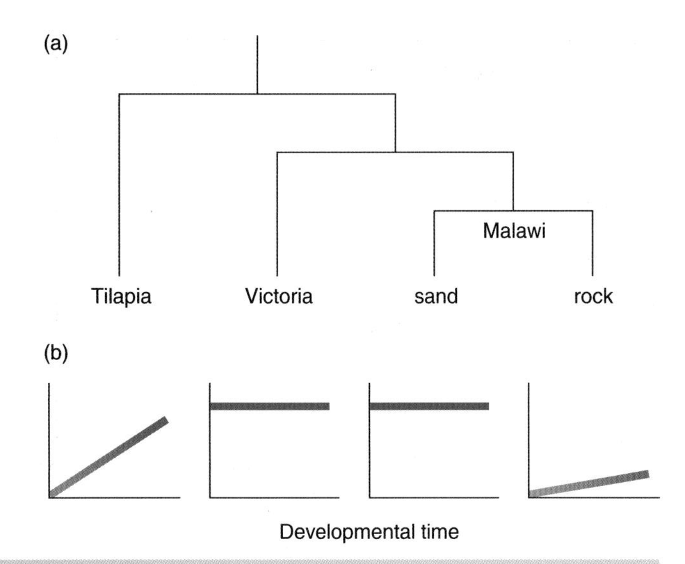

Figure 6.6 Comparison of the tilapia transition of opsins through development (now represented schematically as a line) with the heterochronically-shifted patterns in the other cichlid species studied. (Modified from Carleton *et al.* 2008, *BMC Biology*.)

live in rocky-bottomed areas of the lake, others that are characteristic of sandy areas; and it also included information from species in Lake Victoria. The results of this comparative study showed that some of the lake species had the adult tilapia pattern of opsins from an early developmental stage, while others showed a persistently juvenile pattern (Fig. 6.6). Again, we need to be careful about the difference between absolute and relative developmental time. However, this is not too problematic in this case because these various species have rather similar overall developmental periods.

The selective reasons for the heterochronic shifts that have occurred are not clear. Possibilities include changes in visual sensitivity related to mate choice and clear versus murky water. While hypotheses involving these factors await testing, it is at least possible to confirm that possession of a different combination of opsins affects vision in terms of the peak absorbance wavelength that results from having a particular combination of opsins (Fig. 6.7).

6.5 Heterochrony and Fitness

Whether we are examining a molecular aspect of developmental timing, such as the period of expression of a developmental gene, or a morphological one, such as the timing of onset of reproductive maturity, there is always some variation within each species, and indeed within each *population* of any species, in the exact timing of the aspect concerned. Some of this variation may be **plastic**, in other words resulting from direct environmental (e.g. temperature) effects on development; but at least some of it is **heritable**. Evolution is thus able to convert intraspecific variation into interspecific differences. But is this evolution always adaptive? In other words, is it always selectively driven? This question cannot yet be conclusively answered.

When the Japanese geneticist Motoo Kimura reviewed his theory of **neutral** evolution[27], involving the spread of selectively-neutral variants by **genetic drift**, he was careful not to extend it beyond the realm in which it had arisen: that of the precise DNA sequence of **housekeeping genes** and the corresponding amino-acid sequence of the protein products (usually enzymes) of these genes. He emphasised that in the realm of morphology Darwinian selection held sway, despite his argument that the vast majority of evolutionary changes in the DNA sequences of housekeeping genes were neutral.

However, in the early 1980s when Kimura wrote *The Neutral Theory of Molecular Evolution*, little was known about the evolution of **developmental genes**.
This is important because it raises the question of whether we should expect their

evolution to be (i) similar to that of housekeeping genes (because in both cases we are looking at groups of genes) or (ii) similar to that of morphology (because it is developmental genes and their products that provide the basis of morphology). In the first case we might expect the evolution of developmental genes to be largely neutral; in the second, we might expect it to be largely Darwinian.

There is, though, a third possibility. The DNA *sequences* (both regulatory and coding) of developmental genes may evolve most often due to drift, especially in cases where the altered sequence does not affect either the gene's expression pattern or the function of its protein product. Despite this, the majority of changes that affect *either the expression pattern of the gene or the function of the protein* (e.g. the affinity of a transcription factor for target genes) may be mostly, or even entirely, driven by selection.

This point should be borne in mind through the next three chapters, because it applies equally to the other possible types of developmental repatterning – **heterotopy, heterometry** and **heterotypy**. We should not get into the habit of *assuming* a selective reason for repatterning; rather, we should get into the habit of proposing selective hypotheses and, where possible, testing them – especially given the recent radical proposal[28] that natural selection is relatively unimportant in developmental evolution as well as in the evolution of housekeeping genes.

It seems appropriate to end this section with some examples of heterochrony where the link to fitness is particularly clear. Inevitably, these tend to involve heterochronic shifts in characters related to reproduction. This was the case with the paedomorphic salamanders discussed previously. But rather than expand on that story, we should look at some plant examples, since there have so far been none of these in the present chapter.

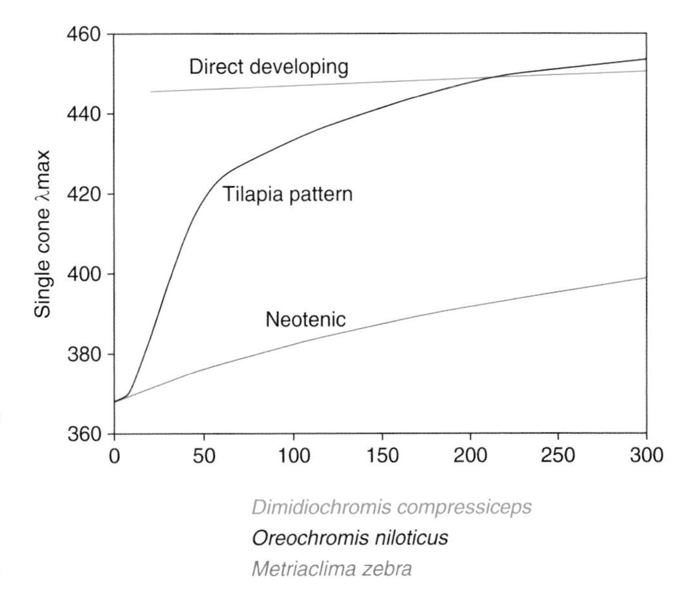

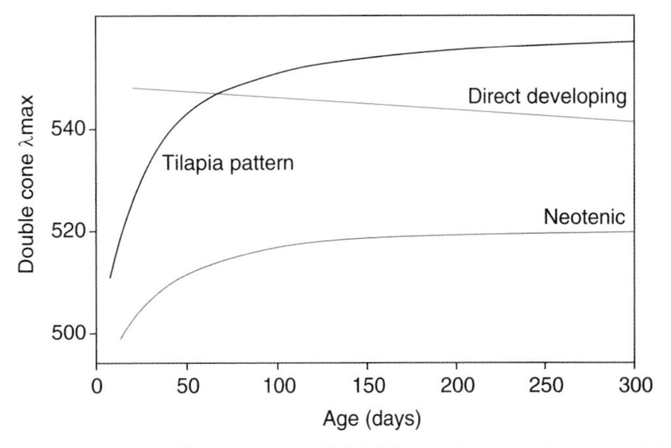

Figure 6.7 Different age-related patterns of peak absorbance wavelength conferred by expression of different combinations of opsin genes at different ages in different species. Top: Pattern for single cones. Bottom: Pattern for double-cones. (The retina contains a mosaic of single and double cones). (Modified from Carleton *et al.* 2008, *BMC Biology*.)

A recent study[29] has shown that some basal **angiosperms** (family **Hydatellaceae,** which are closely related to the waterlilies) have a uniquely early seed-provisioning system: they allocate nutrients to the embryo-nourishing tissue *before* fertilisation. Although apparently unique among angiosperms, this early provisioning is common

(a)

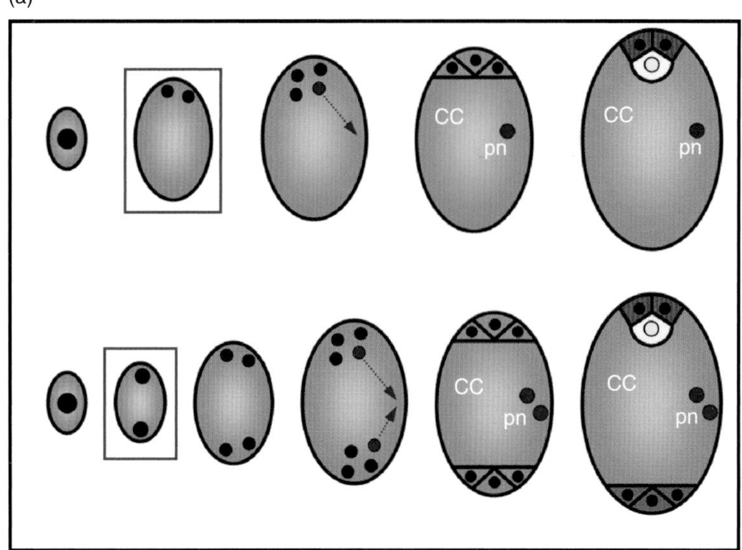

(b)

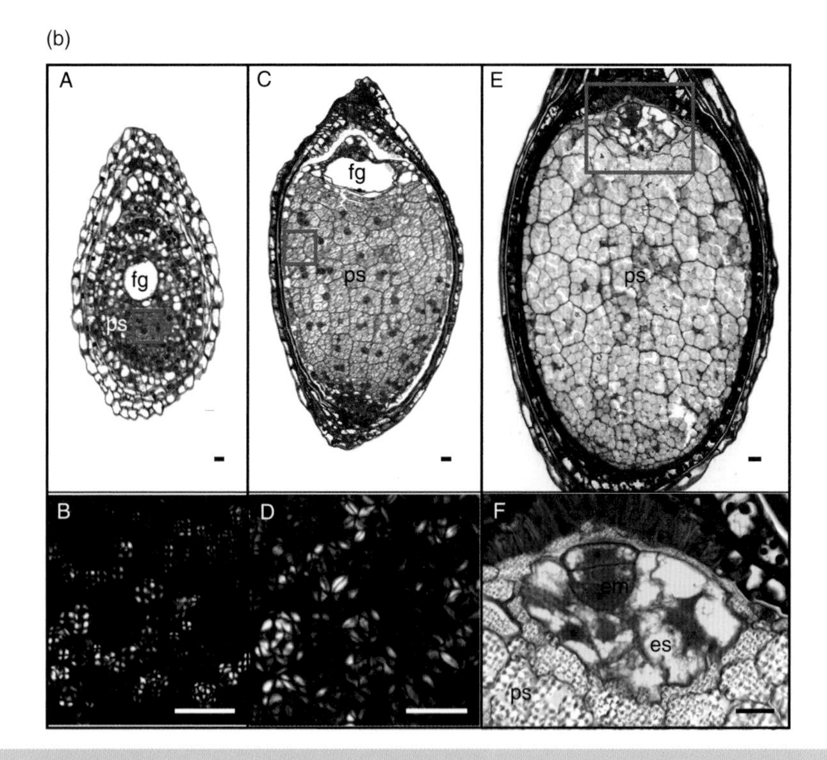

Figure 6.8 (a): Comparison of early development in the basal angiosperm *Hydatella* (top) and in a 'typical' angiosperm (bottom). Egg cell shown in yellow; CC – central cell; pn – polar nucleus. (b): Series of stages in *Hydatella* both before (A – D) and after (E, F) fertilisation. These show the perisperm (ps) packed with starch granules (B, D, F). Scale-bars – 10 μm; fg – female gametophyte; es – endosperm. (From W. E. Friedman, 2008, *Nature*, **453**: 94–97.)

to all **gymnosperms**, so it may represent the primitive, or **plesiomorph**, state. If so, then the heterochronic shift to post-fertilisation provisioning must have been an important element of the transition to the more typical angiosperm condition. One possible selective advantage of this shift is that the parental plant wastes no energy provisioning ovules that do not get fertilised.

Figure 6.8 shows the early development of the female gametophyte in *Hydatella* and in a 'typical' angiosperm. Note that in *Hydatella* there is just one polar nucleus inside the central cell, whereas in a more typical angiosperm there are two such nuclei. This means that, after fertilisation of the polar nuclei by the second sperm nucleus of the pollen, the resulting embryo-nourishing endosperm is diploid in *Hydatella*, whereas it is typically triploid. However, another difference between *Hydatella* and the typical angiosperm condition is that most of the embryo-provisioning occurs via another diploid tissue called the perisperm. This can be seen to be packed with starch granules before fertilisation occurs (Fig. 6.8), that is, before there is an embryo that needs to be nourished.

While the above example involves comparisons between higher taxa, there have been many studies on within-species mutations that cause heterochronic shifts in flowering time – for example, in *Zea mays* (maize), *Pisum sativum* (pea) and *Arabidopsis thaliana* (thale cress). These studies have led to the identification of genes that, when mutated, can lead to heterochronic changes, including the appearance of flowers in juvenile plants – a paedomorphic result. Such changes appear to have been common in the adaptive evolution of angiosperms.

These studies on plant model systems, and equivalent studies on animal models such as the nematode worm *Caenorhabditis*, have led to the use of the phrases 'heterochronic mutation' and 'heterochronic gene'. The former phrase is fine, and is equivalent to the phrase 'heterotopic mutation' of which the familiar 'homeotic mutation' is a sub-category (Chapter 7). But the latter phrase – 'heterochronic gene' is absolutely not fine. If a gene is involved in controlling the timing of some aspect of development in a plant or animal, it can simply be called a 'timing gene'. But a gene in itself cannot be 'heterochronic'. Heterochrony, as we have seen, is a concept in which a *comparison* between different ontogenies is crucial; just as the central subjects of the following three chapters, heterotopy, heterometry and heterotypy, are concepts in which such comparisons are crucial. This comparative nature of these key concepts, and of developmental repatterning in general, should never be forgotten.

SUGGESTIONS FOR FURTHER READING

For an introductory and highly readable account of heterochrony and related issues, see:

McNamara, K.J. 1997 *Shapes of Time: The Evolution of Growth and Development*. Johns Hopkins University Press, Baltimore, MD.

A more advanced (but older) account is provided in:

McKinney and McNamara, K.J. 1991 *Heterochrony: The Evolution of Ontogeny*. Plenum Press, New York.

For the links between heterochrony and other types of evolutionary change in development, such as those described in the next three chapters, see:

Webster, M. and Zelditch, M.L. 2005 Evolutionary modifications of ontogeny: heterochrony and beyond. *Paleobiology*, **31**: 354–372.

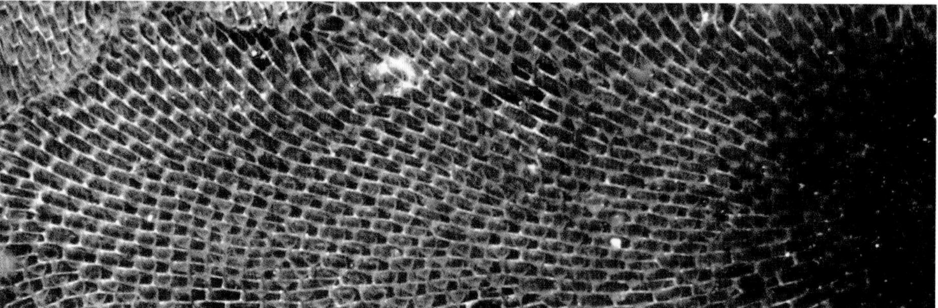

CHAPTER 7 Heterotopy

7.1 What is Heterotopy?

As with **heterochrony** (Chapter 6), we can give an apparently simple definition; in this case, 'changes in the spatial arrangement of development'. But again, on closer inspection, the definition is not as simple as it seems.

Absolute space is no more helpful in attempting to elucidate the nature of heterotopic processes than absolute time was useful in elucidating heterochrony. After all, if we compare 'stage X' in the development of a particular type of animal or plant, closely-related species are often slightly scaled-up or scaled-down in body size, as we saw previously for different species of *Drosophila*. So if we were to superimpose two embryos of different sizes (whether members of closely-related species or even of the *same* species), each small-scale developmental event, such as the position of a boundary between forming structures, for example segments, is slightly displaced in one embryo compared to the other. Such displacements are not regarded as heterotopy – if they were, then 'heterotopy' would be too broad a term to be useful.

We can solve this problem as before (with heterochrony) by restricting the definition to one of evolutionary change in the *relative* occurrence of developmental events – this time their relative arrangement in space. So, size changes that are not accompanied by the repositioning of some developmental structure or process relative to others are excluded.

Thus again we have a type of evolutionary change that is comparative in two different senses. First, it automatically involves comparing the **ontogenies** of two or more creatures – usually but not always belonging to different species; and second, it involves comparing the spatial relationship between two or more developmental structures or processes.

A developmental process in this context can be at any level – from molecular to organismic. However, while the structure of the previous chapter was based on these levels, the structure of the present one will not be. The reason for adopting a different structure here is that increasingly we should be trying to look at a developmental process at several levels, and indeed trying to integrate what we see at those different levels. So in what follows we will approach each example from an initial angle, often a morphological one, but will then spread out to look at the molecular level, and also to look at the fossil record, if there is one, of the heterotopic change we see, and at hypotheses about the possible adaptive significance of the change.

Figure 7.1 Three species of flatfish: plaice, turbot and flounder, all showing the asymmetric design, with the characteristically rotated head, which identifies adults of this group.

7.2 Heterotopic Processes Involving Left-Right Asymmetry

Flatfish: Order Pleuronectiformes

Of today's fish fauna, the vast majority – some 25,000 species – belong to the group called teleosts, or **Teleostii**. This group comprises about 40 orders, one of which, the Pleuronectiformes, contains all species of flatfish, including the familiar plaice, turbot and flounder. These fish have a unique morphology in terms of left-right asymmetry that we tend to take for granted because of its familiarity; yet the origin of this unique morphology, with the head partially rotated in relation to the rest of the body, and both eyes on the same side of the head (Fig. 7.1), represents a fascinating evolutionary problem

The 'typical' teleost, whether a salmon, trout or zebrafish, is approximately bilaterally symmetrical in its external morphology. This symmetry is an ancestral vertebrate feature, which is also shared by most land vertebrates, or **tetrapods**, including ourselves. It contrasts, of course, with vertebrate internal anatomy, which is often markedly asymmetrical. Examples of this asymmetry include the mammalian heart and gut, and the lung asymmetry of snakes, wherein one lung is usually much reduced or even absent. Left-right asymmetry of

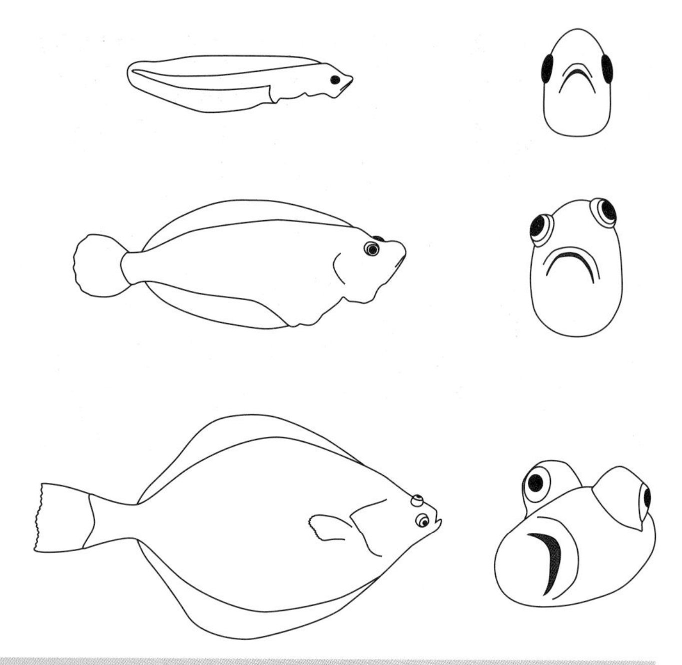

Figure 7.2 As can be seen, early developmental stages of flatfish are bilaterally symmetrical like those of other kinds of fish. The asymmetry characterising flatfish adults arises gradually in larval development by rotation of head structures.

internal organs may have arisen in the **chordate** stem **lineage**[30].

Flatfish embryos (Fig. 7.2) are bilaterally symmetrical. The placement of the eyes is similar to their placement in 'normal fish'. However, at a certain stage in their development, one eye begins to move across the head so that its ends up on the same side as the other one. The end result of the altered developmental process is adaptive given that, during their evolutionary origin, flatfish adopted an ecological strategy of lying on their side on the sea-floor, often camouflaged to look sandy, and sometimes partially covered with sand too. Thus hidden, they are able to strike at unsuspecting prey in a manner that is quite different from the *modus operandi* of other predatory fish. This strategy has clearly been successful: there are more than 500 species of flatfish alive today, so they form a sizeable **clade**; also, it seems that no members of this clade have lost their 'two-eyes-on-one-side' characteristic.

Interestingly, the side the eyes are on is not always the same. Most species have either uniformly left-eyed or uniformly right-eyed individuals, but a few are variable, with some individuals developing one orientation of asymmetry, some the other. The molecular control of eye asymmetry in flatfish is not yet fully understood, but seems to involve a mixture of long-range and short-range signalling. The former includes thyroid **hormone**[31]; the latter includes factors that control the local behaviour of cells, particularly fibroblasts and osteoblasts[32].

What of the nature and timing of the evolutionary origin of this heterotopic developmental pathway? The flatfish design is a relatively recent one. The earliest fossils known are from the most recent geological era – the Cenozoic, which began around 65 MYA with the mass-extinction that killed off the dinosaurs. The nature of the change is such that it looks like it might have originated suddenly, and thus be an example of evolution by **macromutation**, with the original flatfish being a Goldchmidtian 'hopeful monster' (Chapter 2). However, a recent palaeontological study[33] has shown that early flatfish had less asymmetry than those of today, with one eye migrated somewhat from the usual position, but without crossing the body's midline, so that both eyes were still on different sides of the head. One possible **adaptive hypothesis** for this primitive partial eye asymmetry is that flatfish sometimes prop themselves up at an angle using their fins; thus a shifted eye on the underneath might be advantageous compared to a non-shifted one. So a hypothesis of gradual evolution is possible in this case. However, as we will see, it is not in the next.

Before leaving the flatfish example, we should note that it constitutes an example of **recapitulation**. The ancestors of flatfish had symmetrical heads. The embryos of present-day flatfish recapitulate that ancestral state before progressing to their characteristic morphological asymmetry as their development proceeds. This reinforces the point made earlier (in Chapter 2) that de Beer was wrong to conclude that recapitulation had been somehow disproved, whereas Gould was correct when he said that there were many examples in which it could manifestly be seen to happen.

Figure 7.3 A sea-slug or nudibranch. These, like terrestrial slugs, have lost their shells (though note that some slugs have tiny vestigial shells). They can still have a dextral or sinistral body arrangement, but this is less obvious because of the lack of a shell.

Chirality in Snails (Class Gastropoda)

The Mollusca is the second-largest phylum of animals (after Arthropoda), with more than 100,000 extant species. The largest molluscan class is the **Gastropoda**, which includes about 70,000 of these. Although some gastropods have lost their shells at some point in their evolutionary history, or have had them much reduced to the point of being **vestigial**, the vast majority have retained their shells and collectively represent the classic 'snail' body plan. Those that have lost their shells are normally called 'slugs' – whether the land slugs with which we are all familiar or the beautifully-coloured sea slugs, the nudibranchs (Fig. 7.3).

What follows concerns the direction of left-right asymmetry of snails. It can apply to slugs as well, but we will ignore those from now on, because the most conspicuous aspect of this gastropod asymmetry involves the shell. The typical gastropod shell can have either of two types of **chirality** (coiling): dextral or sinistral (Fig. 7.4). The majority of species (~80%) are composed of individuals that are always dextral. Almost all the rest are composed of individuals that are always sinistral. In both cases, 'always' should be interpreted as meaning well over

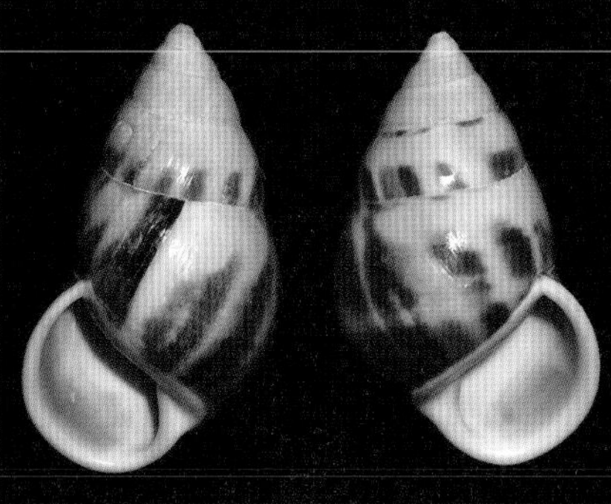

Figure 7.4 Examples of dextral and sinistral gastropod shells. Dextral shells are coiled clockwise when viewed from above, and the aperture is on the right when viewed from the 'front'. The opposites are true of sinistral shells – anticlockwise coiling and an aperture on the left. Top: Different species characterised by different chirality. Bottom: Chirality polymorphism within a species. (Photographs courtesy of Cristina Grande and Nipam Patel).

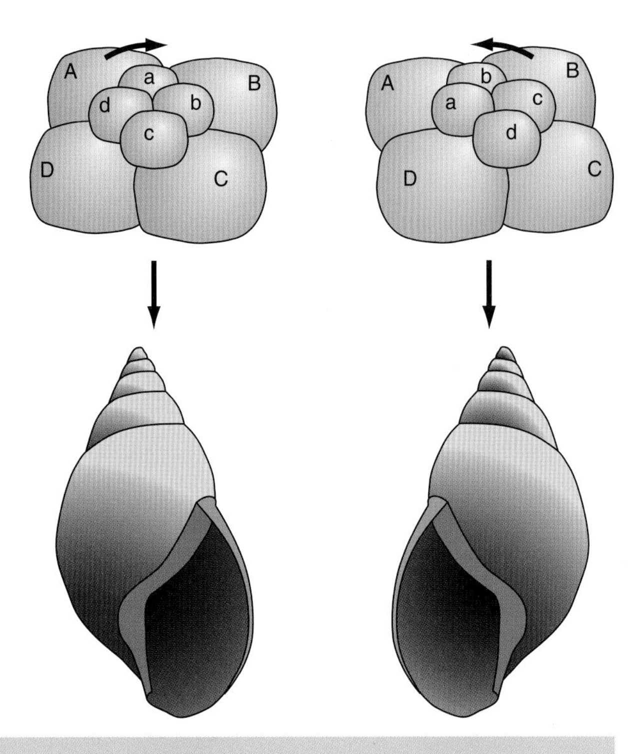

Figure 7.5 The arrangement of cells during early cleavage differs between dextral and sinistral snails. Clockwise rotation of micromeres leads to a dextral shell (LHS), while anticlockwise rotation leads to a sinistral one.

99 per cent – mistakes can and do happen in this respect. A few species (probably <10) are genuinely **polymorphic** – in other words at least some of their populations consist of substantial numbers of *both* chiral types. Examples include the common European pond-snail *Lymnaea peregra* and the land-snail *Partula suturalis* from the Pacific island of Moorea.

The rarity of polymorphic species attests to the small number of occurrences of chirality switches in the history of the gastropods. Another fact that attests to this is the phylogenetic 'clumping' of sinistrality and dextrality. Most families of gastropods consist of uniformly dextral, or uniformly sinistral, species. For example, the families Clausiliidae and Physidae are sinistral, whereas the families Helicidae and Cochlicopidae are dextral. A possible reason for the rarity of chirality transitions in gastropod evolution concerns mating. Although many gastropods are hermaphrodites, they normally cross-fertilise, as noted in Chapter 4. Mating between dextral and sinistral snails seems to be more difficult than mating between snails of the same chiral type.

Chirality affects the left-right asymmetry of soft parts as well as the shell. And its embryological origins can be traced right back to that earliest developmental stage – **cleavage**. Gastropods have spiral cleavage, as shown in Fig. 7.5. As can be seen, dextral and sinistral individuals can already be distinguished at the 8-cell stage by the direction in which the micromeres are rotated relative to the underlying macromeres. Breeding experiments[34] in one of the polymorphic species (*Lymnaea peregra*) revealed that a single gene is responsible for determining which way an individual coils. However, it is not that individual's own **genotype** that matters, but rather the genotype of its mother[35]. This is thus an example of a maternal-effect gene. We previously encountered another such gene (*bicoid* in *Drosophila*). The common theme is that the developmental processes controlled by maternal-effect genes are very early ones. The developmental patterning of most embryos starts out being at least partially under maternal control but, as development proceeds, such effects are phased out and later development is controlled by the embryo's own genes.

Interestingly, just as the flatfish asymmetry example proved that recapitulation happens, the gastropod asymmetry example proves that recapitulation is not a universal law. We can be confident from the breeding programmes on extant polymorphic species that the very first sinistral snail that led from a dextral ancestor to a new sinistral clade carried no outwardly visible trace of its dextral ancestry. In this case, development has been repatterned right from the very beginning, so there is no early embryological 'hangover' from the evolutionary past.

Research on the identity of the genes responsible for switching gastropod chirality is still at an early stage. A decade or so ago, much was known about the genes involved in determining the antero-posterior and dorso-ventral body axes of bilaterian animals, but little was known about the left-right axis. Now, however, all that has changed, especially in vertebrates. Genes involved in the formation of the vertebrate left-right axis include both those that encode secreted signalling proteins, often belonging to the TGF-beta superfamily (e.g. the genes *nodal* and *lefty*), and those that encode **transcription factors** (e.g. *pitx2*). Given the widespread occurrence of many developmental genes across the animal kingdom, often with similar roles, it would not be surprising if the genes involved in gastropod left-right asymmetry turn out to be **homologues** of vertebrate left-right asymmetry genes, despite the deep phylogenetic gulf (>500 MY) that separates these two groups. Indeed, one recent study[36] has implicated *nodal* and *pitx* genes in the determination of molluscan chirality.

Fossil evidence indicates the presence of both dextral and sinistral snails in the distant past – back as far as the **Cambrian** period, before 500 MYA (see Appendix 3 for more on the geological timescale). The phylogenetic history of chirality reversals is probably one of rare occurrences scattered throughout gastropod history. But why should reversal ever occur? What would be the adaptive reason? We should recall that, from a mating perspective, chirality reversals seem to *reduce*, not enhance, **fitness**. Also, it seems that genes **downstream** of those that determine the direction of coiling may become coadapted to stabilising a dextral or sinistral form, and may be less good at achieving such stability if the basic form is reversed.

Here are two hypotheses, and we should recall, as ever, that they are just that. First, it has been proposed that sinistral snails may survive attack by right-handed crabs better than dextral snails do. However, this would not explain a dextral-to-sinistral shift in the terrestrial environment, where the predators are very different. Nor would it explain sinistral-to-dextral shifts.

The second hypothesis involves turning the mating disadvantage 'on its head'. Consider the following scenario: a species of dextral land-snail is expanding its range. As usual, this is a process involving many random events in which individuals come to colonise new areas. Suppose that a single fertilised dextral snail carrying a dominant mutation for sinistrality is accidentally transported to an area where there are no **conspecifics**. All of the offspring of this snail will be sinistral because of the maternal effect. Breeding among these snails may establish a population. Although some dextrals will show up in this population within its first few generations (as progeny of mothers that are **homozygotes** for the dextral allele) they will be in a minority and will therefore be at a disadvantage in terms of mating. Selection will thus be acting against the 'old' dextral allele and may remove it from the new population. Given a sufficient period of isolation, the new population may acquire enough other genetic differences to become incompatible with its parent species, and so we have a new species with reversed coiling.

This **adaptive scenario** is highly improbable. The chances of the migrant snail carrying the appropriate mutant **allele** are vanishingly small. Yet with thousands of species colonising countless new areas over hundreds of millions of years, a vanishingly small probability can become a near certainty.

7.3 Heterotopic Processes Involving the A-P and D-V Axes

We will now look at examples relating to these axes. The first two concern the A-P axis: these involve snake fangs and the vulva of nematode worms. The third concerns a complete reversal of the D-V axis in the evolution of bilaterian animals.

The Evolution of Front-Fanged Snakes

Not all snakes have fangs. But those that do are distributed phylogenetically in a way that suggests that rear fangs represent the original (or **plesiomorph**) character state[37] (Fig. 7.6). Note that the two front-fanged families Elapidae and Viperidae are not **sister-groups**, so it appears that they have evolved their front-fanged state independently. How do fangs manage to relocate to a more anterior position in the evolution of some lineages?

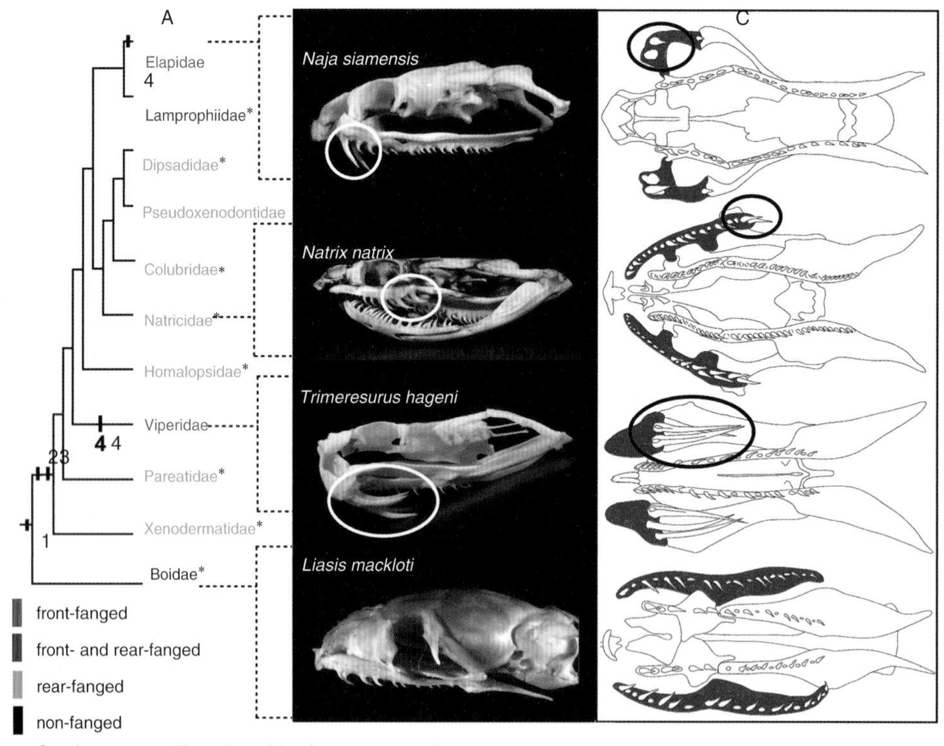

Figure 7.6 A: Snake phylogeny, showing the positions of the two front-fanged groups. As these are not sister groups, the front-fanged condition seems to have evolved twice from rear-fanged ancestors. B: Photographs of the skulls of representatives of four families (lateral view, fangs circled). C: Diagrammatic ventral views of the same four skulls (maxilla in red; fangs again circled). (From Vonk *et al.* 2008, *Nature*, **454**: 630–633.)

Teeth in general, fangs included, are produced from a dental **epithelium**. This is characterised by expression of the gene *sonic hedgehog (shh)*. The fang-forming part of this tissue is located in a posterior position in the upper jaw in both rear-fanged and front-fanged snakes. What happens in the latter is that, during development, **allometric growth** results in a heterotopic shift of the fangs to a more anterior position. Prior to this shift, the anterior part of the upper jaw lacks *shh* expression, so the anterior dental lamina normally found there appears to have been lost.

The evolution of front-fanged snakes is, like the evolution of asymmetric flatfish heads, a relatively recent event – both seem to have occurred in the Cenozoic era (the last 65 MY). It is not hard to come up with an adaptive scenario for the heteropic shifting of fangs to the front of the jaw – in general this would seem to be a better design for a sudden-strike predator. Perhaps in this case the question should rather be why so many snakes have retained the more posterior positioning.

Evolutionary Change in Vulval Position in Nematodes

The vulva is the egg-laying structure of a nematode worm. It is found in hermaphrodites and females (depending on the species) but not, of course, in males. In the model organism *Caenorhabditis*, and many other nematode species, it is situated on the A-P axis at approximately 50 per cent of body length (in the ventral midline). However, in some groups of nematodes it is to be found in a more posterior position, thus revealing heterotopy.

A classic study of this heterotopy was carried out by the biologists Ralf Sommer and Paul Sternberg[38]. They compared *Caenorhabditis* with *Mesorhabditis* (vulva at ~80% body length) and *Teratorhabditis* (vulva at ~95% body length) (Fig. 7.7). The latter two species are much more closely related to each other than either is to *Caenorhabditis*, according to current views on nematode phylogeny[39].

One particularly interesting aspect of this study is that it reveals an evolutionary change in the mechanism of vulval induction. In *Caenorhabditis*, the vulva is formed from three vulval precursor cells (or VPCs); these receive an inductive signal from another cell (the gonadal 'anchor cell') that causes them to proliferate into vulval cells and (ultimately) form the vulva. However, in the other two species, the VPCs migrate posteriorly at an early stage of development, which delays contact between them and the anchor cell. It seems that, in these species, either a different inductive signal is involved, or alternatively a vulval fate is somehow pre-programmed into the cells concerned.

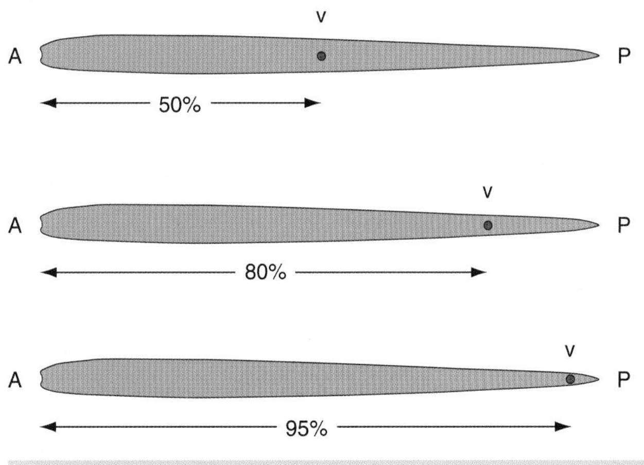

Figure 7.7 The position of the vulva (reproductive opening used for egg laying) on the antero-posterior axis in three nematode genera. From top down: *Caenorhabditis*; *Mesorhabditis*; and *Teratorhabditis*.

Note that the 'delay' mentioned above constitutes a **heterochronic** change. Also, while species such as *C. elegans* with a central vulva have two ovaries (one anterior and one posterior to

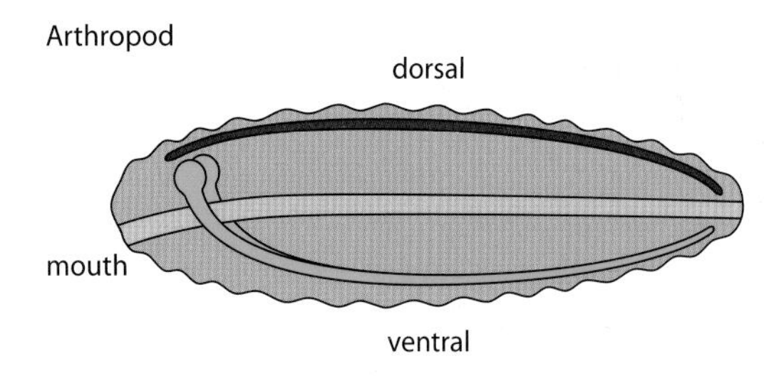

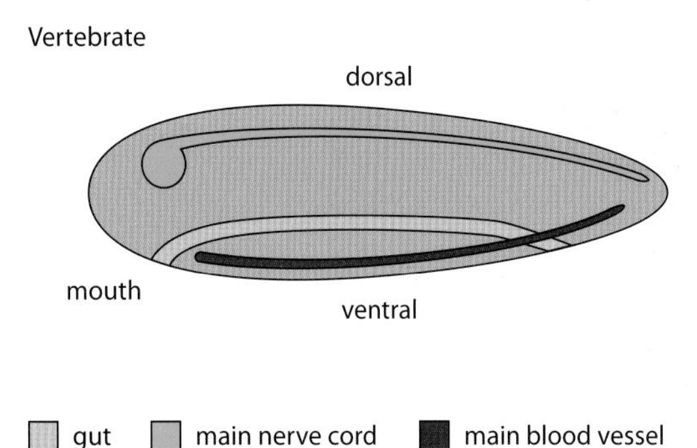

Figure 7.8 Schematic diagrams of the dorso-ventral body layout of vertebrates and arthropods. Note that the nerve cord is dorsal in vertebrates but ventral in arthropods.

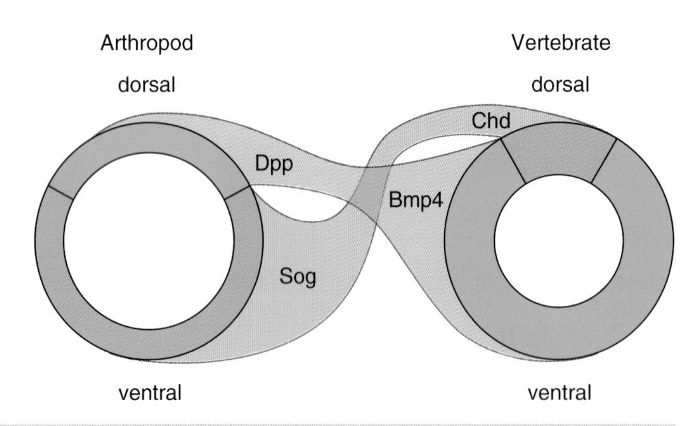

Figure 7.9 Expression patterns of genes belonging to two groups in the dorso-ventral axis of vertebrates and arthropods. Note that each gene-group has the opposite effect (dorsalising/ventralising) in one of these taxa when compared to its effect in the other. (Dpp – Decapentaplegic; Bmp4 – Bone morphogenetic protein 4; Sog – Short gastrulation; Chd – Chordin.)

the vulva), species with a posterior vulva only have a single ovary (anterior to the vulva). This change in the number of ovaries constitutes **heterometry**. So again, as with other examples, we see the compound nature of **developmental repatterning**, which will be explored in more detail in Chapter 10.

Inversion of the Bilaterian Dorso-Ventral Axis

Prior to widespread belief in evolution, and before Darwin's idea of natural selection had been published, a French biologist, Etienne Geoffroy Saint-Hilaire, proposed that the vertebrate body plan was, among other things, an upside-down version of the arthropod one[40]. Structures that are dorsal in vertebrates tend to be ventral in arthropods, and vice versa (Fig. 7.8). For example, the main midline nerve cord in vertebrates is dorsal (enclosed within the spine), whereas it is ventral in arthropods. At the time, many biologists thought this idea rather fanciful, and distanced themselves from it. However, it has become clear in the last 15 years or so that there may be some substance to Geoffroy's claim after all.

This 'substance' lies in the fact that many of the genes involved in dorso-ventral axis formation in vertebrates and arthropods are homologous, but have opposite effects: a gene whose product dorsalises in a vertebrate ventralises in an arthropod, and, once again, vice versa.

Two genes, or groups of genes, have become particularly well-known in this respect[41]. The first involves the genes making bone morphogenetic proteins (BMPs), especially BMP4. In vertebrates, BMP4 has a ventralising effect, whereas in arthropods its homologue *decapentaplegic* (*dpp*) produces a protein that has a dorsalising effect. The second involves the homologous genes *chordin* (dorsalising in vertebrates) and *sog* (ventralising in arthropods). Their products are, perhaps unsurprisingly, BMP/Dpp antagonists (Fig. 7.9).

While Geoffroy's proposal of an upside-down relationship between vertebrates and arthropods has been supported by a type of evidence that he would never have guessed at, there is still a major problem in establishing

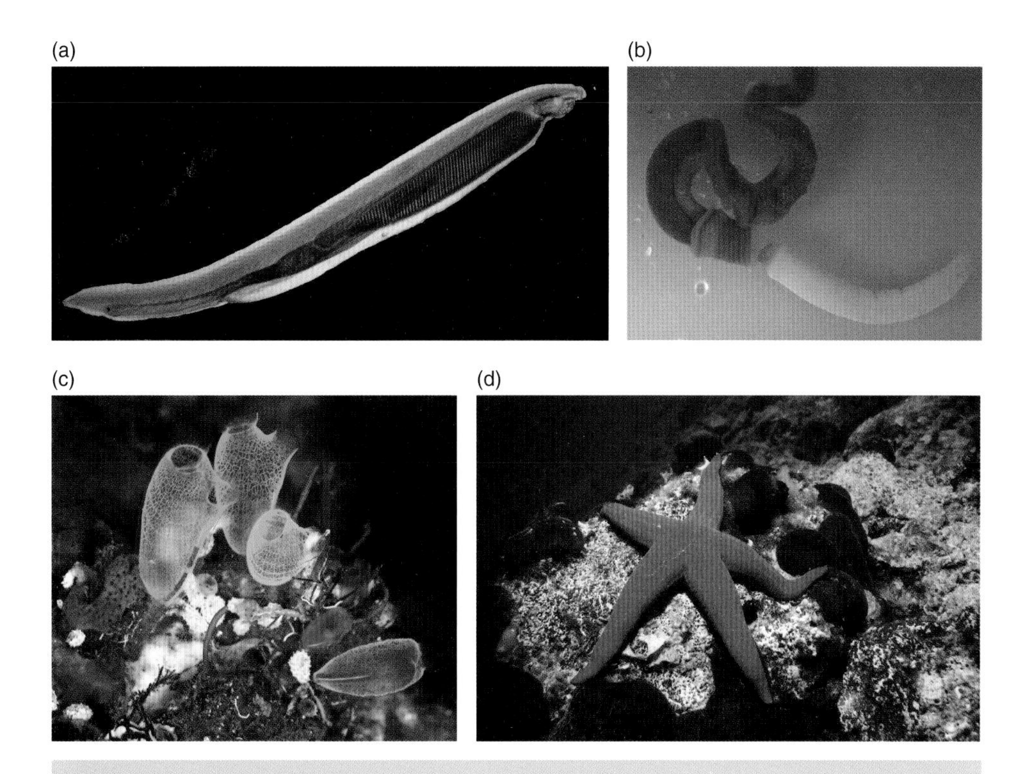

Figure 7.10 Representatives of the four main groups of non-vertebrate deuterostomes: (a) cephalochordates, (b) hemichordates, (c) urochordates and (d) echinoderms. The urochordate shown is a non-colonial form, in contrast to the colonial form seen in Chapter 4. (Hemichordate courtesy of Stephen Green.)

when and how the axis inversion happened. To understand more about these issues, we need to know about the mechanism of D-V axis determination in other bilaterian phyla. It appears that non-arthropod protostomes, such as annelids, have a D-V axis determination system like that of arthropods. However, the situation with non-vertebrate **deuterostomes** is less clear.

There are four main such groups (Fig. 7.10). Within the Chordata are the **cephalochordates** (amphioxus) and the **urochordates** (sea squirts). The former seem to have a vertebrate-style D-V axis, while the latter have such a bizarre body form (at least when adult) that comparisons with them are difficult. Outside the Chordata are the deuterostome sister groups, the **echinoderms** and the **hemichordates**. The former have ceased to be bilaterians. The latter, and in particular the enteropneust worms, have a D-V axis but a diffuse nerve net. Although it is therefore not possible to look at the main nerve cord (since there isn't one), it is possible to look at the genetic determination of the D-V axis in relation to other structures, such as the mouth[42]. Surprisingly, hemichordates seem to have the protosome pattern of gene function (BMP = dorsal). So, it is possible that the evolutionary D-V axis inversion, if it occurred at all, which some still doubt[43], took place specifically in the chordate stem lineage. This is likely to have existed at least as early as the Cambrian (and so >500 MYA), given that chordates can be found in the Cambrian **Burgess Shale** formation.

Given the uncertainties surrounding this important issue, it does not yet seem wise to propose adaptive hypotheses. But perhaps their time will come in the not-too-distant future.

7.4 Other Types of Heterotopy

Figure 7.11 'Chinese lantern' flowers, as can be found in many species of the family Solanaceae. The lantern structure is formed from sepals that are greatly expanded compared to the rest of the flower.

The above examples all relate to what we regard as the three main body axes of bilaterian animals. These axes, however, do not exist in other multicellular creatures, including basal animals such as sponges, and also plants and fungi. So, to demonstrate that heterotopy is a type of developmental repatterning that can occur without reference to such axes, we will now look at an example from the plant kingdom, and specifically from the **angiosperm** family Solanaceae.

Heterotopy and the Origin of the Chinese Lantern Type of Flower

The Solanaceae contains about 3000 species belonging to more then 80 genera. It includes many familiar plants, ranging from tomato to tobacco to the poisonous 'deadly nightshade'. In many cases the flowers are of what we think of as the standard type, with an outer, and inconspicuous, whorl of sepals, surrounding a larger, more colourful, whorl of petals. However, some species in this family, notably in the genus *Physalis*, have evolved a novel flower type in which the sepals are greatly expanded and more colourful, completely enclosing the rest of the flower, and later the fruit (Fig. 7.11).

A recent study has helped to reveal the developmental-genetic basis of the evolution of this unique type of flower[44]. Two genera within the Solanaceae were compared – *Solanum* and *Physalis*; we will concentrate here on one species of each. These are: *Solanum tuberosum* (potato; 'normal' flower type) and *Physalis pubescens* (a species with the Chinese lantern flower type). As is often the case when investigating evolutionary changes in development, it turns out that genes encoding transcription factors are involved. Here, the primary gene of interest belongs to the **MADS-box** family. In the potato, this gene is called *stmads16*, with the *st* standing for the species name; in *Physalis* its homologue is usually referred to as *mpf2*. For simplicity we will refer to both genes here simply as *mads16*.

In the potato, expression of *mads16* is restricted to vegetative tissues – there seems to be no expression in the flower. However, in *Physalis* there is expression of this gene in the sepals, all the way from the small bud stage to the fruit stage (Fig. 7.12). So this is an example of molecular heterotopy – a gene has become expressed in places where it was previously not. The result is morphological heterotopy, with sepals overgrowing other flower structures, and of a sufficiently striking kind that it might be considered to be an example of hereotypy (Chapter 9) or even an evolutionary novelty (Chapter 17).

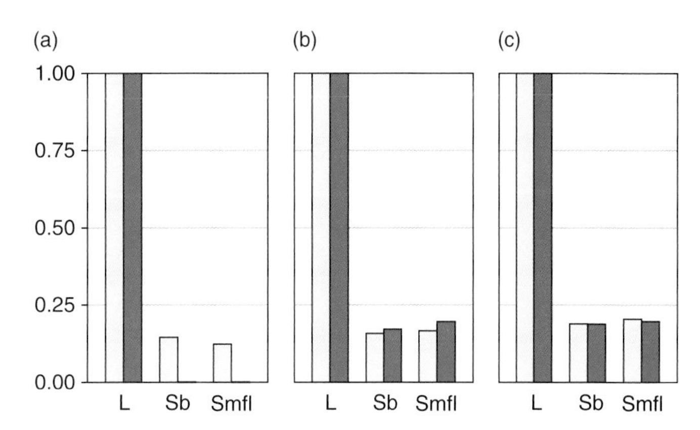

Figure 7.12 Expression (dark blue bars) of the gene *mads16* in (a) potato, which has 'normal' flowers; and (b, c) in *Physalis*, which has Chinese lantern flowers (b – *P. pubescens*; c – *P. peruviana*). Note that expression is restricted to the leaves in potato but extends into the sepals at all stages of flower development in *Physalis*. (Pale blue bars indicate the expression of a related gene, *mads11*.) L – leaf; Sb – sepals from buds; Smfl – sepals from mature flowers. (Reproduced with permission from He and Saedler, 2005, *PNAS*, **102**: 5779–5784.)

Two questions immediately arise. First, how does this gene become expressed in the new place (sepals) in *Physalis*? And second, how does this novel expression pattern lead to the development of a Chinese lantern?

When a gene is expressed in a new place there are two possible reasons for this: either one of the regulatory regions of the gene (e.g. the **promoter** region) has changed so that it is responsive to **trans-acting** molecules already in that new place; or one of those molecules has changed so that it switches on the gene, whereas previously it did not. In the case of the Chinese lantern, it seems to be the former. Comparison of the coding regions of the gene in *S. tuberosum* and *P. pubescens* reveals a high level of similarity (86%), while in the promoter regions the level is much lower (42%).

Turning to the effect of the altered gene in producing the Chinese lantern, various experimental studies have been carried out, including both gene knockdown in *Physalis* and creating a transgenic potato that has the *Physalis* gene. Both of these approaches reveal an effect of the gene on cell proliferation. It seems that the normal *Physalis* gene acts to enhance cell proliferation, thus producing larger sepals.

Of course, as in all similar studies, much is yet to be explained. The fact that sepal cells divide rapidly in *Physalis* will readily explain the existence of larger sepals, but it does not explain the precise form – the lantern – that results. Nevertheless, this is one of the clearest examples of heterotopy at one level (molecular) leading to heterotopy at another (morphological).

Heterotopy in the Evolution of Rodent Teeth

For a final example of heterotopy, we will look at evolutionary divergence in the shape of molars in representatives of two groups of rodents: mice and voles[45]. The species that will serve as representatives are, respectively, the mouse *Mus musculus* and the

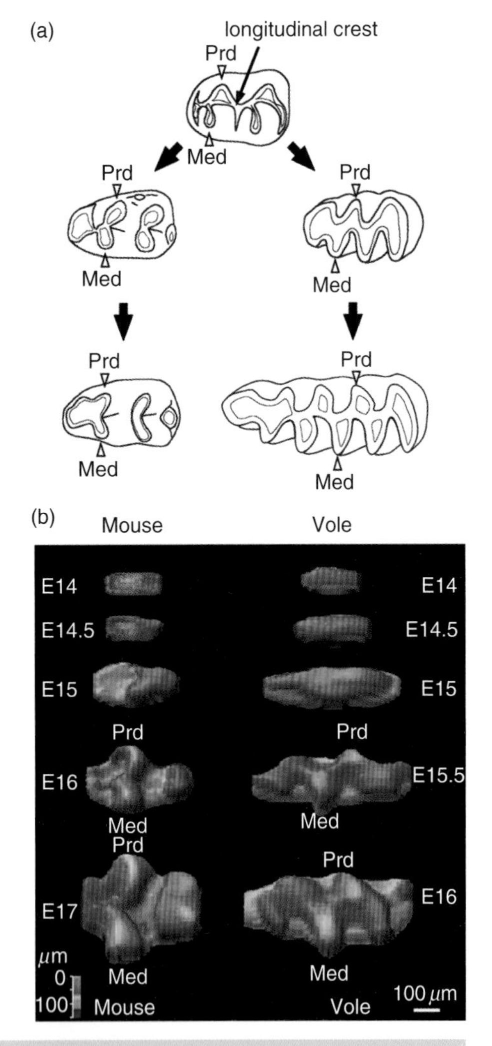

Figure 7.13 Evolution (top) and development (bottom) of the pattern of cusps on the first lower molar teeth of mice and voles. The two cusps shown are the protoconid (Prd) and the metaconid (Med). Numbers show developmental time (e.g. E14 is embryonic day 14). Colour is used to indicate topography. (From Jernvall *et al.* 2000, *PNAS*, **97**: 14444–14448.)

vole *Microtus rossiaemeridionalis*. For simplicity we will refer to these just as *Mus* and *Microtus*. The particular tooth we will focus on will be the first lower molar.

Mammalian teeth, and indeed teeth in general, have a good fossil record: as hard, mineralised structures they are far more often preserved than soft tissue, and indeed considerably more often preserved than bones. So there is much fossil evidence available of the structure of teeth in extinct mammals. The structure of the lower first molar in *Democricetodon*, a possible LCA of mice and voles, or at least a close relative thereof, is known, and can be used as a reference point for comparison with extant species (Fig. 7.13).

Molars have prominent cusps, and the pattern of these can be seen to have altered in both lineages. Specifically, the main two cusps (called the protoconid and metaconid) were diagonally rather than directly opposite each other in the LCA. In mice they have moved so that they are directly opposite. In voles the diagonal relationship has been maintained but more cusps have been added. So, strictly speaking, the type of developmental repatterning that has occurred is predominantly heterotopic in the mouse lineage but predominantly heterometric (Chapter 8) in the vole lineage.

The origins of the different patterns can be visualised in terms of the expression of *sonic hedgehog (shh)*. Recall, from the snake example in Section 7.3, that this gene is generally expressed in tooth-forming tissue. Notice the directly and diagonally opposed expression domains, respectively, in the tooth-forming **epithelium** of mouse and vole (Fig. 7.14).

The developmental repatterning that has occurred in this evolutionary divergence is fairly recent. Like the flatfish and snake examples, it is something that has occurred during the most recent era of geological time, the Cenozoic.

What of a possible adaptive hypothesis? At least in general terms, the construction of such hypotheses should be easier with specific structures such as teeth than with bigger-scale heterotopies such as D-V or L-R axis inversions. Indeed, because teeth are so deeply linked to diet, an adaptive hypothesis framed in terms of dietary changes seems inevitable. However, a specific hypothesis is not so easy. Most mice and voles are herbivorous, eating a wide range of plant material, including foliage, nuts and bark. But the actual diet of any one species is quite variable, depending on habitat and season. Perhaps on average voles eat tougher plant material, leading to selective pressure for molars with larger grinding surfaces, which can be produced by increasing the cusp number. But, as ever, such hypotheses should be treated with caution.

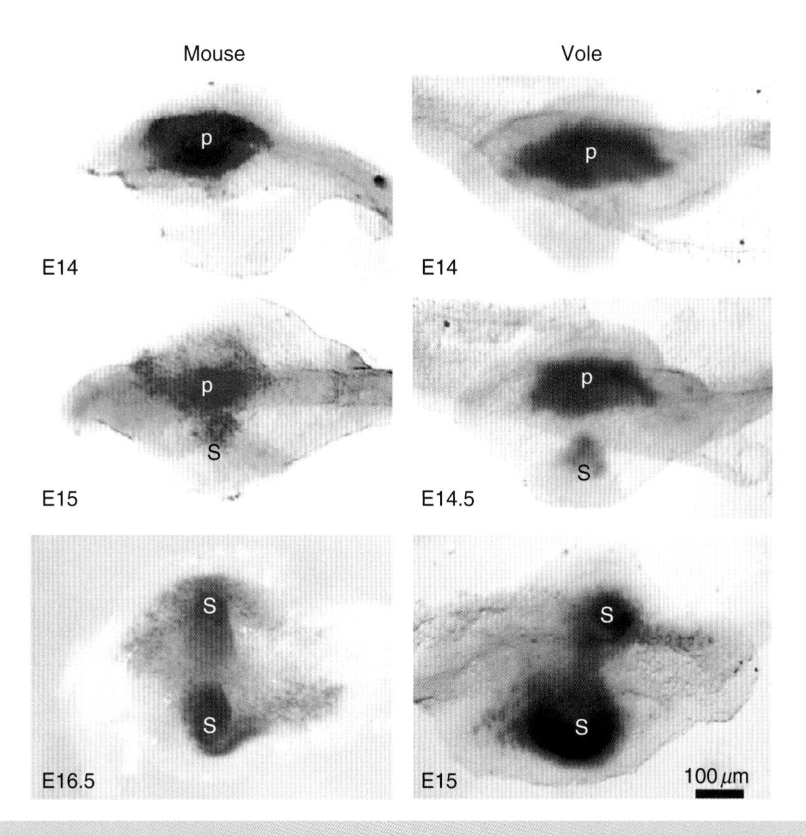

Figure 7.14 Expression patterns of *sonic hedgehog* in mouse and vole first molar tooth-forming epithelium. Although in both species there is initially just a single expression domain corresponding to the primary enamel knot (p, top), there are two by stage E15 including the first secondary knot (S, centre), and subsequently more. The arrangement of the two domains shown becomes shifted diagonally in voles but not in mice. (From Jernvall *et al.* 2000, *PNAS*, **97**: 14444–14448.)

7.5 Concluding Remarks

In this chapter, we have focused on examples of developmental repatterning in which the most conspicuous change was in spatial arrangement, just as in the previous chapter we focused on examples in which temporal changes were most apparent. We will continue this theme of focusing on the most prominent characteristic of particular examples of repatterning in the next two chapters – on **heterometry** and **heterotypy**. However, we are already beginning to see that many instances of repatterning are complex, and that there are usually lots of other changes happening in parallel with the 'main' one. This theme will be picked up in Chapter 10, where the complex nature of most instances of repatterning is the focus of discussion. Finally, there is a link from developmental repatterning in general (and heterotopy in particular) to the origin of evolutionary novelties. The Chinese lantern discussed above is often referred to as a novelty. Another, particularly clear, example of a novelty is the chelonian shell. We will discuss this in the chapter explicitly devoted to novelty, and will see that one of the most important novel things about the chelonian body layout is the heterotopic shift of the pectoral girdle from its usual position outside the ribcage to a new position within.

SUGGESTIONS FOR FURTHER READING

A concise account of heterotopy, along with heterochrony, can be found in chapter 24 of:

Hall, B.K. 1999 *Evolutionary Developmental Biology*, second edition. Kluwer, Dordrecht, Netherlands.

A review article covering heterotopy, again in combination with heterochrony, is:

Zelditch, M.L. and Fink, W.L. 1996 Heterochrony and heterotopy: stability and innovation in the evolution of form. *Paleobiology*, **22**: 241–254.

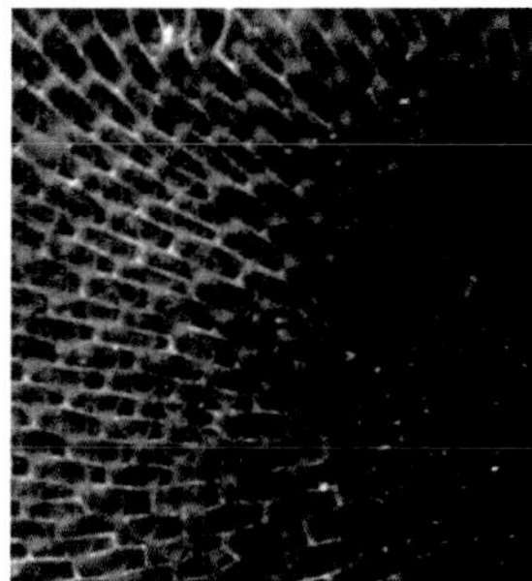

CHAPTER 8

Heterometry

8.1 What is Heterometry?

In contrast to changes in timing, **heterochrony**, and spatial arrangement, **heterotopy, heterometry** refers to situations where the primary change is in neither space nor time, but rather in the amount of something, whether expressed as size, number or concentration. As with heterochrony and heterotopy, the change is comparative in two ways: first, between different **lineages**; and second, between different parts of the same developing organism, whether at the molecular or morphological level. To illustrate the latter point, an **isometric** change in the size of the whole body between one lineage and another would not represent heterometry, but the increase or decrease in size of one organ relative to the rest of the body would.

Of course, if something increases in size, then it will occupy more space and may also take longer to develop. So heterometry may involve spatial and temporal changes too. This is another manifestation of the point that has been made several times already and will be expanded on in Chapter 10, namely that **developmental repatterning** is a complex process, and most examples of it involve multiple types of change. Our strategy here is to look at these individually first, and then to put them together.

The terms heterochrony and heterotopy were both introduced by Ernst Haeckel, who is most famous for his idea of **recapitulation** (Chapter 2). But Haeckel did not come up with a term for altered amount of something in development; so 'heterometry', like 'heterotypy' (next chapter), is of much more recent origin[46].

When dealing with changes in the spatial arrangement of developmental processes, there is an almost limitless array of possibilities. But when dealing with changes in the size of something (e.g. the size of an organ relative to the rest of the body), then there are only two possible types of change: increase and decrease. The structure of this chapter reflects that fact: the next two sections cover increases and decreases, respectively; while the following two sections cover the possibility of changes in both directions, but in different contexts. Each section will be based on one main example.

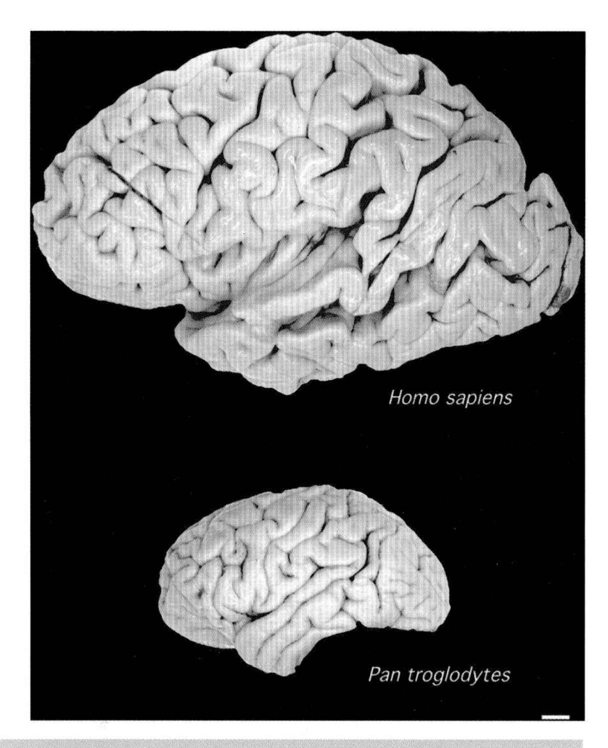

Figure 8.1 Comparison of the sizes of typical adult human (*Homo sapiens*) and chimp (*Pan troglodytes*) brains. In both volume and weight, the difference between the two is about three-fold. The LCA of present-day humans and chimps lived approximately 6 MYA and had a brain size much more similar to chimp than human. Scale bar –1 cm (Reproduced with permission from Todd M. Preuss.)

8.2 Increasing Relative Size

The size of the average adult human brain is about 1350 cm^3. The size of an average adult chimp brain is about 400 cm^3 (Fig. 8.1). Because brain tissue has a weight/volume ratio close to that of water (and hence close to 1), these figures can also be used as approximate brain weights in grams. It is clear from fossil skulls that the last common ancestor (LCA) of chimps and humans had a brain size similar to that of modern chimps. So in the human lineage, brain size has increased more than three-fold.

This increase should be qualified by reference to body size. If, in any lineage, body size increases, then we would expect the size of the organs within that body to increase also. Humans are larger than chimps, but not by much. A typical adult male chimp weighs about 60 kg, a typical adult female about 50 kg. These figures correspond approximately with the *minimum* weights of adult humans of the two sexes. The difference in average weight between adults of the two species is such that humans are very roughly 1.25 times the size of chimps. That would explain an increase in brain size to 500 cm^3 in the human lineage, not to the 1350 cm^3 that is actually found. So increasing brain size in the human lineage is real heterometry, not a mere side-effect of larger body size.

Understanding the developmental-genetic basis of the increase in brain size in the human lineage is still in its infancy. One approach that has been taken is to use information on genes that, when mutant, can cause reduction in human brain size, and to look for **homologues** of these genes in the chimpanzee, and indeed in other great apes also. Two such genes that have been identified are *microcephalin*[47] and *aspm* (*abnormal spindle-like microcephaly-associated*)[48]. In both cases, studies of sequence divergence indicate a possible association either with the size of the brain or with the size of parts of it – in particular the cerebral cortex.

In contrast to the early stage that such *genetic* studies are at, the study of the *evolutionary pattern* of human brain size increase has a long history. Fossilised human skulls (and fragments thereof) have been studied in detail, along with fossils of human post-cranial skeletons, for more than a century. What have such studies told us?

Before considering brain size specifically, it makes sense to look at what is known of human (and proto-human) **phylogeny**. There has been a considerable increase in the number of named species over the last two decades. There are now more than a dozen named species of *Homo, Australopithecus* and *Paranthropus* (with the last genus including some species that were previously regarded as belonging to *Australopithecus*). As new species have been discovered, the proposed pattern of relationships has kept changing. It is likely to continue to do so for some time yet. Because of that, it does not seem useful to hang too much on the precise details of the currently-favoured phylogeny. These may not be necessary anyhow. What we really need, for our present purpose, is an idea of how brain size has changed in relation to some of the more robust features of the phylogeny, and in particular along the **lineage** leading to *H. sapiens*, as follows (Fig. 8.2).

The human and chimp lineages diverged from each other about 6 MYA. In the lineage leading to *Homo*, there was an early radiation of australopithecine species between about 5 and 2.5 MYA. This radiation was not associated with a major increase in brain size. The first species of *Homo, H. habilis*, appeared just over 2 MYA. This is thought to be the first species to have used a complex array of tools. Brain size in *H. habilis* has been estimated to have been about 600 cm^3 – a relatively modest increase over australopithecines. A contemporary of *H. habilis, H. ergaster*, is thought to have had a somewhat larger brain (~700 cm^3), and to have been a more likely ancestor, than *H. habilis*, of the lineage leading to *H. sapiens*.

The most pronounced increase in brain size occurred in the last 2 MY, passing through the 1000 cm^3 mark in *H. ergaster*, and increasing further in *H. heidelbergensis* to about 1200 cm^3 (Fig. 8.3). Of course, the pattern looks different if we (i) consider proportional increases, so that, for example, a change from 500 to 750 is the same as one from 800 to 1200, or (ii) express brain size relative to body size (in which case the increase in *H. ergaster* over *H. habilis* is less noteworthy, because *H. habilis* was small-bodied). Also, it must be

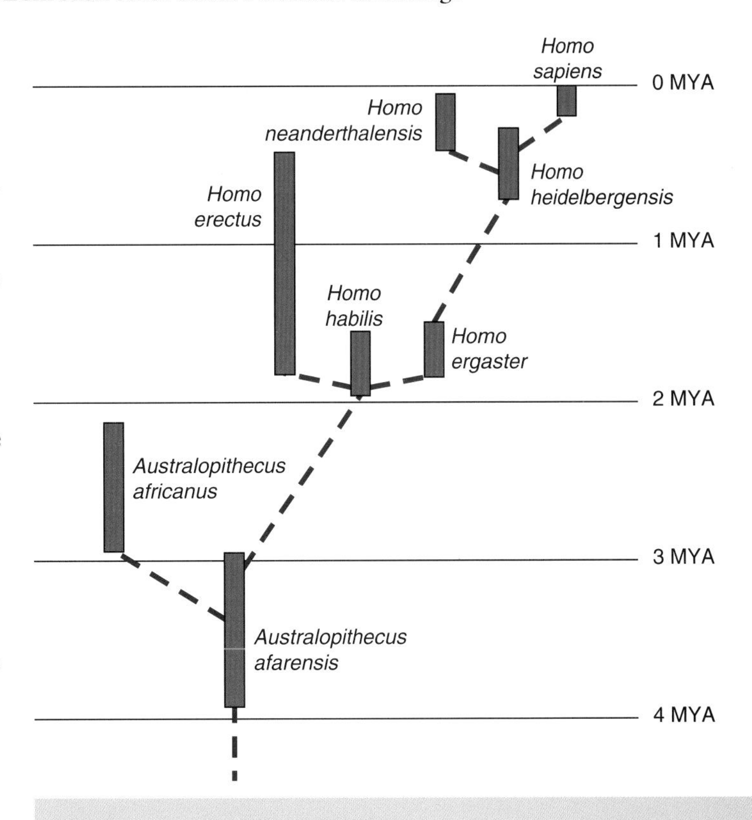

Figure 8.2 Simplified phylogenetic tree of hominin species, with a particular focus on the lineage leading to *Homo sapiens* and its divergences from other lineages.

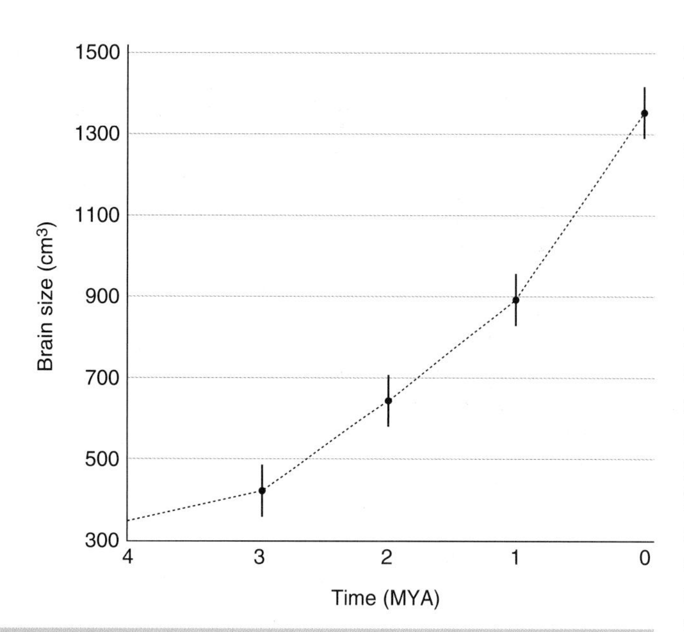

Figure 8.3 Increase in brain size in species of the lineage leading to *H. sapiens* over the last 4 MY. Note that different species often overlapped in time; the apparently simple trend shown needs to be interpreted with this in mind.

remembered that the studies of fossil material usually just estimate total brain size from cranial capacity. There are not robust estimates of the size of particular parts of the brain, such as the cerebral cortex, though we know from comparison of human and chimp brains that humans have a disproportionately expanded cortex.

All the above brain volumes refer to adults. In any species of *Homo* (and indeed in any animal species in which there is a brain), brain size is zero during the earliest stages of **embryogenesis**, and then it has a particular pattern of size increase in the period from the formation of a rudimentary brain (during **neurulation**) until adulthood. As ever, evolution works by altering the **developmental trajectory**. Unfortunately, the amount of information available on juvenile brain size in fossil hominids is limited (though there are several good examples), and information on embryonic brain size in fossil hominids is entirely lacking. However, comparisons between humans and chimps have been made. For example, it has been estimated that human brain size increases by 3.3× from birth to adulthood, while that of chimps increases by about 2.5× over the same period[49].

Finally, what is a plausible hypothesis for the substantial increase in brain size that has occurred in the human lineage? The most obvious is one based on the advances in behaviour that a large brain bestows, whether in tool use or in language, and the advantages that these produce in terms of survival. But another hypothesis is based on sexual selection: preference for mates who have characteristics associated with larger brain size. Such hypotheses are likely to provide fertile ground for ongoing debate, since testing them in relation to any particular period of brain size increase in the human lineage is really not possible.

8.3 Decreasing Relative Size

The relative sizes of specific organs or body parts probably decrease as often as they increase in evolution. Indeed, we could say that this is inevitable because a relative increase in the size of one body part implies a relative decrease in others. However, the *reasons* for relative size changes are often different in cases of one part increasing relative to a composite of the others (as in the human brain) and one part decreasing in relation to a composite of the others (as in the case of **vestigial** structures).

Decreases can be roughly categorised into slight, major and total. Vestigial structures represent the second category, while a shift from presence to absence (e.g.

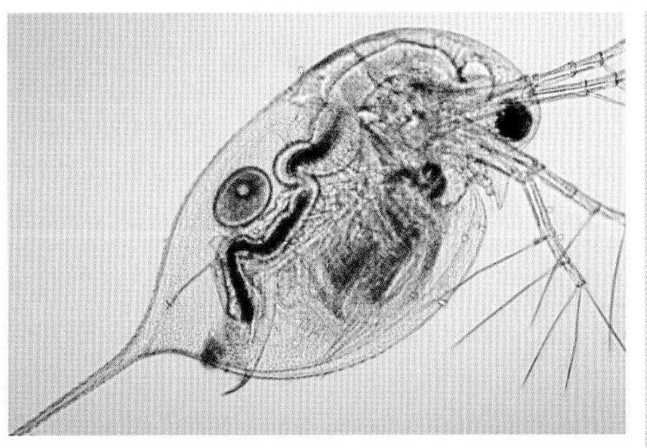

Figure 8.4 Morphological disparity of crustacean body form, as illustrated by species from different crustacean groups. The disparity is much greater than in the insects (which all have the familiar head/thorax/abdomen arrangement), despite the much higher number of insect species.

eyes in many animal lineages) exemplifies the third. Oddly, the loss of something that was previously present, such as a pair of eyes, is often described as a 'secondary' loss. However, this is an illogical usage. If we trace the lineage of an extant eyeless animal (such as a cave-fish: see Section 8.5) back in time and find that its immediate ancestors had eyes, but long before that (**Urbilateria** or earlier) they did not, then the character state of eyelessness is indeed secondary, but the loss that resulted in it is primary, unless there had been a previous loss in the same lineage.

Here, we will focus on an example of a vestigial structure. The commonest textbook example of such a structure is the human appendix; but we will now look at a case in which the entire abdomen has become vestigial. This involves a group of crustaceans – namely the barnacles (Sub-class Cirripedia).

The 70,000 species of Crustacea (itself a sub-phylum of the Arthropoda) exhibit a wide range of body forms. There is no such thing as a 'typical' crustacean, given this incredible morphological disparity (Fig. 8.4). In many familiar crustaceans, such as shrimps and lobsters, the abdomen comprises a significant part of the trunk. However, in some other crustaceans, including crabs and barnacles, it has become markedly reduced or absent.

There are three groups of barnacles, of which one (the encrusting group Thoracica) is much more familiar to most people than the other two (the burrowing group Acrothoracica and the parasitic group Rhizocephala). Despite their very different adult forms, they are grouped together in the Cirripedia because of the similarity of their larvae (Fig. 8.5).

Studies of the **nauplius larva** of a species of parasitic barnacle, *Sacculina carcini*, have revealed the presence of a vestigial abdomen[50], which can be seen through the **expression pattern** of the segment polarity gene *engrailed*. Posterior to the thorax,

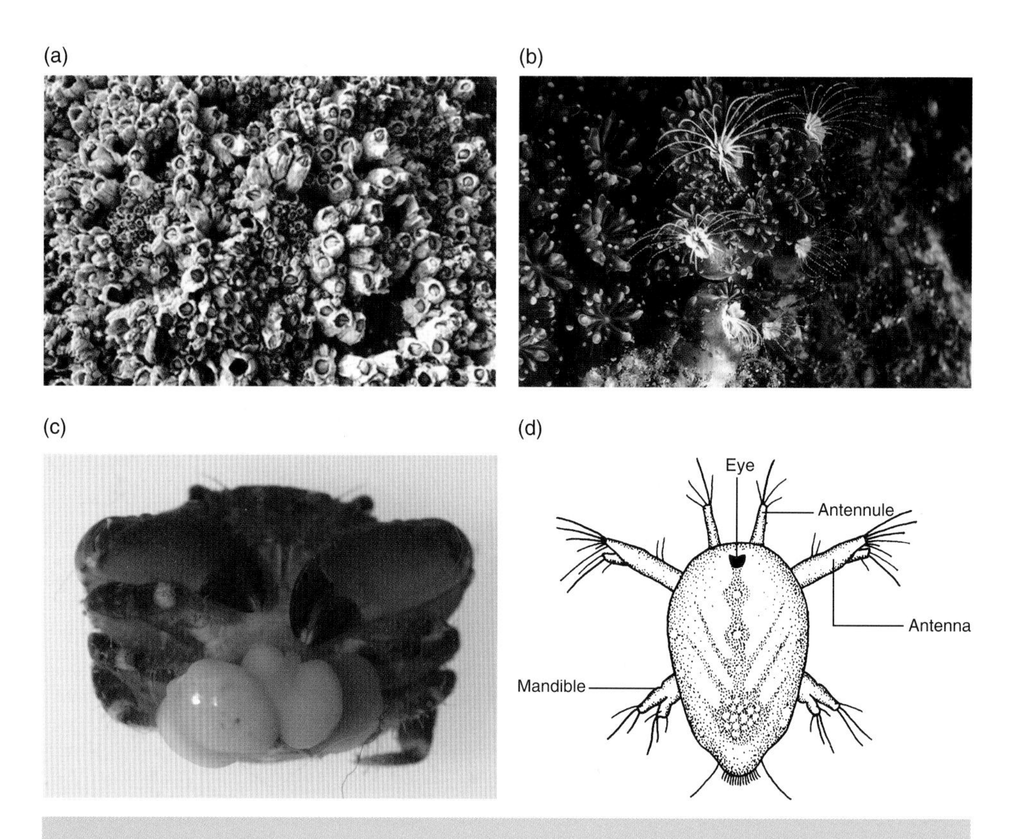

Figure 8.5 The adults of the three groups of barnacles are very different; yet all go through a developmental stage called the nauplius larva. Pictured here are adults of (a) Thoracica (encrusting a rock), (b) Acrothoracica (projecting from a coral) and (c) Rhizocephala (projecting from a parasitised crab); plus a typical barnacle nauplius larva (d).

but also wrapped round at an angle to it, five abdominal stripes of *engrailed* expression can be seen (Fig. 8.6). So, while adult barnacles lack an abdomen, their larvae have a vestigial one. Its small size must be based on changes in genes that are **upstream** of *engrailed* in the segmentation cascade, but details of these changes are not yet known.

There are more than 1000 species of barnacle. Although their close relationship with each other is clear (based on larval and molecular evidence), their position within the overall phylogeny of the Crustacea is less so. One closely-related group is the Copepoda. Copepods show a reduced abdomen with 4 or 5 small segments lacking in appendages. More distantly related crustaceans often have 4, 5 or 6 abdominal segments. So, together with these comparative data, the number of *engrailed* stripes in the nauplius larva of *Sacculina* suggests that the reduction of the abdomen in the Cirripedia has occurred by decreasing the size of all segments (ultimately to zero in the adult), rather than by, for example, a gradual loss of segments from the posterior end.

This evolutionary reduction of the abdomen in the barnacle lineage is very ancient, especially when compared with evolutionary increase in brain size in the human lineage that we examined in Section 8.2, where the changes concerned occurred in the last 2 MY. The fossil record of barnacles extends back at least to the Silurian Period.

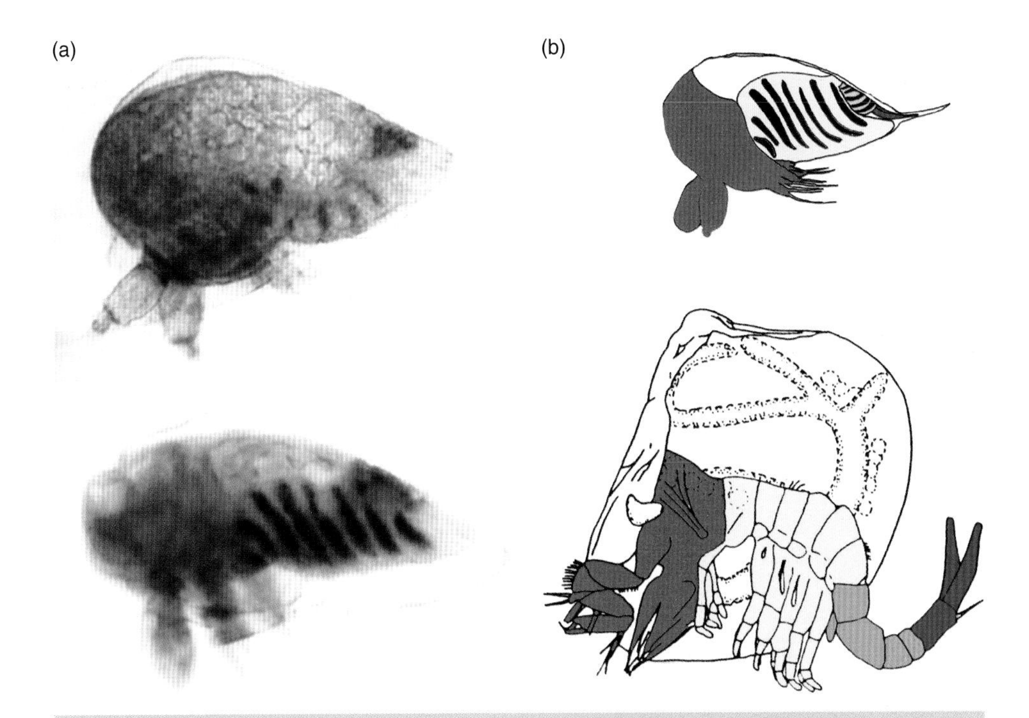

Figure 8.6 Small abdominal expression stripes of the segmentation gene *engrailed*, revealing the presence of a vestigial abdomen in the nauplius larva of a barnacle. (a) photographs of *engrailed*-stained specimens; (b) interpretive diagrams. The dark stain in the top right of the photos shows abdominal *engrailed* expression – see blue/black section of top diagram. (From Gibert *et al.* 2000, *Evolution & Development*, **2**: 194–202.)

Study of the Silurian species *Rhamphoverritor reduncus* (belonging to the Thoracica) from a fossil formation that dates to about 425 MYA, shows that the abdomen was reduced in the larva and absent in the adult[51]. This provides us with a *minimum* age for the origin of the barnacle body form. How much before this the *actual* origin was remains unclear: some fossils from the Cambrian Period have been interpreted as barnacles, though this interpretation is uncertain.

As with many other examples in this chapter and the previous two, we can see here the morphological result of repatterning (in this case negative heterometry), and trace at least part of its developmental-genetic causality. We can also estimate, at least very approximately, when it occurred in terms of geological periods. And again, as with most of the other examples, we can construct a plausible **adaptive hypothesis**, while keeping in mind that such hypotheses are difficult, or in some cases impossible, to test.

Another point that should be borne in mind in relation to adaptive hypotheses is that the adaptive reason for a particular trend – in this case reduction of the abdomen – may change as the trend continues. Something that begins to happen for one adaptive reason can later keep happening for other such reasons. This phenomenon was given the name **exaptation** by Stephen J. Gould and Elizabeth Vrba[52] – it will be covered in more detail in Chapter 12.

With these caveats in mind, we might hypothesise that the initial stages in the reduction of the abdomen in the barnacle stem group hundreds of millions of years ago were not driven by selection associated with the well-established sessile, burrowing or parasitic life-styles that were to emerge eventually. In contrast, the later stages of abdominal reduction may well have been intimately associated with the adoption of those life-styles, in which an abdomen is an unnecessary structure.

8.4 Bi-directional Heterometry

The evolutionary change in brain size in the lineage leading from the earliest australopithecines to modern humans is a good example of a unidirectional trend in size *in that particular lineage*. The italicisation is necessary because in at least one offshoot lineage, the one leading to *H. floresiensis*[53], brain size decreased, even when measured in relation to body size, which also decreased. But, in the lineage leading to *H. sapiens*, as far as we know, there was no period during which brain size decreased. The rate of increase varied, of course, and we saw above that it was most rapid in the last 2 MY. Earlier than that, it was slower. Also, in the later, more rapid phase, there have been proposals – not universally accepted – that the increase occurred in a **punctuated equilibrium** pattern. And a further complication is that *Homo neanderthalensis* had a slightly higher average brain size than *Homo sapiens*. But this does not imply a decrease from Neanderthals to us because they were not our direct ancestors. Rather, after the split in the lineages leading to these two species, brain size (ironically) increased slightly faster in the Neanderthals than in the 'sapients'.

Often, however, when the relative size of some structure or organ changes in evolution, we observe changes in both directions – especially if we look at a long enough span of evolutionary time. To observe such bi-directional heterometry, we will examine variation in the size of eyespots on the wings of butterflies.

Figure 8.7 An example of an eyespot pattern on the wings of a butterfly. The species shown gets its common name, the peacock butterfly, from the presence of these eyespots, which superficially resemble those on the tail of a peacock.

The insect order Lepidoptera, which contains the butterflies and moths, includes about 180,000 species, making it one of the largest orders in the animal kingdom. Rather than butterflies and moths being separate sister **clades** within the Lepidoptera, as used to be thought, the butterflies are now known to be nested within the moth group, thus rendering the latter **paraphyletic**.

Many butterfly species have eyespots on their wings – such as the familiar European peacock butterfly, *Inachis io* (Fig. 8.7). But we will concentrate here on 'the browns' (family Satyridae – or subfamily Satyrinae according to some authors), which includes more than 2000 species, many of which have wing eyespots. We will focus in particular on the African brown, *Bicyclus anynana*, which has been the subject of much study.

There is considerable variation among the 80+ species of the genus *Bicyclus* in eyespot pattern (Fig. 8.8). It is likely that, during the evolution of this genus, there have been many increases and decreases in eyespot size and number. There is also much variation within *Bicyclus anynana*, including a difference between wet-season and dry-season forms with, respectively, larger and smaller eyespots (Fig. 8.9).

Figure 8.8 Variation in eyespot patterns in the butterfly genus *Bicyclus* (family Satyridae). This variation is the result of selection for different eyespot patterns as the genus has diversified from a single stem species. The species are (clockwise from top left): *B. martius*, *B. taenias*, *B. xeneas* and *B. vulgaris*. (Photographs courtesy of Oskar Brattström.)

Figure 8.9 Wet (left) and dry season (right) forms of the species *Bicyclus anynana*, showing the pronounced difference in eyespot size and number. This is an example of seasonal polyphenism (photographs courtesy of Oskar Brattström).

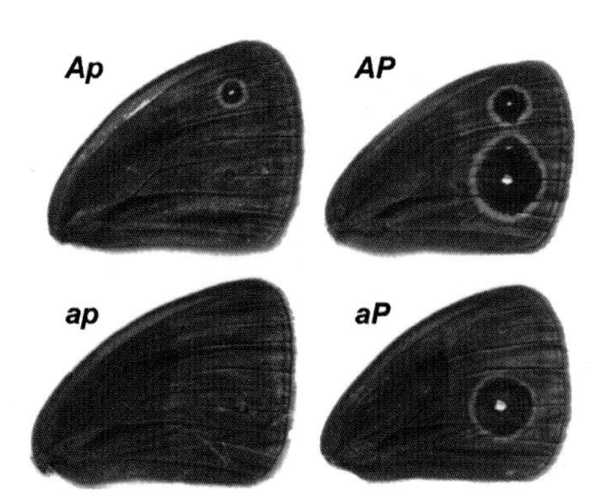

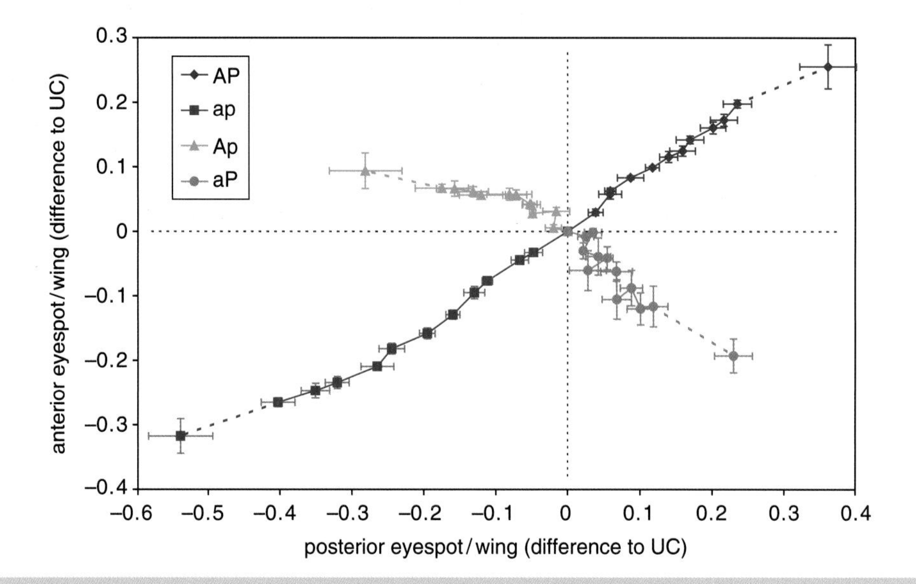

Figure 8.10 Results of artificial selection experiments conducted on eyespot size in *Bicyclus anynana*. Selection was carried out in the laboratory and was aimed at increasing and decreasing the sizes of two eyespots shown (top) on the forewing. A/a – anterior eyespot; P/p – posterior eyespot. Upper/lower case letters refer to selection for larger/smaller eyespots. UC – unselected control. (From Beldade *et al.* 2002, *Nature*, **416**: 844–847.)

Although this seasonal **polyphenism** represents **plasticity**, rather than rapid natural selection, there is also a heritable component of the eyespot pattern. Thus it is possible to artificially select for larger or smaller eyespots, hence reproducing, on a very rapid time-scale, the kinds of evolutionary changes that have occurred naturally in *Bicyclus* in the longer term. The results of artificial selection for increased and decreased size of anterior and posterior eyespots on the dorsal forewing are shown in Fig. 8.10. Clearly, changes are possible in all four directions, though selecting for both eyespots getting bigger or smaller in parallel is in some

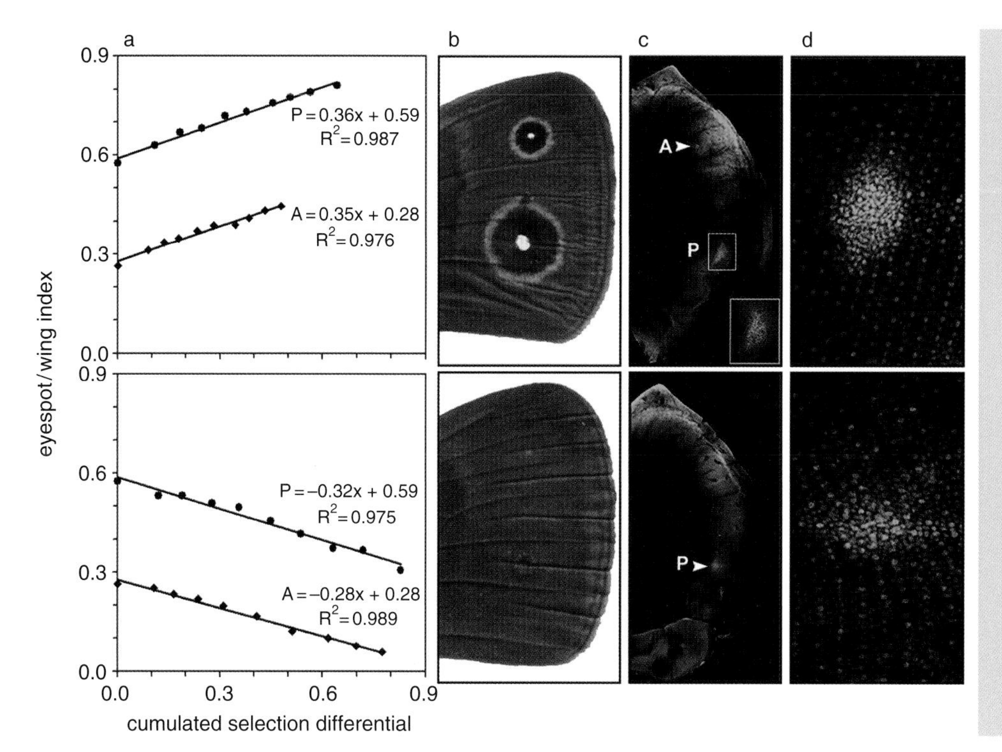

Figure 8.11 Comparison of the sizes of expression domains of the gene *distal-less* in the imaginal discs of late larvae (c, d) that will turn into adults (b) with larger (top) and smaller (bottom) eyespots. Also shown is a representation (a) of the accumulated amount of selection on eyespot size. (From Beldade *et al.* 2002, *Nature*, **416**: 844–847.)

sense 'easier' than selecting for larger anterior spots and smaller posterior ones or the converse pattern[54]. This result has contributed to the debate about the role of **developmental constraint** in evolution, but we will defer consideration of that until Chapter 13.

The pattern of butterfly wing eyespots (and wing pigmentation more generally) has long been a popular system for investigating evolutionary change, but until the 1980s the studies concerned were only able to deal with the adult **phenotypes**. Since then, however, much has been learned about the developmental-genetic basis of eyespots. One gene that is involved in eyespot formation is *distal-less*[55] – a gene first discovered (and named) because of its role in leg formation in *Drosophila*.

Lepidopterans, like dipterans, are **holometabolous** insects in which the adult develops from a series of **imaginal discs** embedded in the **larva**. The cells of these discs proliferate during metamorphosis, producing adult structures. Circular regions of expression of *distal-less* can be seen in late larval-stage wing discs, and these expression domains are larger in larvae that will give rise to adults with larger eyespots (Fig. 8.11).

So, in this study, it is possible to connect gene expression during development with the pigmentation pattern of the adult. It is also possible to connect intraspecific with interspecific variation – the changes produced artificially within one species of *Bicyclus* show how differences among species can arise naturally in longer-term evolution. But what of the adaptive value of eyespots? In other words, when the selection involved is natural rather than artificial, what is its basis?

There is general agreement that the adaptive value of eyespots on butterfly wings relates principally to the ecological interaction between butterflies (as prey) and

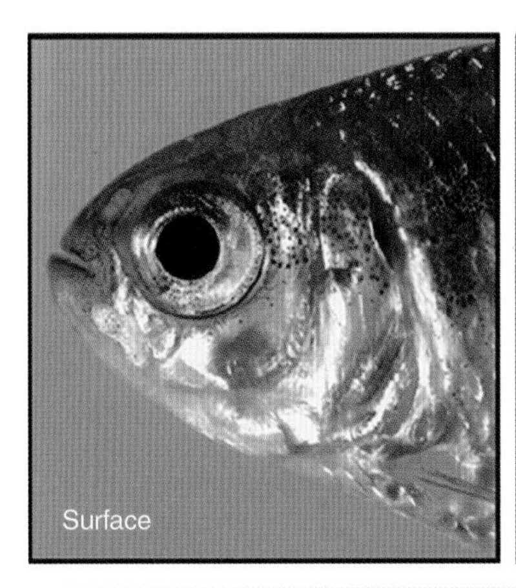

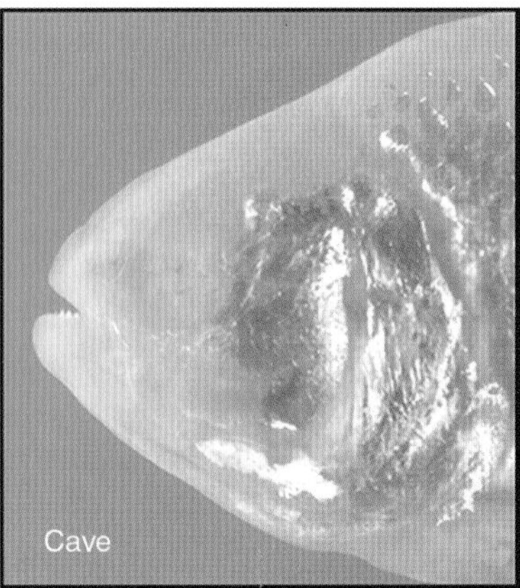

Figure 8.12 Morphology of surface-dwelling and cave-dwelling specimens of the species *Astyanax mexicanus* (Photographs courtesy of William Jeffery).

visually-searching vertebrates, usually birds (as predators). However, the exact reason why eyespots may be advantageous in terms of reducing successful predatory attacks, and hence predator-induced mortality, varies[56]. In cases such as *Bicyclus*, the eyespots are thought to attract predators to peripheral parts of the wing, where the damage sustained is less, and often survivable. However, in other butterflies and moths, much larger eyespots are thought to deter bird predators by resembling the eyes of a creature, such as an owl, that might itself eat the would-be predator – there is a group of butterflies called 'owl butterflies', which display just such eyespots.

8.5 Heterometric Compensation

The above example shows both kinds of heterometry – positive and negative. So does the next one, but in a different context. Before, it was the same character – eyespot size – going up or down in different selected lines in the lab or in different lineages in the wild. Now, we will look at an example in which one character decreases and another one increases. Specifically, we will look at the reduction or loss of eyes in blind cave fish, and the enhancement of other features, such as taste buds, which in a sense compensates for the reduction/loss of eyes.

More than 80 teleost fish species have become obligate cave-dwellers. Also, at least one species, the Mexican cave-fish *Astyanax mexicanus*, has both cave-dwelling and surface-water-dwelling forms (Fig. 8.12). In this species, it seems that many different populations have independently evolved the characteristic phenotype of cave-fish, including reduced/absent eyes and a loss of pigmentation. When crosses are done between geographically separate blind populations, some of the offspring are sighted, indicating that the genes involved in the reduction or loss of eyes are different in those separate populations.

Most of the research conducted so far on this fascinating creature has been focused on the reduction/loss of eyes and pigmentation. But it is known that several characters have increased, while eyes have been reduced. Many of these relate to enhanced olfactory ability. This is logical because in a totally dark environment, in which eyes are redundant as sensory organs, the other senses, and in particular smell/taste, are of greater importance. Here is a list, though not an exhaustive one, of characters which have been enhanced in individuals of the blind cave populations: larger jaw size; more teeth; more taste buds; larger nostrils; and increased size of the olfactory bulb of the brain[57].

The selective value of these enhancements seems clear enough. What is less easy to see, both in this example and in others, is the reason why eyes and pigments are reduced or lost. There are, in general, several competing (though not mutually exclusive) hypotheses for character loss:

1 There may be antagonistic **pleiotropy** – that is, the genes which enhance some features may also reduce others.
2 It may be that, given a limited supply of energy, it is advantageous for a developing organism to divert energy away from making an unnecessary structure so that it can be used in the development of other structures.
3 Any organ – whether an eye or something else – is subject both to attack by parasites and to possible pathological conditions, such as tumour formation. If the organ has zero benefit, these possible problems constitute negative selection.
4 Even if none of the above applies, the genes responsible for a redundant organ or structure will, like other genes, be subject to mutation. The vast majority of mutations are detrimental or selectively neutral rather than advantageous. In the absence of positive selection, such mutations may accumulate.

With regard to the genes involved in vertebrate eye development, two key players[58] are *sonic hedgehog* and *pax6*. Developmental studies have shown that the expression domain of *shh* in the anterior midline of cave-fish embryos is enlarged. Since *shh* signalling has a negative effect on *pax6*, the upregulated *shh* expression is at least partly responsible for the reduction/loss of eyes, because *pax6* controls the development of the optic cup.

Although many of the obligate cave-dwelling teleost species evolved their characteristic phenotype many millions of years ago, the cave populations of *Astyanax* are of relatively recent origin. The first Mexican populations – surface-dwellers – appeared after the emergence of the Panama isthmus some 3 MYA. So the cave populations are younger than this, though exactly how much younger is not yet clear.

Many other examples of heterometric compensation are known. For example, in some beetles horns can be produced – these are often restricted to males, and are thought to confer an advantage in terms of sexual selection. However, when horns are present, neighbouring structures in the head, including eyes, are reduced in size. Often, the phrase 'trade-off' is used in this context.

The compound nature of developmental repatterning has been repeatedly emphasised through the last three chapters, and will be expanded on in Chapter 10. One remark about it which is appropriate here is that heterometry will nearly always imply some degree of heterotopy, because if something is increased or decreased, then it will normally take up more or less space. An exception to this is where the increase/ decrease is in the concentration of some developmental agent, such as a **hormone** travelling via the bloodstream, where the volume of blood remains the same.

In contrast to the expectation that heterometry would usually involve some degree of associated heterotopy, there need not be any associated heterotypy (altered type: Chapter 9). In elongate geophilomorph centipedes, for example, closely-related species often have different ranges of segment number. However, if one species typically has (say) four more segments than another, the extra segments are not of different types. Indeed, the centipede segmental body plan is often referred to as having homonymous segmentation, because most of the segments are more or less the same as each other.

SUGGESTIONS FOR FURTHER READING

Since heterometry is a term with a very recent origin, unlike heterochrony and heterotopy, both of which were coined by Ernst Haeckel in the 19th century, you will not find many items of literature with this term in the title. However, you most certainly *will* find many interesting sources giving more detail on the examples of heterometry discussed herein, even though they will not use that term.

The evolution of human brain size, and indeed human evolution in general, has a rich literature, some of it very accessible to students. Examples are:

Lewin, R. 2005 *Human Evolution: An Illustrated Introduction*, fifth edition. Blackwell, Malden and Oxford.
Wood, B.A. 2005 *Human Evolution: A Very Short Introduction*. Oxford University Press, Oxford.

With respect to the evolution of butterfly pigmentation, the larger field in which studies of eyespot evolution belongs, a good review is:

Beldade, P. and Brakefield, P.M. 2002 The genetics and evo-devo of butterfly wing patterns. *Nature Reviews Genetics*, **3**: 442–452.

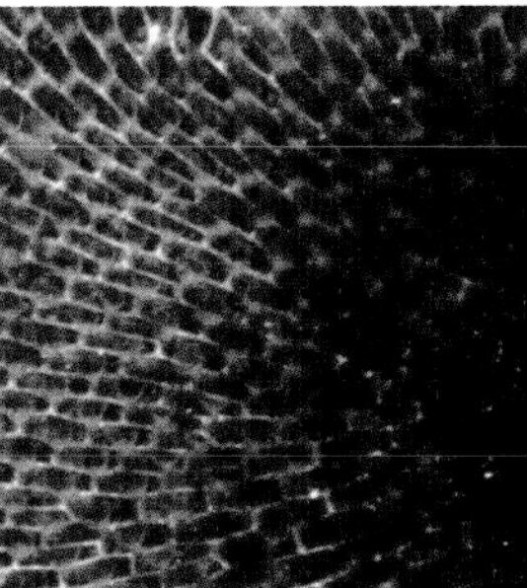

CHAPTER 9
Heterotypy

9.1 What is Heterotypy?

This particular category of **developmental repatterning** is, in a way, complementary to the other three, all of which involved the same 'thing' being produced at different times, in different places or in different amounts. But heterotypy involves an altered 'thing', which is at least potentially produced in the same time, place and amount as its predecessor.

Heterotypy is perhaps easiest to picture when this 'thing' is a protein. So we will start at the molecular level and work our way upwards. A good molecular example of heterotypy, which has been much studied from an evolutionary perspective, is the variant form of haemoglobin that causes the disease sickle cell anaemia in humans. In terms of developmental cascades, haemoglobin is a protein that falls right at the end: other genes **upstream** of it cause it to be switched on in certain cells (red blood cell precursors; recall that the mature red blood cell of most mammals has no nucleus and thus no genes). In the next section we will look at heterotypy in a gene that is more **upstream**, and hence is a developmental, rather than a **terminal target**, gene.

The adult human haemoglobin molecule is a tetramer, consisting of two alpha and two beta polypeptide chains (Fig. 9.1). These are produced by the alpha and beta haemoglobin genes, which in humans are on different chromosomes (16 and 11).

These genes, along with other human haemoglobin genes (those for embryonic and foetal haemoglobin) appear to have arisen by duplication and divergence from a

Figure 9.1 The structure of the adult human haemoglobin molecule. As can be seen, the molecule is a tetramer made up of two alpha chains and two beta chains. There are also iron-containing haem groups shown in green.

single original haemoglobin gene. Their polypeptide products thus have much similarity in sequence, and are of approximately, but not exactly, equal length. The alpha and beta chains in the human haemoglobin molecule are 141 and 146 amino acids long. In people suffering from the disease sickle cell anaemia, the alpha chain is normal and the beta chain is altered in just one place – position 6 – where there has been a substitution of a valine for a glutamine residue. This single change of one amino acid for another has a major effect. The mutant haemoglobin, often called HbS, forms quasi-crystalline rods, which distort the shape of the cell from its normal wheel-shape to an elongate sickle-shape (Fig. 9.2). These cells are less good at transporting oxygen than normal ones, and many affected individuals die as a result.

Note that this problem is caused by the wrong *type* of amino acid. The variant haemoglobin molecule is present in the same place as its normal equivalent (the red blood cells), for the same period (throughout life after it replaces foetal haemoglobin) and in approximately the same amount. So this is clearly a case of heterotypy.

Sickle cell anaemia normally features in population genetics texts rather than those of evo-devo. The reason is that this system provides one of the best examples of the population-level phenomenon of **heterozygous advantage**. In malarial regions of the world, notably parts of sub-Saharan Africa, heterozygotes are resistant to malaria, because their red cells provide a less good home for the malarial parasite *Plasmodium* than do normal red cells. Also, they have enough normal haemoglobin beta to transport oxygen in the blood. So, where malaria is endemic, the **heterozygote** is the fittest of the three **genotypes**, with the normal **homozygote** second-fittest and the HbS homozygote the least fit. Such heterozygote advantage constitutes a form of stabilising (as opposed to directional) selection, and this can maintain a **polymorphism** indefinitely in a population.

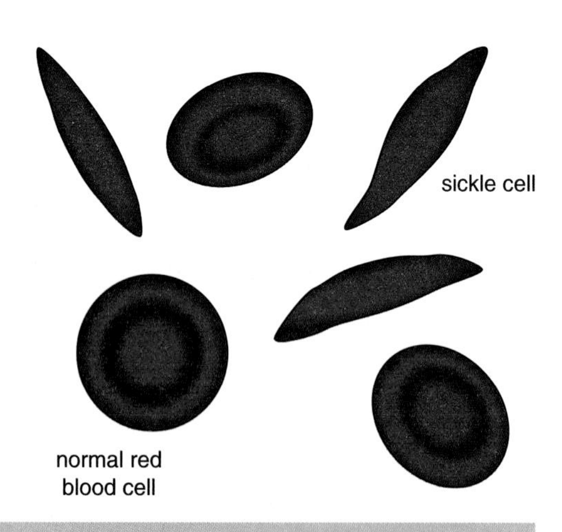

Figure 9.2 The shapes of normal and sickle-type human red blood cells.

If, instead of comparing normal haemoglobin beta with HbS in humans, we compare normal haemoglobin beta (or alpha) with their equivalents in other species of mammal, we see many cases of amino-acid substitutions. All of these are examples of heterotypy. Indeed, at the molecular level in general, what produces heterotypy rather than the other three types of repatterning is any change in the coding, rather than the regulatory, region of a gene. Such heterotypic changes at the molecular level may, however, go on to produce other types of change (e.g. heterochronic ones) at the morphological level. This point will be expanded on in Chapter 10.

9.2 Altered Products of Developmental Genes

To appreciate the difference between heterotypy and other sorts of developmental repatterning at this level, it is necessary to have at least an outline picture of how developmental genes interact to regulate each other's activities. We will now build such a picture (on the foundations laid in Chapter 3).

One approach to building this picture is to focus on a three-step cascade of interacting developmental genes, and in particular on the 'middle' gene in this sequence. From the perspective of this gene, the other two can be thought of as **upstream** and **downstream**, or as controller and target. It is true that in many cases any particular developmental gene may participate in different regulatory cascades, in different tissues or body regions, or at different times. But we will ignore such complications here.

So, focusing on a single cell, and on the interactions between these three genes, which we can simply call A, B and C for now, there is a series of interactions, as shown in Fig. 9.3. What is illustrated is a system in which genes A and B make **transcription factors**, while gene C makes a secreted protein. It is easy to imagine other combinations. The choice does not matter, because any combination will serve to show the main points.

Most **eukaryote** genes have a coding region, an upstream **promoter** region, which includes the transcription start-site, and other regulatory regions, sometimes called enhancers and silencers, depending on whether their effects are usually positive or negative. As noted previously, these regulatory regions can be 'upstream' or 'downstream' from the coding region, sometimes by distances of several kilobases, and may also be within the **intron** sequences that are interspersed with the coding **exons**, and are transcribed but not translated. In Fig. 9.3, one regulatory region per gene is shown for the sake of simplicity. The quote marks around 'upstream' and 'downstream' above are intended just as a gentle reminder that these terms used in the genetic sense (with regard to the direction of transcription) mean something very different than when they are used in the developmental sense (with regard to a cascade of gene-protein interactions): see Glossary.

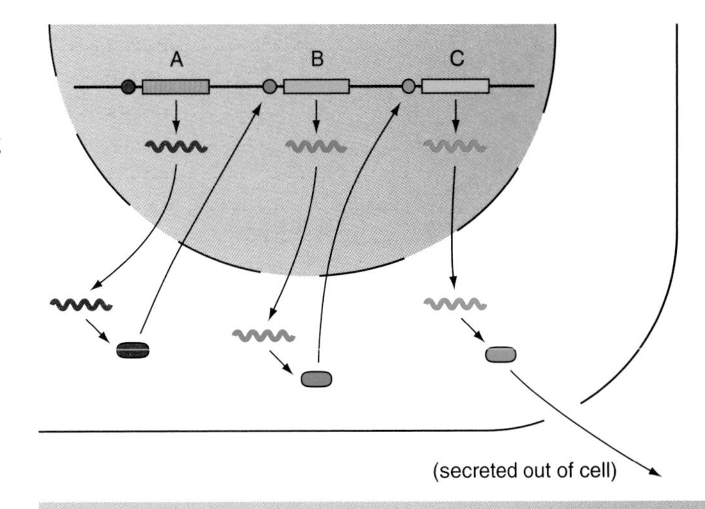

Figure 9.3 Generalised cascade of three interacting genes and proteins. Genes A and B make transcription factors; gene C makes a secreted protein.

These regulatory regions all fall into the category of **cis-regulators**. In general, the prefix '*cis*' means 'on the same side'; and in gene regulation it means 'on the same chromosome', usually with the additional proviso that, although it might be a few kilobases away, it is not *very* distant – for example, being at the other end of the chromosome.

The idea behind a *cis*-regulatory region is that it works via some mechanism that requires chromosomal continuity. Such a mechanism is not hard to envisage for a promoter region that is close to the coding region. We can imagine a large molecule such as **RNA polymerase** II, which transcribes protein-coding genes, binding to the promoter region and then moving along the coding region, catalysing the synthesis of heterogeneous nuclear RNA (hn RNA or pre-mRNA – the pre-intron-removal stage).

But how does a cis-effect work in the case of a regulatory region that is a few kilobases away? One mode of action for these long-distance cis-regulatory effects is through conformational changes in the **chromatin** within a section of chromosome. These may include looping of the DNA so that stretches of it, which are normally not in contact, come together.

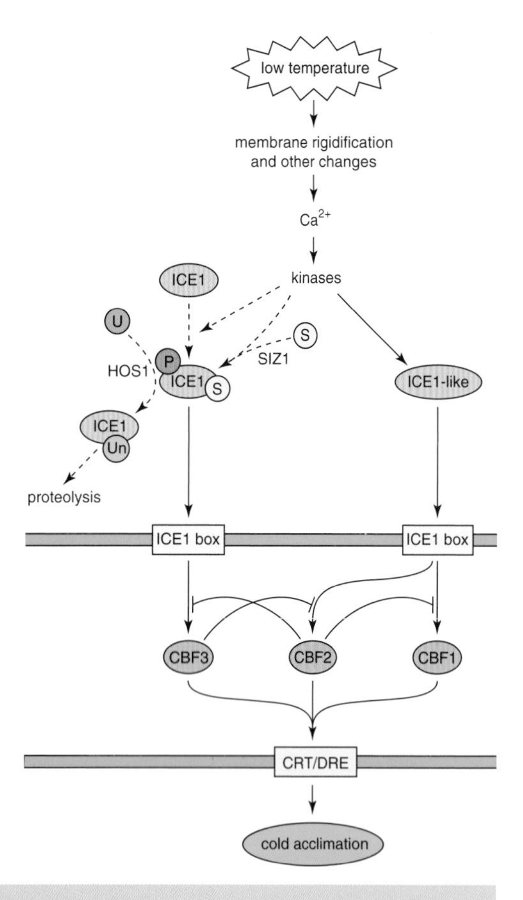

Figure 9.4 The transcriptional network that underlies cold-responsiveness in *Arabidopsis*. It starts with calcium ions and ends with transcription factors (CBFs) regulating gene expression. There are far more intermediates than shown in this simplified version. (Modified from Chinnusamy *et al.* 2007, *Trends in Plant Science*, **12**: 444–451.)

In contrast to cis-regulation, there is also **trans-regulation** – meaning a regulatory effect where a gene's expression is influenced by something that is not physically part of the same section of chromosome. An example is a transcription factor – a protein that derives ultimately from another gene (say, on another chromosome entirely), that was made by translation of that gene's mRNA in the cytoplasm, that migrates into the nucleus, and that attaches to the *cis*-regulatory site of the gene in focus. In the context of Fig. 9.3, the protein product of gene A is a *trans*-regulator of the expression of gene B. In turn, gene B's product is a *trans*-regulator of gene C.

It will be clear from the above that the phrases '*cis*-regulation' and '*trans*-regulation', which are often found in the literature, are misleading. Gene expression is regulated by an *interplay* between *trans*-acting agents and the *cis*-acting control regions to which they bind. So 'the norm' is for gene regulation to be both *cis* and *trans* all at once.

So far, this story has just been about a developmental process in a single organism. But we are interested ultimately in extending the story to visualise evolutionary changes in such processes. In particular, for the present chapter where the focus is on heterotypy, we are interested in how a change in a coding region produces developmental repatterning. We saw a change in the coding region of the haemoglobin beta gene in the previous section but, while this had physiological consequences, it did not have developmental consequences because that gene is at the end of a developmental cascade of interacting genes and proteins, not in the middle. So now it is time to look at another instance of an altered gene product, but in a more upstream gene (a real equivalent to one of the abstract genes shown in Fig. 9.3).

In *Arabidopsis*, as in other flowering plants from temperate regions, there is a process called cold-acclimation, in which exposure to low temperatures has the effect of increasing tolerance to subsequent freezing. It has been shown that several hundred genes are involved in this process[59]. Some of the molecular details are now known[60], and are shown in Fig.9.4.

A mutation in the coding region of one of the genes involved in 'cold-signalling' can have a marked effect on the result, in terms of the plant's ability to develop normally at low temperature. Specifically, a mutation in the gene *hos2* produces the result[61] seen in Fig. 9.5. The mutation is a single base substitution – C to T at position 541 from the start site. This changes the code GCT to GTT and thus inserts a valine residue in place of an alanine one.

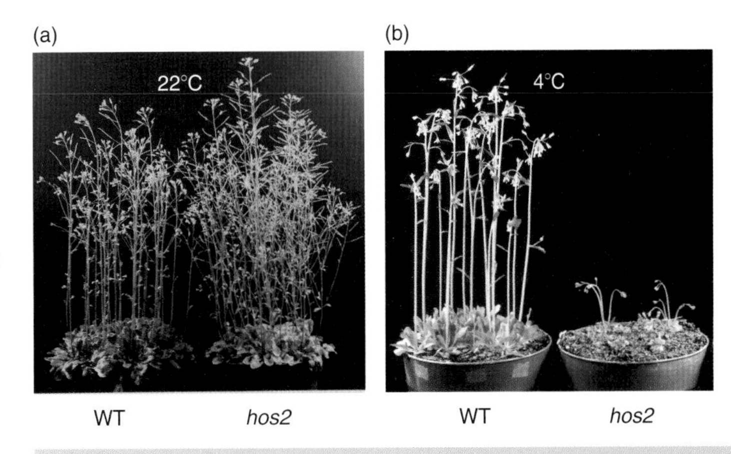

Figure 9.5 The developmental effect of the *hos2* mutation on *Arabidopsis* plants grown in the cold (b). Note the lack of a comparable effect of *hos2* when the plants are grown under warmer conditions (a). WT – wild-type. (From Xiong *et al.* 2004, *The Plant Journal*, **40**: 536–545.)

The effect on the plant shown in Fig. 9.5b suggests that the normal role of the Hos2 protein in cold acclimation is a positive one. However, the *hos2* mutant has been shown to up-regulate the CBP transcription factors (Fig. 9.4), which would indicate that the wild-type version of the gene represses them. This is the opposite of what would be expected. The explanation of this paradox is still awaited, but may lie in other effects of *hos2* on different components of the cold acclimation system. Regardless of the answer to this puzzle, the effect of a single amino-acid substitution in a coding region – i.e. minimal heterotypy of gene product – is to produce organism-level developmental repatterning, which includes a large change in the size of plants grown in the cold, as seen in Fig. 9.5.

9.3 Altered Pigmentation

A uniform colour change in an animal or plant provides another kind of example of heterotypy. We will now look at the colour of the shell of the land-snail *Cepaea nemoralis*, a member of the **gastropod** family Helicidae. As noted in Chapter 4, pigmentation in *Cepaea* has been much studied by ecological geneticists, primarily in the period from the 1950s to the 1970s. The shell pigmentation pattern exhibits many polymorphisms. These include the 'base colour' of the shell, and the presence or absence of a series of dark bands, which may also vary in number (Fig. 9.6). The adaptive significance of the variation in overall pigmentation pattern is thought to relate largely to predation by birds, with different shell morphs being **cryptic** to varying degrees against particular backgrounds[62]. However, it is also related to climate: darker shells have been shown to take up heat at a higher rate than paler ones, and to equilibrate at a higher internal temperature[63].

Unlike pigmentation patterns on butterfly wings (Chapter 8), the developmental genetic basis of shell pigmentation in *Cepaea* is not yet known. However, the formal genetics are understood, in terms of numbers of **loci, alleles** and their **dominance**.

Figure 9.6 Different colour morphs of the land snail *Cepaea nemoralis.* As can be seen, some shells also have dark bands superimposed upon the paler base colour. Most natural populations of this species are polymorphic for both base colour and banding pattern.

For example, yellow and pink unbandeds differ at a single locus; and the allele conferring pink pigmentation is dominant over that for yellow.

Unlike marine gastropods, which have indirect development via a planktonic **trochophore larva**, development in land-snails is direct, without a larval stage. So what hatches from the egg is a miniature snail. The shell of the hatchling is soft compared to that of later developmental stages, and is referred to as a protoconch. There may be differences between this and the rest of the shell in both micro-sculpturing and pigmentation. However, after hatching, the shell colour and the presence/absence of bands do not change: if a shell starts off yellow, it ends up yellow. We know this because a shell is a kind of preserved record of its own development, given the way it grows (Fig. 9.7). In *Cepaea*, the time from hatching to the cessation of growth at the adult stage is approximately three years.

Whatever the developmental genetic basis, the difference between pink and yellow unbanded shells represents heterotypy. As far as we know, nothing has changed in terms of timing, positioning or number. We simply have a difference in the type of pigment. Of course, the conclusion about which type of developmental repatterning is involved depends on precisely which comparison (between which different shell morphs) we make. The difference between pale pink and dark pink is **heterometry**; while the difference between five-banded and single-banded is both heterometry and **heterotopy**. As ever, developmental repatterning is complex: this will be the main theme for the next chapter.

Figure 9.7 A gastropod shell is effectively a preserved record of its own growth. The shell pictured has had two growth periods in successive summers punctuated by cessation of growth in the intervening winter. This punctuation can be seen as a discontinuity in the shell (top-left region of picture).

9.4 Altered Morphology and the Origin of Novelty

The term heterotypy means 'altered type'; the term novelty means 'new type'. There is no clear line of distinction between these two. Rather, there is a spectrum of magnitudes of alteration in type, somewhere along which we perceive a structure as new, and hence we call it an evolutionary **novelty**. Smaller changes we just perceive as modifications on an existing theme. Exactly where on this spectrum the transition from a modification to a novelty is perceived is a subjective issue – different biologists will not agree on the placement of this point of transition.

Because of this difficulty, it is hard to know what to deal with here, and what to defer to Chapter 17. We will focus on just one example here: the alteration of the anterior-most trunk segment in centipedes to form a pair of poison claws (Fig. 9.8). This choice is based partly on the fact that the evolutionary change concerned is from one type of segment to another type, in contrast to the origin of the turtle shell (Chapter 17), where it is not a case of changing one type of shell to another, but rather producing a shell in a **lineage** that started off without one.

Figure 9.8 The poison claw segment of a centipede. This type of segment is unique to this particular arthropod class (photograph courtesy of Michel Dugon).

We have already used centipede examples in other contexts (e.g. **heterochrony**: Chapter 6) and have already seen the pattern of their phylogeny. However, in relation to the evolution of the poison claw segment, this 'internal' phylogeny (of the order Chilopoda) is not particularly important, because all of the 3,000+ **extant** centipede species, and all extinct ones for which we have sufficient fossil evidence, possess poison claws. There are no known cases of loss of this defining character of the group. No animals outside the Chilopoda possess a similar poison claw segment. So this kind of segment appears to have arisen only once in evolution and, at least as yet, has persisted in all the lineages that have descended from the original 'ur-centipede'.

What *is* important in relation to the origin of the poison claw segment is the 'external' phylogeny of centipedes; in other words their relationship with other taxa, especially of myriapods, and the identification of which of these taxa is the **sister-group** of the Chilopoda.

There is not yet consensus on this issue. In fact, it is not even certain that the Myriapoda (normally thought of as an arthropod subphylum) is a **monophyletic** group, though this seems more likely than not. Assuming that it is monophyletic, the two most favoured patterns of relationship among its four constituent classes – centipedes, millipedes, pauropods and symphylans – are shown in Fig. 9.9. In one case, centipedes are the outgroup to the rest of the myriapods, whereas in the other they are the sister-group of the symphylans. One piece of evidence in favour of the former pattern is the position of the reproductive structures, with centipedes being the only 'opisthogoneate' (posterior reproductive openings) group of the four.

Whichever pattern of relationship is the true one, the lineage leading to centipedes evolved a poison claw segment, whereas the lineages leading to the other three myriapod groups did not. This relates to a dietary difference between centipedes and the other myriapods. Centipedes are universally carnivorous, while all the other myriapods are detritivores. Of course, such sweeping statements about the feeding ecology of groups containing hundreds or thousands of species are risky. Some juvenile centipedes may be partly herbivorous or detritivorous, especially before their poison claws are well-formed. And doubtless many grazing millipedes accidentally ingest tiny invertebrates buried in the decaying plant matter that is their main food. There are also a very few species of millipede that seem to be genuinely carnivorous. However, these are very minor exceptions to a surprisingly general rule. So it is clear that the origin of the poison-claw segment and the origin of a predatory ecology were related.

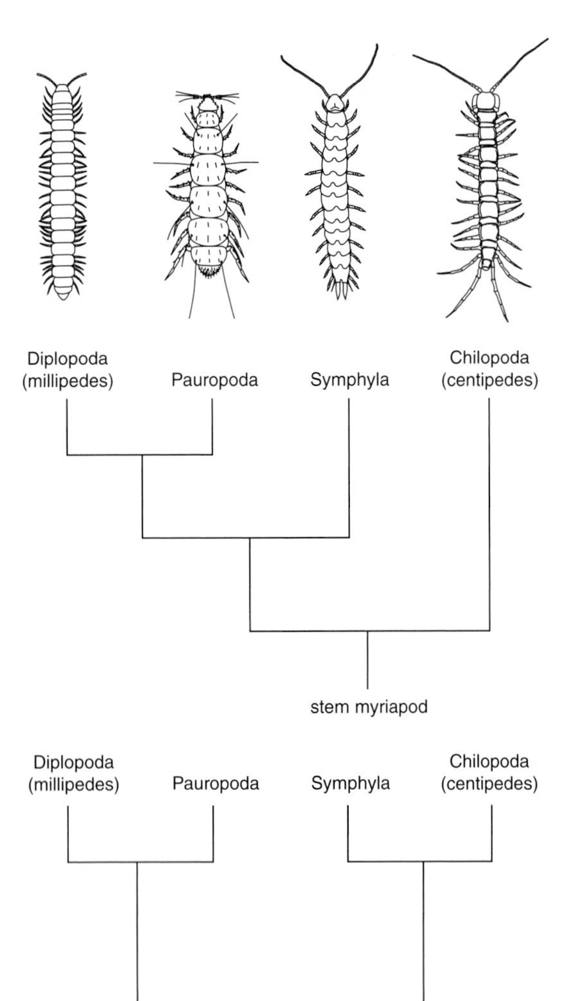

Figure 9.9 Two hypotheses of myriapod phylogeny. Centipedes are the only opisthogoneate group; the other three are progoneate. These terms refer to whether the reproductive openings are at or near the back or the front of the animal (in that order).

Unfortunately, the fossil record of myriapods is poor. The group is entirely terrestrial and its evolutionary origin is associated with an invasion of the land. It seems that myriapods were among the first land animals, and that their first appearance was not long after, and was no doubt enabled by, the invasion of the land by plants. In terms of geological periods, both the earliest known land plants and the earliest known myriapods[64] are from the Silurian Period (c. 443–417 MYA). The earliest fossils that are unequivocally centipedes[65] are from the subsequent Period, the Devonian, and date to about 375 MYA. So we can bracket the evolutionary origin of the centipede poison-claw segment with these dates, but cannot be any more precise than that. Whether the poison claws evolved early or late within this period, and whether the segment concerned was transformed rapidly or slowly, remains unknown.

It is hard, perhaps impossible, to produce heterotypy at the level of morphological structures (as opposed to the level of gene sequence or pigmentation) without other sorts of developmental repatterning occurring too. How can a structure become altered in type if it remains the same in its position in the body, its period of development and its size? However, this difficulty has been with us in various guises from Chapter 6 onwards: few cases of repatterning are 'pure' in the sense of belonging exclusively to one of the four types.

Interestingly, at the level of gene expression, the poison claw segment has a unique **Hox gene** code[21]. In other words, the combination of Hox genes expressed in this segment (three of them) is different from that of any other segment in the animal (Fig. 9.10). This expression pattern must have arisen via heterotopy – in the form of changes in the antero-posterior zone of expression of certain Hox genes compared with the LCA of centipedes and other myriapods.

At the morphological level, though, there is no significant heterotopy, as the segment has not migrated along the antero-posterior axis. Also, there is no significant *early* heterochrony: the poison-claw segment is indistinguishable from neighbouring ones during early **embryogenesis**, as shown in Fig. 9.11. Its

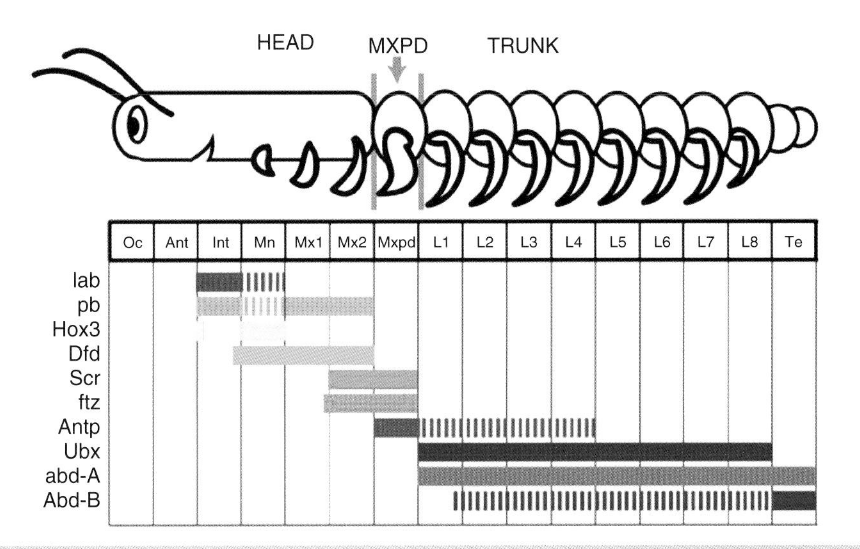

Figure 9.10 Diagram of the combinations of Hox genes expressed in different segments of a centipede. Note that the poison-claw segment is characterised by a unique three-gene pattern of expression. Poison claws are sometimes called maxillipeds (Mxpd). In the Mxpd segment, the three Hox genes *scr*, *ftz* and *antp* are expressed. This particular combination is not found in any other segment. (From Hughes and Kaufman, 2002, *Development*, **129**: 1225–1238.)

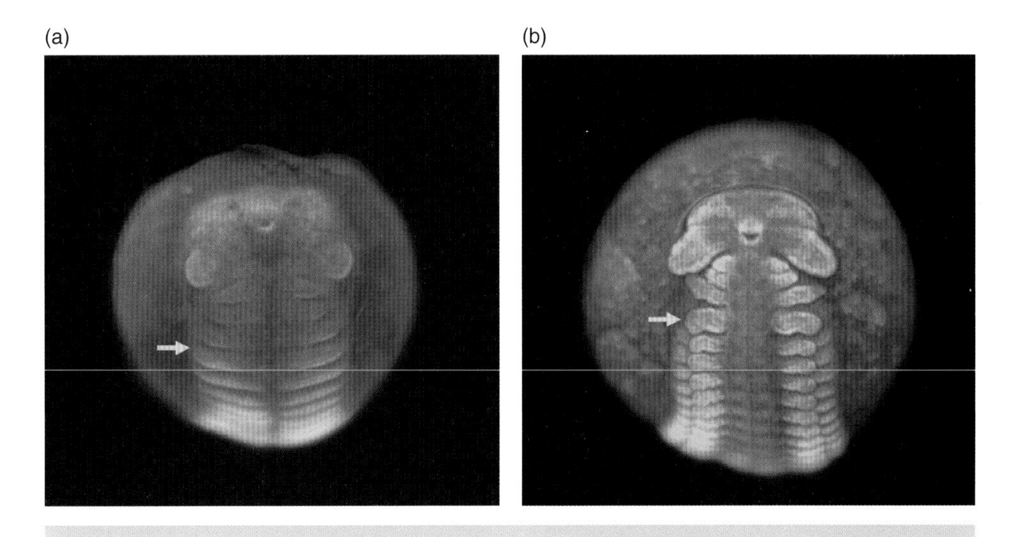

Figure 9.11 Early centipede embryos showing (a) a stage at which the poison claw segment is no bigger than others around it and (b) a later stage in which it is beginning to be larger than those others (see arrow) (photographs courtesy of Michel Dugon).

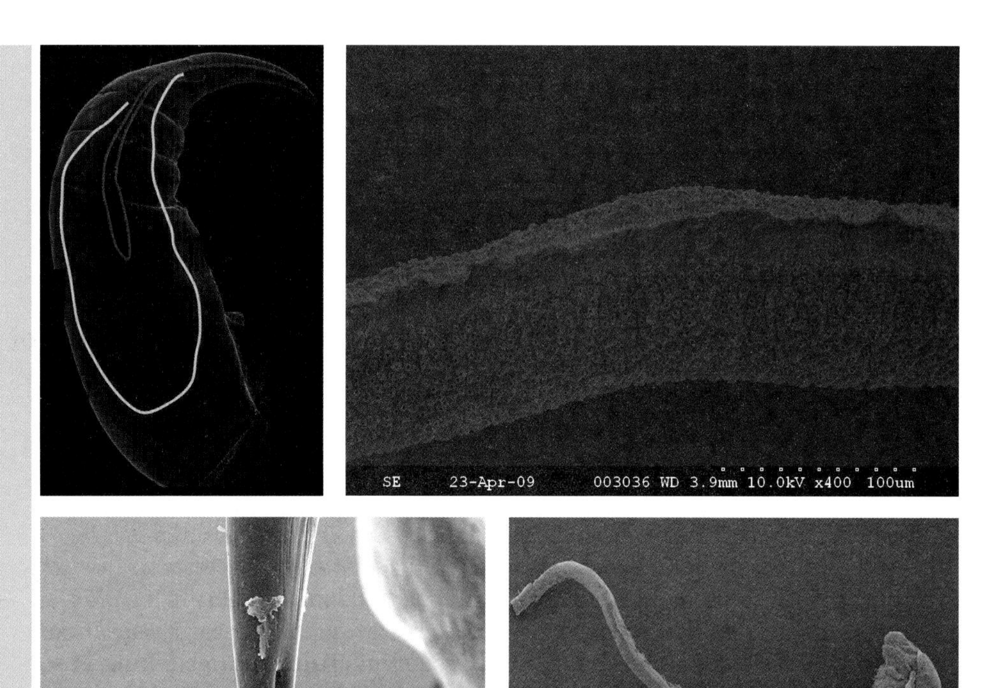

Figure 9.12 The internal structure of an adult centipede's poison claw. Note the poison gland (yellow) and duct (red) through which the poison passes (top left), and the syringe-like tip used to inject the poison into prey animals (bottom left). Also shown are the dissected-out duct (top right, high magnification) and gland + duct (bottom right, lower magnification). (Photographs courtesy of Michel Dugon).

development has not been speeded up relative to others during early development, despite the fact that it ends up bigger. But ultimately the size (especially of the appendages) does increase, which implies later-stage heterochrony, and by definition constitutes positive heterometry.

In functional terms, though, what is important about the poison claws of centipedes is their ability to inject toxin into prey animals. This is achieved by a system of poison-secreting cells that surround a lumen, the calyx, which leads to a tube that ends at a syringe-like opening near the tip of the claw (Fig. 9.12). This cell type presumably did not exist in the LCA of centipedes and other myriapods. Its appearance was an important element in the evolution of the whole poison-claw system. And it reminds us that heterotypy at the level of the cell is important, in additional to heterotypy at the levels of gene and organism: indeed it connects these two with each other.

9.5 The Origin of New Cell Types

Often, particular groups of multicellular organisms are associated with a particular cell type that seems to be unique to them, and so must have evolved from some other cell type at or near the point of origin of the group concerned. An example from a

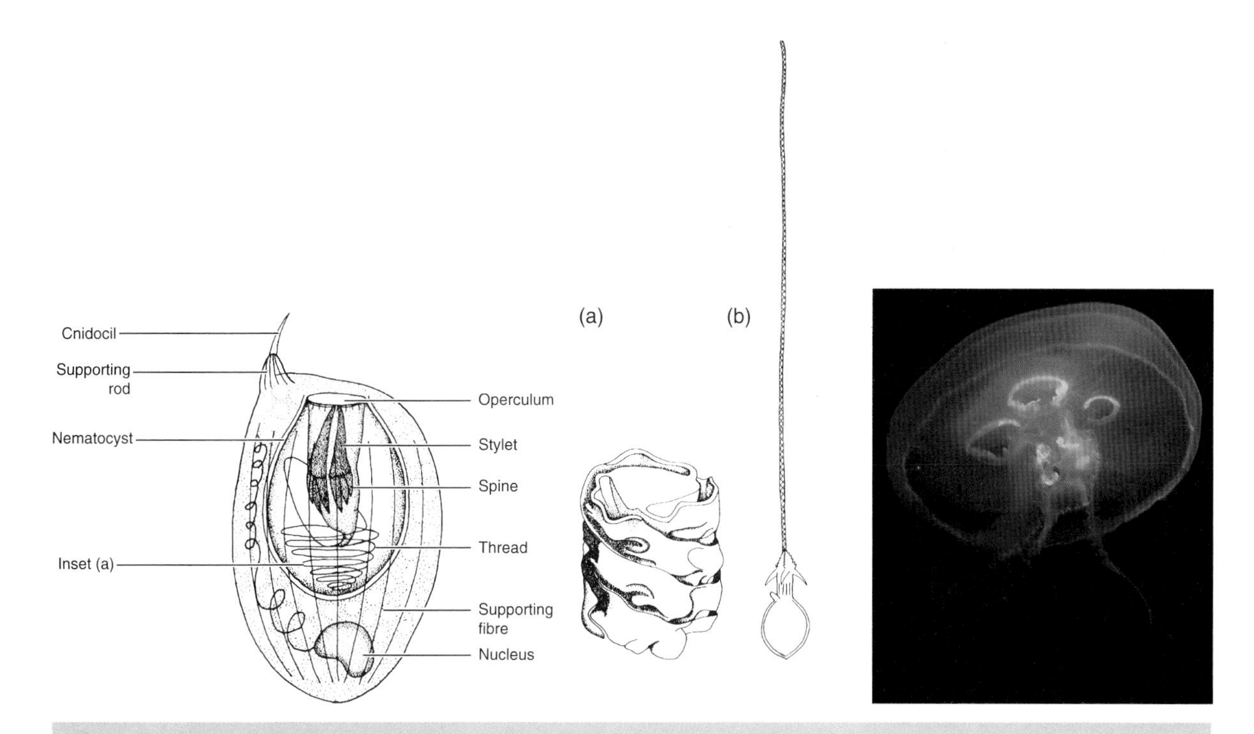

Figure 9.13 The stinging cell, or nematocyte (left), possessed only by members of the phylum Cnidaria, for example, jellyfish (right). The left-hand picture of the nematocyte shows several features, including the sensory cnidocil, which, when touched by prey, causes eversion of the 'thread' (or 'tube'), which delivers toxins to the prey. (a) Coiled thread, high magnification; (b) Everted thread, lower magnification.

basal animal group is the cnidarian nematocyte. The phylum **Cnidaria** contains about 10,000 species of jellyfish, sea anemones, hydroids and corals. These have stinging cells called nematocytes (or cnidocytes), which are used to poison prey organisms (Fig. 9.13). No other animal group possesses such cells.

Another example of the association between possession of a particular cell type and membership of a particular animal group is that between endothelial cells and vertebrates[66]. The heart chambers and blood vessels (both large and small) of vertebrates are lined with a layer of endothelial cells (Fig. 9.14). Even the most primitive extant vertebrates, the lampreys, have endothelial cell linings of their blood vessels and hearts. In contrast, endothelial cells are not, in general, found in the circulatory systems of invertebrates. However, it appears that the cephalopod molluscs, such as the squid and octopus, may provide an exception to this general rule[67]. This may be a case of **convergent** evolution of a closed circulatory system and a cell type that constitutes a part of it.

The split between the vertebrates and the other chordate lineages occurred at least as early as the Cambrian (>525 MYA; Chengjiang site in China). So the vertebrate endothelial cell must have originated at least as long ago as this. It seems to have been retained in all vertebrate species, presumably due to its positive influence on the overall

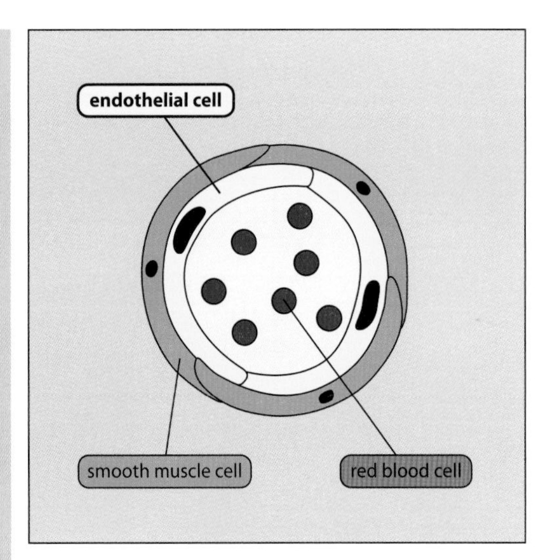

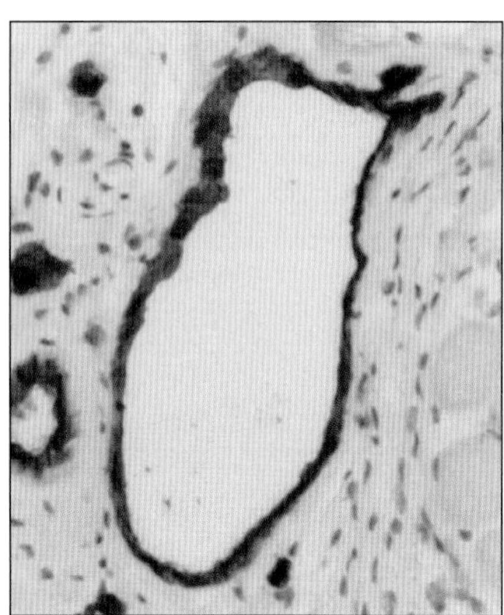

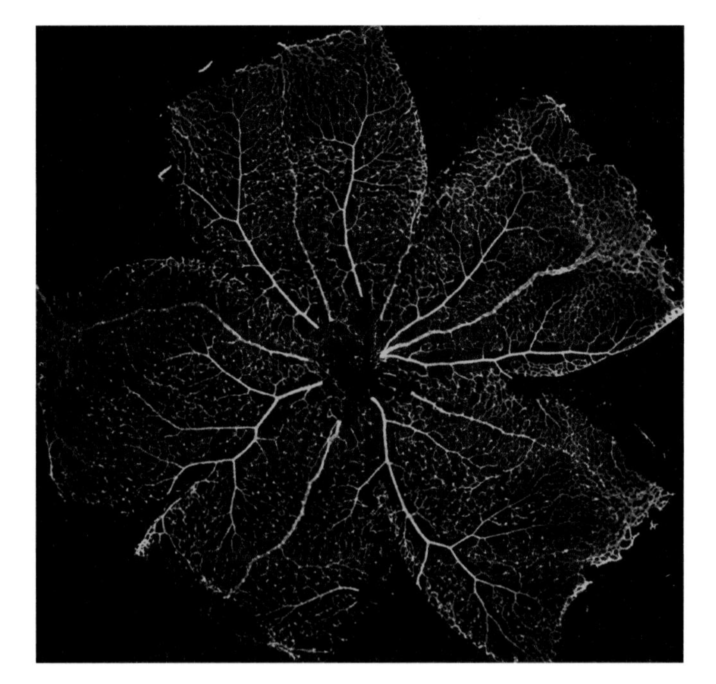

Figure 9.14 Endothelial cells lining blood vessels in a mouse. This lining arrangement applies to all vertebrates but not to non-vertebrate chordates. Top left: Diagrammatic cross-section of vessel. Top right: Actual cross-section (endothelial cells stained brown). Bottom: Panoramic view of the network of vessels in a mouse retina (endothelial cells stained green) (photographs courtesy of Helen Arthur).

integrity of the vasculature. It is associated with two other evolutionary events, one morphological, the other molecular. The former is the origin of hearts with two or more chambers; the latter is the proliferation of a new protein family – the Xins[68] – that appears to be essential for the correct development of all vertebrate hearts.

The same problem of finding a line to separate the appearance of 'altered types' and the origin of 'novel types' that we encountered in relation to morphology also besets our interpretation of the evolution of cell types. Endothelial cells did not suddenly appear *de novo* in the vertebrate lineage, but rather must have evolved from some previously-existing type of cell. It has been proposed that endothelial cells may have originated, in evolution, from another (mobile) type of blood cell called an amoebocyte, versions of which are found much more widely in the animal kingdom than just in the vertebrates.

SUGGESTIONS FOR FURTHER READING

As with the previous chapter, you will not find many references to the material of this chapter under the chapter title word – heterotypy. This is because, like heterometry, this word has a very short history in evolutionary biology, unlike heterochrony and heterotopy, both of which were introduced in the 19th century. However, you will nevertheless find good reviews of altered types of parts of organisms, for example new cell types.

For a discussion of the origin of a particular new type of cell in evolution, see chapter 5 in:

Gerhart, J. and Kirschner, M. 1997 *Cells, Embryos and Evolution*. Blackwell, Malden, MA.

For a more general discussion of the origin of new cell types in both evolution and development, see the following review article:

Vickaryous, M.K. and Hall, B.K. 2006 Human cell type diversity, evolution, development, and classification with special reference to cells derived from the neural crest. *Biological Reviews*, **81**: 425–455.

Also, note that since 'altered type' and 'new type' are not clearly separated from each other, the further reading in Chapter 17 on evolutionary novelties is also relevant here.

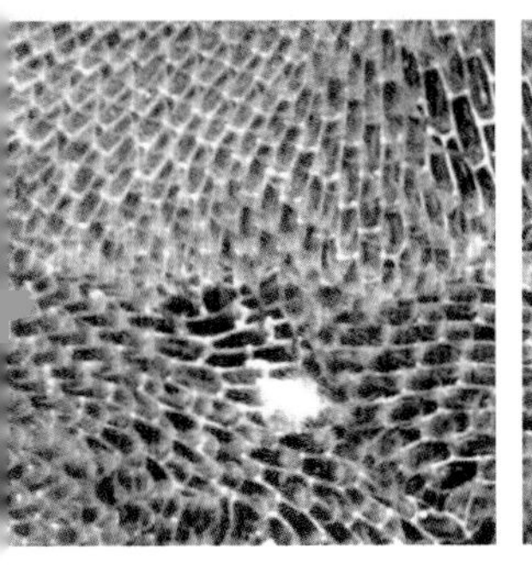
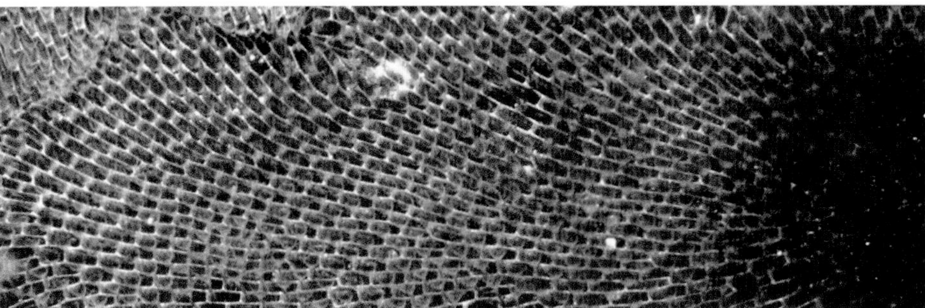

CHAPTER 10
The Integrative Nature of Repatterning

10.1 Repatterning is a Complex Process

Throughout the last four chapters, our approach, when dealing with each case study, was to focus on the 'main' kind of repatterning – change in timing (Chapter 6), spatial arrangement (Chapter 7), amount (Chapter 8) or type (Chapter 9). However, we also noted in each of those chapters that repatterning is rarely 'pure' in the sense of consisting of one kind *only*. We should expect it to be the norm rather than the exception that instances of **developmental repatterning** will involve two, three or even all four kinds of change, especially if we look at how the repatterning is manifested at different levels of organisation – gene, cell, developmental pathway and resultant morphology.

In the present chapter the focus is, in a sense, reversed. Instead of trying, as far as possible, to focus on the 'main' kind of repatterning, and thus to simplify matters, we will focus on complexities of various sorts. This new focus will include examining the compound nature of many instances of developmental repatterning. But it will also include other elements. And the first of these (Section 10.2) tells a cautionary tale: however useful, indeed *necessary* for the purpose of study, is a structured approach in which we can distinguish between **heterochrony, heterotopy, heterometry** and **heterotypy**, the developmental system of an animal or plant does not necessarily make such a distinction. Sometimes, as we will see, organism-level development can respond in the same way to different kinds of molecular-level repatterning.

10.2 Different Kinds of Repatterning can Produce a Similar Result

We will examine a study that emphasises this point using the Sonic Hedgehog signalling system in vertebrate embryos that we examined in Chapter 3 (recall, in particular, Fig. 3.16). In the vertebrate lineage, **gene duplication** events have produced a family of at least three *hedgehog* genes – *sonic, indian* and *desert hedgehog*. The first of these, which we will now discuss further, has been the most studied.

Sonic Hedgehog signalling is involved in many developmental processes in vertebrates, including the following: early development in the midline dorsal region involving the **notochord** and **neural tube** (as we saw in Chapter 3); determination of left-right asymmetry[69]; limb development; formation and maintenance of the gastro-intestinal tract[70]; and formation of teeth (as we saw in the case of both snakes and rodents in Chapter 7). And this list is not exhaustive. So, clearly, Sonic hedgehog (Shh) signalling is of vital importance in the development of vertebrates of all types, from **teleost** fish such as the zebrafish (which has undergone an additional gene duplication, giving *shh-a* and *shh-b*) to humans and other mammals.

We will focus on the dorsal midline signalling role of Shh. Here, cells of the notochord secrete Shh protein, which forms a concentration gradient in the vicinity – the further from the notochord the lower the concentration. Immediately dorsal to the notochord is the forming neural tube, whose cells will differentiate in various ways. For example, ventral neural tube cells become 'floorplate cells', while lateral ones give rise to motor neurons (Fig. 3.16). The conventional wisdom is that neural tube cells react to Shh concentration in a threshold manner.

Recently, however, it has been shown that increased duration of exposure to Shh protein can have the same effect as increased concentration[71]. In other words, molecular heterometry and heterochrony in this system can produce similar results. This conclusion was arrived at through an experiment with chick embryos, as follows:

Using neural tube tissue in culture, its cells were exposed for a fixed time period to varying concentrations of Shh in one experiment and to a fixed concentration for varying time periods in another. The response looked for was not at the cellular level (as described above) but rather at the molecular level. In particular, the number of neural tube cells expressing two **transcription factor** genes that are switched on by Shh was monitored. The results are shown in Fig. 10.1. Note that in addition to the positive effect of Shh on both transcription factors, Olig2 and Nkx2.2, the latter also has an inhibitory effect on the former (Fig. 10.1a). The pattern of change in the number of positive cells for each transcription factor reflects this, because when the level of Nkx2.2 gets high enough, this has a negative effect on the level of Olig2 (Fig. 10.1b).

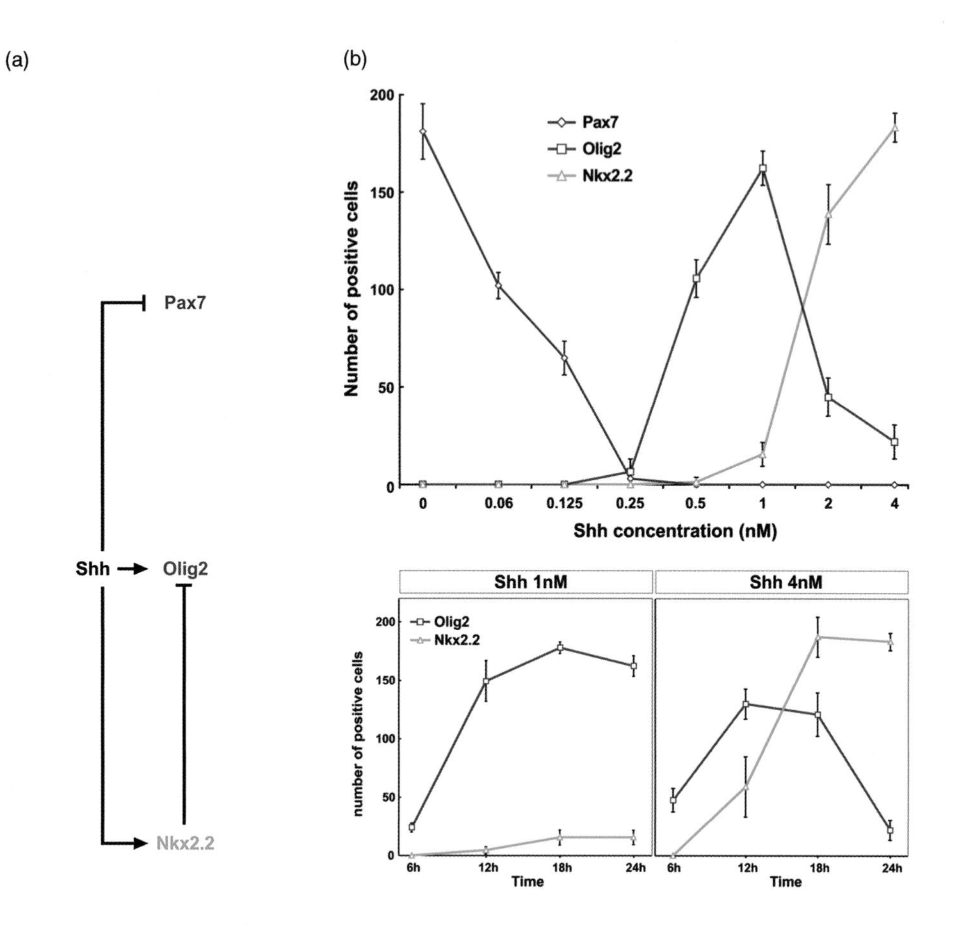

Figure 10.1 Genes downstream of Shh signalling can be influenced by both the concentration of Shh and the duration of exposure to it. (a) The pattern of activating and inhibiting effects. (b) Numbers of positive cells for downstream genes in relation to concentration and time. See text for further explanation. (Reproduced from Dessaud *et al.* 2007, *Nature*, **450**: 717–720.)

Figure 10.1b deserves close scrutiny. Note the different *x*-axes – concentration in one case and time in the other. The only overlap in the two experiments is the treatment wherein neural tube cells are exposed to a 4 nM concentration of Shh for 24 h; and the results of the two runs of this treatment are similar, as expected (RHS of top, and bottom right, graphs). More interesting is the cross-over point where the ascending number of Nkx2.2-positive cells overtakes the declining level of Olig2-expressing ones. This occurs at an Shh concentration of about 1.5 nM under the fixed exposure time of 24 h (first experiment); but it occurs at a time of about 15 h in the fixed concentration of 4 nM (second experiment). So, using this feature as one way of looking at the response to Shh, the combinations of 4 nM for 15 h and 1.5 nM for 24 h produce similar results.

A complication here is that, beyond a certain point, cells seem to become desensitised to Shh signalling. Note that in the 'variable duration' experiment, the number of Nkx2.2-positive cells at 24 h is more or less the same as it is at 18 h, despite the earlier rapid rise in the number of positive cells. Clearly, the dynamics of such systems are highly complex.

Finally, we should note that this example strains the definitions of heterochrony and heterometry somewhat. When these terms were defined earlier (Sections 6.1 and

Section 8.1), the context was two different **lineages**. Yet here the context is two different experiments. We will shortly (Section 10.6) look at the different kinds of causality of developmental repatterning.

10.3 Compound Repatterning at a Single Level of Organisation

One of the major developmental and morphological changes that occurred in the evolutionary lineage that led to mammals, as it departed from a **sister-group** that for the moment we will call the reptile/bird **clade**, was the modification of the jaw joint[72] and, associated with this, the origin of the mammalian ear ossicles (or ear-bones). These are the three small bones (malleus, incus and stapes) that transmit sound from the eardrum to the inner ear in humans and other mammals. The human jaw joint (Fig. 10.2), and the mammalian jaw joint in general, involves an articulation between the cranium and the lower jaw (mandible). At first sight, there would seem to be no reason why there should be a connection between these two things – one is a component of our hearing system, the other of our food-acquisition system.

However, over a period of more than a century, many studies have been carried out which link these two things, including studies of comparative morphology, embryology, palaeontology and, most recently, comparative developmental genetics. We will now examine the results of these studies, which reveal that this evolutionary change involved at least three kinds of developmental repatterning – heterotopy, heterometry and heterotypy.

In mammal embryos, the malleus and incus originate from the same structure as the mandible and maxilla – the first pharyngeal arch. During **embryogenesis**, they separate from the jaw region and migrate into the area that will become the ear. This suggests an evolutionary origin from jaw bones.

Extant reptiles, and probably the stem **amniotes** from which they arose, have a single ear ossicle – the stapes – and a different arrangement of jaw bones than that found in mammals (Fig. 10.3). Notice that, in this figure, there is a small bone called the articular that is part of the lower jaw, and a bigger one called the quadrate that is part of the upper jaw. These are the bones that evolved into the malleus and incus of mammals (the articular becoming the malleus, the quadrate becoming the incus).

As evolutionary changes go, this one has an excellent fossil record. Many stem-group 'mammal-like

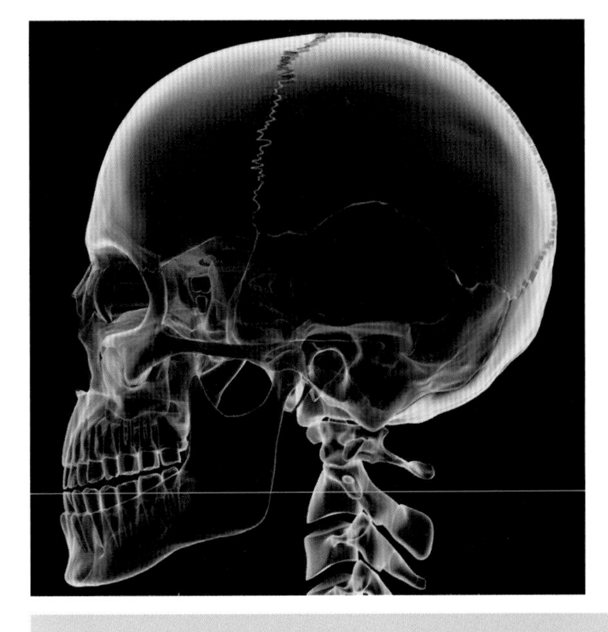

Figure 10.2 The human jaw joint as an example of the mammalian condition. The lower jaw (mandible) articulates not with the upper jaw (maxilla) but rather with a cranial bone called the temporal.

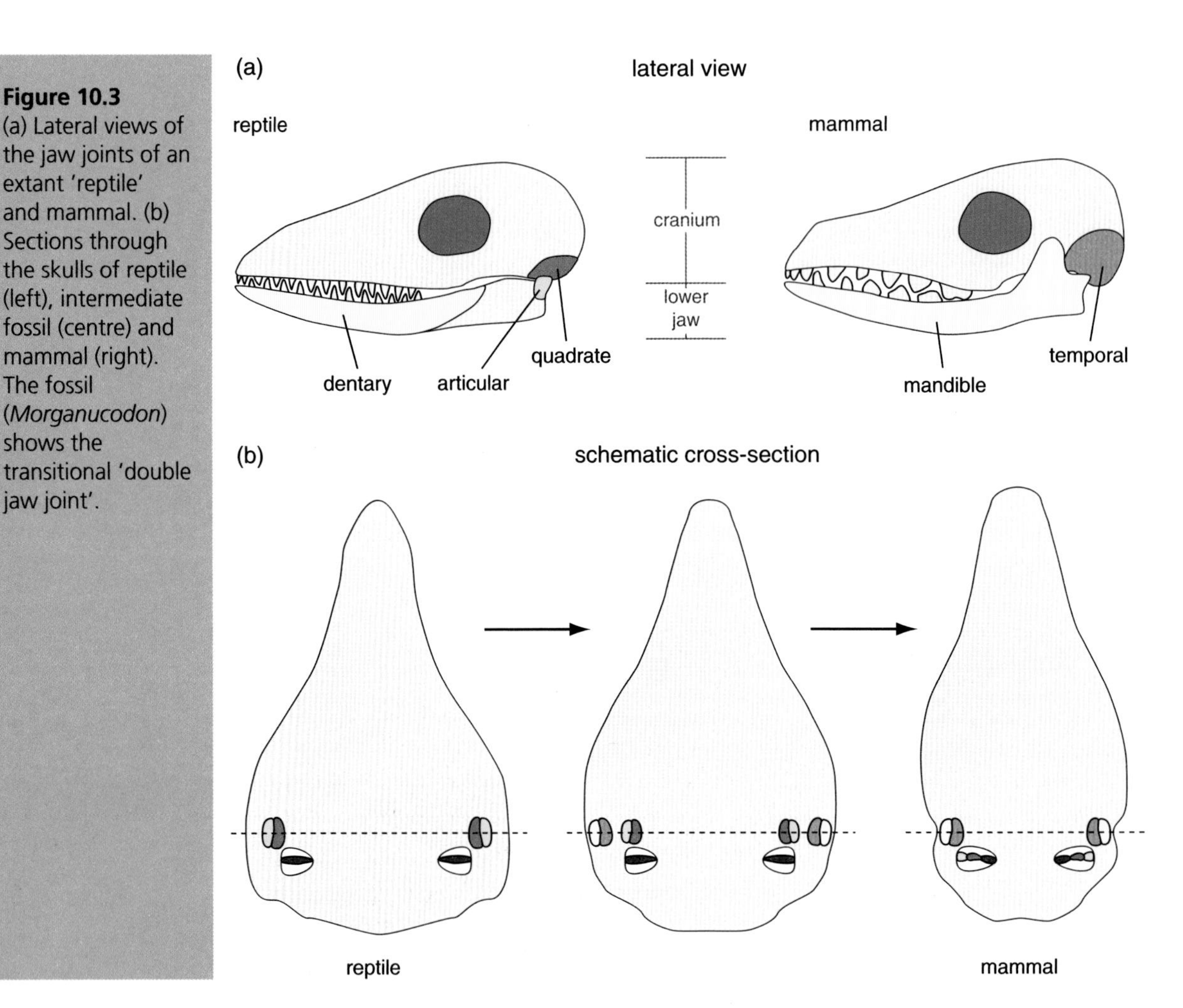

Figure 10.3
(a) Lateral views of the jaw joints of an extant 'reptile' and mammal. (b) Sections through the skulls of reptile (left), intermediate fossil (centre) and mammal (right). The fossil (*Morganucodon*) shows the transitional 'double jaw joint'.

reptiles' and early mammals are now known, showing various stages of the transition from jaw bones to ear bones. A key point in this transition is represented by the fossil *Morganucodon*, which had a double jaw joint (Fig. 10.3b, centre): it had evolved direct contact between the 'main' upper and lower jaw bones, which was the beginning of a mammal-type jaw joint; but it still had, coaxially with this, a joint involving the articular and quadrate. As in many cases of duplicate structures (or genes), one of them is 'free' to evolve in new directions because the other can do the job required – in this case jaw articulation – alone. This is indeed what happened. The articular and quadrate lost their role in the jaw and became the malleus and incus of the middle ear. The phylogenetic position of the morganucodontids in relation to other 'mammal-like reptiles' is shown in Fig. 10.4. The transition occurred approximately 200 MYA.

As noted earlier, this example of developmental repatterning in evolution involves a combination of heterotopy, heterometry and heterotypy. The bones concerned have shifted their position relative to other structures, have decreased in size and have taken on a new type of role – in hearing rather than in feeding. There may well have been an element of heterochrony too. And a similar compound change

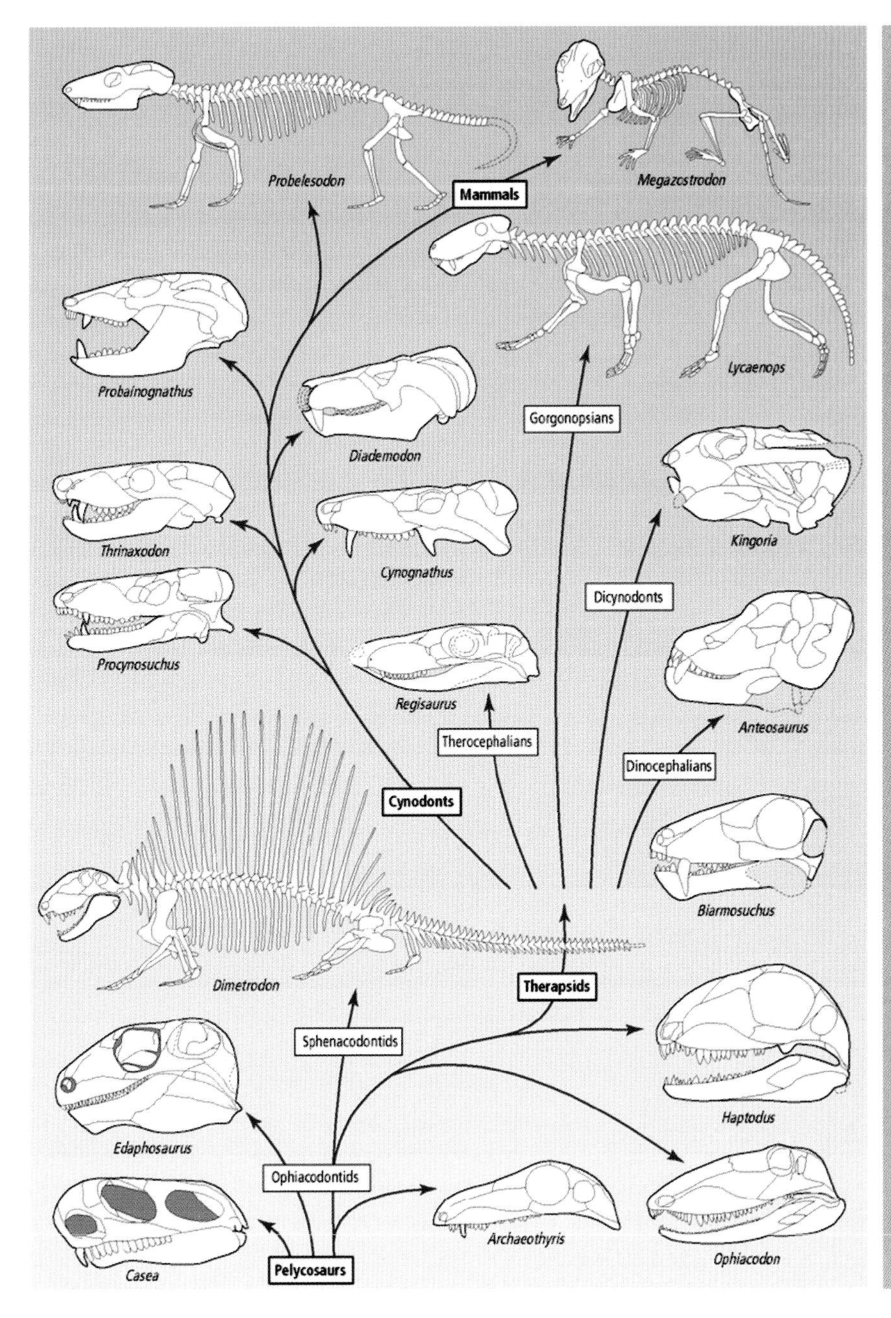

Figure 10.4 Evolutionary diversification of the 'mammal-like reptiles', showing the position of the Morganucodontids, here represented by *Megazostrodon*, a relative of *Morganucodon* (top right). (From M. Ridley, *Evolution*, 3rd edition, 2004.)

occurred earlier in the case of the stapes. Although already an ear ossicle in early members of the reptile/bird clade, the earliest amniotes had a large stapes that helped to support the braincase. Early amniotes are thought not to have had an eardrum, so ear ossicles, had they existed in those creatures, would not have been able to function as they were able to later.

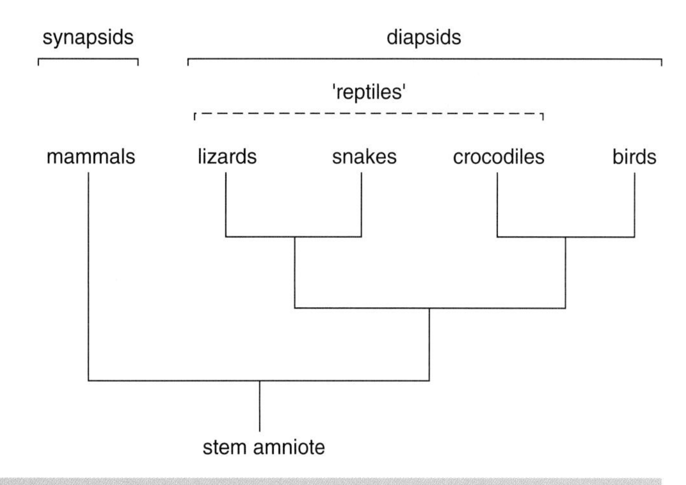

Figure 10.5 A simplified phylogeny of the Amniota. All groups shown are monophyletic except 'reptiles'.

With regard to the selective basis for the shift from articular and quadrate to malleus and incus, there are several related hypotheses, all to do with improved hearing, as might be expected. The earliest mammals are thought to have been small nocturnal insectivores in which acute hearing would have been a distinct advantage. Whether the evolution of the mammalian ear ossicles improved amplification, frequency range or impedance balancing (in relation to the transition of sound from a gaseous to a liquid medium) is unclear. Indeed, two or even all three of these may have been improved simult aneously.

Up to now, this story has been told using a minimum of unfamiliar names of the **taxa** involved. The main emphasis in this chapter is on developmental repatterning and not on taxon names. However, ultimately we need to be able to place instances of developmental repatterning in their correct position on a phylogenetic tree (Chapter 11); and taxon names, while sometimes off-putting, are useful in this latter exercise. So a brief digression will help at this point.

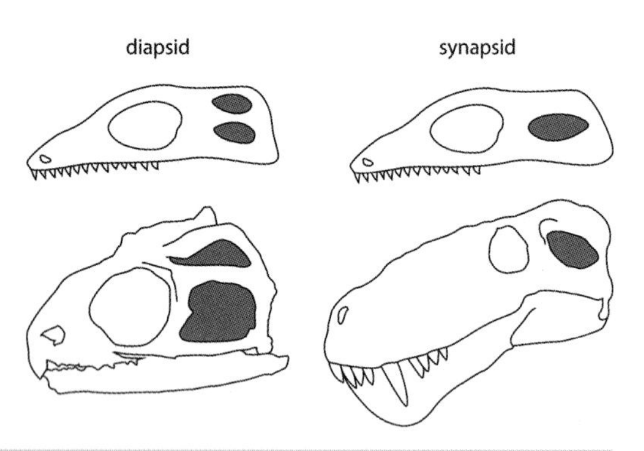

Figure 10.6 Diapsid and synapsid skulls, showing the post-orbital holes (dark grey) that give these types of skulls their names. Top: Simplified diagrams. Bottom: Specific examples – an extant diapsid (tuatara) and an extinct synapsid.

The earliest land vertebrates were probably still dependent on water for part of their life-cycle, such as modern amphibians, whose simple eggs would desiccate if not submerged or heavily lubricated. An important evolutionary **novelty** that allowed independence from water was the amniote egg: hence the name **Amniota** for all the descendants of the stem-group amniote; this taxon includes what are popularly known as reptiles, birds and mammals. However, 'Reptilia' is a **paraphyletic group**. The cladistic classification of the Amniota that is now usually accepted is shown, in simplified form, in Fig. 10.5. As can be seen, this involves an early split between the group that led to mammals (Synapsida) and the group that led to all other amniotes (Diapsida). The birds, though a **speciose** clade (Aves), are nested well within the Diapsida as shown. The terms synapsid and diapsid refer, respectively, to a single or a double hole in the skull (Fig. 10.6). The situation is complicated, however, because of secondary changes; for example, snakes are 'diapsids' without two such holes, just as they are 'tetrapods' without legs. Also, the turtles belong to a group with no such hole – anapsids – whose position in the tree is still not clear; hence their omission from the figure.

Since the purpose of this example was to show that different kinds of repatterning can occur simultaneously at the *same* level of organisation (morphological), we will not look at the molecular changes involved in the evolution of the ear ossicles. However, these are beginning to be understood.

10.4 The Kind of Repatterning can Change Between Levels of Organisation

We now turn from an example where different kinds of repatterning occur at the same level of organisation to an example in which different kinds of repatterning occur at *different levels* of organisation. The example upon which we will concentrate involves the pattern of small hair-like projections called trichomes on the dorsal cuticle of *Drosophila* **larvae**. These are non-sensory structures. Their function may relate to the locomotion of the larvae; certainly the larger ventral trichomes, which are generally referred to as **denticles**, have such a function. Since the larvae of most *Drosophila* species inhabit a semi-fluid medium, often rotting fruit, external projections give a degree of 'grip' and thus enable more rapid movement. It seems likely that different species, with different preferred larval microhabitats, need different amounts of grip in order to move effectively.

Of the 2000-plus species of the genus *Drosophila*, the most intensively studied include the 'model' species *D. melanogaster* and its closest relatives *D. sechellia*, *D. mauritiana* and *D. simulans*. A recent study focused on the different trichome patterns in these four species[73], and in particular on the difference between *D. sechellia* and the other three (Fig. 10.7).

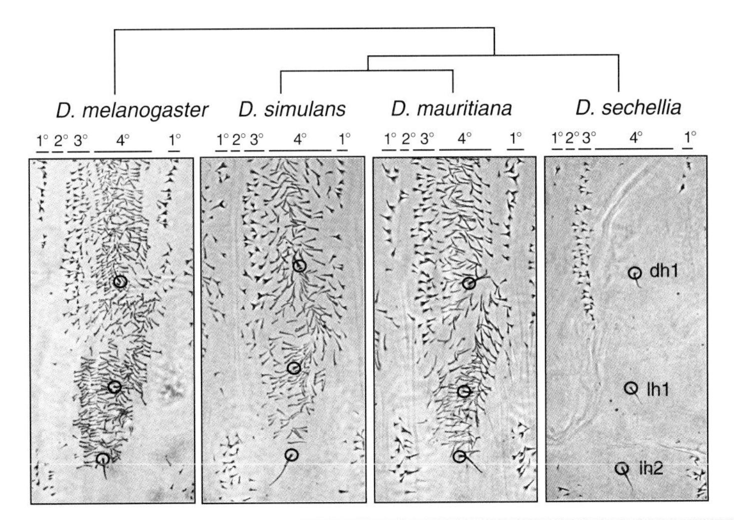

Figure 10.7 The phylogeny of species of the *melanogaster* subgroup of *Drosophila,* with larval trichome patterns of each species illustrated. Notice the 'naked' appearance of *D. sechellia.* 1–4: positions of different types of cell; dh – dorsal hair, lh – lateral hair. (Reprinted with permission from McGregor *et al.* 2007, *Nature*, **448**: 587–590).

As can be seen, *D. sechellia* has far fewer dorsal trichomes than the other species, and is **derived** as opposed to **basal** – so, at the morphological level, this is a case of negative heterometry. But what are its underlying cellular- and molecular-level causes? At the cellular level, trichomes are produced by epidermal cells forming extensions that are filled with the cytoskeletal protein actin. So, microscopic observation of fewer trichomes per segment implies fewer epidermal cells producing such extensions. This reduction is in turn caused by reduced expression, in *D. sechellia*, of a gene called *shavenbaby* (*svb*), which lies at a crucial nexus of gene interactions that lead to the development of trichomes[74] (Fig. 10.8).

Sophisticated genetic experiments, involving transgenic flies, demonstrated that the 'naked' pattern of *D. sechellia* requires the *D. sechellia* versions of three *cis*-regulatory regions of the gene. The crucial sequence differences have not yet been specified, but the *D. sechellia* and *D. melanogaster* regulatory regions are known to be between 3 and 5 per cent different in their sequences.

Although the full details of the interaction between these different regulatory sequences and their **trans-acting** partners (**transcription factor** products of

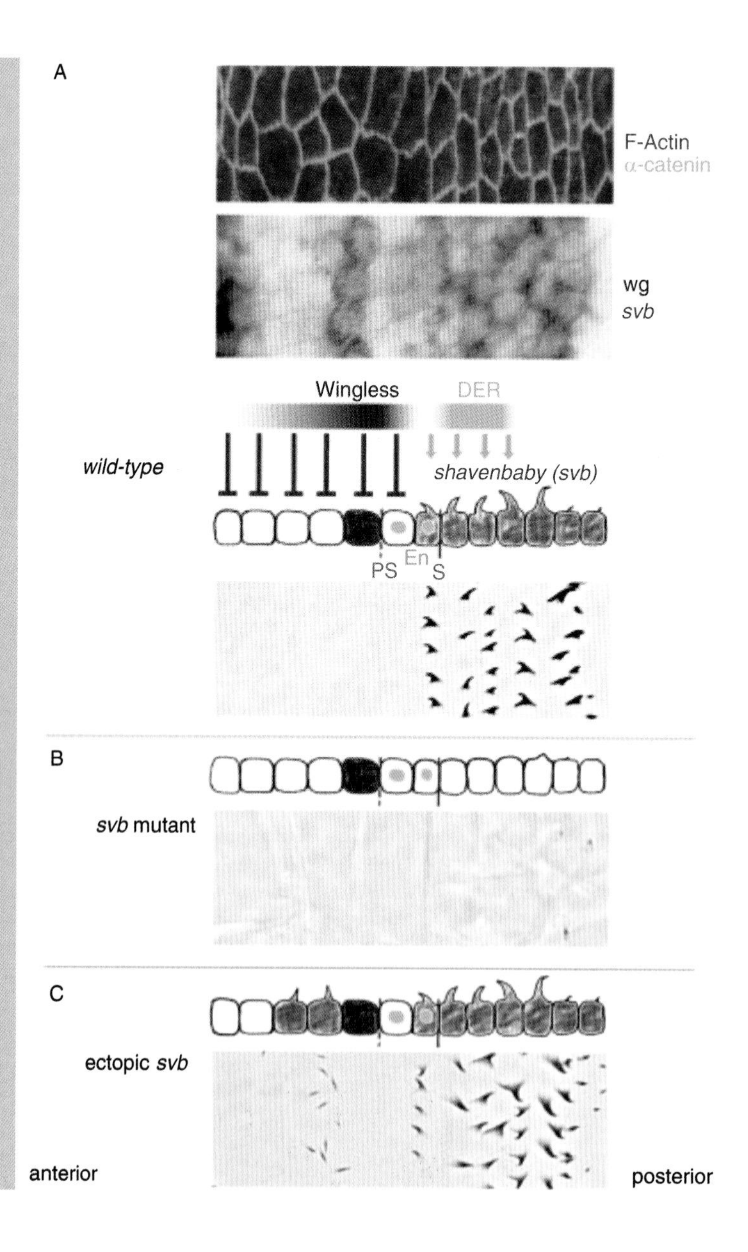

Figure 10.8 The role of the gene *shavenbaby* in the development of trichomes. A, staining for the products of several important genes, including *svb*; and (below) negative and positive effects of wingless and DER on *svb*, resulting in trichome formation having a particular distribution (RHS). B, *svb* mutant, with trichomes absent. C, ectopic expression of *svb*, resulting in addditional areas of trichome formation (LHS). DER – *Drosophila* Epidermal Growth-factor Receptor. (From Payre, 2004, *Int. J. Dev. Biol.*, **48**: 207–215.)

upstream genes) are not known, it seems that the *D. sechellia* regulatory sequences have been evolutionarily altered in such a way that the *svb* gene is not turned on by the factors that *do* turn on *svb* in *D. melanogaster*, and that this is specifically the case in particular groups of cells in each segment. A possible series of changes in regulatory regions in the *D. sechellia* lineage, which could account for the 'naked' appearance of the dorsal parts of its larval segments, is shown in Fig. 10.9.

Evolution of a ***cis*-regulatory** region of a gene that turns its activation by *trans*-acting molecules into a lack of activation (or even a repression) by those same molecules can be thought of as heterotypy, because the interaction has changed from + to 0 (or from + to –). So here we have molecular-level heterotypy causing morphological-level heterometry.

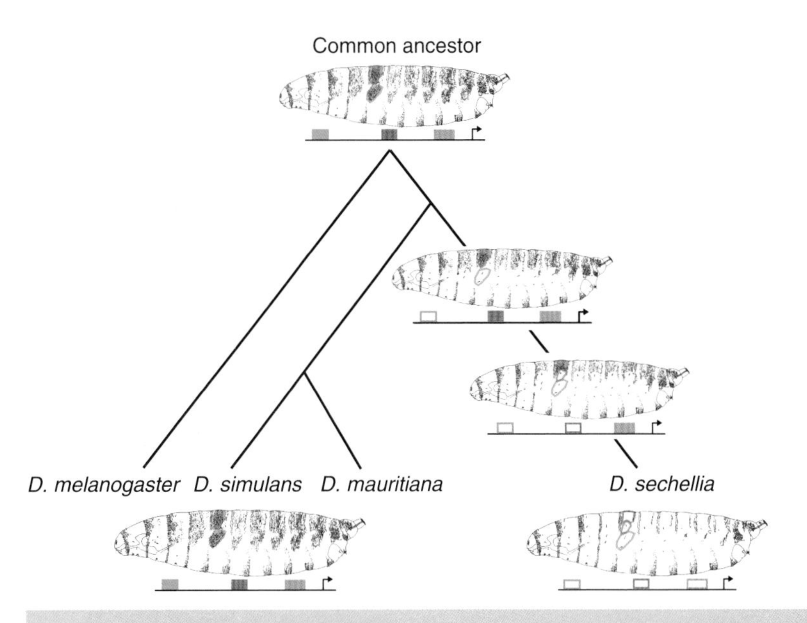

Figure 10.9 Hypothesis of the way in which the trichome pattern may have evolved, with a particular focus on the evolution of the 'naked' pattern of *D. sechellia*. Changes in 3 regulatory regions of the gene (coloured rectangles) may have been responsible for reducing or eliminating trichomes in the dorsal segmental parts outlined. (From McGregor *et al.* 2007, *Nature*, **448**: 587–590.)

10.5 Categories and Subcategories of Repatterning

We started our tour of the ways in which development can be evolutionarily repatterned in Chapter 6, with the best-known kind of repatterning, namely heterochrony. But we deliberately ignored much of the terminology associated with somatic-reproductive heterochrony as archaic and unhelpful. There is absolutely no need to produce an equally-elaborate scheme of newer terms to cover all the different kinds of repatterning (heterochrony included) that have been the subject of this chapter and the previous four. However, we do need clarity of thought on the relationships between different kinds of repatterning, and thus the formulation of an overall *outline* classification of these kinds will be useful. We already have the briefest of such outlines in the recognition of 'the four heteros', but we can go a little further than that, as follows.

First, we can ask the question: is there a fifth possible kind of reprogramming? Strangely, the answer is 'maybe'. Recall that 'the four heteros' all involved changes that were comparative in two senses: one lineage compared with another; and the timing, spacing, size or type of one 'thing' (whether gene product or structure/organ) compared with others. If we dispense with the need for the second type of comparison, we can see a kind of repatterning that simply involves the *re-scaling* of development. If, for example, in a descendant lineage compared with its ancestral one, development has been altered so that total developmental time is longer and the resultant adult bigger (and thus occupying more space), the changes are all absolute rather than relative. However, while such repatterning is theoretically possible, it is unlikely to occur in practice because the growth of multicellular organisms is almost always **allometric** to some degree. Therefore, repatterning taking the form of perfectly **isometric** changes in scale must have been very rare in evolution, and indeed may not have happened at all.

Second, we can ask: is it of value to recognise a limited number of sub-categories within each of 'the four heteros'? The answer in this case is almost certainly 'yes'. Indeed, we have done this already in some cases, so what follows is merely a reminder, and a completion, of this recognition process.

Note the outline classification shown in Fig. 10.10. Heterochrony is simply classified into the three categories of **somatic**, somatic/reproductive and reproductive. The various old terms given in Chapter 6 just for reference (**neoteny**, etc) are all sub-categories within somatic/reproductive heterochrony (and are not shown here). Somatic heterochrony is when the development of one part of the body, or one organ system, is accelerated or retarded relative to others. Reproductive heterochrony is an equivalent for the reproductive system as opposed to the soma. This might at first seem odd as there is just one reproductive system, but of course parts of it may be accelerated or retarded at the cellular level relative to others.

Heterotopy is also divided into three categories, but on a very different basis, as expected. One category is 'movement heterotopy' where some structure moves relative to others, as in the case of the movement of bones from jaw to ear in the

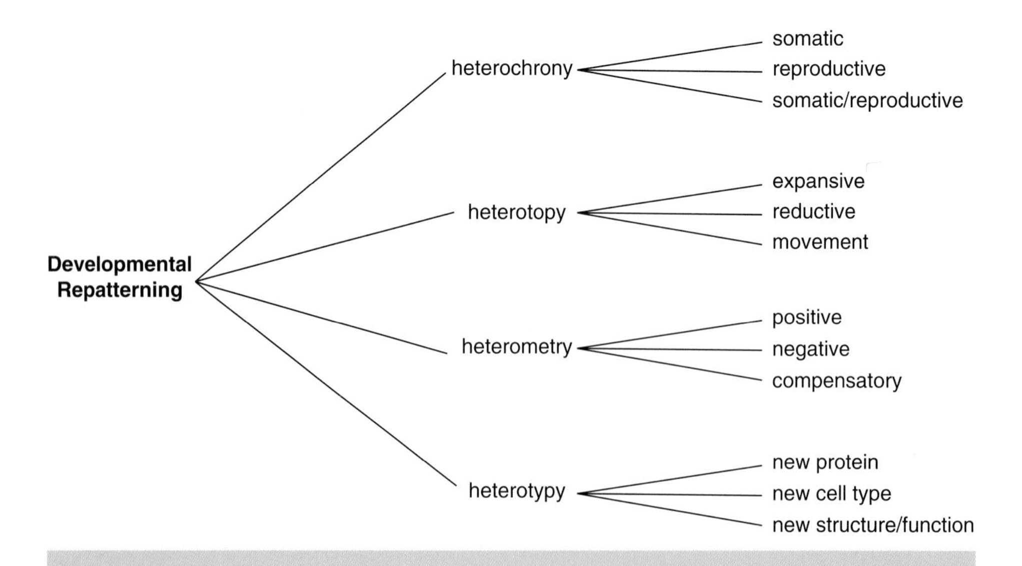

Figure 10.10 Categories and subcategories of developmental repatterning.

evolution of early mammals. There is also expansive heterotopy, where the spatial domain of something – gene expression, morphological structure, or often both – is enlarged. There are several ways in which this can happen, including the **ectopic expression** of a gene. The opposite of expansive heterotopy is reductive heterotopy, where the overall position of something relative to the rest of the body changes through diminution. These last two categories of heterotopy necessarily involve hetrometry too; whether they are better seen as one or the other depends on the individual case.

Heterometry is easy to split into three categories: upward (positive); downward (negative); and both, but in different systems (compensatory). The blind cave-fish examined in Chapter 8 revealed compensatory heterometry in that some features (e.g. to do with chemoreception) had increased, compensating for the reduction or loss of eyes.

Heterotypy is more difficult. There is an almost infinite range of possibilities for the form taken by a 'new type'. However, we can at least consider all these possibilities to fall into three level-based categories: new proteins (and other molecules); new cell-types; and new types of structure/function. These were exemplified, in Chapter 9, by a new type of haemoglobin, the origin of the endothelial cell, and a new type of segment (the poison claw segment in centipedes).

Any classification of the sort shown in Fig. 10.10 is really just an aid to clear thinking about the processes involved. If, in any particular case, it does not seem to be doing this job well enough, then it can be modified. For example, the molecular, cellular and morphological levels used as categories of heterotypy could also be used instead of, or as well as, the categorisations shown in the figure for the other three types of repatterning.

10.6 The Causes of Repatterning

In a long-term evolutionary process, such as the evolution of the mammalian ear ossicles, the developmental repatterning involved is nearly always compound. This is partly because many of the individual changes have themselves been compound, but partly also because many changes have been accumulated over a long period of time. Although the causal nature of each of these changes is largely unknown, we inevitably view it as originating in gene mutation, because non-**heritable** changes are of no consequence for evolution.

In shorter periods of evolutionary time, developmental repatterning is again often compound in nature, and caused ultimately by mutation. Sometimes, as in the case of the evolution of trichome patterns on *Drosophila* larvae, it is even possible to pin down the mutational changes involved to a considerable level of detail – in that case, to three regulatory regions of the *shavenbaby* gene.

When looking at the developmental repatterning caused by a single mutation, whether one that is involved in evolution (e.g. the one causing **chirality** switches in gastropods) or one that is not (e.g. antennapedia in *Drosophila*), the link between the mutation and the repatterning is particularly clear. Sometimes, the earliest cellular changes that initiate the repatterning of development right through to the adult are known – as we saw for the gastropod chirality example in Chapter 7.

We have only looked at one case in which repatterning was not caused by a mutation – the experiment involving different concentrations of Shh protein and different durations of exposure to it. The best way to interpret such experiments is that the experimenters are mimicking the effects of mutations either in the *shh* gene or, more likely, in one or more of its upstream controllers.

So, the question that arises is as follows. In the course of evolution, is there any way in which developmental repatterning can be produced other than by gene mutation, broadly defined to include all changes in DNA base sequence, whether large or small, genic or non-genic, and involving nuclear or organelle genomes? The answer is 'yes', for two reasons.

First, it has become clear in the last decade that patterns of DNA methylation, which do not involve changes in base sequence, yet which can be associated with genes being turned on or off, can be inherited. As noted in Chapter 5, the study of such processes is now generally described as **epigenetics** (though this word was previously used in a very different way).

Second, while repatterning caused solely by environmental factors, such as the production of winged aphids when population density is high, is not inherited, the ability to respond in such a way to environmental variation is. Thus natural selection can favour the ability of a developmental system to be **plastic**. This whole area – the evolution of developmental **reaction norms** – is covered in Chapter 16. It is now regarded to be of considerable importance. Of course, its causality must lie ultimately in the genes, but in this kind of evolution the environment plays a more active role in repatterning than in a 'conventional' example, where there is no plasticity.

SUGGESTIONS FOR FURTHER READING

The two main examples of compound developmental repatterning dealt with in this chapter involved: the evolution of the mammalian ear ossicles; and the evolution of bristle patterns in *Drosophila*.

The first of these can be put into a fuller context by considering the evolution of the mammals in general. This subject is dealt with by:

Kemp, T.S. 2005 *The Origin and Evolution of Mammals*. Oxford University Press, Oxford.

The second can also be put into the fuller context of the use of *Drosophila* as a model system for studies in evolutionary biology in general, as dealt with by:

Powell, J.R. 1997 *Progress and Prospects in Evolutionary Biology: The Drosophila Model*. Oxford University Press, New York.

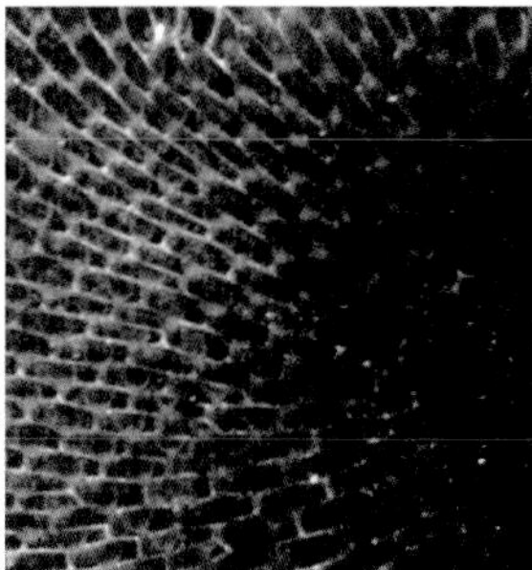

CHAPTER 11

Mapping Repatterning to Trees

11.1 Pattern, Process, Homology and Trees

Although Charles Darwin's key contribution to evolutionary theory – natural selection – was in the area of evolutionary process rather than evolutionary pattern, the sole picture included in *The Origin of Species* is an evolutionary tree, or **phylogeny** (Fig. 11.1). Darwin saw all sides of the evolutionary picture. However, since its Darwinian foundations, evolutionary theory has tended to become polarised, in the sense that particular schools of evolutionary thought have concentrated very much on either pattern or process. The '**modern synthesis**' of the mid-20th century, stemming from the seminal works of Fisher, Haldane and Wright, was focused on process, while the **cladistics** school, stemming from the work of the German biologist Willi Hennig[75], was focused on pattern.

As a scientific discipline progresses, fragmentation of this kind is inevitable. In this respect, evolutionary biology is no different from any other. To deepen our knowledge of the mechanics of natural selection, or of the pattern of relationships among particular taxonomic groups, the appropriate focus is necessary, even if, for the school of biologists concerned, it involves a neglect of evolution's 'other side'.

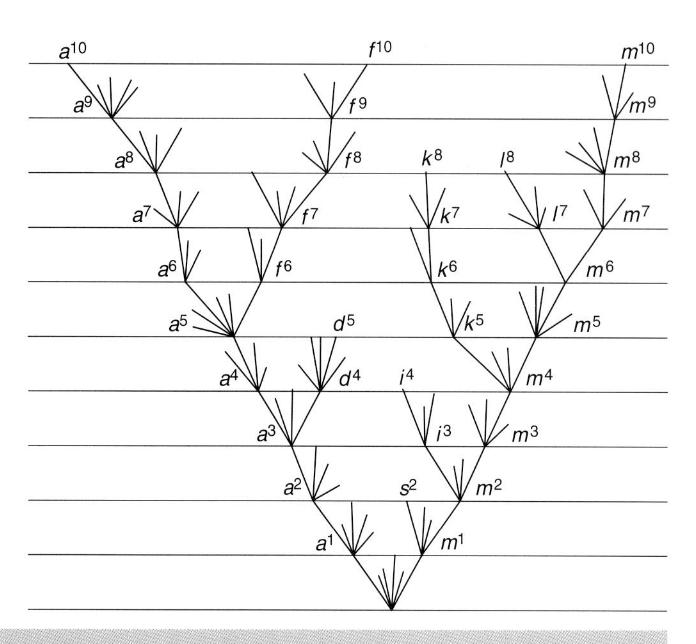

Figure 11.1 The diagrammatic evolutionary tree that appears in Darwin's *Origin of Species*, and is, famously, the only picture in the entire book.

While the splitting of different disciplines to study different aspects of evolution was inevitable, and has to a large extent been productive and thus 'good', there has also been a darker side to it. A small minority of the more extreme proponents of each particular approach have been rather dismissive of the other. For example, some supporters of the modern synthesis wrote off studies of evolutionary relatedness as mere stamp-collecting. For example, the British biologist couple, the Medawars, said[76] that it is often the case that 'nothing of any importance turns on the allocation of one ancestry rather than another.' They took this view because they saw natural selection as something that could explain *all* evolutionary changes, regardless of particular patterns of relationship. In contrast, some cladists regarded natural selection as impotent in terms of its predictive power. They took the view that because it could potentially explain everything, it explained nothing. The American cladist Donn Rosen[77] declared that there was 'no need to placate the ghost of neo-Darwinism; it will not haunt evolutionary theory for much longer.'

We may either look at the 'silent majority' of each school, and so see a positive endeavour aimed at understanding evolutionary pattern on the one hand and process on the other, or look at the extremists of either side and see a narrow, negative approach to evolution that is hostile to other approaches. But whichever way we look at it, the fragmentation of scientific disciplines – in this case evolutionary biology – ultimately paves the way for renewed synthesis. That is, eventually the bigger picture re-emerges, and when it does so it has greater depth than before.

Evo-devo represents the emergence of this renewed synthesis. Perhaps that is overstating the case, in that the 'bigger picture' is not yet clear, but it is becoming increasingly so. Thus we might more accurately describe present-day evo-devo as the *beginning* of the emergence of renewed synthesis. What has been discovered through the evo-devo studies of the last three decades has been truly amazing; but there will be many more amazing findings to follow.

A crucial evolutionary concept that has not yet been discussed in detail here is **homology**. This term refers to characters (whether at the molecular or morphological level) whose similarity between two **taxa** indicate their presence in the last common ancestor (LCA) of those two taxa; and hence their common descent, with different modifications in each descendant **lineage**, from that LCA. The term homology was in fact introduced by the English biologist Richard Owen[78] in the 1840s (i.e. pre-*Origin of Species*) to distinguish between similarities that were in some sense 'true' as opposed to those that were 'superficial'; but the post-*Origin* interpretation of these as similarities due to shared ancestry on the one hand and evolutionary **convergence** on the other has been adopted ever since.

Evo-devo now gives us an opportunity to bring together pattern and process. The *modus operandi* is as follows. We acknowledge three things. First, that natural selection is the mechanism producing adaptive change at the population, and ultimately the species, level. Second, that at the individual level, the processes bringing about the changes upon which natural selection subsequently works are **mutation** (*sensu lato*) and **developmental repatterning**. Third, that all living organisms are related to each other (assuming that there was only a single 'successful' origin of life on our planet), and that there is thus an overall tree of relatedness, sections of which can be focused in on to give a more detailed picture of relatedness within one or another taxonomic group (e.g. vertebrates, insects, **angiosperms**).

Having acknowledged these three things, we focus on a particular part of the overall tree of life, whether big (e.g. the animal or plant kingdom) or small (e.g. a particular order or family). We take the tree of this group that has been reconstructed using cladistic methods. We then attempt to pinpoint on this tree the occurrence of particular developmental repatterning events. This endeavour should reveal the distinction between homology and **homoplasy**. We try to characterise the nature of each repatterning in terms of **heterochrony**, etc. (as described in the last five chapters) at both molecular and morphological levels, and also to construct hypotheses as to why each repatterning might have been selectively favoured.

This whole endeavour takes on board not only evo-devo *sensu stricto* but also a whole range of other evolutionary disciplines, ranging from **phylogenetics** to population biology to palaeoecology. Its outcome should be (and already is in some cases) a much more complete picture of the evolution of organismic form than has ever been available before. In the next sections of this chapter, we will look at specific examples involving both animals and plants.

11.2 The Origin(s) of Animal Segmentation

Three animal phyla are acknowledged to contain largely or exclusively animals with a segmented **body plan**: the annelid worms, the arthropods and the chordates (Fig. 11.2). No arthropod has yet been discovered in which all traces of segmentation have been lost at all stages of the life-cycle, though some parasitic forms have lost *adult* segmentation, such as the parasitic barnacles we saw in Chapter 7. A few annelids do appear to have completely lost segmentation, such as the one shown in Fig. 11.3, though it remains possible that these are **basal** and thus primitively unsegmented. In any event, they represent less than 1 per cent of all annelid species. No vertebrates have lost segmentation, though not all basal chordates are segmented.

In relation to this last point, the recent discovery[79] that the **urochordates** (sea squirts) rather than the **cephalochordates** (amphioxus) are the closest relatives of the vertebrates makes matters more complex because, while amphioxus shows segmentation of its musculature, the sea-squirt tadpole seems not to, and the adult

Figure 11.2 One representative of each of the three large phyla with a segmented body plan. They are: (a) an earthworm (Annelida); (b) a millipede (Arthropoda); and (c) a snake (chordata).

Figure 11.3 An unusual annelid. The specimen shown belongs to the phylum Annelida (segmented worms), but seems to have lost all traces of segmentation (photograph with permission from Peter Wirtz).

clearly is unsegmented. So, even within the chordates, there is uncertainty over whether segmentation arose once (and was lost in sea-squirts) or twice (in amphioxus and in vertebrates).

The situation regarding animal segmentation becomes more complex again when we consider some animal groups outside those already mentioned, which seem not to be segmented in a general way but do show partial segmentation in the sense of having structures (internal and/or external) that are serially repeated along the antero-posterior axis. These include the cuticular rings of kinorhynchs, the multiple shell plates of polyplacophoran molluscs, and the nephridia (kidneys) and gills of monoplacophoran molluscs (Fig. 11.4) – both of these latter groups being considered as basal within overall molluscan phylogeny.

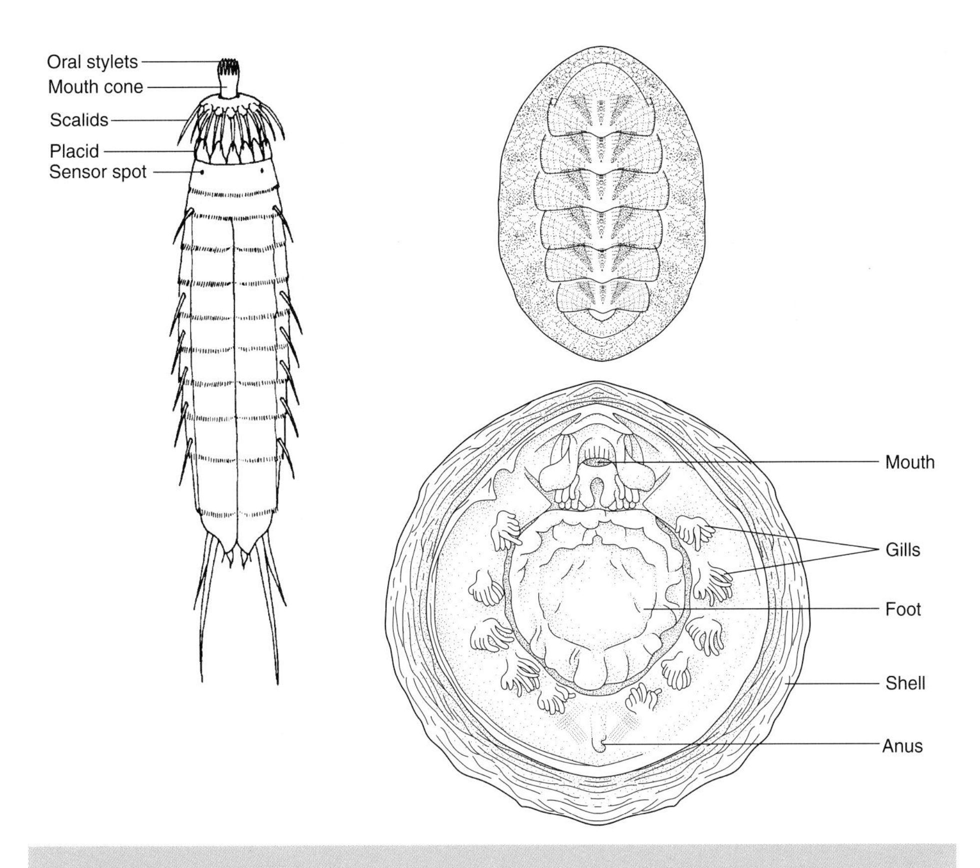

Figure 11.4 Animals that have serial repetition of structures along the antero-posterior axis but are not regarded as having true segmentation. Top (left): A kinorhynch, with multiple rings of cuticle. Top (right): A polyplacophoran mollusc (a chiton; dorsal view) with multiple shell plates. Bottom: A monoplacophoran mollusc (ventral view), with serial repetition of some internal structures.

For the moment we will ignore both these partially-segmented groups and the basal chordates. We will focus on the three large segmented groups – the annelids, arthropods and vertebrates, and ask the question: did 'true' or 'complete' segmentation evolve on one, two or three occasions (Fig. 11.5)?

In the 1960s and 1970s, evolutionary biologists thought they knew the answer[80]: 'two'. This was because the annelids and arthropods were thought to be evolutionary **sister-groups**, and thus closely related, whereas the 'supergroup' that they supposedly formed was thought to be only very distantly related to the vertebrates. The fact that in annelids and arthropods segmentation is 'overt' (i.e. externally visible), whereas in vertebrates it is not (because it is an essentially mesodermal rather than ectodermal phenomenon), seemed to clinch the 'two origins' hypothesis.

One of the biggest changes in our view of animal relationships came with the recognition of two large **clades** of bilaterian **protostomes**, each including several

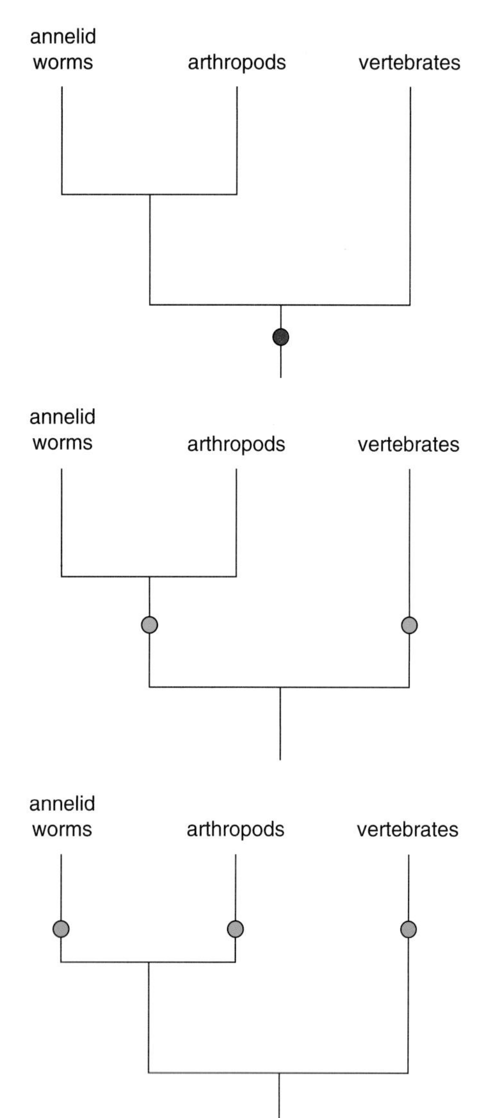

Figure 11.5 Did segmentation originate once (red), twice (blue) or three times (green)? The previously accepted view (twice) has given way to a debate between supporters of the other two views.

phyla: the **Lophotrochozoa** and the **Ecdysozoa**[81]. These are shown in Fig. 11.6, which is essentially a more detailed version of Fig. 11.5. Under the new view of large-scale relationships, which has remained robust in the face of many challenges over more than a decade, the annelid-arthropod **sister-group** relationship is seen to be false, thus withdrawing one of the main reasons for supporting a 'two origins' hypothesis of segmentation over either of the other hypotheses.

The developmental genetics of segmentation has now been studied in all three of the main segmented groups – though more extensively in arthropods and vertebrates than in annelids. There is an accumulating body of resultant comparative data at our disposal, which should potentially help to choose between the single, double and triple origin hypotheses. Yet so far no clear choice has emerged, because the data can be interpreted in different ways.

Information on the developmental genetics of segmentation arrived first for those of the various **model organisms** (Chapter 3) that were segmented: the fruit-fly *Drosophila* as a representative of the arthropods, and the various vertebrate model systems – mouse, chick, frog and zebrafish.

The developmental genetic basis of segmentation in *Drosophila* had been reasonably well established[20] by the late 1980s. This was described earlier (Chapter 4); but it might be useful if we remind ourselves here of the beginning and end of the process. Since segmentation takes place along the antero-posterior axis, the first step is the establishment of that axis. This is done via several localised anterior and posterior determinants. One of the main head determinants is Bicoid mRNA/protein, which has been made by the maternal *bicoid* gene in the egg's nurse cells. After the formation of the A-P axis, it is progressively divided and subdivided by the segmentation-gene cascade until the final number of rudimentary segments is formed, each of which is characterised by a transverse stripe of expression of the segment polarity genes, such as *engrailed*. All rudimentary segments are formed at about the same time.

It was assumed by many biologists that this system, with minor modifications, would apply across the whole of the phylum Arthropoda. However, this assumption turned out to be wrong.

The process of **somitogenesis** in vertebrates had been studied in detail long before the genetic work on the process of segmentation in *Drosophila*. However, lacking the necessary information on the relevant developmental genes, these early vertebrate studies were largely restricted to being descriptive. But an exception to this was an attempt, in 1976, to provide a model of how the process might work. This is called the 'clock and wave-front model' because it involves the interaction between a

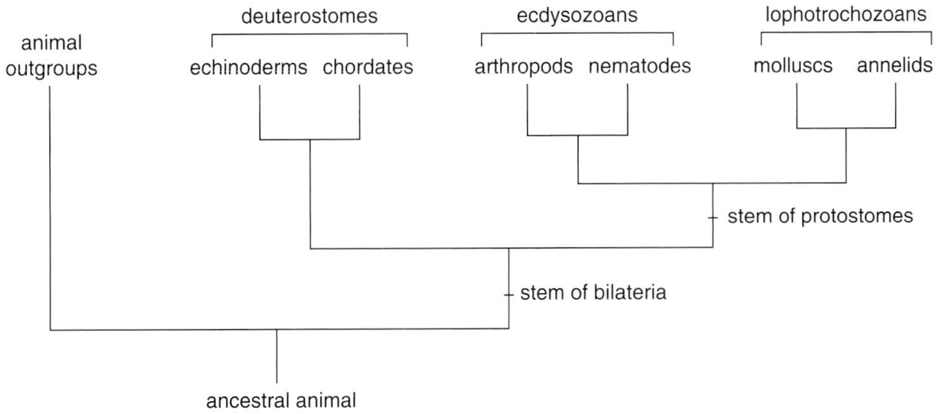

Figure 11.6 A phylogeny of the animal kingdom, showing the division of the protostomes into two groups, Ecdysozoa and Lophotrochozoa. These group names translate as 'moulting animals' (e.g. arthropods) and 'animals with a lophophore feeding structure and/or a trochophore larva' (e.g. annelid worms). Although the names relate to developmental/morphological features, the two groups were discovered through molecular studies.

pulsating 'clock' of some sort, now known to involve the periodic expression of certain genes, controlling the production of segments by interacting with an anterior-to-posterior migrating wave, now known to involve the expression of other genes (Fig. 11.7). One of the central elements of the somitogenesis clock is Notch-Delta signalling. This behaves in the way described for signalling systems in general in Chapter 3; and many downstream genes are caused to 'pulsate' by the periodic Notch-Delta activity.

The early descriptive studies on which the clock-and-wavefront model was based were largely conducted on somitogenesis in the mouse and the chick. As in the case of extrapolation from *Drosophila* to arthropods, there was a tendency to extrapolate from mouse and chick to all vertebrates. In the vertebrate case, further comparative studies have suggested that the extrapolation was not so unreasonable, though clearly many of the details of the process, including the identities of some of the genes involved, do vary from one vertebrate system to another.

Until very recently, the processes of arthropod segmentation and vertebrate somitogenesis looked to have little in common. This favoured independent evolutionary origins. Although less was (and still is) known about segmentation in annelids, their wide phylogenetic separation from both arthropods and vertebrates suggested that segmentation originated three times (at least) in the animal kingdom; and thus that the segments (or somites) of the three main segmented groups of animals were not homologous.

It has become clear, however, that the *Drosophila* system is not representative of all arthropods. Indeed, some of its components are not even true of all insects. As noted previously, the role of *bicoid* as a head-determinant is now known to be a recent

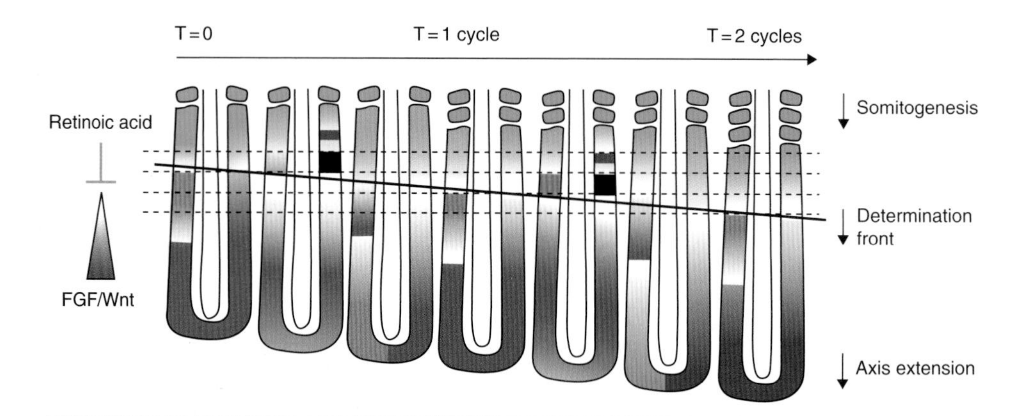

Figure 11.7 One version of the 'clock-and-wavefront' model of vertebrate somitogenesis. Top – time measured by clock cycles; Bottom – left and right bands of paraxial mesoderm (separated at the anterior end, joined at the posterior) in seven successive embryonic stages. The determination front (diagonal line) represents the boundary between RA (green) and FGF/Wnt (purple) signalling, and the threshold for cells becoming competent to form segments. Orange (shown in LHS of each stage only) – wave of clock expression. A somite becomes determined (black, RHS) when a clock pulse sweeps through cells that have become competent as the wavefront moves posteriorly. Somites (green blocks) form in the paraxial mesoderm in antero-posterior order, as shown. (From Dequéant and Pourquié, 2008, *Nature Reviews Genetics*, **9**: 370–382.)

evolutionary innovation characterising only a sub-group of the **Diptera**. And the approximate simultaneity of segment formation in *Drosophila* is not found in the **short-germ** insects, such as the Orthoptera (crickets and grasshoppers). These exhibit anterior-to-posterior segment formation, like vertebrates.

But the most interesting recent finding has involved the comparison of the *Drosophila* system not with other insects but rather with non-insect arthropods, and in particular with the sub-phyla Chelicerata (arachnids and their allies) and Myriapoda (centipedes, millipedes and their relatives). These represent the most distant possible points of comparison, given the now-accepted close relationship between insects and crustaceans.

Periodic expression of Notch-Delta signalling has recently been discovered in a spider (recall Fig. 6.3) and in a centipede (Fig. 11.8). Since this signalling does not contribute to the formation of segments in *Drosophila*, but does to the equivalent process in vertebrates, a very different view emerged of the ancestry of segmentation. Perhaps *Drosophila* is a very **derived** insect, which has departed a long way from the ancestral arthropod mechanism of segmentation, possibly for reasons connected with its rapid, **long-germ** development. Perhaps spiders and centipedes have departed less from the ancestral arthropod segmentation mechanism; in which case that mechanism has some significant similarities with its vertebrate counterpart. Perhaps the segments of animals belonging to distantly-related phyla are homologous after all.

The question we are approaching here is the nature of the LCA of all bilaterian animals. In particular, was this creature, which has been dubbed the **urbilaterian**, segmented or not? So far, we have seen the probable number of origins of segmentation move historically from two to three to just one. If this last option is correct, then the urbilaterian was segmented (Fig. 11.9).

However, this conclusion may be premature because of the phenomenon of gene **co-option**. We will examine this in more detail in Chapter 15. A brief description will suffice here. An ancestral animal may have had within its genome a suite of genes that could potentially be used to drive a segmentation system, even though that animal is not itself segmented. The genes may have played entirely different roles in the ancestor; but there may be something about them that predisposes them to be used later (in a descendant) to form segments.

If this is so regarding the urbilaterian, then that ancient animal may have had the relevant genes but have been completely unsegmented. One possibility is that it may have looked something like one of today's (unsegmented) acoel flatworms (Fig. 11.9). And, of course, intermediate possibilities exist too. Some biologists have suggested that it is too simple to think of an animal as being either segmented or not; rather they point to the examples of partial segmentation noted earlier (e.g. kinorhychs) and take the view that we should think of particular structures or organ systems as being segmented or not, rather than apply such a distinction to the whole animal.

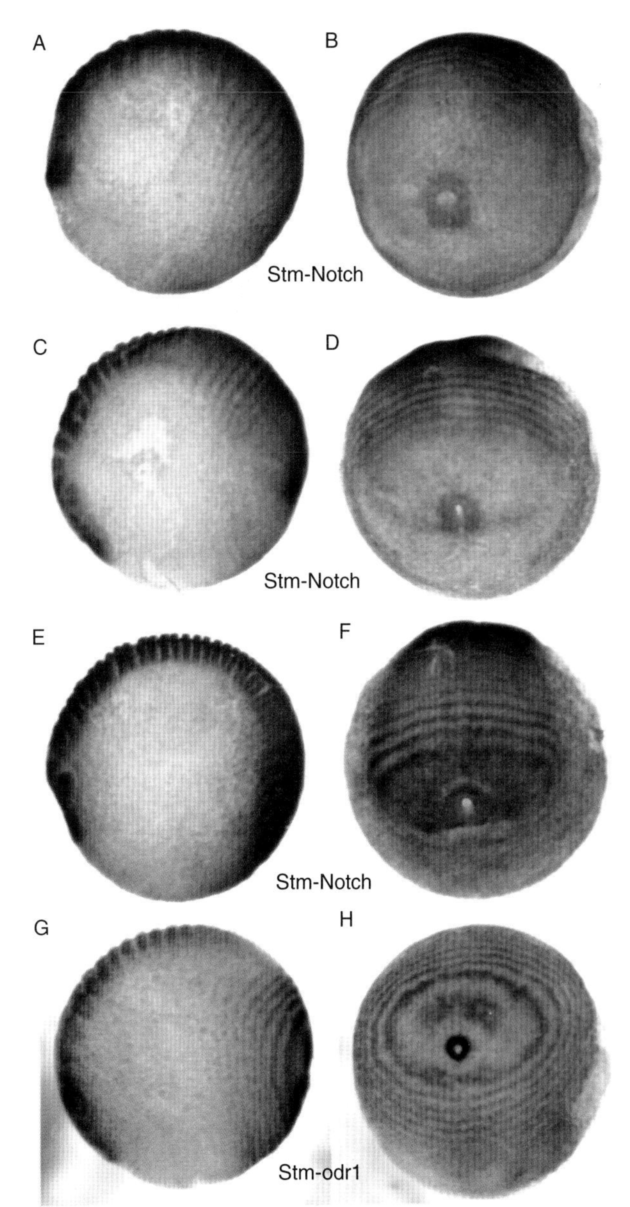

Figure 11.8 A to F: Waves of expression of *notch* in a centipede (*Strigamia maritima*; Stm). These, along with similar waves of expression in spiders (Fig. 6.3) indicate a role for Notch-Delta signalling in the segmentation of some arthropods (including stem arthropods?) that is somewhat similar to its role in vertebrate somitogenesis. G, H: Expression of another segmentation gene, *odd-skipped related 1*.) (From Chipman and Akam, 2008, *Developmental Biology*, **319**: 160–169.)

11.3 The Vertebrate Fin-to-Limb Transition

Since it is not yet clear how many times animals became segmented, a single origin, and thus homology of all animal segments, remains one of the possibilities. Regarding the invasion of land by animals, however, no such possibility exists. It is abundantly clear that animals have invaded the land many times; thus the evolution of adaptations

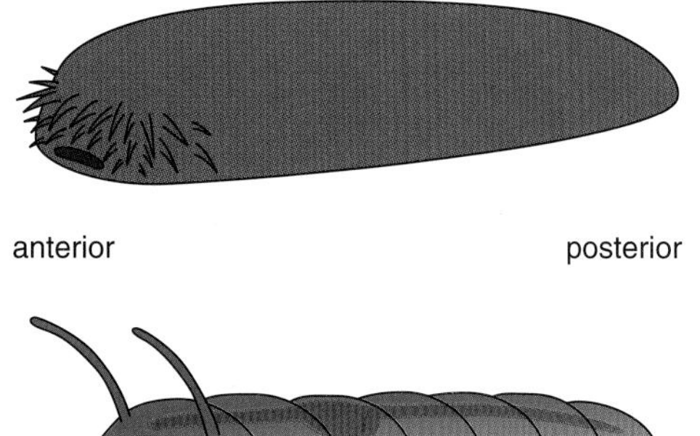

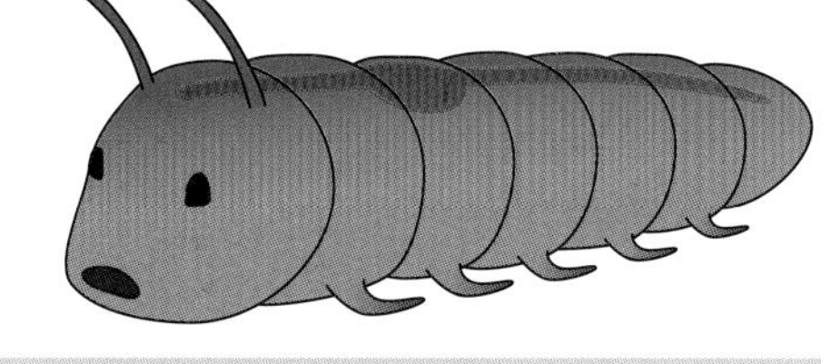

Figure 11.9 Two views of the structure of the original bilaterally-symmetrical animal, or urbilaterian: Top: Relatively simple and unsegmented, like today's acoel flatworms. Bottom: Complex and segmented.

for life on land, including methods of breathing air and of walking, have occurred independently in different groups and are convergent rather than homologous.

Regarding walking limbs evolving from swimming appendages, this has happened separately in vertebrates and arthropods. However, while it has also happened separately in various arthropod groups (at least once in each of the four arthropod subphyla, and probably more often than that), it seems to have occurred just once in the vertebrates. This single transition – from the fins of fish to the walking legs of **tetrapods** – has been the subject of much study, especially by palaeontologists. As a result of several important fossil discoveries made over the last few decades, we have a reasonably clear picture of the groups involved and of the timing of the transition. We also have relevant comparative developmental-genetic information from living fish and tetrapods. And we have hypotheses – though not more than that, as will shortly become clear – on the adaptive nature of the transition.

The earliest known tetrapods lived approximately 375 MYA, in the Devonian period. Their closest living relatives are the lungfish and coelacanths (Fig. 11.10). But many Devonian fossils are known that are intermediate between today's lungfish and today's tetrapods, as can be seen from the

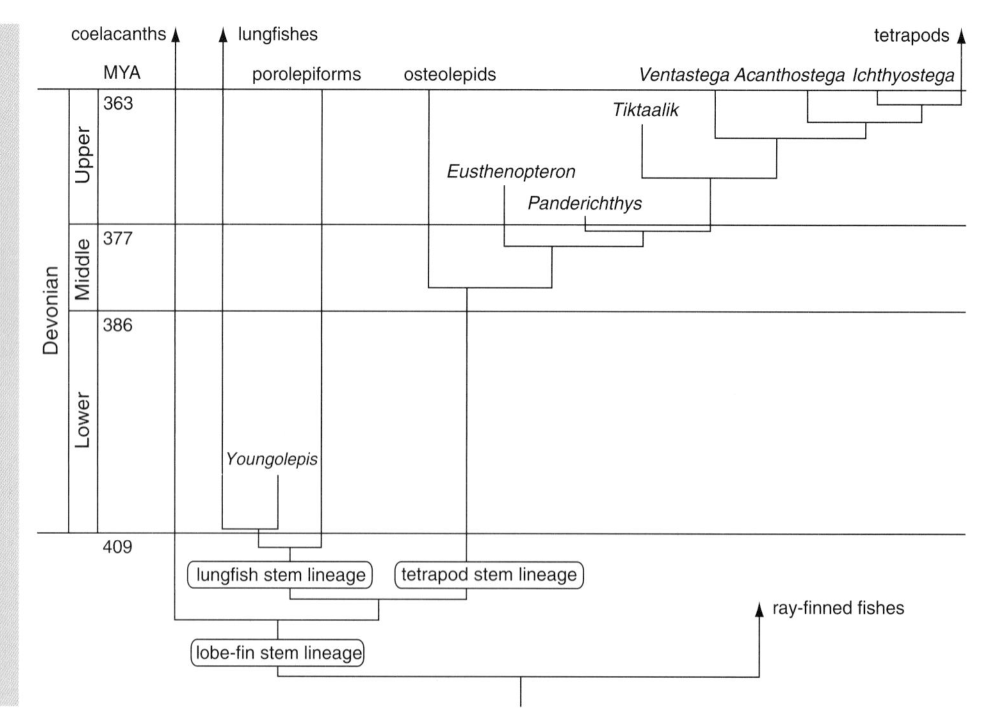

Figure 11.10 Phylogeny of lobe-finned fishes, showing the origin and sister-groups of the tetrapods. Upward arrows indicate extant groups; all others are extinct. (Modified from:Clack, J.A. 2002 *Gaining Ground: The Origin and Evolution of Tetrapods.* Indiana University Press, Bloomington, IN.)

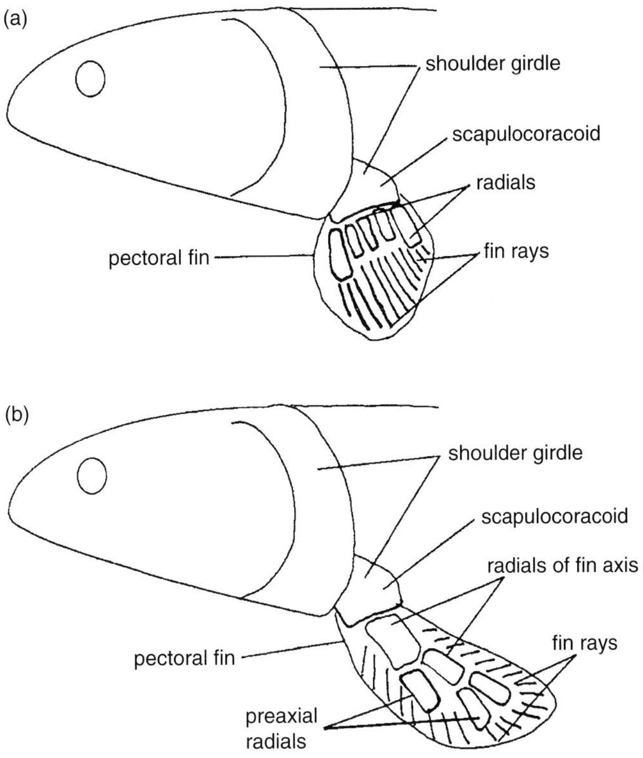

Figure 11.11 Diagram of the different skeletal supports of the paired fins of (a) ray-finned and (b) lobe-finned fishes (courtesy of Jenny Clack.)

figure. These reveal details of the transition from aquatic to terrestrial vertebrate morphology. Many characters changed in this transition, including the structure of the skull and trunk, but we will focus here on the limbs.

All of the groups named in Fig. 11.10 belong to the Sarcopterygii or lobe-finned fishes. The sister-group to this, the Actinopterygii or ray-finned fishes, contains all the other bony fishes, including familiar forms such as salmon, mackerel and zebrafish. The skeleton of a ray-fin has, as the name suggests, a series of fin rays that support the soft tissue of the fin (Fig. 11.11a). In contrast, the skeleton of a lobe-finned fish has more robust skeletal elements supporting a thicker, fleshier (and hence lobe-like) fin, as shown in Fig. 11.11b.

By looking at a relatively small number of Devonian fossils, we can see the transition from fish fin to tetrapod limb. To avoid the potential problem of becoming immersed in too much detail, we will focus on just a few fossil genera (Fig. 11.12), especially

- *Eusthenopteron* (a fish)
- *Panderichthys* (a fish)
- *Tiktaalik* (transitional; has been dubbed a 'fishopod')
- *Acanthostega* (a tetrapod)
- *Dendrerpeton* (a tetrapod)

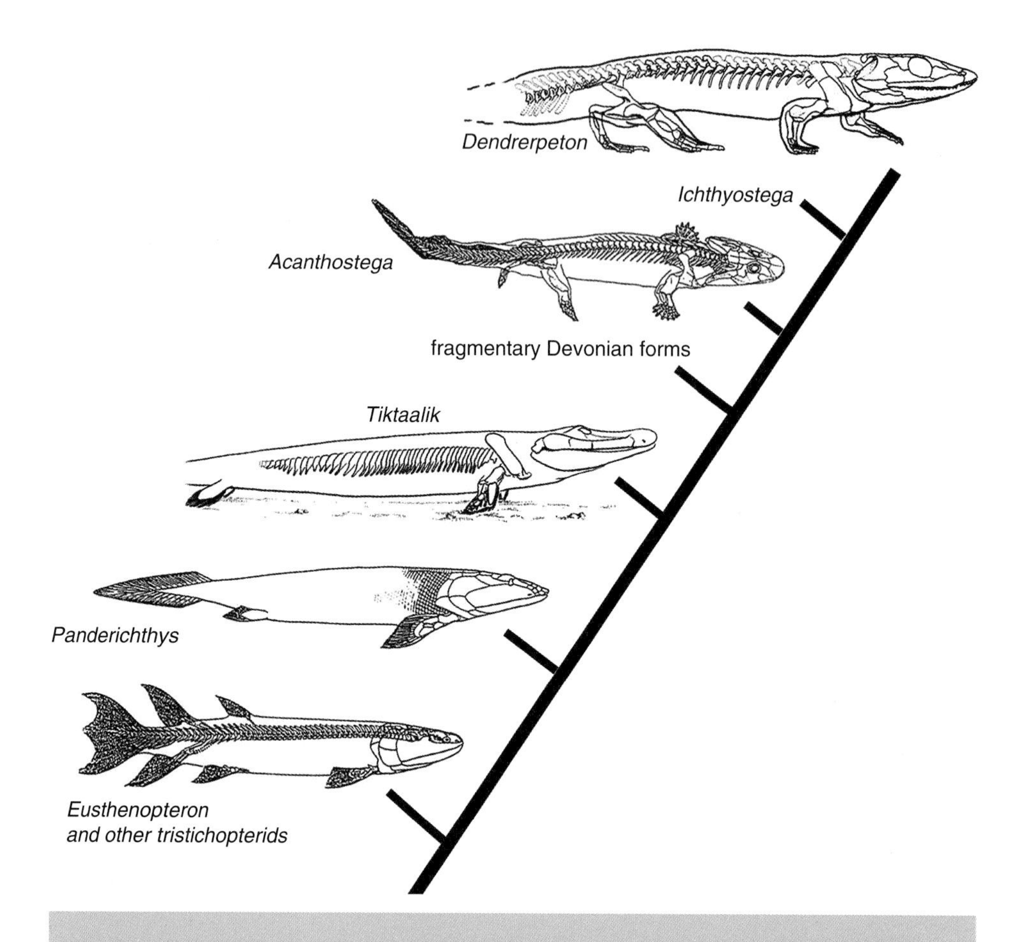

Figure 11.12 Cladogram showing selected stages in the fin-to-limb transition, from *Eusthenopteron*, which clearly has fins, to *Dendrerpeton*, which clearly has limbs and is thus an early tetrapod. The most intermediate-looking form is *Tiktaalik*, which has been dubbed a 'fishopod' (courtesy of Jenny Clack).

Note that the term 'fish' is being used here in its casual sense. Technically, from a cladistic perspective, we tetrapods are all 'fish', because the Tetrapoda is a sub-clade of the larger clade Sarcopterygii, the lobe-finned 'fish'.

As can be seen from Fig. 11.12, when appendages with clearly recognisable digits first arose, there were up to eight digits per appendage (in *Acanthostega*) rather than the five that might have been expected on the basis of the commonly accepted notion of a pentadactyl limb, which accurately describes the five-digit state found in humans and many other (but far from all) present-day tetrapods. The number of digits reduced both in early lineages, such as the one represented by *Dendrerpeton*, and in many later ones, including the extreme case of the mammalian lineage leading to the perissodactyls, such as the horse, which has a single large digit forming the hoof.

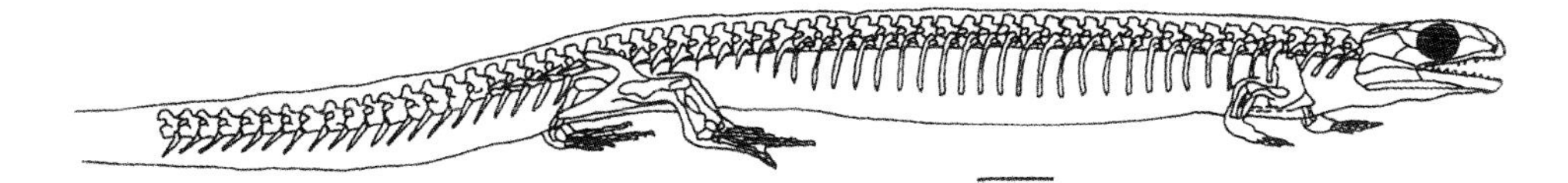

Figure 11.13 A reconstruction of the complete skeleton of *Westlothiana*. This creature, which lived in the Early Carboniferous, appears to have had a fully terrestrial mode of life (courtesy of Jenny Clack).

The early tetrapod fossils are surprising in another way too: it seems that they were not terrestrial. Various lines of evidence suggest this conclusion, including their possession of webbed feet. Although it is hard to reconstruct the ecology of animals that lived about 375 MYA, one view is that they inhabited water-bodies that were densely vegetated and that they used their rudimentary legs to 'paddle' around in this thick vegetation. However, we need to be cautious here, because we have begun to enter the territory of **adaptive scenarios**. Alternatives to the selective advantage of early limbs as paddles include: ability to walk on land when the original pond habitat dried up; ability to do so but only to walk to another pond; ability to burrow into the mud when the pond dried up to avoid desiccation prior to its re-filling; and, most radically, ability to walk, several million years earlier than previously thought, in shallow, tidal marine environments[82].

All the fossils discussed so far have been from the late Devonian, with the exception of *Dendrerpeton*, which is from the Carboniferous. The boundary between the Devonian and Carboniferous periods is about 360 MYA. The Carboniferous is one of the longer geological periods, with its base at about 300 MYA, and can be divided into the Early and Late Carboniferous (or the Mississippian and the Pennsylvanian, which are the corresponding American terms).

Unfortunately, the fossil record of most of the Early Carboniferous (360–320 MYA) is poor, so we know little of how tetrapods evolved for much of this time. However, a significant discovery of fossil tetrapods was made in the 1980s at a disused quarry near Edinburgh. These fossils are from the closing stages of the Early Carboniferous. They include *Westlothiana* (Fig. 11.13) and other genera that are thought to have been fully terrestrial animals.

Sometime between the late Devonian aquatic tetrapods and the 'late Early Carboniferous' terrestrial ones, several important evolutionary events must have taken place. These include the evolution of a truly terrestrial existence by *some* tetrapods, one of the bases for which is the amniote egg possessed by all reptiles, mammals and birds, but also the split between the stem lineage of this group (the **Amniota**) and its sister group, which led to today's amphibians.

Up to the first mention of eggs in the preceding paragraph, this account has been unfolding as a story of adults. This is because adults fossilise best, juvenile stages less well, and eggs/embryos usually not at all. However, as has been stressed from the outset of this book, the only way that evolution can make a new kind of adult is by deflecting the course of development. So it would be interesting to know some of the

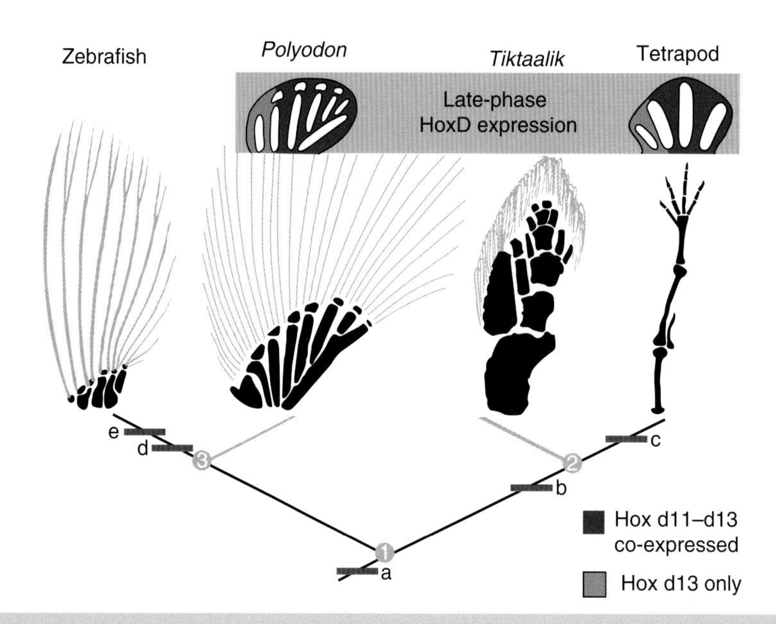

Figure 11.14 Expression of certain HoxD genes in fins and limbs. The genes *hoxd11–d13* are co-expressed in the posterior part of the fin/limb in both cases, while only *hoxd13* is expressed in the anterior part. The fish used in this study was the paddlefish *Polyodon*; the tetrapod used was the chick, *Gallus*. a–e: A series of evolutionary changes of which a and e relate directly to Hox genes. a: Acquisition of late-stage *hoxd* expression. e: loss of this expression. (From Davis *et al.* 2007, *Nature,* **447**: 473–476.)

developmental changes that were involved in producing the morphological changes in adults that have been pictured earlier in this section. We can get some idea of these changes by comparing the developmental genetic basis of fin and leg formation in present-day fish and tetrapods.

Two recent studies have made just such a comparison, importantly using basal (non-teleost) fishes (paddlefish and shark). These studies showed a remarkably similar pattern of late-stage expression of some of the HoxD genes between paddlefish/shark and tetrapod (chick/mouse). Specifically, the genes *hoxd11–13* are all co-expressed in the posterior part of the fin/foot, whereas in the anterior part, only *hoxd13* is expressed (Fig. 11.14). This is just one of many aspects of appendage patterning that have been retained in these long-separated lineages (though it has been lost in teleosts, such as the zebrafish). Its importance lies in its rejection of the hypothesis that the wrist/ankle and digit region of the tetrapod limb is an evolutionary *addition* and has no counterpart in fish.

Of course, while some aspects of developmental-genetic patterning have been conserved between basal fish and tetrapods, other aspects of patterning must have changed – otherwise we would not observe the morphological differences that we do. So, what has changed? One possibility is that the development of the tetrapod limb, with its well-developed digits, is due to an elongation of the second phase of

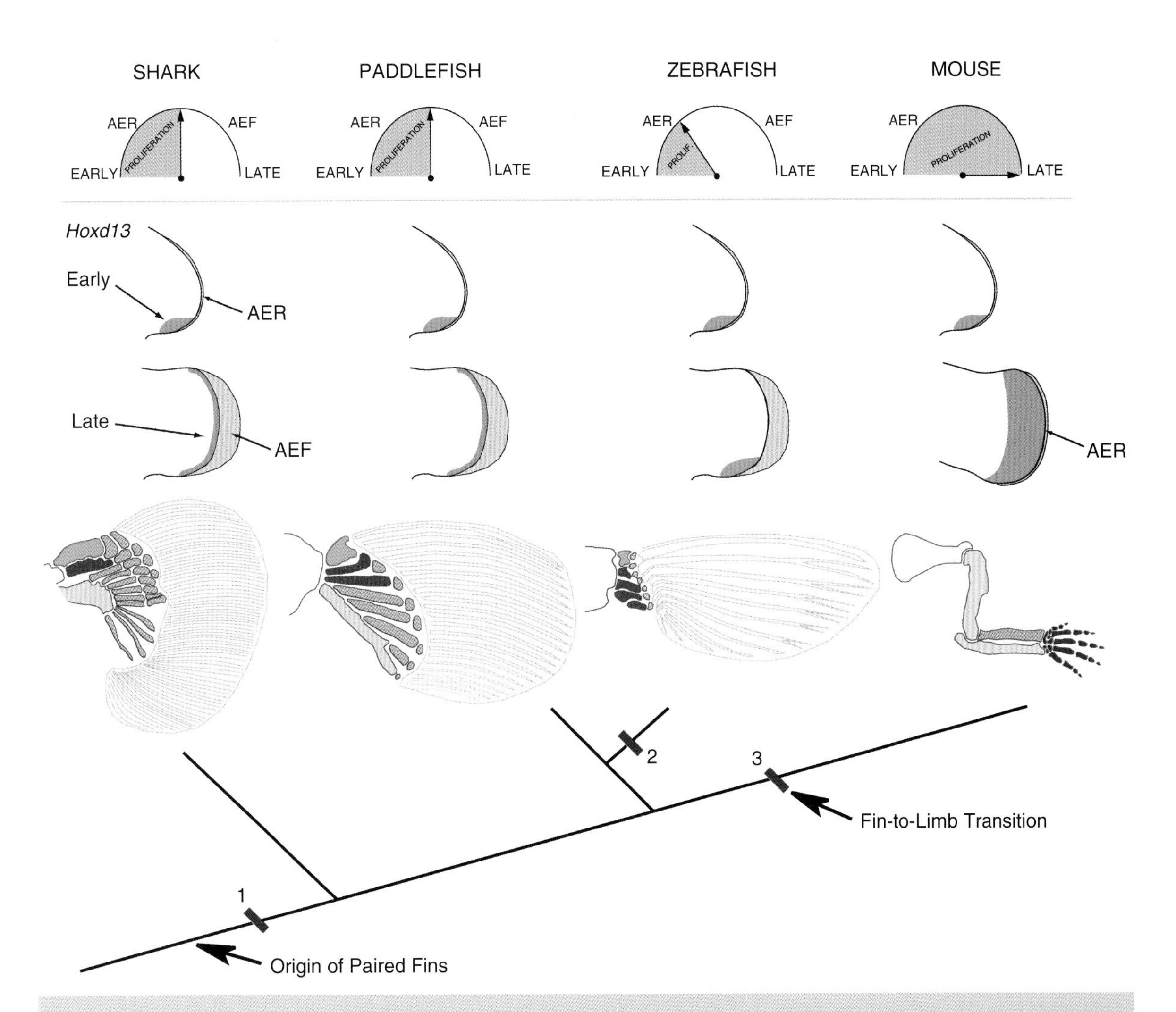

Figure 11.15 One possible developmental-genetic basis for the production of elongate limbs in tetrapods is shown here: extended late-phase expression of *hoxd13*. Note the inclusion of the three groups in Fig. 11.14 (though with tetrapods represented by the mouse this time); and the addition of gene expression data on the shark. 1,2 – gain and loss of late-phase *hoxd13* expression (same as a and e in Fig. 11.14). 3 – extended late-phase expression, associated with the development of elongate limbs. Top – possible changes in the timing of the switch from the cell-proliferative AER to the non-proliferative AEF (Apical Ectodermal Ridge/Fold). (From Freitas *et al.* 2007, *PLoS ONE*, **2**: e754.)

HoxD gene expression, perhaps associated with an extended phase of cell proliferation at the distal end of the developing limb (Fig. 11.15). The next few years are likely to see further progress in our understanding of the combination of conservation and change in gene expression patterns associated with the evolution of the tetrapod limb.

11.4 The Origin of Flowers

Like the origin of walking limbs in vertebrates, the origin of flowers in plants occurred just once; thus all flowers are homologous. Indeed, this particular origin is one of the most significant events in plant evolution. The dominance of the land vegetation of the Carboniferous by non-vascular plants such as tree-ferns has given way to dominance of the current land vegetation by **angiosperms**. This dominance is seen in most forest ecosystems, such as the tropical rainforests, and also in most herbaceous ecosystems such as prairies, steppes and savannahs. The biggest single exception to this generalisation is the dominance of conifers in the vast boreal forest that sweeps across northern North America and Eurasia.

It may be helpful to pause at this point and ask the deceptively simple question: what is a flower? In addition to the more familiar examples (rose, daffodil, waterlily and so on) there are many other very different forms (Fig. 11.16). These include the Chinese lanterns that we looked at in Chapter 7, and the rather inconspicuous flowers of grasses. In terms of their function, flowers can be male, female or have both sexual parts represented. The size of flowers varies enormously, as does their occurrence in single or multiple (composite) forms. Some of these features are variable, even within narrow taxonomic groups such as genera, making it difficult to identify the ancestral flower type, even if we can identify the most basal angiosperm group, which itself has not yet been established with certainty.

A recent phylogenetic tree based on molecular data (Fig. 11.17) suggests that the most basal living angiosperm may be the genus *Amborella*, consisting of a single known species from the islands of New Caledonia, northeast of Australia. Another angiosperm group that is thought to be basal is the Nymphaeales (including the waterlilies and the family Hydatellaceae, which was discussed in Chapter 7). With regard to fossil evidence of angiosperms, there are fossil flowers, flower parts and pollen dating back to about 130 to 140 MYA. However, as usual, fossil evidence only indicates the minimum age of a group: its actual origin may have been much earlier. In the case of angiosperms, analysis of molecular data has suggested a possible origin about twice as long ago as the first fossils, but this estimate should be treated with extreme caution, especially since many 'molecular clock' estimates of the origin of animals now seem to have been very misleading.

A 'typical' flower, inasmuch as there is such a thing, is shown in Fig. 11.18a. There are four whorls of structures, some or all of which may have evolved from modified leaves. From outer to inner, the structures are: sepals (not visible in the figure), petals, stamens (male) and pistil (female). However, in some angiosperms, including many that are thought to be basal, these structures are arranged in spirals rather than circles. One piece of evidence for the flowers-from-leaves theory is that mutation of a gene called *leafy* in the plant model system *Arabidopsis* causes replacement of the flowers by leaves. It is known that this gene regulates those of the **MADS box** group that encode **transcription factors** and are involved in the development of flowers (recall the ABCDE model of flower development that was illustrated in Fig. 11.10 in Chapter 5).

Figure 11.16 A small sample of the great diversity in the forms of flowers.

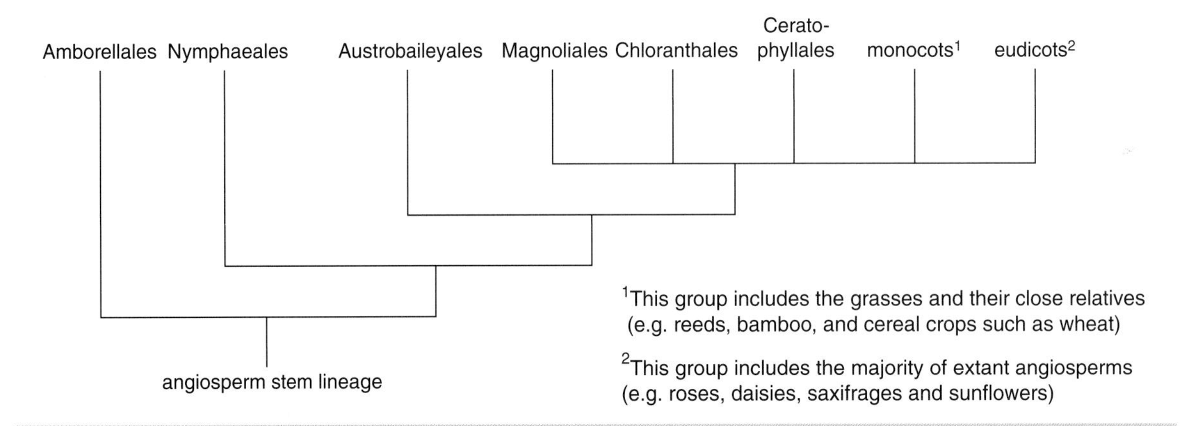

Figure 11.17 A phylogeny of the angiosperms based on analysis of molecular data. This suggests that *Amborella* is the most basal extant angiosperm, and thus of particular interest in relation to the evolutionary origin of flowers.

Figure 11.18 A 'typical' flower (a) and, in contrast, (b) the rather inconspicuous small flowers (left male, right female) of *Amborella*. (Courtesy of Sangtae Kim, Matyas Buzgo and Professor Pamela Soltis.)

The flowers of *Amborella* are small (just a few mm across) and rather inconspicuous (Fig. 11.18b). The petals and sepals are so similar that they have been referred to jointly as tepals. Because any particular *Amborella* flower only develops one set of sexual structures fully, a flower cannot self-pollinate. Ecological studies have shown that *Amborella* pollination in nature is caused by both wind and insects.

The selective advantages that were involved in the origin of flowers are not clear, though those involved in many later modifications of flower structure in particular lineages, for example for pollination by particular insects, are very clear indeed. Given the flower's many parts, different (but related) forms of selection were probably involved in the origin of each. A suitably cautionary note on which to end this section is the title of a paper that appeared in *Nature* a couple of years before the time of writing this chapter (see Suggestions for Further Reading): 'After a dozen years of progress the origin of angiosperms is still a great mystery.'

11.5 General Conclusions on Repatterning and Selection

In this chapter we have been dealing with major evolutionary events, each of which has undoubtedly taken place in stages over a long period of time, even though this is clearer in some cases than others. There is one main conclusion to be drawn from these events about developmental repatterning, and another to be drawn about selection/adaptation.

First, repatterning. Although some of the more minor instances of repatterning that we looked at in Chapters 6 to 9 were characterised by a predominance of one of the four types of change (**heterochrony**, **heterotopy**, **heterometry** and **heterotypy**) with perhaps minor components of the others, major repatternings are probably always compound. Of the three main examples looked at in this chapter, the compound nature of the change is probably easiest to see in relation to the vertebrate fin-to-limb transition. We examined this at both morphological and molecular levels; and at both of these the repatterning can be seen to be compound. Changes in both temporal and spatial expression of some of the Hox genes are involved in the process (hence heterochrony and heterotopy); and new kinds of structures emerge (the wrist/ankle and the digit; hence heterotypy). Although we did not explicitly encounter heterometry, it is hard to believe that it is not involved somewhere. So, one message of this chapter backs up the main message of the last: the integrative (or compound) nature of developmental repatterning.

Now, selection. In the flower example we noted that the selective advantage of the first flowers is hard to determine, but that many subsequent changes in flower structure are much easier to interpret in selective terms, including cases of adaptation to pollination by certain insects. If the *Amborella* flower-type is indeed similar to the first flowers (which is by no means certain), then those flowers were hardly a great visual attractant to insects, even if they may have sometimes been insect-rather than wind-pollinated. And in the limb example we noted that limbs that could later be used for walking in completely terrestrial habitats were probably initially evolved for paddling, either in densely-vegetated bodies of fresh water or in shallow marine waters. In both cases, what we have is thus **exaptation**. An evolutionary change begins going in a particular direction for one selective reason, associated with one form of adaptation, but later continues in the same general direction for quite another selective reason. Such exaptation is an important topic and will be discussed further in the next chapter.

SUGGESTIONS FOR FURTHER READING

For an excellent account of ideas and information on the origin(s) of animal segmentation see:

Chipman, A.D. 2010 Parallel evolution of segmentation by co-option of ancestral gene regulatory networks. *BioEssays*, **32**: 60–70.

For an accessible (and personal) account of the fin-to-limb transition from a palaeontological perspective, see:

Clack, J.A. 2002 *Gaining Ground: The Origin and Evolution of Tetrapods*. Indiana University Press, Bloomington, IN. (second edition *in press*).

For a review of the origin of the angiosperms, covering both progress in this area and remaining uncertainties, see:

Frohlich, M.W. and Chase, M.W. 2007 After a dozen years of progress the origin of the angiosperms is still a great mystery. *Nature*, **450**: 1184–1189.

PART III

The Direction of Evolution

'The internal structure of organisms has directly influenced the avenues of phylogeny.'

Lancelot Law Whyte, *Internal Factors in Evolution*, 1965

In this, the third part of the book, we focus on the issue of what mechanisms steer the course of evolution in certain directions rather than others. As the great American palaeontologist G.G. Simpson once said: 'The various major schools of evolutionary theory have arisen mainly from differences of opinion as to how evolution is oriented.' The prevailing view in Simpson's time, the mid-20th century, was that the direction of evolution was primarily determined by Darwinian natural selection. Is this still the prevailing view today? To put it another way, has the advent of a whole new approach to evolution – evo-devo – changed the prevailing view of how evolution is oriented? This question will now be answered in stages, as follows:

Chapter 12: Natural selection is alive and well in the evo-devo era, and is responsible for adaptation as envisaged by Darwin and many others since. But an evo-devo approach urges special attention to the phenomena of coadaptation and exaptation.

Chapter 13: The nature of the developmental process imposes 'constraints' on the possible directions of change that evolution can produce. Or, put another way, developmental bias leads to increased probability of change in some directions and decreased probability of change in others. Thus development itself can influence the direction of evolution.

Chapter 14: An overview of developmental genes is followed by a more detailed look at some key examples. Ways in which these genes can evolve are examined and we note that it may be 'easier' to produce adaptive change by altering regulatory rather than coding sequences of developmental genes.

Chapter 15: Developmental bias now shows up in another guise. It is 'easier' to co-opt old genes for new roles than to invent new genes. Gene co-option is seen to be one of the most important new concepts to have emerged from evo-devo research.

Chapter 16: The environment is not merely a 'sieve' that selects the more fit from the less fit. In many cases, the environmental conditions under which development takes place affect the character states of the developing, and later the adult, organism: a phenomenon known as phenotypic plasticity. Although a plastic response is not inherited, the ability to make such a response often is; in such cases, natural selection acts not on fixed characters but rather on 'developmental reaction norms'.

Chapter 17: Here we look at origins: of species, evolutionary novelties and body plans. Some of these origins may involve an especially important role for developmental bias.

Chapter 18: Some lineages have evolved towards massively increased complexity, either of body form or of life-cycle. How they have done so is a fascinating issue. However, we must be careful, when concentrating on these increases in complexity, not to over-generalise. There is no law of increasing complexity in evolution. Rather, some lineages exhibit complexification, others exhibit simplification, and others simply exhibit *diversification*, without any change in the level of complexity of the resultant creatures. When complexity does increase, duplication-and-divergence of parts is an important mechanism.

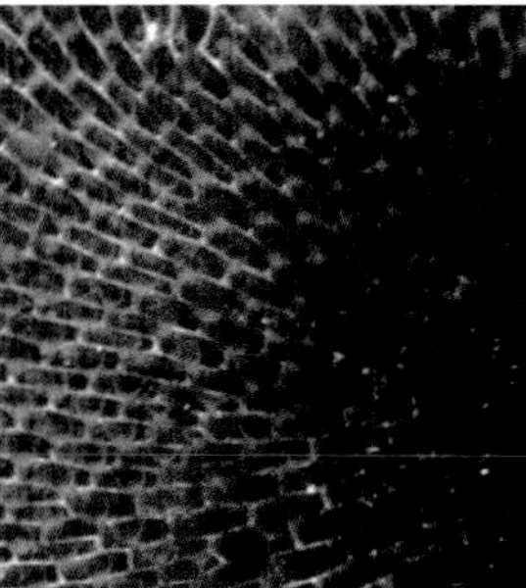

CHAPTER 12
Adaptation, Coadaptation and Exaptation

12.1 Natural Selection on a Continuously Variable Character

In Chapter 4, we examined selection on discrete variants that comprise a **polymorphism** in a population, such as pale and **melanic** moths. Here, we will examine the ways in which natural selection can act on a character that exhibits continuous variation in a population, such as human height. This kind of variation is usually thought of as exhibiting a normal distribution of character values in the population. While this is a useful, simplified way to picture the variation in such a character, it is often only true of a subset of a population; the whole population usually has a more complex distribution. For example, in most human populations, height is normally distributed only if we look at a subset of the population, such as adult males or adult females.

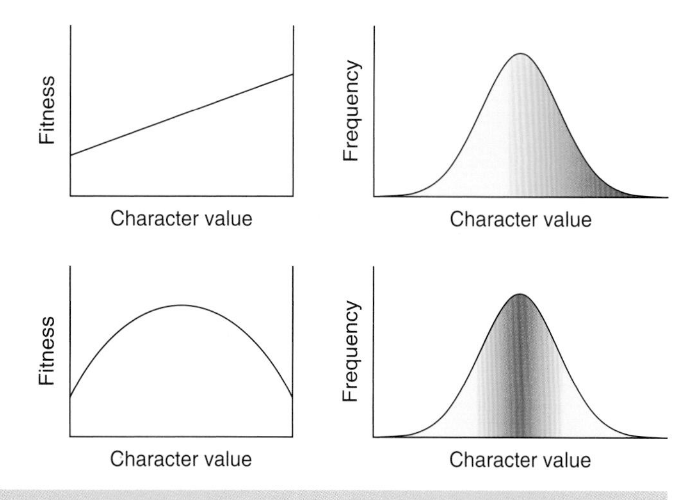

Figure 12.1 Two types of selection acting on a continuously-variable character: directional (top), which causes evolutionary change; and stabilising (bottom), which acts to maintain current character values. In the right-hand pictures, fitness is indicated by the colouration: Redder – more fit; yellower – less fit.

Thousands of characters exhibit continuous variation in a typical natural population of any given species. How selection works, in any particular case, depends on how variation in **fitness** relates to variation in character value. There are two main possibilities: (i) fitness increases (or decreases) with character value; (ii) the fitness maximum is in the centre of the distribution and drops off to both sides. These correspond to directional and stabilising selection (Fig. 12.1). The former shifts the mean character value in the population, and so causes evolutionary change; the latter acts to maintain the current mean value. As will be apparent, they are broadly equivalent to directional and balancing selection acting on polymorphic variation; albeit one of the names is different.

Of course these two 'main possibilities' of ways that selection can act on continuous variation are just that. There are many more complex ways in which fitness could map to character value. Variants at both extremes of the distribution could be fitter than those in the middle, leading to 'disruptive selection', which is probably much rarer than its stabilising and directional counterparts. And indeed there could in theory be multiple fitness peaks and troughs as character value increases, though it is often hard to picture a biological reason for that sort of situation.

While the moth *Biston* is the classic **case study** of directional selection acting on polymorphic variation, Darwin's finches provide the equivalent classic for selection acting on continuous variation – especially on the size and shape of the birds' beaks. In addition to the vast primary literature on the subject, there are several books entirely devoted to this small group of rather drab birds.

The main gist of the story is well known, and only needs to be told in outline. Most of Darwin's finches (about 13 species, though recent molecular work questions the exact number) occupy the Galapagos Islands (Fig. 12.2), situated off the west coast of South America, with one species living on Cocos Island, which is several hundred kilometres to the northeast of the Galapagos archipelago. The group appears to be **monophyletic** and to have arisen by repeated **speciation** of an ancestral finch species, dispersing members of which initially arrived from the South American mainland.

Island archipelagos provide a particularly fertile geographical setting for speciation events (Chapter 17), probably due to a combination of two main factors: different ecological conditions on different islands; and restricted movement between islands. *Drosophila* fruit-flies have undergone many speciation events in the Hawaiian archipelago, producing in that case not just 10 to 15 species but rather hundreds of them.

The **phylogenetic** relationships of Darwin's finches[1], as determined by studies based on the sequence of portions of their mitochondrial DNA, are shown in Fig. 12.3. This tree confirms many previous views on relatedness, including the existence of monophyletic groups of ground finches and tree finches. Note, however, that the

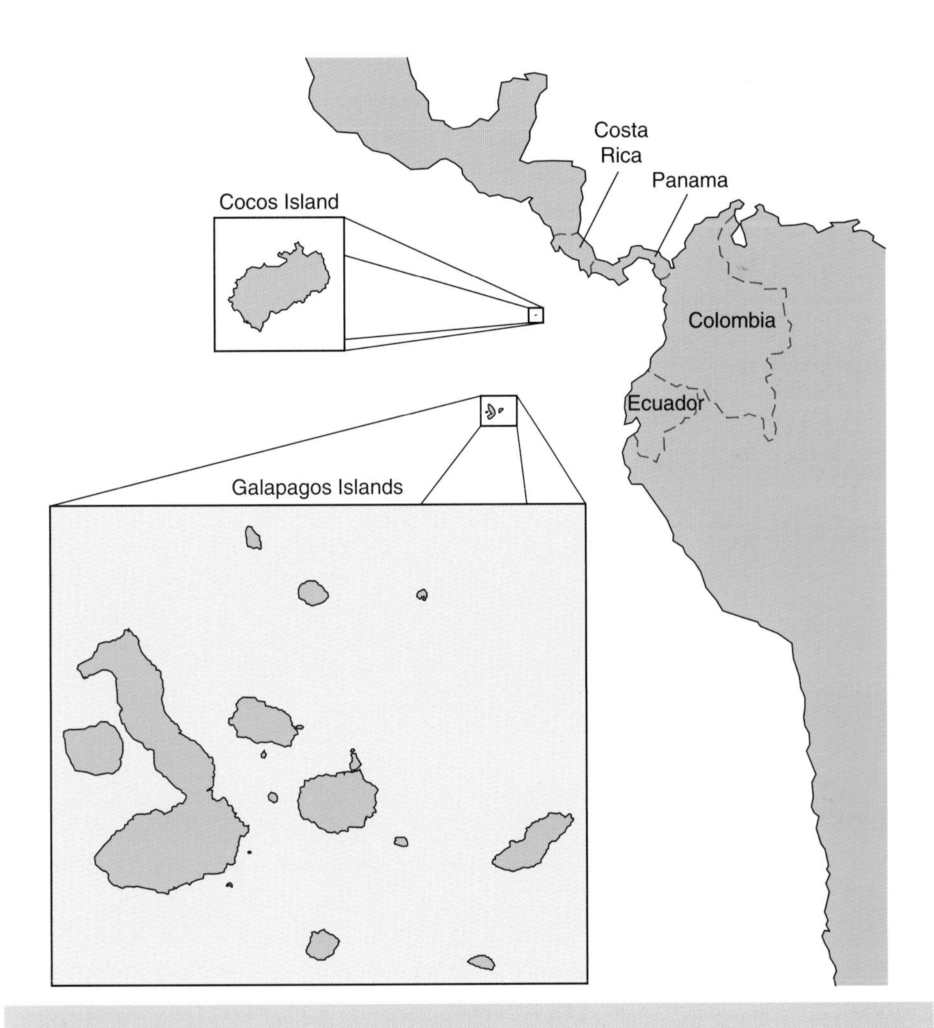

Figure 12.2 Map of the Galapagos archipelago, home to all but one of Darwin's finches; and Cocos Island, where the other member of this species group is found.

Cocos finch is not the **outgroup** of the overall **clade**; this suggests that the colonisation of the Galapagos from the mainland was not via Cocos Island.

Scientific interest in Darwin's finches has generally been focused on the size and shape of the beak (see LHS of Fig. 12.3). This is because these characters have an obvious link with diet, and that in turn has an obvious link with specific habitats, and sometimes with specific islands. The series of speciation events that produced this group of finches is thought to have occurred largely because of inter-island migration followed by adaptation to a particular array of food sources present on the island concerned. It is also possible that competition among some pairs or groups of the finch species has provided an additional dimension to the selective process; though this can be regarded as part of the same overall phenomenon, providing that such competition acts through joint use of a common food supply. Looking at Fig. 12.3, it can be seen that there is variation in the following beak characters: overall size, length, depth, pointedness and curvature. Most

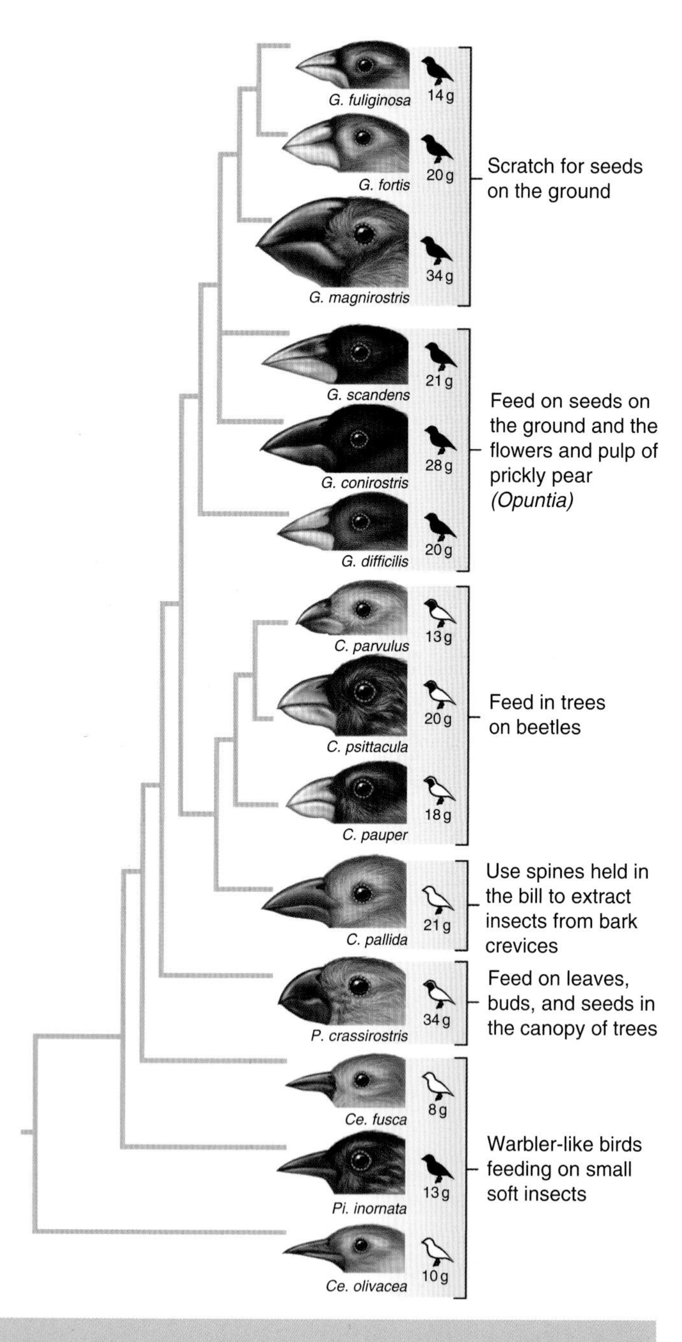

Figure 12.3 Phylogeny of Darwin's finches, showing variation in beak shape and size; and that the Cocos finch (*Pinaroloxias inornata*) is nested within the clade rather than being the outgroup. (From Townsend *et al.* 2008, *Essentials of Ecology*. Blackwell Publishing Ltd.)

interspecific comparisons show compound changes that involve at least two of these characters.

In one sense, the story so far is the opposite of that of *Biston*, because in that case we were looking at intraspecific differences that might one day contribute to a speciation event but have not yet done so; whereas in the case of Darwin's finches, we are looking at the result of accumulated directional selection in the past, which has indeed contributed (along with selection on other characters) to the divergence of different species. However, careful study of populations of individual Galapagos finches, notably of *G. fortis*, has also shown intraspecific variation in beak size that can be correlated with changes in seed size and hardness caused by changing climate. So in this case study we can actually see both intra- and inter-specific dimensions of the selection that is involved.

When the British ornithologist David Lack wrote his book *Darwin's Finches* (a phrase that he coined) in the 1940s, many of the above components of the story were known. However, since then, two further components (in addition to the molecular phylogenetics noted above) have begun to be studied, and both of these are of key relevance to understanding the role of selection acting on continuously variable characters in general. One is genetic, the other developmental, though in the end these two components will merge together.

In terms of its genetic basis, the beak size of a bird, like other continuously variable characters in other creatures, is only partially heritable. However, the **heritability** of beak size in *G. fortis* has now been estimated, and the value is very high – about 0.95, meaning that about 95 per cent of the variation in the population studied is due to genetic rather than environmental (e.g. nutritional) causes. This genetic component is likely to be due to polymorphism at many genes, though for the most part we do not yet know which genes they are. The genes contributing to a continuously variable (or 'quantitative') character are called **quantitative trait loci**, or QTLs for short. Determining the number and identity of these in any particular case is difficult; but notable successes in this regard have now been achieved in some other systems (Fig. 12.4).

Because there are many QTLs contributing to the variation in a quantitative character, such characters are often described as **polygenic**. But equally, any one gene may contribute to two or more phenotypic characters – a phenomenon known as **pleiotropy**. Both these situations are common, and indeed should be regarded as the norm. They are contrasted with the one-to-one relationship between genes and phenotypes, which was the simple situation we started with, in Fig. 12.5.

With regard to new *developmental* information on the beaks of Darwin's finches, two important findings have been produced by a research group led by American developmental biologist Clifford Tabin. These concern expression of the genes that make the proteins BMP4 (bone

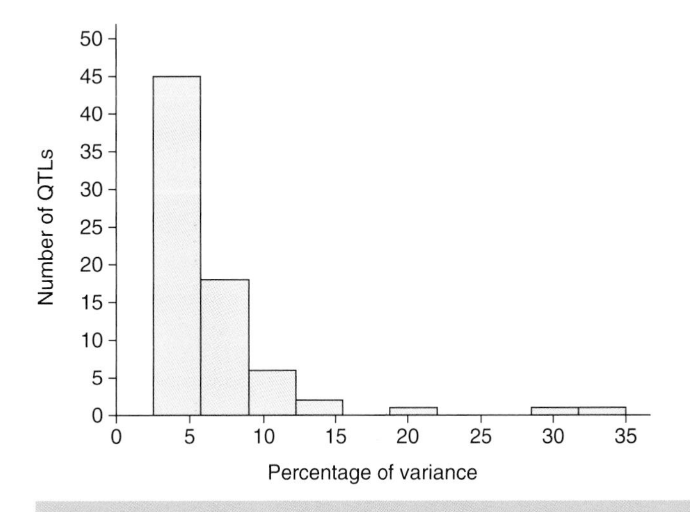

Figure 12.4 Mapping the quantitative trait loci (QTLs) underlying the variation in growth form among tomato plants. In this particular study, 74 QTLs were detected underlying the variation in 11 morphological characters. As can be seen, most loci have only a very small effect on the variation, while a few loci have much larger effects. (Source: Tanksley, S.D. 1993, *Ann. Rev. Genet.* **27**: 205–233.)

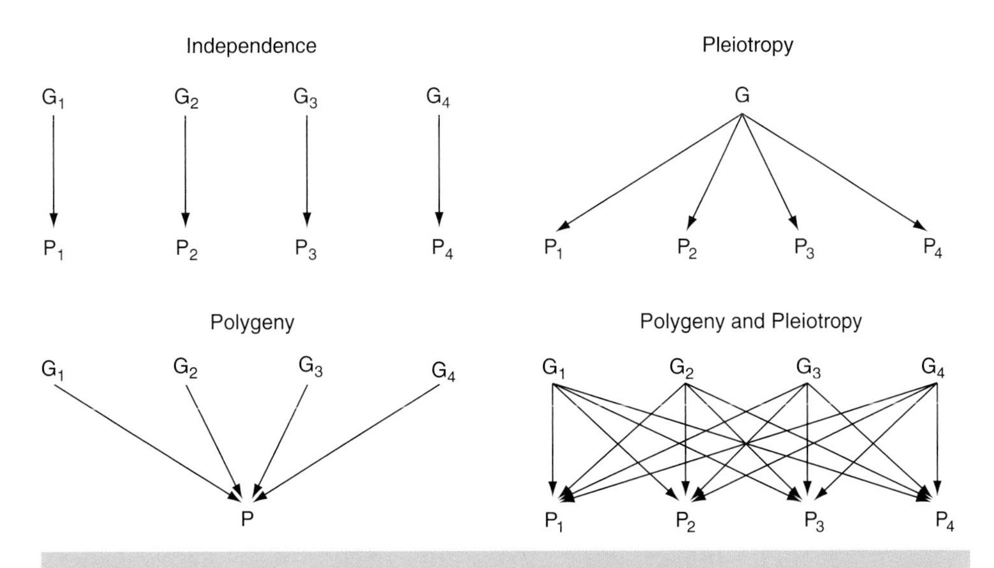

Figure 12.5 Possible relationships between genes and phenotypes: independence (one-to-one); pleiotropy; polygeny; and a combination of polygeny and pleiotropy. The last of these is probably the commonest.

morphogenetic protein 4) and calmodulin (*cal*cium-*modul*ated prote*in*), both of which are known to be important **developmental genes** in relation to many characters in many different creatures – for example, the BMP4 homologue in *Drosophila*, Dpp (Decapentaplegic) helps to establish the dorso-ventral body axis, as we saw in Chapter 3.

What the Tabin group have shown is that expression of *bmp4* is associated with deep and broad beak forms[2], while *calmodulin* expression is associated with long, pointed beaks[3]. Experimental manipulation of the expression of these genes in chick embryos confirmed the predicted changes in beak morphology that arose from the comparative gene expression work on Darwin's finches.

The link between these new genetic and developmental findings will probably turn out to be that *bmp4* and *calmodulin* are two of the QTLs that contribute to the overall variation in beak characters within each of the finch species. However, whether one or both of these genes are among the most important of the QTLs is not yet clear, and indeed this may well be dependent on the particular species under investigation.

12.2 Natural Selection on Two Characters; and the Idea of an Adaptive Landscape

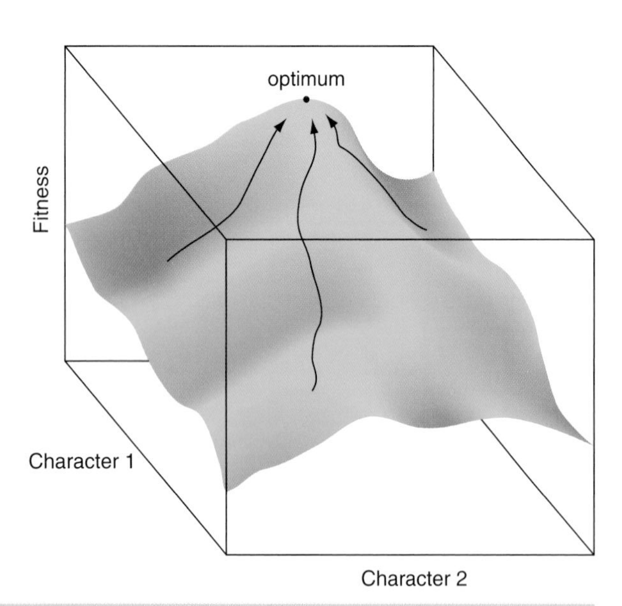

Figure 12.6 The adaptive landscape model of selection acting on combinations of the values of two characters. The arrows indicate that selection will move character values to the highest point on the landscape – the 'local optimum'.

As already noted, it is unlikely that natural selection acts independently on each individual character. After all, a multicellular creature is an integrated whole, not a series of only loosely connected characters or traits. We will now try to imagine how selection might act jointly on two (continuously-variable) characters. This is still too simplistic, but it is a step in the right direction.

Selection acting on combinations of the values of two characters can be considered using the **adaptive landscape** concept, introduced by the American pioneer of population genetics, Sewall Wright[4] (Fig. 12.6). Here, the values of characters 1 and 2 (whatever they may be) form the base of the landscape, while fitness is represented by the vertical axis. Fitness variation thus corresponds to the topography of a real landscape in this sort of representation. If there is no variation in fitness across different combinations of character values then the landscape is flat – as in a plain or steppe. Alternatively, if there is much variation, then the landscape is mountainous, with a series of fitness peaks separated by lower-lying valleys or troughs.

Using this form of representation, any population will evolve towards a fitness peak, as shown in the figure. However, there are three complications to this apparently simple picture that need to be mentioned. First, given that environmental change over time is the rule rather than the exception in nature, it has been suggested that a 'seascape' would be a better analogy than a landscape, because the fitness peaks may move (like the crests of waves) as the environment changes. Second, what happens after the population climbs a particular fitness peak will depend on several things, including the extent of its own variation and the structure of the landscape, in particular the degree to which the different peaks are separated by deep troughs – i.e. whether the landscape is smooth (hilly) or more rugged. Third, the adaptive landscape model is sometimes taken to equate fitness with ecological adaptation; this may not be warranted, as we will see in the following section.

12.3 Developmental and Functional Coadaptation

The two classic case-studies of natural selection, the peppered moth (Chapter 4) and Darwin's finches (Section 12.1), although different in the nature of the variation available for selection to act upon (polymorphic versus continuous), have something important in common: they both involve a type of directional selection producing adaptation that is environment-specific. Melanic moths have a selective advantage in soot-polluted environments, but a selective *disadvantage* in clean, rural ones. Also, although the comparison is less straightforward because it is an interspecific one, finches with 'chunky' beaks have an advantage when it comes to eating hard seeds, while they would no doubt be at a disadvantage if forced to compete with pointed-beak finches on a (hypothetical) island where cacti provided the sole food source.

This common feature of these two classic case-studies is apt to give rise to an unwarranted equating, in people's minds, of increased fitness and increased environment-specific adaptation. It is true that these two things are often linked; but it is equally true that they need not be.

In a standard textbook on evolution, the most likely subject to be encountered at this point would be sexual selection, for example that which is thought to have resulted in the production of the peacock's tail. Since mating is important in all environments, such products of sexual selection are generally not environment-specific. Rather, they convey increased fitness to their bearers regardless of the precise environmental conditions. However, since sexual selection is adequately dealt with in standard evolution texts, the subject that we will now concentrate on is not sexual selection but rather something that has been labelled (unfortunately, as we will shortly see) 'internal selection'.

To understand what is meant by this phrase, we need to think about two separate aspects of adaptation: the standard kind, referring to the degree to which an animal or plant is adapted to its ecological *milieu*; and a second, less often discussed, kind, referring to the degree to which the interacting parts of an animal or plant – at both morphological and molecular levels – are **coadapted** to each other. We will consider, in order to make this distinction, some aspects of the morphology of bears and foxes. In particular, we will contrast two traits – the scale of the body and the pigmentation of the fur.

Figure 12.7 An arctic fox and, for comparison, a typical fox from a temperate zone region such as Ireland. The environment-specific nature of the adaptation (in fur colour) is clear in this case.

Everyone knows that polar bears are white. Arctic foxes are also typically white (Fig. 12.7), though the colour of their fur changes on a seasonal basis, being whitest in winter, browner in the summer months. We interpret the whiteness, in both cases, as being the result of natural selection for camouflage – that is, it makes these predatory animals less visible to their prey in a snowy environment. This is probably correct; after all, if the fur colour was adapted to maximise heat uptake, which might also be useful in such cold climates, it would be black. The problem is solved by these animals having especially thick but pale fur, the former property aiding insulation, the latter aiding prey-catching ability.

No-one doubts that the whiteness of polar predators is an environment-specific adaptation. Temperate-zone bears and foxes are typically some shade of brown. But now we will broaden the picture to consider how to, in a sense, 'turn a fox into a bear' or, more generally, how to make one type of carnivorous mammal into another in the course of evolution.

Notwithstanding the more general use of the term carnivorous, meaning meat-eating in any group of animals, the Carnivora is a taxonomic order of mammals that includes bears and foxes, together with dogs, cats, weasels and many others. The order has two main subdivisions, the Feliformia (cat-like) and the Caniformia (dog-like). All are **monophyletic** groups. Within the Caniformia, bears and foxes are not **sister-groups**, so the idea of evolution 'turning a fox into a bear' is somewhat fanciful. But evolution gradually turning one kind of caniform into another is quite the opposite: an oft-repeated fact.

Suppose that the ancestral form is small, and what evolves from it is scaled up in body size. Although it is obvious that such an overall change must involve many co-ordinated alterations in body form, this rarely receives the attention it deserves. To put it another way, coadaptation of body parts receives far less attention in standard evolutionary biology texts than does adaptation to specific environments. This disparity of treatment has no basis in fact: both processes are very important in evolution.

In the context of scaling up a caniform mammal, if the initial steps create some imbalance, such that, for example, the skull is scaled up more than the rest, then later changes that correct this imbalance will be selected for. These latter changes will improve fitness in all environments and thus are best considered as improving the level of coadaptation. There is a notable contrast here with the evolution of camouflaged colouration, which improves adaptation in specific environments but probably leaves the level of coadaptation unchanged.

Despite the importance of this distinction, many intermediate scenarios can be imagined in which an evolutionary change may be adaptive in all environments but more so in some than in others. A useful way to picture all the possibilities ranging from (i) adaptation to a single set of environmental conditions to (ii) coadaptation that is equally advantageous in all environments, is to use **trans-environment fitness profiles** (Fig. 12.8).

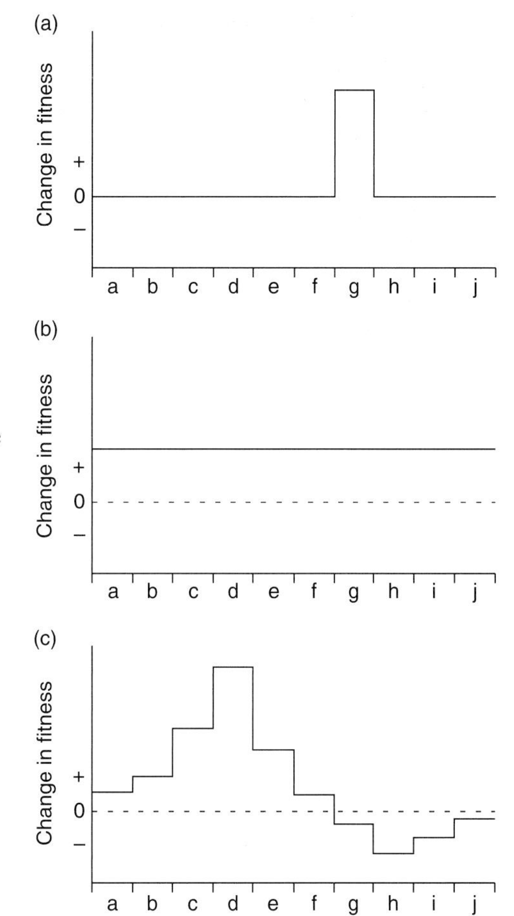

Figure 12.8 Trans-environment fitness profiles. This form of representation can be used to illustrate variation in the degree of selective advantage possessed by some developmental/phenotypic change. At one extreme (a) is an advantage unique to one environment; at the other (b) is an advantage that is equal in all environments. Most actual adaptations are probably in between these extreme cases (c).

Not only can the two contrasting caniform mammal evolutionary changes be illustrated in this way, so too can other contrasting changes in other taxa. For example, it is clear that the evolution of melanic moths and the evolution of moths whose wing-thorax joints have improved co-ordination, and thus can fly better, could be represented as contrasting patterns in Fig. 12.8. This kind of general representation of evolutionary changes, the trans-environment fitness profile, deserves to be much more widely used than it has been to date.

Now, back to that label of 'internal selection'. The person who championed the ideas discussed in this section back in the 1960s was the Scotsman Lancelot Law Whyte, who was a 'polymath', writing important works in many fields ranging from psychology to physics and engineering, rather than being specifically a biologist. In a series of publications, Whyte called the process leading to coadaptation by two names: first[5], *developmental selection*; and, later[6], *internal selection*. There were good reasons for both of these choices, and yet both were ultimately flawed.

Whyte started by using 'developmental selection', because he was thinking not of coadaptation of the adult but rather of the embryo. Any particular embryonic stage needs to be coadapted in two different ways: functionally in its own right (like the adult) but also temporally, in the sense that it must connect in an integrated way with the stages before and after it, so that the overall process of **embryogenesis** flows successfully between stages.

Whyte later switched to using 'internal selection', because that included the adult as well as the developmental stages that preceded it. However, while this was a more inclusive (and therefore in at least one sense better) term, it was misunderstood by some neo-Darwinians of the time to indicate that the selective process was itself somehow internal to the organism. This mis-interpretation led to Whyte's ideas being criticised by evolutionists (even those who were interested in 'coadapted gene complexes') and to a lack of acceptance of his main point. This is a classic case of a bad name leading to the neglect of a good idea.

In order to ensure that directional selection leading to coadaptation of body parts no longer languishes at the fringes of evolutionary theory, it is better if we use another term to describe the kind of non-environment-specific selection that Whyte had in mind. A better term is: *selection for internal integration*. Note that this is a name for the process, while coadaptation is a name for its outcome. Whatever term we use though, the pictorial approach that is embodied in Fig. 12.8 applies. This is perhaps indeed a case where 'a picture is worth a thousand words'.

12.4 Morphological Geometry and Selection

Many of the studies that have been carried out on the evolution of the beaks of Darwin's finches, including those described in Section 12.2, used simple linear measures of the beak, notably length and depth. While much headway can be made using such metrics, any morphological structure, whether in a bird, in another kind of animal, or in a plant, can also be characterised by more complex forms of measurement that more adequately reflect its complex structure. One such form of measurement involves the use of 'landmark' features. We will now look at an example of using landmarks – this is based on the mandible (the lower jaw) of mice.

Mice belong to the mammalian order Rodentia, which has about 2500 species (see Fig. 12.9 for some examples). Within this, mice (and rats) belong to the subfamily Murinae, which includes more than 500 of these. Rodents are characterised by just two pairs of upper and lower incisor teeth, which grow throughout life (this growth being counteracted by gnawing). Rodents in general, and mice in particular, are very widespread in their distribution. Even just taking the single mouse genus *Mus* (to which the **model animal** *Mus musculus* belongs), its distribution extends across most continents and many islands. This makes it very different, in terms of its geography, from Darwin's finches, which, as we saw previously, are very geographically restricted.

The study of *Mus* upon which we will focus, however, is one that was laboratory-based, and is more about predicting responses to natural selection than about describing and analysing them after they have happened (as was the case with *Geospiza*). This is an interesting and unusual approach.

A diagram of the *Mus musculus* mandible is given in Fig. 12.10. As can be seen, there are 11 landmark features characterised, with 1 and 11 representing the base of the incisor. In a study of these features, carried out by Swiss biologist Chris Klingenberg[7], their tendency to co-vary was determined in the form of both genetic and **phenotypic** covariance matrices (more on these in Chapter 13). Covariance of characters, especially when it has a large heritable component, is important in relation

(a)

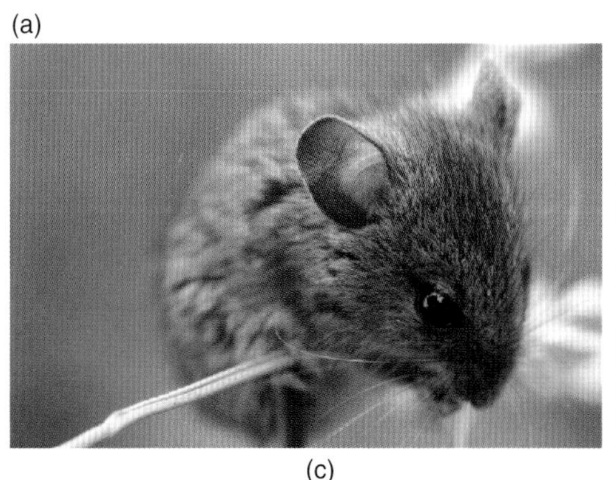

(b)

(c)

Figure 12.9 Three rodent species. The first two, a mouse and a rat, are closely related and belong to the family Muridae. The third, a porcupine, is much more distantly related. Rabbits (not shown) are sometimes mistakenly thought to be rodents, but in fact belong to a different mammal order – the lagomorphs.

to how they respond to selection. In general, the higher the covariance, the greater the correlated response in landmark j when selection is applied to landmark i.

Figure 12.11 shows the results of two kinds of simulated selection, labelled A and B. In A, selection to alter the angle of the incisors produces a predicted response in the two landmarks involved in that character; but also correlated responses in many others (e.g. landmarks 4, 8, 9 and 10). In B, selection to increase the depth of the posterior part of the mandible produces a predicted response in the main landmarks concerned (4 and 7) and, again, a correlated response in others (e.g. 1 and 10). Measurement of the direct and correlated responses showed that the direct response to selection was the smaller of the two in A, but the larger of the two in B.

This contrast is interesting because A was an arbitrarily determined form of selection (on the part of the experimenter) while B was based on the actual pattern of mandible shape divergence between mice and rats. Further arbitrary and non-arbitrary forms of selection reinforced this message – the former were

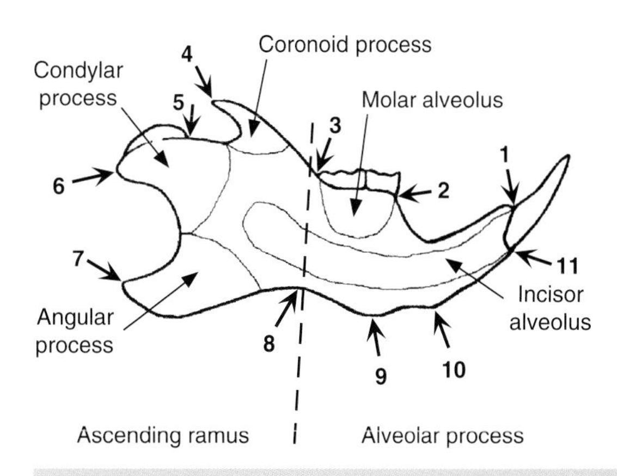

Figure 12.10 A mouse mandible, showing a series of 'landmarks' (numbered; these are all precise points) whose positions can be tracked in real or simulated selection experiments. The major parts of the mandible (areas rather than points) are also named. (Reproduced with permission from Klingenberg and Leamy 2001 *Evolution*, **55**)

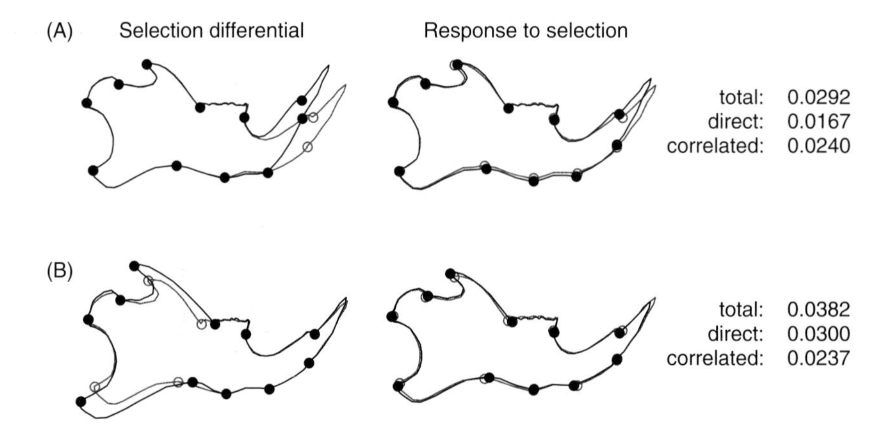

Figure 12.11 Simulated selection on mouse mandibles. A: 'arbitrary' simulated selection on the position/angle of the incisors. B: simulated selection based on the difference between mouse and rat mandibles, and thus on the kind of natural selection that may have been involved in their divergence from a common murine ancestor. Left – selection imposed; centre – response to selection; right – estimates of the relative magnitudes of direct and correlated responses. Pre-selection morphology – open circles and grey lines; selective differential/response – solid circles and black lines. (Reproduced with permission from Klingenberg and Leamy, 2001 *Evolution*, **55**)

always characterised by indirect responses being greater than direct ones, while the latter were always characterised by the opposite pattern. The authors of this study interpret this difference as suggesting that actual evolution of mouse mandibles may be influenced by the nature of the covariance matrix – i.e. by *the structure of the available variation* – as well as by the nature of the selection applied.

12.5 Long-term Evolution and Exaptation

The main examples that were used in Sections 12.1 and 12.4 both involved short-term evolutionary changes that led to limited diversification. It may seem odd to classify them together in this way, because the selective processes discussed in Darwin's finches took place over a few million years, perhaps about 5 MY (the approximate age of the Galapagos Islands), and produced about 13 species; while the radiation of murine rodents (in mandibles and other characters) has taken much longer and produced many more species. Nevertheless, despite these differences, both examples of evolutionary change are, in the grand scheme of things, not very long-term or large-scale. They would be referred to as examples of **macro-evolution**, as distinct from what the American palaeontologist G.G. Simpson[8] called **mega-evolution** – major evolutionary changes such as the vertebrate fin-to-limb transition discussed in Chapter 11.

A difference between short- and long-term evolutionary processes is that in the former the kind of natural selection involved often remains broadly the same throughout the process, while in the latter the kind of selection involved may alter radically as the evolutionary change progresses. This statement might well be questioned on the basis that even within 200 years the selection *for* melanism in *Biston* turned into selection *against* melanism (Chapter 4). However, it was still selection relating to the same thing: the degree to which the different forms were **cryptic** or conspicuous to visually-searching predators. So this is still the same *kind* of selection – what some would call the same selective agent – but acting in reverse. In long-term evolution, what we often find is that the selective agent itself changes or, to put it another way, that the reason why a structure is adaptive to its possessor changes as the structure is elaborated.

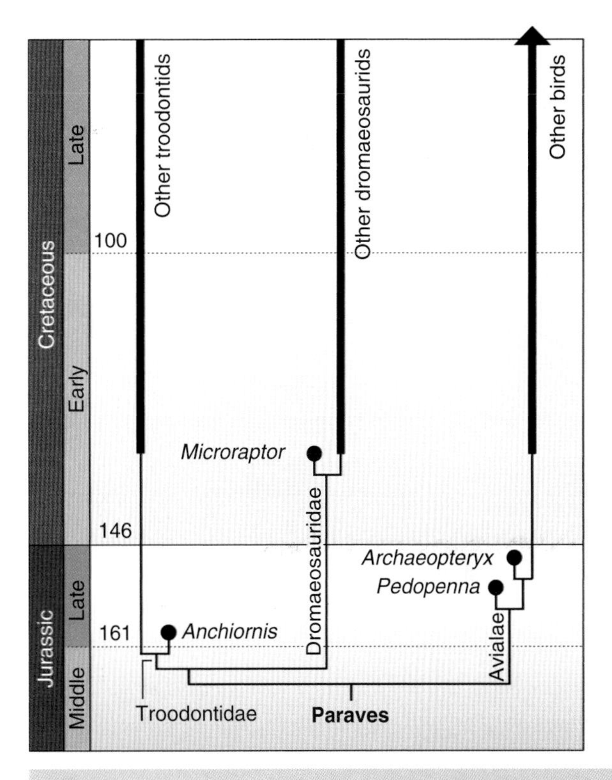

Figure 12.12 The birds are most closely related to, indeed may be said to be a clade within, the theropod dinosaurs, which include dromeosaurs and others. Feathers have been found in several species of non-avian theropods, indicating that the origin of feathers pre-dates the avian stem lineage. Note that the birds are the right-hand lineage in the picture; other lineages are non-avian theropods, including the dromeosaurs (centre). The earliest-known feathered theropod is now *Anchiornis*, which, as can be seen, pre-dates the iconic fossil *Archaeopteryx*. Numbers (LHS) are millions of years ago (MYA). (Reproduced, with permission from Witmer, L.M. 2009, *Nature*, **461**.)

This is probably one of the most important features of long-term evolution, and it provides the answer to the often-asked question of why the first stages in the evolution of a structure, which much later on can be seen to be adaptive, occurred at all. A structure that first arose for a particular adaptive reason but later became advantageous for a quite different reason has been called an **exaptation** by Stephen Jay Gould and Elizabeth Vrba[9]; they also called the second stage of the process leading to it by the same name. In other words, the ultimately exapted structure has arisen from an initial phase of adaptation followed by a second phase of exaptation.

A good example of exaptation in long-term evolution is the origin and elaboration of the feather. Initially thought to be unique to birds, this **epidermal** structure has now been found in other members of the theropod dinosaur group to which birds belong[10] (Fig. 12.12). Feathers have not, at least to date, been found in any other groups of dinosaurs.

It is fairly certain that feathered dinosaurs, such as some dromeosaurs (Fig. 12.12), did not fly. If that is true, then feathers did not originate as an adaptation for flight. The most likely initial advantage of proto-feathers is that they helped to keep the animal's body well insulated. Their more complex structure than that of the flat reptilian scales from which they probably evolved (Fig. 12.13) would certainly provide better insulation. But once proto-feathers had evolved, and some of their bearers

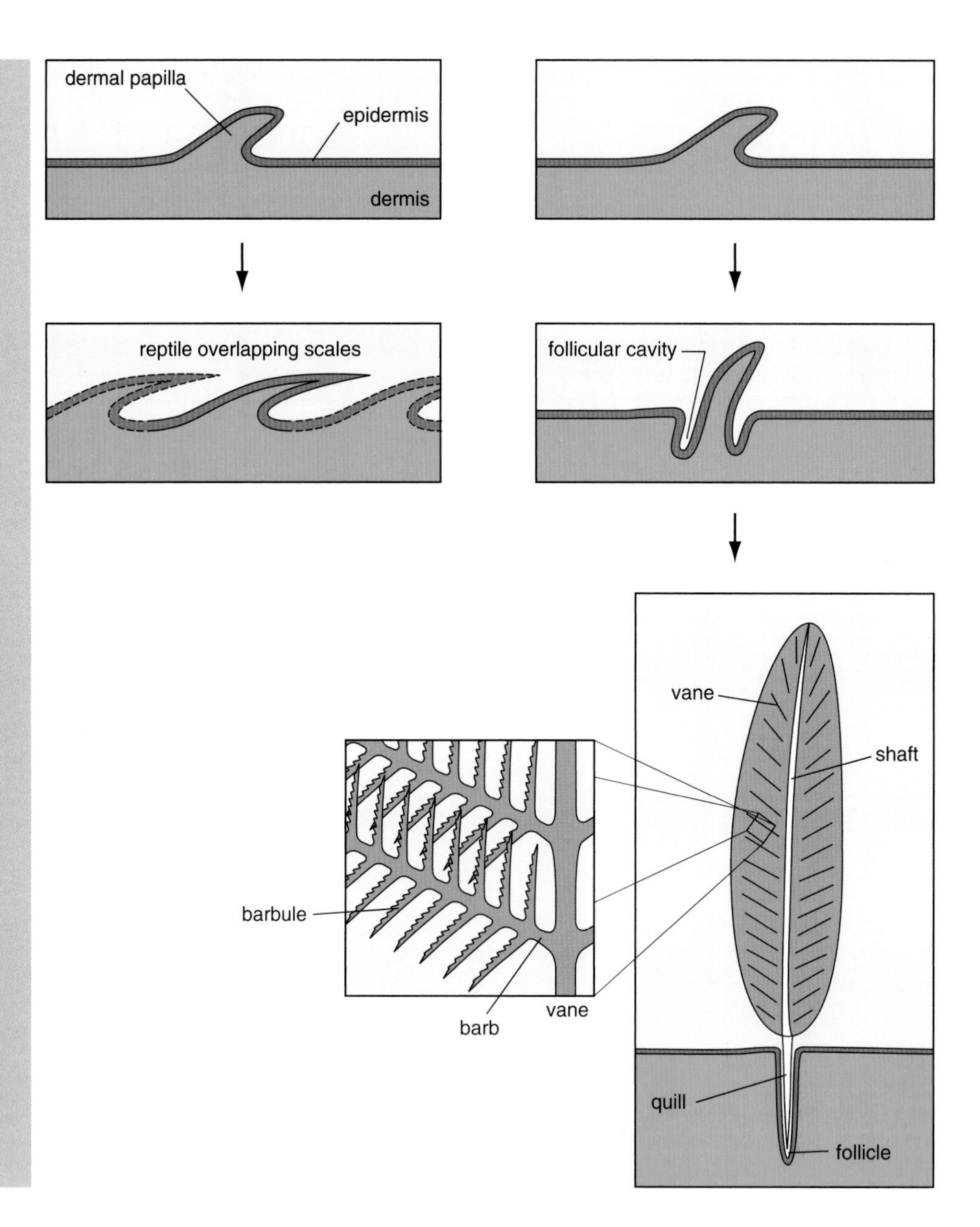

Figure 12.13 Comparison of the structure of feathers and the reptilian scales from which they have evolved. The pictures show both the outside layer of skin (epidermis) and the underlying layer (dermis). Note the significantly increased complexity of the avian feather, compared to the reptilian scale.

began to fly (or at least to glide), the structure of feathers came under a new form of selection related to improved flight ability.

We have now completed our journey through the mechanics of natural selection, from a beginning in the realm of ecological adaptation in 'simple' (or apparently so) characters to a more complex finishing point incorporating pleiotropy, polygeny, adaptive landscapes, trans-environment fitness profiles, morphological geometry, coadaptation and exaptation. We have thus arrived at a reasonably complete understanding of the selective process; albeit yet more can be understood, especially in quantitative terms, by consulting advanced texts on population genetics. But is this crucial process of natural selection evolution's only (or 'main' as Darwin postulated) driving agency? This is the question to which we now turn.

SUGGESTIONS FOR FURTHER READING

A recent textbook on evolution (of 800+ pages) is particularly extensive (and advanced) in its discussion of natural selection, with several chapters being devoted to this subject. It is thus a good source of further reading in this area:

Barton, N.H., Briggs, D.E.G., Eisen, J.A., Goldstein, D.B. and Patel, N. 2007 *Evolution*. Cold Spring Harbor Laboratory Press, Cold Spring Harbor, New York.

For a recent book on Darwin's finches, see:

Grant, P.R. and Grant, B.R. 2008 *How and Why Species Multiply: The Radiation of Darwin's Finches*. Princeton University Press, Princeton, NJ.

An unusual but fascinating and provocative book on the importance of selection related to organismic integration (and so to internal coadaptation rather than external adaptation) is as follows – but be warned that it is hard to get hold of these days:

Whyte, L.L. 1965 *Internal Factors in Evolution*. Tavistock Publications, London.

For those wishing to access a book-length account of landmark features and the geometry of morphology, see:

Bookstein, F.L. 1991 *Morphometric Tools for Landmark Data: Geometry and Biology*. Cambridge University Press, Cambridge.

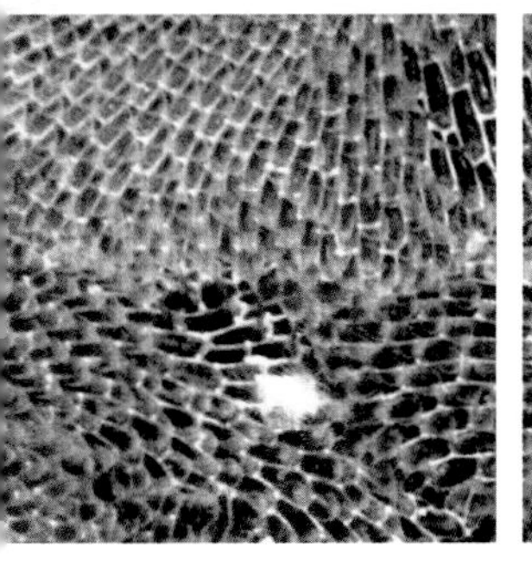
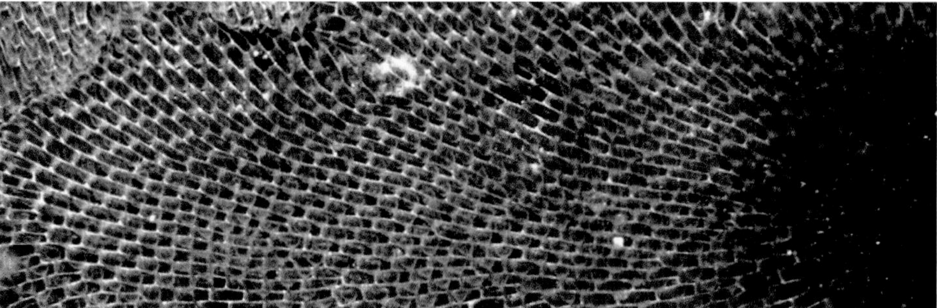

CHAPTER 13
Developmental Bias and Constraint

13.1 A Key Question about Evolution's Direction

The examples looked at in previous chapters showed evolution going in certain directions – from one kind of pigmentation to another in moths, and from one shape of beak to another in birds. The explanation of the directions was given in terms of natural selection: **fitness** differences determined that, in each case, evolution went in one particular direction. And in the case of the peppered moth, when fitness differences reversed so too did the direction of evolution. These observations could be taken to support a view that selection is the sole determinant of the direction evolution takes in any particular lineage.

However, there are other observations that suggest this may not be the whole story, and that the direction of evolution is determined by at least two factors: natural selection at the population level; and development (together with heritable variation in it) at the individual and family levels. We began to examine some developmental

Figure 13.1 A manatee and a sloth. These are the only two groups of mammals that break the 'rule' of seven cervical vertebrae.

aspects of beak shape in the last chapter, and we noted in Chapter 1 that selection acted not only on adult beaks but on the whole **developmental trajectory** that starts, in each individual bird, with no beak at all, and ends with one of adult proportions (Fig. 1.3). Now, we will build on the developmental part of the story and see whether development itself may help to determine evolution's direction.

We will start with a contrast involving the structure of the neck in birds and mammals. These are both large **clades**, containing about 10,000 and 5000 species, respectively. Neck length varies greatly in both groups, and this variation is probably due, for the most part, to natural selection. But the type or direction of change at the developmental level that produces the variation in neck length that is apparent to an observer is very different, as follows. Birds make long necks by adding cervical (neck) vertebrae; mammals, in contrast, make long necks by elongating their cervical vertebrae.

Consider, to begin with, the giraffe – whose neck has become the clichéd example for explaining the difference between Darwinian and **Lamarckian** evolutionary mechanisms. We will now use it as a starting point to explain another kind of difference between evolutionary mechanisms, in terms of their developmental basis.

Giraffes have the same number of cervical vertebrae as humans: seven. Moreover, almost all 5000 species of mammal have seven neck vertebrae, regardless of the length of their neck. The only known exceptions (Fig. 13.1), among extant mammals, are the three species of manatee, which have six cervical vertebrae, and sloths, a group of six species in which the number of cervical vertebrae varies between six and nine.

The situation in birds is very different. Long-necked birds have evolved more than once – think, for example, of the rather distantly related swans, flamingos and rheas (Fig. 13.2). Swans typically have 23 to 25 cervical vertebrae, flamingos typically 18 to 20. These figures contrast markedly with those for short-necked birds, which usually have 13 or 14 cervical vertebrae. Thus the highest number found in the group is approximately double the lowest one.

There are two possible reasons for this difference between the ways in which birds and mammals make long necks: (i) natural selection; and (ii) what has been called

Figure 13.2 A swan, a flamingo and a rhea. Each of these belongs to a group of birds that has (independently of the others) evolved especially long necks. Unlike long-necked mammals, such as the giraffe, long-necked birds have many additional cervical vertebrae.

developmental constraint[11]. We will regard these as alternative hypotheses for now, though they are not necessarily mutually exclusive, and will consider the degree to which each is supported by the available evidence.

A good place to start is the sloth group, and in particular those species with eight or nine cervical vertebrae. Many accounts of these interesting creatures point out that the extra cervical vertebrae allow greater flexibility of the neck. While this is probably true, it seems hard to believe that these are the only few species out of 5000 or more species of mammal in which such flexibility would be advantageous. Thus while a selective hypothesis might work for sloths, it is very poorly supported by the constancy of neck vertebra number in all the other species.

And so to the alternative hypothesis: developmental constraint. It is possible that there is something about the mammalian developmental system that precludes (or virtually so) the production of variants with a number of cervical vertebrae that is different from seven, but readily permits the production of variants in which the cervical vertebrae have different lengths. Although we do not yet know the mechanistic basis for such a 'bias', the available evidence supports this hypothesis much better than its alternative, selective, one.

If the developmental constraint hypothesis is correct, then the direction of evolutionary change is determined not solely by natural selection but by a combination of selection and constraint, in the following manner. Whether neck length goes up or down in a particular lineage is determined by selection; but whether these changes are achieved by changing the number or length of the cervical vertebrae is determined by the nature of the developmental system, and in particular by the readiness with which it allows the production of different kinds of variant – this is sometimes referred to as '**evolvability**'.

There is, however, a complication in the form of an intermediate hypothesis. Recall that, in the previous chapter, we distinguished between environment-specific adaptation and its internal equivalent – selection for developmental and (more generally) organismic integration, i.e. **coadaptation**. The extreme constancy of mammalian cervical vertebra number argues against the first of these but not necessarily against the second. In humans, it has been noted that there is a link between vertebral abnormalities, in particular the transformation of cervical vertebrae (without attached ribs) into quasi-thoracic ones (with attached ribs), and the occurrence of cancer. Many embryos affected in these ways die[12]. If this phenomenon is general across the mammals, then it may be that variant numbers of cervical vertebrae can be produced, but are usually unfit at an early stage for internal reasons, and so do not survive to post-embryonic, and ultimately adult, stages, to 'test' whether they would be fit for external reasons. However, note that while this approach may provide a basis for selection against *fewer* neck vertebrae, it does not provide a basis for selection against *more* of them.

'Developmental constraint' and 'developmental selection' are two of the most important ideas in the field of evo-devo. Both have a long history, pre-dating the explosion of 'genetic evo-devo' in the 1980s, but both connect with the genetic manifestation of the subject. For example, **Hox genes** are important in the specification of vertebral type and possibly also in the production of the often-associated tumours in human embryos. Both ideas are associated in a major way with the debate about the relative importance of external and internal factors in evolution (one of Stephen Jay Gould's so-called eternal metaphors[13]), and in particular whether what is often called a pan-selectionist view of evolution – but would be better called a 'pan-externalist' one – is sufficient. If development is important in interacting with selection to determine evolution's direction, as it now seems, then a pan-externalist view is clearly *not* sufficient.

Given the importance of this debate, it is crucial that we have a terminology that clarifies rather than obscures the key issues. We saw in the previous chapter that there were problems with the term 'developmental selection' (and the associated one 'internal selection'). Now we will see that there are also problems with the term 'developmental constraint'. However, just as there are problems, there are also solutions.

13.2 Making Sure the Question is about Processes, not Terminology

In 1991, a paper[14] appeared in the journal *Trends in Ecology and Evolution* with the wonderful title – 'Ontoecogenophyloconstraints? The chaos of constraint terminology.' The point the authors were making was this: the general idea that evolution is subject to various kinds of constraint is important but has been hindered by a poor, and inconsistently used, terminology. The 'Onto' with which their title begins is short for 'ontogenetic', i.e. developmental, and this is the area upon which we will focus here. However, it is worth noting that there are other kinds of constraint acting upon evolution too, One of these, represented by the 'phylo' in the above title, is phyletic constraint. The idea here is that where you can go, in evolutionary terms, is constrained by your starting point. For example, it is possible to make an insect that is adapted to utilising the dung of large mammals, but the type of insect that you get as a result of selection favouring the use of this particular resource depends on the features of the ancestral insect from which a new 'dung insect' is produced. Hence dung beetles and dung flies have some similar attributes but are very different animals (Fig. 13.3).

Figure 13.3 A dung fly and a dung beetle. These result from convergent evolution to a dung-utilising niche in dipterans and coleopterans. Although the selection involved in the two cases probably had much similarity, the different starting points mean that the results are strikingly different animals. This is often referred to as phyletic constraint (Dung fly photograph courtesy of Steve Marshall.)

Probably the most famous article ever written in the area of evolutionary constraints is the 1979 **spandrels** paper[15] by the American evolutionary biologists Stephen Jay Gould and Richard Lewontin (Fig. 13.4). These authors stated that organisms are

> so constrained by phyletic heritage, pathways of development and general architecture that the constraints themselves become more interesting and more important in delimiting pathways of change than the selective force that may mediate change when it occurs.

Note again the appearance of other kinds of constraint, including phyletic, alongside the developmental variety on which we are concentrating here. More crucially, note the potential impact of this short quote on evolutionary theory. Constraints, if they exist, may deserve to be ranked 'up there' along with natural selection as major evolutionary forces.

However, there is a problem with the word constraint itself: it is too negative. 'Developmental constraint' implies that development's evolutionary role is only to preclude the appearance of certain variants. Although Gould recognised this danger, his attempt to get around it largely failed. He pointed out that constraint could be interpreted in both positive and negative ways[16]. But to exemplify the positive interpretation he was forced to give what most people would now regard as an archaic usage: someone saying at a meeting that they felt 'constrained to speak', meaning that, as we would say nowadays, they felt 'compelled to speak'.

Like species, language evolves; but it rarely evolves backwards. So, attempting to revive a usage from a long-gone century is doomed to failure. Rather, we have to come up with a different word, one that is more readily interpreted in both positive and negative ways. The best such word that has been suggested so far, though there may be a better choice yet to come, is bias[17] – specifically, in this context, **developmental bias**.

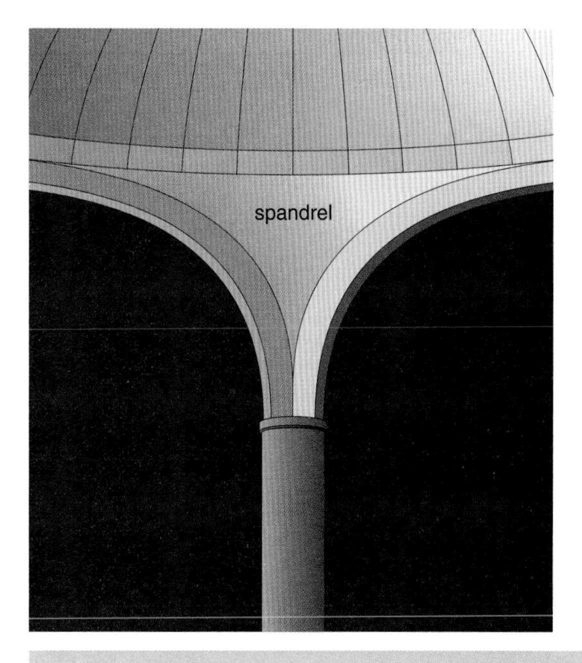

Figure 13.4 Diagram of a spandrel – an architectural structure of 'curved triangular' form that is an inevitable consequence, in a building, of the joining, at a right-angle, of two arches underneath a dome. Some structures in animals and plants may, like spandrels, be inevitabilities rather than adaptations.

Another landmark article[11] in this area was published in 1985 by the English evolutionary biologist John Maynard Smith (and a long list of co-authors). Its main title was 'Developmental constraints and evolution'. The authors of this paper defined developmental constraint as 'a *bias* on the production of variant phenotypes…', but they did not go as far as suggesting the replacement of 'developmental constraint' with 'developmental bias', as is evidenced by their title. That suggestion was made later[17], in an article published in 2001, along with the

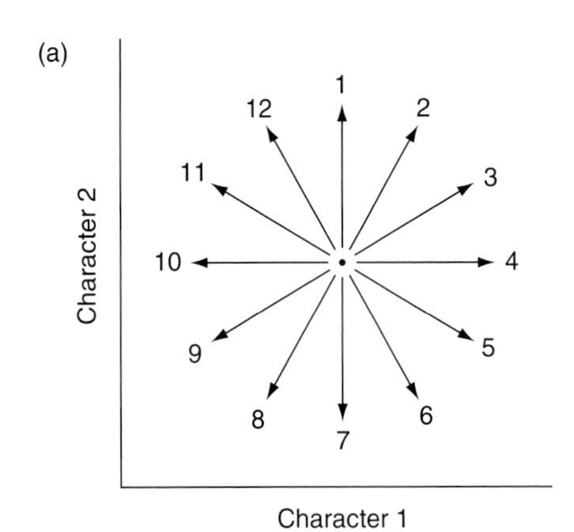

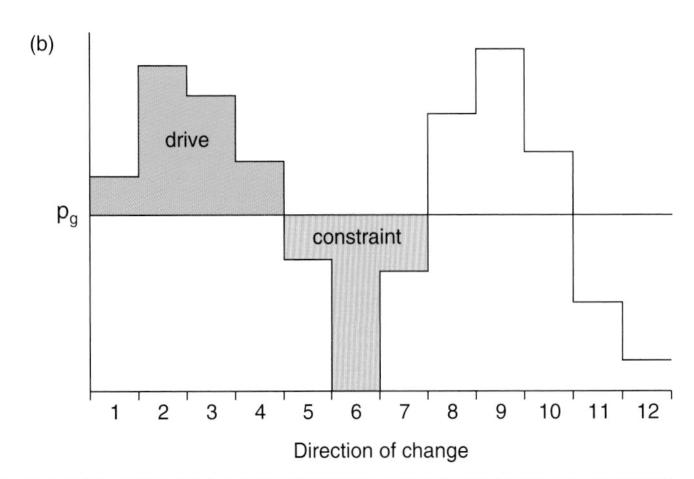

Figure 13.5 (a): Twelve possible directions of compound change in two-dimensional character space, each involving different relative amounts of change in characters 1 and 2. (b): Two possible ways in which the probability of generating the 12 kinds of change might vary (or not): the line of equiprobability (lack of developmental bias) and a 'flight of steps' indicating very different probabilities (presence of developmental bias in both positive and negative forms – respectively, developmental drive and developmental constraint).

introduction of another term, **developmental drive**. The argument in favour of this new terminology is as follows.

Consider two characters and ways in which they might vary (Fig. 13.5). This can either be treated as an abstract argument where the characters may be of any sort and at any level – from molecular to morphological. Alternatively, it may be visualised with two particular characters in mind – for example, the lengths of the forelimbs and hindlimbs of tetrapods. Some readers will no doubt prefer one approach, others the alternative one.

In Fig. 13.5a, there are 12 possible compound directions of change, in terms of the combinations of directions of change in the two characters. (The number 12 is arbitrary – we could just as easily look at 100 possibilities, but the picture would become too cumbersome.) Of the 12 combined directions, we can classify them as follows: numbers 2, 3, 8 and 9 represent positive co-variation of the characters because both go in the same direction – either up (2, 3) or down (8, 9). Numbers 5, 6, 11 and 12 represent negative co-variation: when one character's value goes down, the other goes up. Finally, numbers 1, 4, 7 and 10 represent independent variation in which one character changes in value while the other does not.

Many pairs of characters tend to positively co-vary. Taking the tetrapod limb example, when forelimbs get longer so too do hindlimbs. This is most obviously true in extreme cross-species comparisons (like that of a mouse and an elephant) but it is probably also true of most small-scale comparisons, including intraspecific ones. Humans with longer arms generally also have longer legs. But this is a statistical argument rather than an absolute one, so it is best considered in terms of probabilities (Fig. 13.5b) Here we see two patterns. The first is the horizontal line of equiprobability in which all of directions 1 to 12 are equally likely to occur. The second is the line taking the form of a series of upward and downward steps, representing variation in the probability of generating variants in each of the 12 directions, from any given starting point. The first line represents the absence of developmental bias, the second its presence.

Notice that in the developmental bias line in Fig. 13.5b, the highest four probabilities are for directions 2, 3, 8 and 9, which are those representing positive co-variation. So this abstract approach is compatible with the example of tetrapod limbs in which we expect longer fore- and hind-limbs to usually go together. However, note also that some cases of negative co-variation are also possible, though less frequent – for example, number 12. These correspond to cases such as *Tyrannosaurus rex*, where the hindlimbs are massive but the forelimbs are tiny (Fig. 13.6).

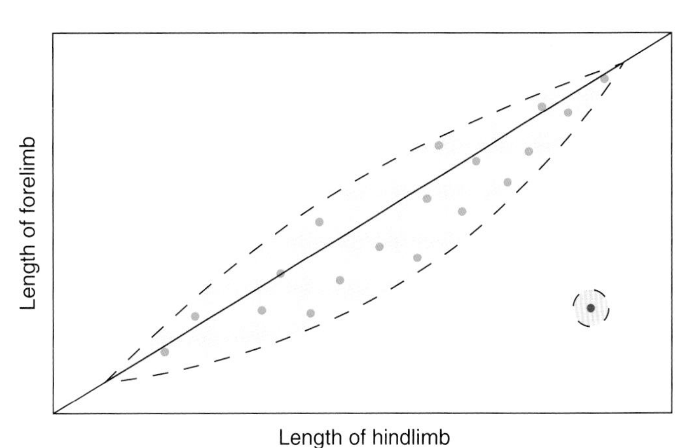

Figure 13.6 In tetrapod vertebrates, the lengths of the fore- and hind-limbs usually vary together, as in the contrast between a mouse and a horse (main grouping of points, each representing a species, in green). However, in some tetrapods the co-variation has been 'broken', and one pair of legs is very small compared to the other, as in *Tyrannosaurus rex* (outlier).

Now, back to the importance of having an adequate terminology that serves to help rather than to confuse. We can construct such a terminology as follows:

1. Any line on Fig. 13.5b, other than the line of equiprobability, represents developmental bias, because some forms of change are more probable than others.
2. Where the probability of a particular form of change is low (under the line of equiprobability), we can call this developmental constraint.
3. Most constraint is relative, though there is one case in the figure (number 6) where a form of change is impossible – this can be called absolute constraint.
4. Where particular forms of change have high probabilities, such as those for positive co-variation, we can call this developmental drive.
5. Therefore, developmental bias is a cover term, which includes both drive and constraint.
6. All of these refer to 'local' bias (whether drive or constraint), because they relate to how a given developmental system can be altered in the short term to produce variants that are only slightly different from the starting point.
7. It is also possible to have grander-scale 'global' constraints. For example, Stephen Jay Gould pointed out that, while it is possible to have large groups of organisms that can move by walking or flying, there is no group characterised by moving using wheels ('Kingdoms without Wheels' – a book[18] chapter). This is probably because wheels cannot be produced by a cell-based developmental system (whereas they can easily be produced in a factory).

One very important message emerges from this approach, and in particular from Fig. 13.5b, which shows only two lines out of a potentially infinite number. All of the lines not shown represent forms and degrees of bias. In contrast, there is just one line of equiprobability (the one shown) representing lack of bias. Therefore it is a logically inescapable fact that *developmental bias is the rule rather than the exception in nature.*

However, this conclusion is only the first part of a non-pan-externalist view of evolution. The second conclusion that we need to reach is that, in addition to being very common, developmental bias does actually affect the direction of evolutionary change rather than being over-ridden by natural selection. We will address this second point in Sections 13.5 and 13.6. Meanwhile, there are two other connected points that deserve mention, as follows.

First, the adoption of an adequate terminology is only a beginning. Ultimately, we would like to understand *the causal developmental mechanisms* underlying biases that we observe at the level of adult morphology – such as the sizes of forelimbs and hindlimbs in tetrapods. For example, we can ask the question: is positive co-variation in the sizes of forelimbs and hindlimbs due to genetic variation in the production of growth hormones, and if so, how does that work in terms of affecting cell proliferation rates in different parts of the body? We will look at genetic and developmental aspects of developmental bias in Chapters 14 and 15.

Second, in Fig. 13.5b, cases of independent variation of characters (numbers 1, 4, 7 and 10) were shown to be associated with a mixture of drive (three cases) and constraint (a single case). But this pattern is just one of many possibilities. How often in fact is one character able to alter without causing changes in at least one other character? This question is considered further in the next section.

13.3 Dependence versus Independence of Different Characters

This section could have been entitled 'Charles Darwin versus Alfred Russel Wallace', because the two great pioneers of the concept of natural selection held very different views: Darwin[19] was impressed by the extent to which characters often co-varied (and frequently used the phrase 'correlation of growth'); Wallace[20], on the other hand, emphasised that characters could often vary independently of each other.

It is clear that, taking all possible combinations of characters across all species, instances of both correlated and independent variation can be found. But the relative prevalence of the two, and patterns in this prevalence related to which characters are chosen, are both important issues that we cannot yet quantify and so are key issues for future research.

The difference between Darwin's and Wallace's emphases in relation to the degree to which characters are dependent on, or independent of, each other has not received the attention it deserves. Although a textbook is not the place for a detailed investigation of this difference, a brief reference to relevant comments made by Darwin and Wallace will help to reinforce the message, as follows.

Chapter 5 of *The Origin of Species*, entitled 'Laws of Variation', contains a section called 'Correlation of Growth'. Here, Darwin says:

> I mean by this expression that the whole organisation is so tied together during its growth and development, that when slight variations in any one part occur, and are accumulated through natural selection, other parts become modified. This is a very important subject, most imperfectly understood.

There is a marked contrast between the sentiment expressed in the above quote from Darwin and the one below from Wallace. Specifically, the following quote is from the

Concluding Remarks section of chapter III, 'The Variability of Species in a State of Nature' in his 1889 book, the main title of which is, rather self-effacingly, *Darwinism*. Here, Wallace emphasises the importance, to the theory of evolution by natural selection, of the existence of variation within species living in nature (as opposed to under domestication). After commenting on this importance, he goes on:

> 'Yet more important is the fact that each part or organ varies to a considerable extent independently of other parts.'

The degree to which different characters co-vary can now be visualised against the background of the **adaptive landscape**, a concept introduced in Chapter 12. This approach has been adopted by proponents of evo-devo, but also by quantitative geneticists. Strangely, the ways in which the two groups visually represent character correlation and adaptive landscapes can be remarkably similar; yet the ways in which members of the two groups interpret their pictures *appear to be* (but may not actually be) quite different, as we will shortly see.

13.4 Evo-Devo Meets Quantitative Genetics

Scenarios in which populations with and without developmental bias (the evo-devo term) or character co-variation (the quantitative genetics term) exist in an environment characterised by a particular adaptive landscape are shown in Fig. 13.7.

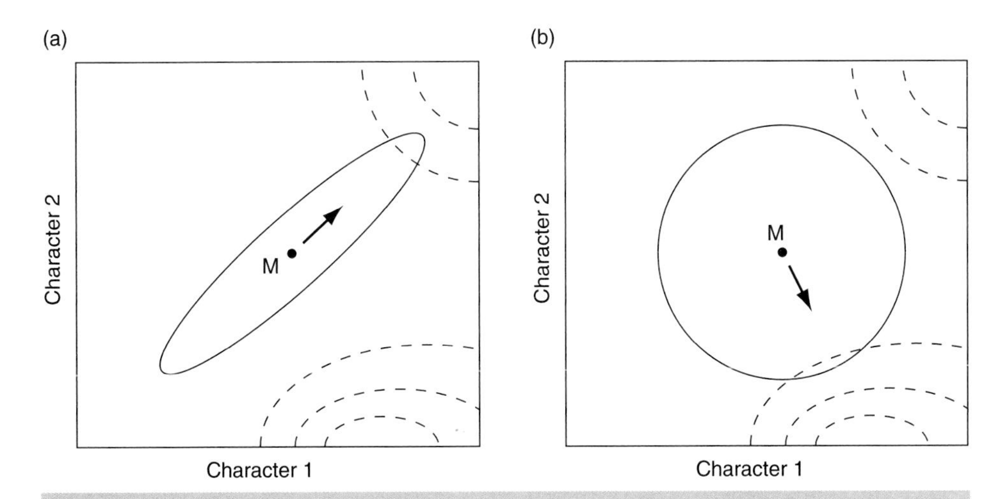

Figure 13.7 Populations with identical joint mean values (M) for characters 1 and 2 evolving in identical adaptive landscapes with the only difference between the left-hand (a) and right-hand (b) scenarios being the presence versus absence of developmental bias/ character correlation. Note the completely different evolutionary directions produced by this single difference. Solid line: extent of variation. Dashed lines: fitness contours.

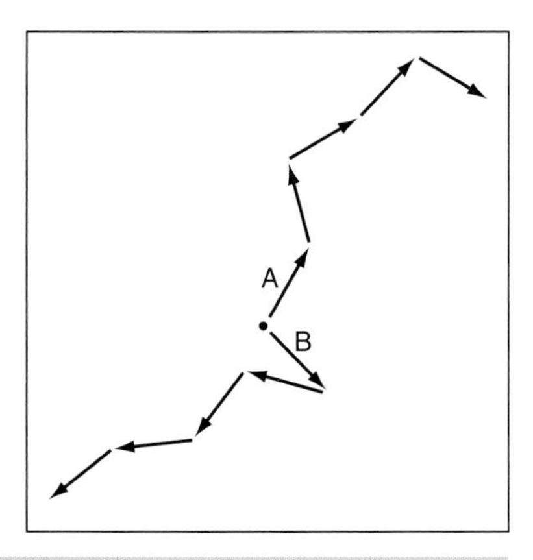

Figure 13.8 Possible long-term divergence in morphology caused by the initial divergence shown in Fig. 13.7. The A and B labels on the initially divergent directions here correspond to (a) and (b) there.

Importantly, although the two adaptive landscapes are identical, and although the populations' mean values of both characters are also identical, the presence or absence of bias in the variation of the characters will lead to evolution taking very different directions. Furthermore, after the two populations have climbed different adaptive peaks then, depending on the nature of the landscape extending around those peaks, the two populations may diverge further, as exemplified in Fig. 13.8. Ultimately they may end up, in the distant evolutionary future, with radically different morphologies.

The picture shown in Fig. 13.7 can be found in both evo-devo and quantitative genetics papers. As noted earlier, there is no difference between the proponents of the two disciplines in how to picture variation and evolution against the background of an adaptive landscape. The difference, if there is one, lies in how to interpret the evolutionary process so pictured. But is there really a difference in interpretation? Let's investigate this issue.

To do this, we need to examine a concept from quantitative genetics called the variance-covariance matrix (Fig. 13.9). This summarises the amount of genetic variation in each character within a population (down the diagonal) and the degree of genetic co-variation between pairs of characters (all off-diagonal parts of the matrix). Variation is measured as the *variance* and co-variation as the *covariance* – statistical terms that will be familiar to some readers but not others – however, for the latter group, the use of 'variation' and 'co-variation' will readily suffice.

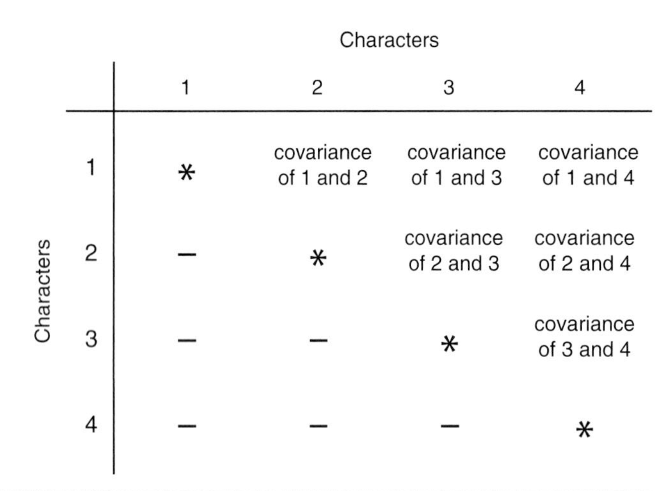

Characters	1	2	3	4
1	*	covariance of 1 and 2	covariance of 1 and 3	covariance of 1 and 4
2	–	*	covariance of 2 and 3	covariance of 2 and 4
3	–	–	*	covariance of 3 and 4
4	–	–	–	*

Figure 13.9 The variance-covariance matrix of quantitative genetics. This provides a table of measurements of the degree to which various characters co-vary in a population. Here, there are (arbitrarily) four characters, labelled 1 to 4. Note that the phenotypic covariances of pairs of characters can be broken down into genetic and environmental components, just as can the variances of individual characters, which are sometimes given along the diagonal indicated here by asterisks.

Note that up to now we have concentrated on phenotypic variation – for example, in Figs 13.5 and 13.7; also in Chapter 12. Now, however, we are considering genetic variation. The two are not the same because most phenotypic characters are only partially **heritable**, as we saw when discussing beak size in Darwin's finches. Note also that, while we previously considered possible *developmental* causes for character correlation in relation to tetrapod limb lengths, the G matrix is typically used by quantitative geneticists to represent characters in adults, without reference to their developmental origins.

These points help us to see more clearly the nature of the difference between the evo-devo and quantitative genetics approaches to evolution. It is not so much a difference of interpretation but rather a difference of focus, as follows.

The quantitative genetics approach links phenotypic variation to its genetic but not its developmental basis. One quantitative geneticist's response[21] to the Gould and Lewontin paper's advocacy of developmental constraint was: 'the genetic variance/covariance matrix of quantitative genetic theory measures developmental constraints.' However, proponents of evo-devo would argue that it does no such thing; rather it measures the end results of such constraints (or biases) in terms of adult characters. To measure the developmental biases/constraints themselves, it is necessary to understand the nature of the developmental mechanisms concerned.

The evo-devo approach is complementary in its strengths and weaknesses: it involves an interest in understanding the developmental mechanisms underlying the phenotypic variation, but is less focused on separating the genetic components from the non-genetic, while acknowledging the distinction between them. There are other differences too. For example, the evo-devo (some would say *eco*-devo) concept of developmental **reaction norms** (Chapter 16) blurs the distinction between genetic and non-genetic components of phenotypic variation. Also, quantitative geneticists stress that the structure of the G matrix itself can change in evolution, while many proponents of evo-devo suspect that there may in turn be limits to the extent to which this is possible.

The existence of these differences of focus should be considered in a positive light in relation to current evolutionary theory. Our goal should be to build on the strengths of each and, when sufficient progress in understanding each has been made, to put them together so that we have a more synthetic theory. In other words, it is not a case of evo-devo *versus* quantitative genetics, but rather evo-devo *plus* quantitative genetics. There is much exciting research to be done in this combined area over the coming decades. It may not be an exaggeration to say that this research, together with the conceptual synthesis that will emanate from it, will come to be seen as one of the major advances in the history of evolutionary biology. Time will tell.

13.5 Developmental Bias and 'Routine' Evolution

It is time to return to concrete examples. In this section we will look at three examples involving 'routine' evolution; in the next we will look at an example that is decidedly non-routine in that it involves the production of a very bizarre body form.

Butterfly Eyespots

Some very interesting artificial selection experiments were carried out on the eyespots of the butterfly *Bycyclus anynana* by Paul Brakefield's research group in Holland. There were two main series of experiments – the first focusing on eyespot size[22], the second on eyespot colour[23]. We will look at these in turn.

We began to examine the selection experiments on eyespot size in Chapter 8, and saw that selection was possible in all four directions, but easier in some than in others (Fig. 8.10; see also more generalised version in Fig. 13.10). Specifically, it is easier to select for both the anterior and posterior eyespots to get bigger, or for both to get smaller.

It is harder, but still possible, to select for converse changes in the two pairs of eyespots (anterior bigger, posterior smaller and *vice versa*). These results should be interpreted in relation to the nature of the developmental process producing the eyespots – in particular the fact that it exhibits a form of bias that leads to a strong positive correlation between the sizes of the anterior and posterior eyespots, as can be seen in Fig. 13.10.

Since it is *possible* to select for the big/small and small/big combinations, it is clear that, in the laboratory, artificial selection *can* overcome the effects of developmental bias. But in nature the equivalent question is: *will* natural selection overcome bias? This is a much harder question to answer. In fact, the answer depends on the relationship between the structure of the available variation and the structure of the adaptive landscape. In Fig. 13.7, we saw a scenario in which natural selection was unable to overcome developmental bias, and so the latter played a major role in determining the direction of evolutionary change. Now, in Fig. 13.10(left), a different scenario is contrasted with that, in which developmental bias does not contribute to the direction of evolution. Clearly, both outcomes are possible. We do not yet have sufficient information to be able to estimate their relative frequency.

The second set of artificial selection experiments, on eyespot colour, produced a very different result from those on eyespot size. Again, two pairs of eyespots were used. Again, four directions of selection were attempted, this time involving increasing the amount of gold colour in the eyespots at the expense of black, and *vice versa*. The results of the experiment are also shown in Fig. 13.10(right). Unlike the selection on eyespot

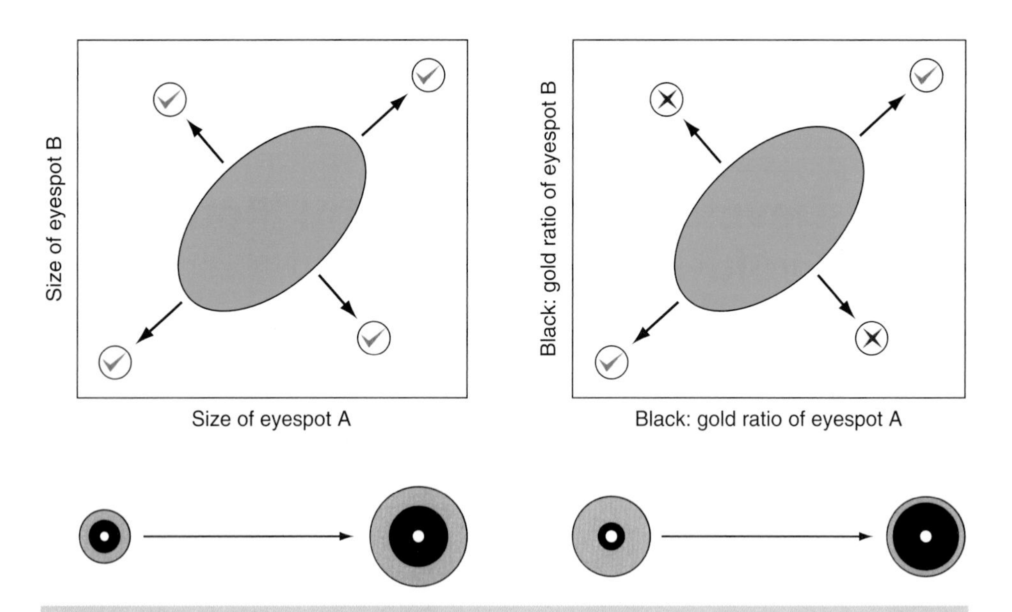

Figure 13.10 Left: The pattern of co-variation in two different eyespot sizes in the butterfly *Bicyclus anynana* and the success of attempts to select artificially in four directions. (Note that this is a diagrammatic version of Fig. 8.10) Right: Parallel diagram for the pattern of co-variation in colour between two different eyespots and the result of attempts to artificially select, again in four directions. Note the different results of the size and colour experiments.

size, which was able to make one pair of eyespots bigger, the other smaller, selection on colour was unable to make one pair of eyespots golder, the other blacker. This can be interpreted as a lack of ability of artificial selection to overcome the effect of developmental bias in this case, despite the fact that the degree of colour co-variation in the eyespots concerned was no stronger, at least at the phenotypic level, than that for size co-variation.

Both the contrast between the two artificial selection experiments and the contrast between artificial and natural selection argue for caution in drawing a general conclusion about the importance of developmental bias in determining the direction of evolution. Nevertheless, science is all about searching for general principles that subsume many specific facts. Perhaps examining some other characters in other species will help in this respect.

Centipede Segments

There are more than 3000 extant species of centipede. These belong to five taxonomic orders, two of which we contrasted in Chapter 4: Lithobiomorpha and Geophilomorpha. This time we will focus entirely on the latter. As noted earlier, the number of leg-bearing segments in the Geophilomorpha varies from 27 to 191. However, what was not mentioned earlier is that in all 1000+ species of geophilomorphs the number (or numbers) of leg-bearing segments is always odd[24] (Fig. 13.11). So, while almost all odd-numbered character states between 27 and 191 are found in one or more species of the group, not a single species is characterised by an even number of leg-bearing segments. In other words, the numbers 28, 30, 32.....188, 190 are all missing. Very occasionally, an aberrant specimen with an even number is found, but such specimens are so rare that a typical natural population of a geophilomorph species contains none of them.

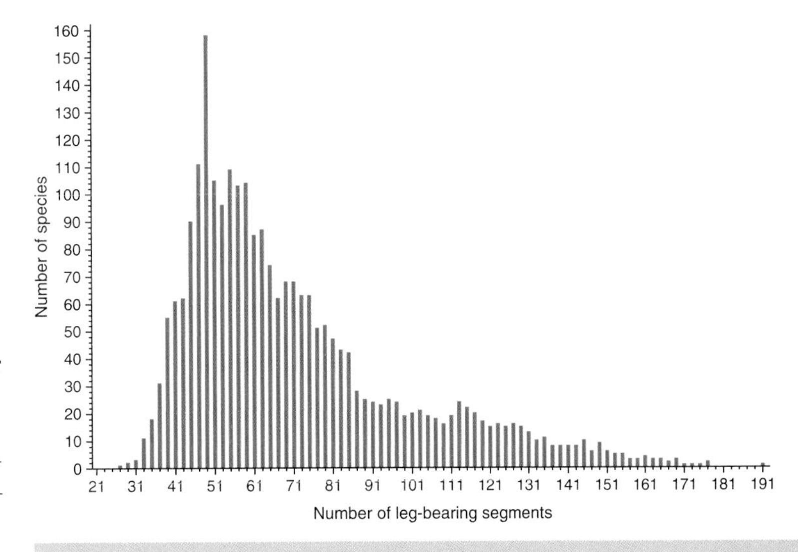

Figure 13.11 The distribution of leg-bearing segment numbers in the centipede order Geophilomorpha, which comprises over 1000 species. Note that almost all odd-numbered character states between 27 and 191 segments are found in one or more species; while even-numbered character states are never found (except as occasional aberrations of individual specimens). (Redrawn with permission from Minelli and Bortoletto, 1988, *Biol. J. Linn. Soc.*, **33**: 323–343.)

There are two possible reasons for this unusual distribution of character states: natural selection and developmental bias. The former – selection against all even numbers – seems highly implausible. The latter – a mechanism of segment development that incorporates a bias against the production of even numbers – is more plausible, and indeed has recently been characterised. It involves a switch from gene expression patterns corresponding to the production of

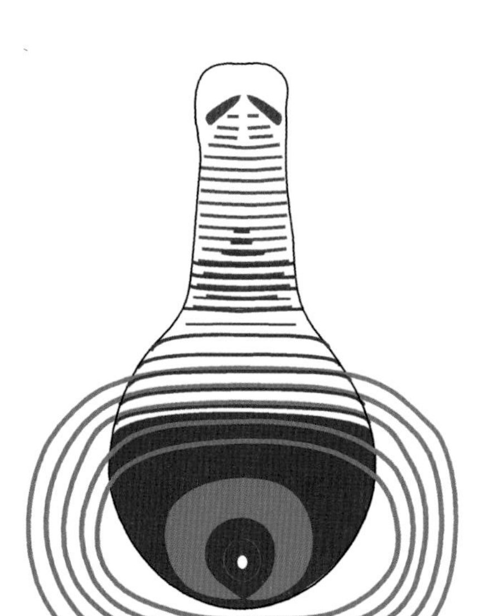

Figure 13.12 Cartoon of segment formation in the centipede *Strigamia maritima*, as seen through the expression patterns of three genes – *engrailed* (red), *caudal* (blue) and *even-skipped* (green). It appears that initial double-segment periodicity of *caudal* gives way to single-segment periodicity in later development, through intercalation of secondary stripes of expression between primary ones. This may explain the all-odd pattern of character states found in nature (Reproduced with permission from Chipman, A. *et al.* 2004, *Evolution & Development*, 6).

double-segment units to patterns that involve single segments[25] (Fig. 13.12).

There are two ways to interpret this developmental mechanism and its link with final segment number. First, if a secondary stripe of expression of the gene *caudal* is intercalated between *every* pair of primary ones, then the resultant number will always be odd. Second, we might choose to interpret the picture shown in Fig. 13.12 more broadly, as evidence of a mechanism in which segment formation has a multiplicative/divisive component rather than being simply additive (adding segments one at a time from the head backwards). In this interpretation, if it is a case of double-segments being formed first and then divided into single segments, such a mechanism will always produce an even, rather than odd, number of segments. Although this would seem to be the opposite of what is observed, the poison-claw segment (which we discussed in Chapter 4) was probably evolutionarily derived by modification of a leg-bearing segment. If it is included in the number of 'trunk segments', then that number is always even and is between 28 and 192. Either way, we have a potential mechanism for the developmental bias involved.

Of course, while the exact number of segments in a centipede is partly determined by developmental bias, it is partly also determined by natural selection. In general, the number of segments is highest in the most subterranean species. So in some lineages evolving a more completely subterranean existence we might expect selection to cause an upward trend in segment number, but developmental bias to affect the modal number of segments that is eventually arrived at. This combination of the roles of selection and bias is reminiscent of our starting point – the different ways in which mammals and birds produce long necks.

Angiosperm Leaf Arrangements

The Canadian biologist Brian Goodwin[26] noted that about 80 per cent of **angiosperms** have a particular one (out of three possible) phyllotactic, or leaf-arrangement, patterns (Fig. 13.13), and interpreted this as an example of the role of developmental bias in evolution. He states that:

> ...we get an interesting conjecture: the frequency of the different phyllotactic patterns in nature may simply reflect the relevant probabilities of the morphogenetic trajectories of the various forms and have little to do with natural selection.

While this is one possible interpretation of the observed frequency distribution of the three kinds of leaf arrangement, with one being much commoner than the other two, such a distribution is not strong evidence for an influence of developmental bias. The distribution could equally well be interpreted as a result of natural selection. After all, the vast majority of birds have wings adapted for flying, while only a few, such as

Figure 13.13 The three main types of leaf arrangement patterns in angiosperms: spiral, distichous and decussate. About 80 per cent of species have the spiral pattern. The reason for this preponderance may lie in selection, constraint, or both. (Reproduced, with permission, from Brian Goodwin, *How the Leopard Changed its Spots,* Touchstone Books, 1994.)

Figure 13.14 Although there are many species of chelonians, two of which are shown here, they all have a body plan in which the ribs are fused with the dorsal part of the shell (the carapace) and the shoulder girdle is inside the ribs, unlike the usual vertebrate arrangement (as in humans) where the shoulder girdle bones are outside of the rib-cage.

penguins, have wings adapted for swimming. But this very asymmetric distribution is most readily interpreted in terms of the action of natural selection.

This example thus serves as a useful cautionary message. Goodwin may be right in his claim that developmental bias plays a role in determining the frequency of different phyllotactic patterns in nature; but equally, he may be wrong. To resolve this issue in the future, we need to know more about the developmental processes involved, and also to map the phyllotactic patterns onto a robust and complete **phylogeny** of angiosperms, when such a phylogeny becomes available.

13.6 Developmental Bias and the Origin of Evolutionary Novelties

This section connects with Chapter 17, where the origin of novelties will be discussed more widely. The issue of how to define a novelty will be dealt with there. Here, we will focus on a single example of a structure that is so obviously novel, from an evolutionary perspective, that a definition is hardly required: the shell of chelonians (turtles).

The Chelonia is a **monophyletic** group with about 300 extant species, all characterised by a particular body layout that is very different from that of all other tetrapods in several respects, including possession of a shell and associated features such as a shoulder girdle that is inside, rather than outside, the ribcage (Fig. 13.14). The evolutionary origin of the Chelonia is obscure, despite some interesting recent fossil finds[27], and so it is not possible to identify with certainty the appropriate **sister-group**. But this in fact does not matter, because any reasonable choice of sister-group implicates the same evolutionary changes in the Chelonian stem **lineage** – the production of the features noted above from a more typical tetrapod design in which the shoulder girdle is in its normal external position. A recent paper[28] on turtle development sheds light on how this change in design might have occurred.

Figure 13.15 Degrees of limb reduction in skinks. The family Scinkidae contains hundreds of species. As shown, some skinks have 'normal' legs; some have reduced legs (compared to body size); and some are entirely legless (© blickwinkel/Alamy).

It is interesting to compare the single (or possibly double: see **glyptodonts**) origin of shelled tetrapods with the multiple origins of elongate legless ones. There are estimated to have been more than 50 such origins. The snakes constitute the most obvious legless group, but there are many other groups of legless lizards, and there are also the caecilians – a group of legless amphibians. Why is it apparently so easy for a tetrapod lineage to evolve leglessness and yet so hard to evolve a shell?

The answer to this question lies in the nature of the **standing variation** present in all populations, which we looked at briefly in Chapter 4. It is probably true that all populations of all tetrapod species that possess legs have variation, however slight, in the length of their legs relative to their overall body size. Given the existence of that variation, it is not hard to see how selection can progressively reduce leg length from 'normal' to 'short' to zero. Indeed, the lizard family Scinkidae shows all stages in this process (Fig. 13.15).

In contrast, how many tetrapod populations possess variation in the degree to which the shoulder girdle is external as opposed to internal? Depending on how we choose to measure such variation, the answer is either 'none' or 'almost none'. So, the relative frequencies of origins of these two types of tetrapod body plan may well lie not in the relative frequencies with which the selective advantages of a shell and the loss of legs occur, but rather in a bias in the developmental system wherein some kinds of variation are common, others exceedingly rare. This line of argument, which will be developed further in Chapter 17, inevitably leads to the issue of what role, if any, **macromutations** have in evolutionary change.

SUGGESTIONS FOR FURTHER READING

For a classic paper on developmental bias/constraint, see:

Gould, S.J. and Lewontin, R.C. 1979 The spandrels of San Marco and the Panglossian paradigm: a critique of the adaptationist program. *Proc. Roy. Soc. Lond. B*, **205**: 581–598

For a recent review of the subject, see:

Brakefield, P.M. 2006 Evo-devo and constraints on selection. *Trends Ecol. Evol.* **21**: 362–368.

For an account of the potential importance of developmental bias in evolution written for a general audience, see:

Arthur, W. 2004 *Biased Embryos and Evolution*. Cambridge University Press, Cambridge.

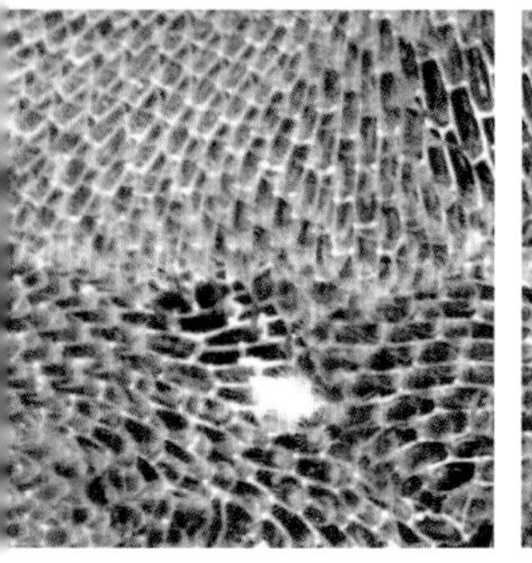

CHAPTER 14
Developmental Genes and Evolution

14.1 The Direction of Evolution at the Developmental/Genetic Level

The direction that evolution takes, in the case of any particular character in any particular **lineage**, can be considered at many levels. We started at the level of the adult **phenotype** when considering mammalian necks, lepidopteran pigmentation, bird beak shapes and centipede segments in the preceding chapters. However, we also began to look at the direction of change at the developmental level that underlies the direction of phenotypic change. In the case of mammalian versus avian necks, we noted that while natural selection determines the direction 'up' or 'down' in terms of overall neck length, it is more likely that a difference between the developmental systems of birds and mammals is responsible for one group lengthening their necks by adding cervical vertebrae, the other by elongating those already present.

This contrast at the developmental level can be thought of either as *directions* or as *types* of change. The latter term is better, as it is more inclusive. At the level of a particular adult phenotypic character, there are often two obvious opposing *directions* of change: bigger/smaller, longer/shorter; paler/darker (though of course more complex changes are also possible). But underlying each of these pairs of opposing directions may be multiple *types* of developmental and genetic change.

For example, there are at least three developmental ways to bring about evolutionary neck elongation. The birds and mammals exhibit the two most obvious ways of making longer necks, but a third type of **developmental repatterning** could, at least in theory, also be employed: thickening of the inter-vertebral discs, without elongation of the vertebrae themselves.

Directional phenotypic evolutionary transitions can be underlain by multiple types of change at the genetic level also. For example, a lineage that evolves a more complex body form may do so by (i) having a greater number of developmental genes; (ii) having a greater number of controlling elements such as micro-RNAs; or (iii) having a more complex pattern of interaction involving the *same* number of genes and controlling elements. We will examine evolutionary complexification further in Chapter 18.

In the previous chapter we began to consider the important question of whether the nature of the developmental process may influence the direction of evolutionary change. As we pursue this question at genetic and molecular levels in this chapter and the next, we can reformulate the question as follows. Are some types of developmental change more easily achieved than others? If so, does this affect the direction of evolution? Much hinges on what is meant by 'ease' in this context. There is a link here, with the concept of **evolvability**, championed in the context of evo-devo by the American biologists John Gerhart and Marc Kirschner[29], and considered in a recent paper[30] to be the 'proper focus' of evo-devo. Basically, an 'easier' change equates with a greater degree of evolvability.

One interpretation of the 'ease' of producing a particular type of developmental repatterning is the frequency of occurrence of variants which produce that type of repatterning. If the variants concerned are sufficiently frequent that they are normally found in the **standing variation** of a population, then evolution of that type of repatterning will generally be 'easy', though with the caveat that this may not be true of a variant causing antagonistic **pleiotropy**. If, in contrast, the frequency of occurrence of the required variants is so low that they are almost never found in the standing variation, then the type of repatterning concerned will not be easy at all. This connects back to the comparison of the different ease of evolving legless and shelled vertebrates with which we concluded the previous chapter. But now we want to connect all these related concepts – ease, evolvability, bias and constraint – with developmental genes. So the first step is to understand the nature of these genes, building on the brief introduction to them given in Chapter 3.

14.2 Developmental Genes: An Overview

The **genomes** of animals with large complex bodies have turned out not to be as large as was initially thought. In the case of humans, a typical guesstimate of the number of genes in the genome, before the human genome project was carried out, was 100,000. Now, following completion of the project, the answer turns out to be only a little more than a quarter of that figure. However, 25,000+ genes is still a very large number. Most animals and plants seem to have genome sizes in the range 10,000 to 30,000 genes.

Developmental versus Regulatory Genes

In Chapter 3, we divided genes into three major classes, while acknowledging that there was some overlap between them. The three classes are: **housekeeping genes** (e.g. those making routine metabolic enzymes); **terminal target genes** (e.g. the genes for haemoglobin and myoglobin); and **regulatory genes** (e.g. those making **transcription factors**). We noted that it is the last category that is most crucial in governing developmental processes.

However, there is a complication: equating regulatory gene with **developmental gene** is overly simple, for the following two reasons. First, a gene whose main job is to regulate the level of expression of a housekeeping gene is by definition regulatory, but it may have no effect on development. Second, a gene making an enzyme that itself has no direct role in development may be 'developmental' in an indirect way if, for example, the role of its enzymic product is to make a developmental agent (e.g. a steroid **hormone**) from a non-developmentally-active precursor molecule.

So we now recognise three types of developmental gene:

1 those producing transcription factors;
2 those producing the components of **signalling pathways** that ultimately activate (or repress) transcription factor genes; and
3 a catch-all third category of 'indirect developmental genes'.

This last category is probably very large and heterogeneous. Having acknowledged its existence, we will focus now on the other two, which are more manageable, and yet still problematic because they can be subdivided further in different ways.

Transcription Factors

One way to subdivide transcription factor genes is by the nature of the motif in their protein product that binds to the DNA of their **downstream** target genes. The main four such motifs (Fig. 14.1) are: helix-turn-helix, helix-loop-helix, zinc finger and leucine zipper. **Homeobox**-containing genes constitute a subset of the first of these in that their homeodomain products are of the helix-turn-helix type. Note that all four of these protein motifs are characteristics of the *secondary* structure of the proteins concerned – the three-dimensional shapes, such as helices, that are formed by particular stretches of the protein molecule.

Within one of these four categories, the proteins involved can be further subdivided by their *primary* structure: the sequence of amino acids, particularly in the conserved region corresponding to the DNA-binding motif. For example, **homeodomain**-containing proteins can be hierarchically clustered according to the similarities of their 60-amino-acid homeodomain regions. This is broadly equivalent

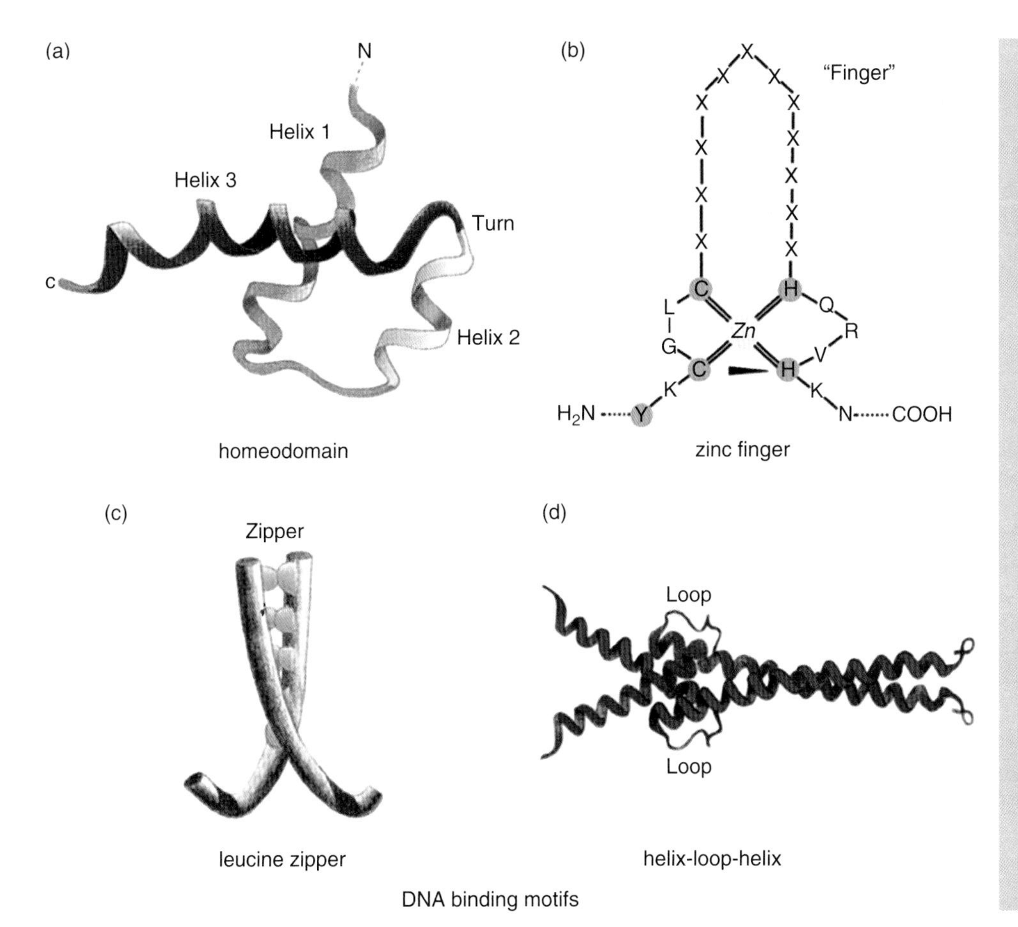

Figure 14.1 The four main classes of transcription factors, based on their secondary structures: (a) helix-turn-helix, (b) zinc finger, (c) leucine zipper and (d) helix-loop-helix. The homeodomain transcription factors fall into the first of these four classes. (Reproduced with permission from Carroll, S. *et al.* 2005, *From DNA to Diversity*, Blackwell.)

to a hierarchical clustering of the genes themselves (Fig. 2.10) based on the nucleotide-sequence of their homeoboxes. For example, the **Hox gene** subgroup (Section 14.3) is characterised by possession of a particular type of homeobox – the *antp* type.

So far, so good. However, suppose that we wish to classify genes not by the primary and secondary structures of their protein products but rather by their developmental roles, for example in a particular developmental process such as segmentation. Two groups of genes involved in segmentation in *Drosophila* are the pair-rule genes (one of the groups involved in determining segment number: refer back to Fig. 3.16) and the Hox genes, which confer **segment identity**. The relationship between a motif-based group and a role-based group may or may not be simple. The pair-rule group is heterogeneous and includes genes with products containing homeodomain, zinc finger and helix-loop-helix motifs (examples being *paired*, *odd-skipped* and *hairy*, respectively). However, the Hox group is homogeneous in that all Hox genes, whichever part of the antero-posterior axis they are expressed over, and whichever segments they characterise, are homeobox genes and thus make proteins with helix-turn-helix homeodomains.

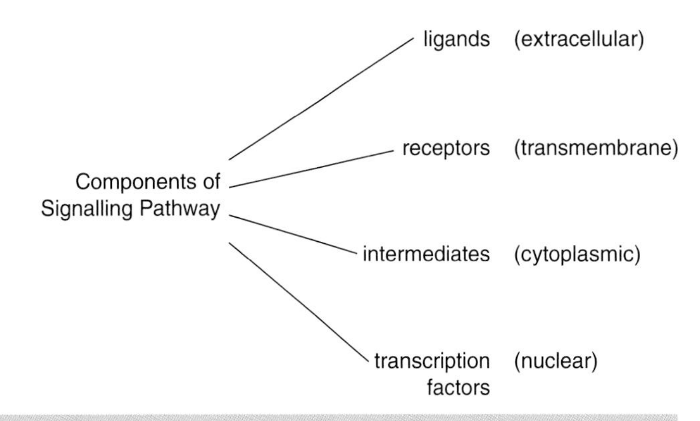

Figure 14.2 One way to classify the proteins of a signalling pathway: by their physical location. Thus we have the extracellular ligands, which attach to transmembrane receptors, which interact with cytoplasmic proteins, some of which migrate into the nucleus and have gene-switching effects. Members of the last class are by definition transcription factors, and can thus be classified as in Fig. 14.1.

Signalling Proteins

A similar problem is found in trying to classify signalling proteins: again, they can be classified in different ways, and the resultant classifications may not coincide. We already saw, in Chapter 3, that there are different signalling pathways, such as Hedgehog and Wnt. So one way to classify particular developmental genes is in terms of which pathway their protein product participates in (but with the complication that it is sometimes more than one).

A very different (indeed perhaps orthogonal) way to classify the genes concerned is in terms of *where* in a pathway their product functions, for example whether it is a mobile **ligand** moving from one cell to another, a membrane-bound receptor to which the ligand attaches in a target cell, or one of the several proteins that are usually employed within the target cell to get the signal from the receptor through the cytoplasm and the nuclear envelope to the target genes (Fig. 14.2). Taking any class of proteins defined in this way, it could be subdivided into a series of types. For example, transmembrane receptors can be relatively simple molecules that have an extracellular, a transmembrane and an intracellular domain; or they can be more complex, as in the case of the 'seven-pass' and 'twelve-pass' types, where the molecule criss-crosses the membrane seven or twelve times (Fig. 14.3).

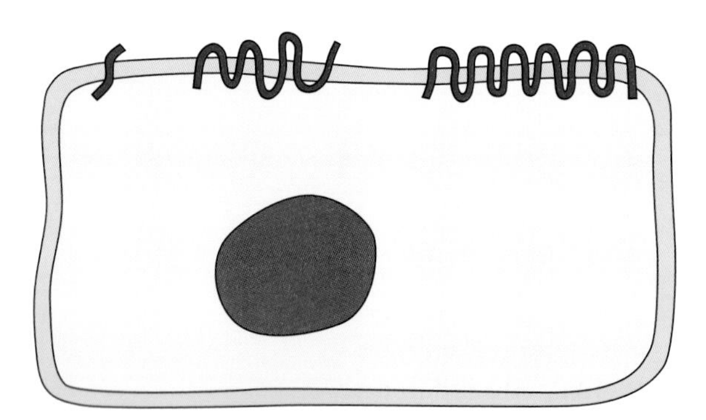

Figure 14.3 Simple (single-pass) and complex (7-pass; 12-pass) transmembrane proteins. Although these three types are common, there are many other such proteins that pass through the membrane a different number of times. The ways in which these proteins work in signalling is very complex.

Reasons for Classifying Developmental Genes

There is a parallel here with classifying species. A systematist will want groupings of species to relate to shared ancestry and closeness of relationship – hence the use of genera. In contrast, an ecologist may group species according to their ecological role; hence the use of 'guilds' – groups of species using the same resources and in similar ways. Sometimes, genus and guild will correspond, but often they will not. Each concept is valid; the choice between them depends on the aims of the investigator.

The same is true of ways of classifying developmental genes. From the perspective of a developmental biologist working on one of the **model organisms** – say *Drosophila* or *Arabidopsis* – the developmental role of a protein is the most important thing about it and the gene that makes it. However, from an evo-devo perspective, additional things become important. These include the evolutionary origin of the gene/protein concerned, perhaps through **gene duplication**; and its subsequent divergence, in both sequence and expression pattern, in the various evolutionary **lineages** that ramify out from the LCA that possessed the duplicated gene, as can be assessed from comparative studies of the extant 'growing tips' of those lineages.

It is clear that, in the course of evolution, the function of a developmental gene's product may change. We noted earlier that in the higher **Diptera**, including *Drosophila*, an important anterior determinant is the Bicoid protein. However, the protein Orthodenticle is used by most insects for that purpose. Also, the gene *fushi tarazu* (*ftz*) acts as a pair-rule gene in *Drosophila*, but it is physically within one of the *Drosophila* Hox complexes (the Antennapedia complex), and has sometimes been described as an 'escaped Hox gene' in *Drosophila*, because in many other animals it does indeed have a Hox-type function (i.e. helping to determine segment identity: see Section 14.4) rather than a pair-rule function. And the *Drosophila* gene *zerknult. It* (*zen*) shows a similar pattern: it resides within the Hox complex but does not behave as a typical Hox gene, specifying A-P segment identity. Yet in non-insect arthropods and in vertebrates (where the homologous gene is *hox3*), *zen* does indeed have a Hox-type function. These are just three examples out of many.

So, from an evo-devo perspective, we need to think not just about what *general* role a 'developmental protein' has, such as being a transcription factor, but also what *specific* role (or roles) it has, in terms of being part of the developmental machinery for making segments, legs or the heart in an animal, or for making roots, leaves or flowers in a plant. Examples of these specific roles are the subject of the next section.

14.3 Developmental Genes: Examples

In a typical **bilaterian** animal, there are four body axes – antero-posterior, dorso-ventral, left-right, and (in the limbs) proximo-distal. There are also several types of organs present across *most* of the Bilateria, such as a heart (even if only a rudimentary pulsatile tube), a brain (even if only a smallish ganglion or pair/group of ganglia at the anterior end of the animal) and one or more pairs of eyes (even if these are just simple ocelli). However, in some groups within the Bilateria, the axes and organs mentioned above may be absent. Most obviously, perhaps, extant **echinoderms** are 'bilaterians' that have lost the bilateral symmetry that gives the group its name, along with some typical bilaterian organs, such as the brain.

We will now select one body axis, the dorso-ventral one, and one organ, the eye, as examples to look at in more detail. We will examine both transcription factor genes and genes producing secreted proteins, and also interactions between them.

Establishment of the Dorso-ventral Axis

There is a key transcription factor gene involved in the formation of the dorso-ventral body axis in *Drosophila* and other insects. Appropriately, its name is *dorsal*. However, like many genes, its name is in a sense the opposite of its developmental role, which in the

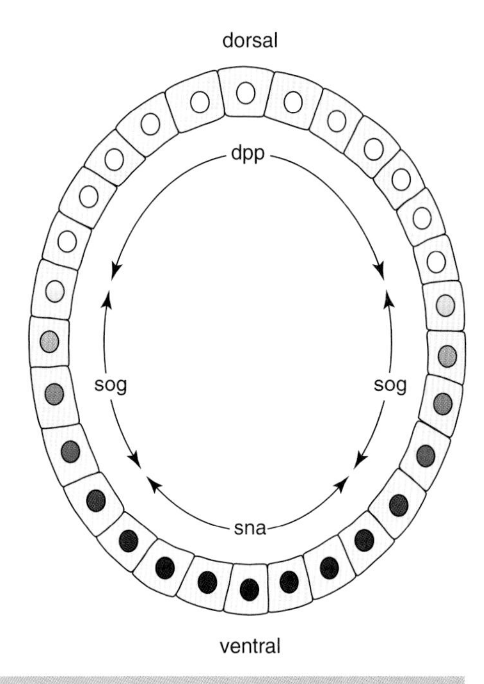

Figure 14.4 Some genes/proteins involved in establishing the dorso-ventral axis in a *Drosophila* embryo. The most upstream of those shown is the gene *dorsal*, whose protein product is transported into the nucleus maximally in ventral cells (dark blue), and minimally (or even not at all) in dorsal ones (white). Many downstream genes respond to this gradient, including *decapentaplegic*, *short gastrulation* and *snail*.

case of *dorsal* is a ventralising one. This apparent contradiction arises because of the common practice of naming genes after the effects of mutations in them: mutations causing loss of function often result in a phenotype that is the opposite of the normal one. If a mutation destroys a ventralising function, then the mutation itself will be dorsalising.

Figure 14.4 shows the pattern of Dorsal protein distribution in a cross-section of a *Drosophila* embryo, and some of the **downstream** genes that are affected by it. The first thing to notice is that the D-V gradient of Dorsal protein is not so much in its amount but rather its location – in the nucleus or the cytoplasm[31]. As a transcription factor, Dorsal can only switch on downstream genes when it is in the nucleus; and this is the case only in the ventral part of the embryo, with a peak of nuclear localisation around the ventral midline.

This particular type of gradient illustrates the fact that it is not just the control of gene expression that matters. Control of many subsequent stages between expression and protein function may also be important. So, following the control of expression, control of splicing, export to the cytoplasm, translation, post-translational modification of the protein, and interactions of the protein with others, are all stages at which the developmental effects of a gene may be regulated. In the case of the Dorsal protein, its location, in the sense of nuclear-v-cytoplasmic, is controlled by interaction of Dorsal with Cactus, a protein that can bind to Dorsal and prevent its movement into the nucleus. Phosphorylation of Cactus, a form of post-translational modification of that protein, prevents it binding to Dorsal. This happens in the ventral region of the embryo; hence the nuclear location of Dorsal in that region.

Several genes are downstream of *dorsal* in the cascade that leads to full establishment of the D-V axis. These, which we looked at from a comparative perspective in Chapter 7, include *decapentaplegic* (*dpp*), *short gastrulation* (*sog*) and *snail* (*sna*). As can be seen from Fig. 14.4, the first is expressed dorsally, while the other two are expressed ventro-laterally and ventrally, respectively. Dpp is a secreted signalling protein of the TGF beta super-family; Sog is also a secreted protein, of a type that has a cyteine-rich (CR) domain, and it acts to inhibit Dpp signalling; finally, Snail is a transcription factor with a zinc finger motif.

Clearly, all of these genes are intermediaries in a long chain of events that leads to the formation of structures and organs that are appropriate to the dorsal and ventral regions of the body. For example, the gene *tinman* is downstream of Dpp signalling, and is involved in heart formation, which in *Drosophila* and most other invertebrates is a dorsal structure. However, Tinman is a homeodomain-containing transcription factor, so its role is to control the expression of yet further downstream genes, some of which are actually involved in the physical process of building heart tissue. As usual in examining developmental cascades, we can only see a part of what is a very long and complex series of gene/protein interactions.

As we saw in Chapter 7, the vertebrate **homologues** of the *Drosophila* D-V axis genes have opposite developmental effects[32], in that those which dorsalise in *Drosophila* ventralise in vertebrates and *vice versa*. This D-V axis contrast provides evidence that the 19th-century French biologist Geoffroy Saint Hilaire was right when he proposed, in 1822, that vertebrates are 'upside-down' invertebrates[33].

Development of Eyes

Again, we will start with the situation in *Drosophila* and then broaden out to consider other groups. A key gene in the development of eyes in *Drosophila* is *eyeless*, the name of which, like that of *dorsal*, is derived from the effect of mutations in it. *Eyeless* is a homeobox-containing gene, and so makes a transcription factor of the homeodomain type. This transcription factor activates downstream genes with the ultimate result that eye tissue is formed. Activation of *eyeless* in parts of the body where it is normally switched off results in production of eye tissue there (Fig. 14.5). Expression in 'the wrong place' is referred to as **ectopic expression**, with 'ectopic' meaning the same as it does when used to describe mis-placed pregnancy in humans.

In vertebrates, there is a homologue[34] of *eyeless* called *pax6*. *Pax* refers to the presence of the 'paired box' sequence, which is a conserved region producing a protein domain with DNA-binding capability, like the homeodomain, but roughly twice as long, with about 130 rather than 60 amino acids. In fact, *eyeless* and *pax6* both have homeobox *and* paired-box sequences. Functional studies in vertebrates suggest that Pax6 has two or more functions in eye development, one of them being to directly regulate the gene producing the light-sensitive visual pigment Rhodopsin. It has been shown that the homeodomain of the Pax6 protein binds to the **promoter** region of the gene that produces Rhodopsin-1, a transmembrane protein found in light-sensitive cells of the retina.

Does the existence of homologous genes in insects and vertebrates controlling the production of eyes mean that those eyes are themselves homologous, despite their completely different structures and mechanisms of action (camera-type eye versus compound eye)? Questions of this kind have surfaced repeatedly as evo-devo studies have revealed similarities in the developmental-genetic control of structures that were previously thought to be **homoplastic** (i.e. the result of **convergent** evolution) rather than homologous. The answer may lie in the LCA having some rudimentary version of the organ concerned – such as light-sensitive pigment spots in the case of eyes – and having genes of certain types controlling the development of the 'primitive' structure, a controlling role that can then be elaborated as the structures themselves become elaborated in subsequent evolution.

14.4 The Hox Genes

The Hox genes have become something of a *cause celèbre* in evo-devo. **Homeotic** transformations, in which a body part appears in a place where it would normally not be, were documented in the 1890s by the English biologist William Bateson[35]. Studies

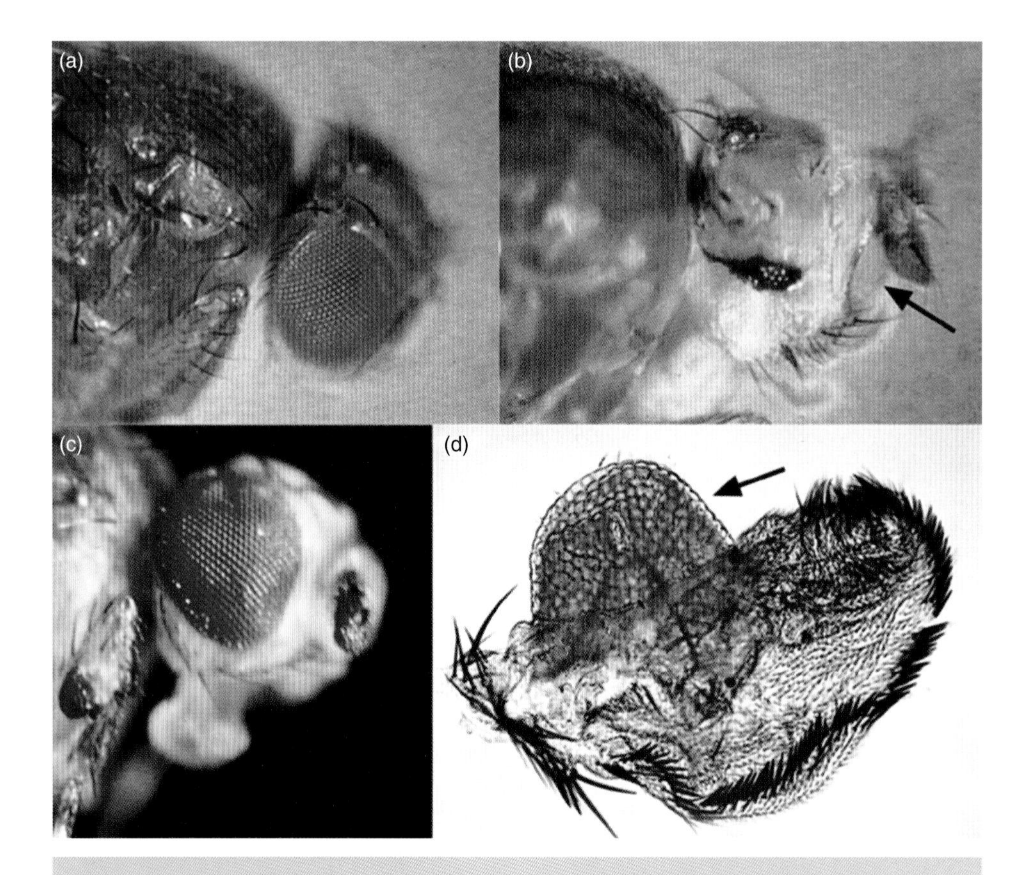

Figure 14.5 Mutant phenotypes of *Drosophila* resulting from mutation and/or ectopic expression of the gene *eyeless*. As can be seen, areas of eye tissue can develop in various abnormal places on the body. (a) Normal eye; (b) Eye absent (arrow); (c) eye present in normal place, but other patches of eye tissue also formed in incorrect places; (d) a large patch of eye tissue (arrow) formed ectopically on the wing. (Reproduced with permission from Carroll, S. *et al.* 2005, *From DNA to Diversity*, Blackwell.)

of homeotic mutants of *Drosophila*, such as those with halteres transformed into wings or antennae into legs, led the American geneticist Ed Lewis to identify the two gene complexes involved in the 1960s and 1970s – the Antennapedia and Bithorax complexes[36]. These were jointly referred to as HOM complexes because they both contained genes that could cause homeotic mutant forms. In the 1980s, studies of mice revealed that there were complexes of homologous genes that had similar roles; these were referred to as Hox complexes, because it was by then known that the genes involved all had homeobox sequences. Now, studies have been conducted on many other animal groups, revealing that all animals have Hox genes. And that name – Hox – is now used in general, even though it is confusing to those first studying the subject, because Hox genes are only a small subset of the much larger category of homeobox-containing genes.

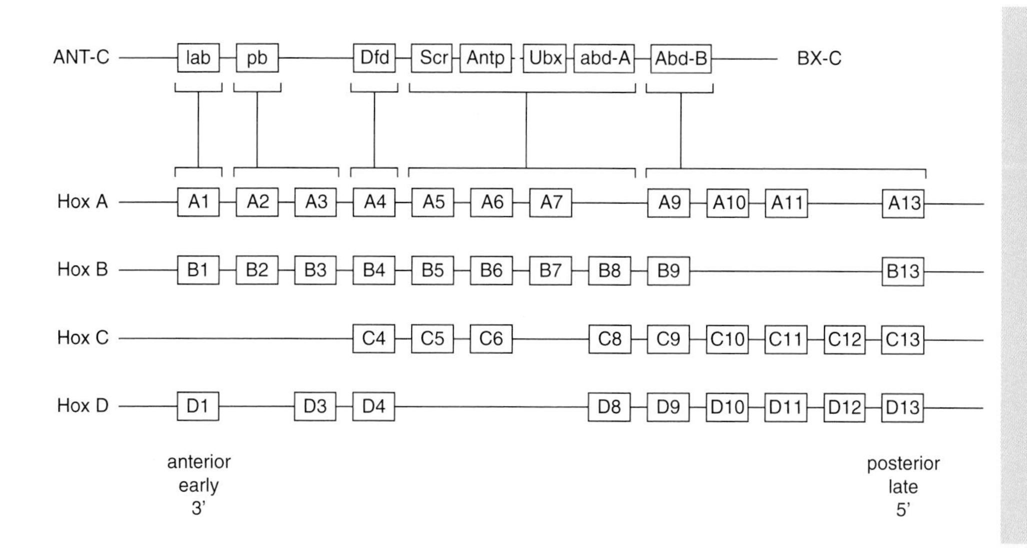

Figure 14.6 The Hox gene complexes of *Drosophila* (top) and mouse, showing the relationships between individual Hox genes. Note that genes to the left of the complex are expressed more anteriorly in the embryo, and usually at an earlier stage, than those to the right.

A Two-Way Comparison: Flies and Mice

The fly and mouse Hox genes, and the pattern of homology between them, are shown in Fig. 14.6. The most reasonable interpretation of the evolutionary changes that have occurred in the two **lineages** is as follows. In the lineage leading to *Drosophila*, an original single cluster with about eight genes became split in two. In the lineage leading to mice the cluster underwent (i) addition of genes, especially in the AbdB group, by individual **gene duplication** events; (ii) two rounds of duplication of the entire cluster; and (iii) loss of some genes within each cluster, probably due to redundancy. The net result of these changes has been to increase the number of Hox genes to 39.

The most remarkable thing to emerge from studies of the expression patterns of the Hox genes in flies and mice has been **colinearity**. The antero-posterior expression zones of the genes reflect their order along the chromosome (Fig. 14.7). Genes expressed at the anterior end of the animal are at the 3′ end of the cluster. This suggests that there is some form of co-ordinated ***cis*-regulation** of the genes that requires physical proximity and a particular linear ordering. In addition to the spatial colinearity just described, in mice there is also temporal colinearity[37] because the anterior segments are formed first. So there is a three-way colinear pattern involving chromosomal order, spatial expression domains and temporal expression periods. The temporal dimension is absent in *Drosophila* because it forms its segments more-or-less simultaneously.

A Multi-Way Comparison: Many Animal Phyla

Many other animals have now been screened for Hox genes. One surprise that has come out of this work is that most **teleost** fish have 7 or 8 Hox clusters, compared with the mammalian 4. It seems that in the lineage leading to teleosts there was

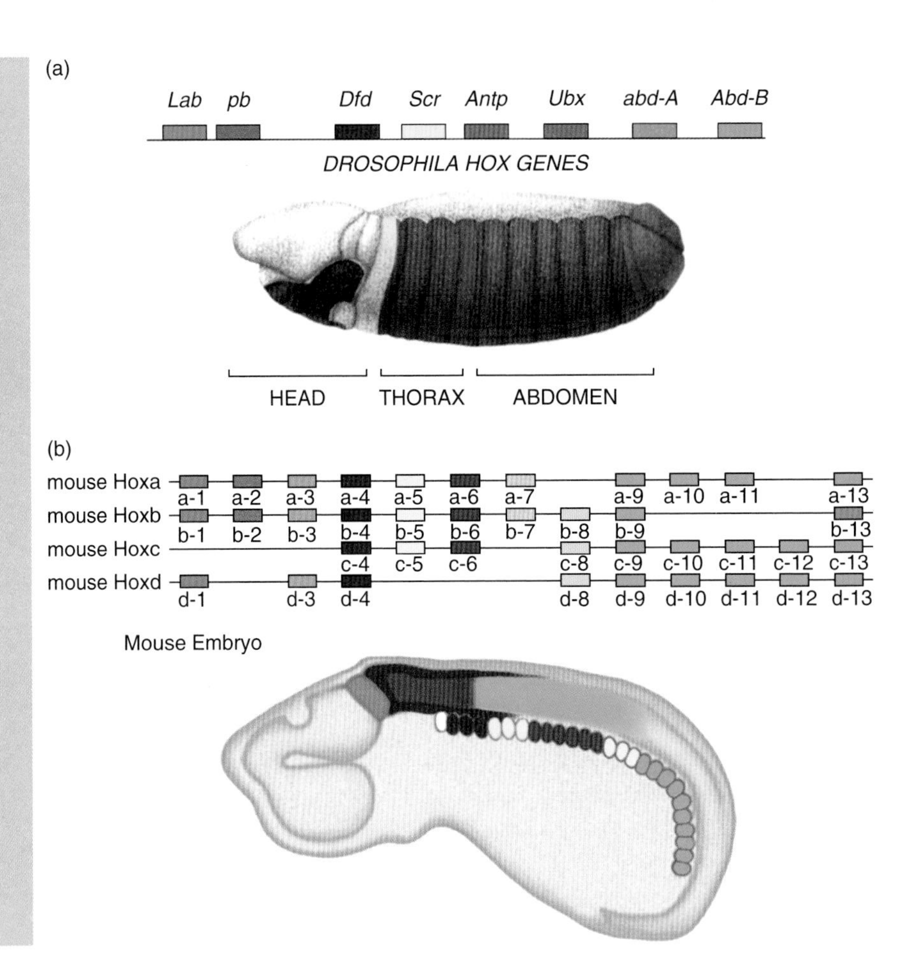

Figure 14.7 The spatial expression zones of the Hox genes along the antero-posterior axis of a fly (a) and a mouse (b). Note the correspondence with the chromosomal order of the genes, as seen both here and in Fig. 14.6. This is referred to as colinearity. (From Carroll, S. *et al.* 2005, *From DNA to Diversity*, Blackwell.)

another round of duplication (from 4 to 8), followed by the loss of one of the 8 clusters. This duplication appears to have been one involving the whole genome rather than being restricted to the Hox clusters; and it was followed by extensive further gene losses, again most likely because of redundancy, so that the typical teleost has 48 Hox genes compared to the typical mammalian number of 39.

Another surprise that has emerged from extending the study of Hox genes to other animal groups is that in some of them the clusters have disintegrated, with Hox genes scattered seemingly at random over the chromosomes, so there is no longer colinearity because one of the linearities has disappeared. This is the case, for example, with the **urochordate** *Oikopleura*.

The Swiss biologist Denis Duboule has classified Hox clusters into four types[38]: organised (O), disorganised (D), split (S) and atomised (A) (Fig. 14.8). The difference between organised and disorganised is two-fold. In organised clusters, all genes are transcribed in the same direction; also, there are no non-Hox genes interspersed with the Hox genes (as there are in the *Drosophila* Antennapedia complex, for example), with the result that the cluster is more compact than a 'disorganised' one. 'Split cluster' is a self-explanatory term, though we now know that the *Drosophila melanogaster* split does not apply to all insects; and in some other groups splits can occur in different places than between the genes *antennapedia* and *ultrabithorax*. At first sight, the term 'atomised cluster' seems illogical, since if the genes have become scattered across the chromosomes there is no such thing as a cluster. But the term is useful to refer specifically to cases where there *was* a cluster in an ancestor and it has *become* atomised in the descendant. The pattern of occurrence of the four types of cluster, O, D, S and A, is shown in Fig. 14.9 for various bilaterian groups.

The spatial colinearity found in both *Drosophila* and mice is unlikely to have arisen independently. Thus it seems safe to say that the LCA of the Bilateria had this arrangement. **Basal** animal groups that split off early from the lineage leading to Bilateria have Hox genes but not in clusters; for example, the **cnidarian** *Hydra*. However, it does not seem sensible to refer to the *Hydra* arrangement as an atomised cluster, but rather just as a dispersed group of genes.

The link between the evolution of the Hox genes and the evolution of the body forms of the animals in which they are found is not simple. The original comparison of *Drosophila* and mouse suggested that more Hox genes might be associated with a more complex morphology. But the inclusion of other groups, notably teleost fish, made clear that there is certainly not

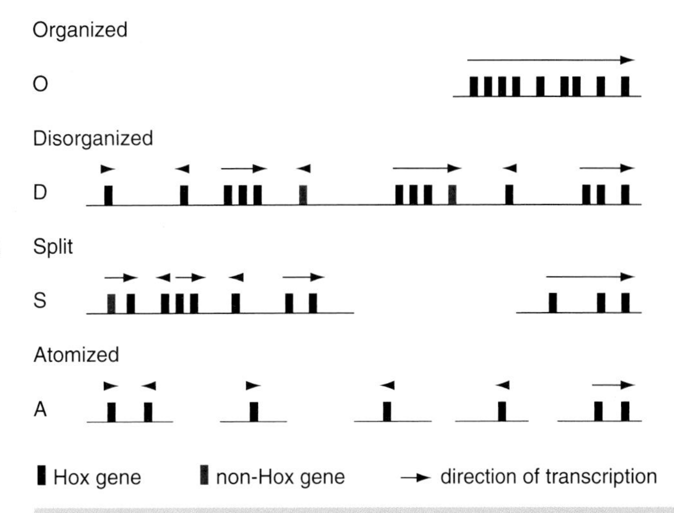

Figure 14.8 Four different types of Hox gene cluster: organised (O), disorganised (D), split (S) and atomised (A). (Redrawn with the permission of Denis Duboule.)

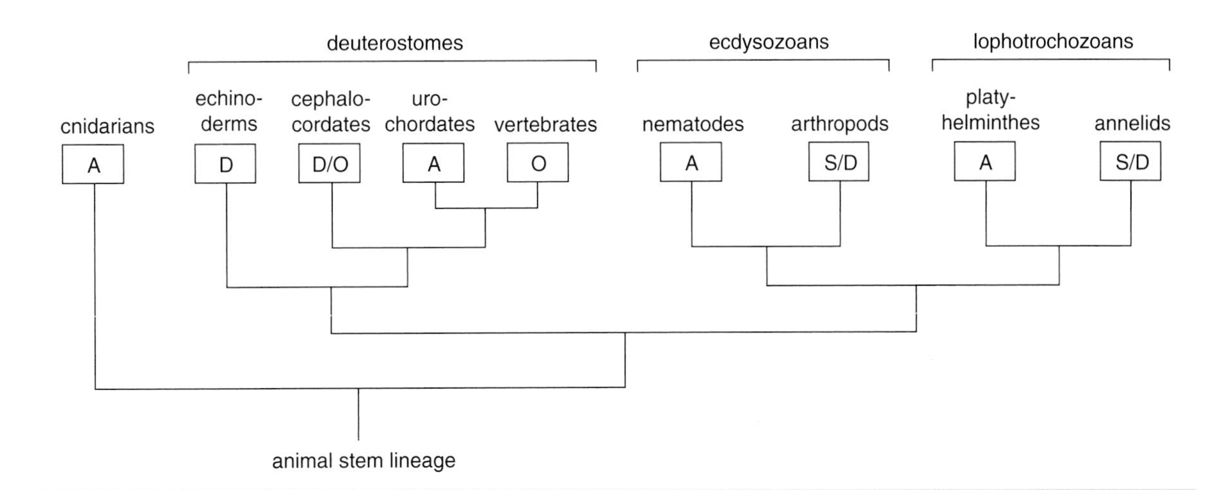

Figure 14.9 The phylogenetic distribution of the four types of Hox gene cluster across several animal phyla (A,D,O and S). (Redrawn with the permission of Denis Duboule.)

a simple relationship between these two things. Hox clusters have doubtless sometimes changed or remained the same due to natural selection – respectively directional and stabilising. However, there is probably also much phylogenetic inertia. For example, why do echinoderms still have a non-atomised cluster (Fig. 14.9) when they cannot have colinearity as they have no antero-posterior axis? When there is directional selection, it is perhaps most often likely to be of the 'internal' variety discussed in Chapter 13. That is, it is likely to be related to improved developmental (or more generally organismic) integration, rather than being related to adaptation to particular environments.

14.5 Gene-Level Forms of Developmental Bias and Coadaptation

A major problem with the subject of molecular developmental genetics is that it is too complex. There are so many genes and proteins involved in each developmental process; and there are so many developmental processes even in any one organism, such as *Drosophila*, that students approaching the subject for the first time can easily feel swamped in detail. If we deal with the subject in a comparative way, as is necessary to see what insights it can give us into evolution, then the situation becomes even worse. There are millions of species of multicellular creatures, and any cross-species comparison will reveal both similarities and differences in the genetic basis of a shared developmental process, with the relative proportions of these being broadly related to the phylogenetic distance between the creatures being compared.

As noted previously, science is all about trying to find general principles that subsume many individual 'facts'. But when the facts become too numerous, the task of finding those general principles becomes daunting. How do we deal with this problem? The answer is three-fold, as follows:

First, it is helpful to take a selective approach so as not to become buried in too much information. Hence the focus on just a few developmental processes and groups of genes in the previous two sections. But equally, too selective an approach may mean that important details and comparisons that might help in the search for general principles are omitted. Obtaining the right balance of selectivity and inclusiveness is difficult, and what is the best balance for any one investigator (in research), or for explanation to any one audience (in teaching), may not be so for another one.

Second, it is essential to keep asking general questions; and it is important exactly which ones to ask. One general question that we should definitely ask is: do some types of developmental-genetic changes arise more readily than others; if so, which; and does this have an effect on the direction of evolution? We will 'dissect' this question shortly.

Third, it is important to relate the above two things, namely (i) the degree of selectivity of approach and (ii) the nature of the general questions being asked, to each other. The importance of doing so can be seen by comparing three general theories (two right and one wrong) produced by three well-known biologists – Charles Darwin, Gregor Mendel and Hugo de Vries.

Darwin[19] accumulated information on many types of animals and plants, so his general theory of natural selection had a broad underlying platform of evidence. Mendel[39], in contrast, based his general theory of particulate inheritance on experiments conducted on a single species, of the pea genus *Pisum*. So the theory was riskier, because it might have turned out that, for example, animals, or even other types of plants, inherited their characteristics by completely different mechanisms than peas. As we now know, they do not; and although some other groups of organisms, such as bacteria, are **haploid** and so fail to obey Mendel's laws, they still have particulate genes.

De Vries[40] operated in a similar way to Mendel: he focused on a particular plant – in his case the evening primrose *Oenothera*. Because he found discrete variants appearing spontaneously, he developed his 'mutation theory' of evolution, arguing that evolution occurred in jumps at the phenotypic level rather than gradually as proposed by Darwin, and that the jumps were based on large-effect mutations of the same sort as Mendel had worked on in peas. As with Mendel, this general theory based largely on studies of a particular species was risky. But unlike Mendel, de Vries was unlucky: *Oenothera* is a genus in which the chromosomes behave in an unusual way, and the discrete phenotypes that de Vries observed were based on this behaviour and not on mutations, in the conventional sense, at all.

Bearing in mind this cautionary message, we now return to general questions that can be asked in the field of gene-level evo-devo, and specifically to the one noted above, namely:

Do some types of developmental-genetic changes arise more readily than others; if so, which; and does this have an effect on the direction of evolution?

This question, which relates to both **developmental bias** and adaptation or **coadaptation** at the gene level, only makes sense if we can contrast specific different 'types' of developmental-genetic change. Here are two such contrasts that can be made:

1 the contrast between achieving developmental repatterning by altering the regulatory regions of a gene versus altering the coding region[41]; and
2 the contrast between **co-opting** an existing gene to perform a new role versus 'inventing' a new gene, say by duplication of an existing one followed by divergence of the duplicated copies.

These are not the only such contrasts that can be made; and a good mental exercise is to try to think of others. But they are the two contrasts that we will focus on here, with the first one being dealt with in the following section, the second one in Chapter 15.

14.6 Changes in Regulatory versus Coding Regions of Genes

As we saw in Chapter 3, the typical **eukaryote** gene can be thought of as having two functional components: coding and regulatory. Neither of these corresponds to an integral stretch of the gene. Coding regions are broken up by **introns**; and regulatory regions may be close to the coding regions, as is the case with promoters, or separated from the coding region by considerable distances, as is usual for the regulatory regions called **enhancers**.

Even without considering post-transcriptional modifications such as intron splicing, gene regulation is a very complex process. Many things happen before the enzyme **RNA polymerase** attaches to the DNA at the transcription start site and begins the business of making the RNA transcript that will ultimately lead to the production of the corresponding protein.

The simplified case of a gene with a single enhancer is shown in Fig. 14.10a. Even in such a simplified case, a transcription factor does not act alone to switch on the gene. At minimum, each employs one co-activator, whose job is to open up the local **chromatin** from its more condensed form associated with lack of gene expression, by acetylase action. More often, a developmental gene will have several separate enhancers, as shown in Fig. 14.10b. What is typically found is that different enhancers, or different combinations of them, interact with different transcription factors, and are used in the expression of the gene in different tissues or regions of the embryo.

This diversity of enhancers may have important consequences for the relative ease of evolution in regulatory versus coding regions of the gene. A change in the coding region that alters the amino acid sequence of the protein is likely to have the same effect in all places and at all times where/when the gene is expressed. In contrast, a change in the DNA sequence of one particular regulatory region may only affect one place/time of expression.

To illustrate this point in a particular context, suppose that the gene under consideration itself makes a transcription factor and is somewhere in the middle of a developmental cascade. It is switched on by the transcription factor product of an **upstream** gene, but its own product then switches on a **downstream** gene. There are probably hundreds of such 'mid-cascade' genes in most animals and plants. Suppose also that the gene in focus has *two* separate developmental roles, and that its expression in each of these is controlled by a different enhancer.
Mutational changes in the enhancer regions will each then affect just one of the gene's roles; whereas mutation in the coding region will potentially affect *both* of its roles – with the 'potentially' referring to whether the mutation concerned affects the function of the protein.

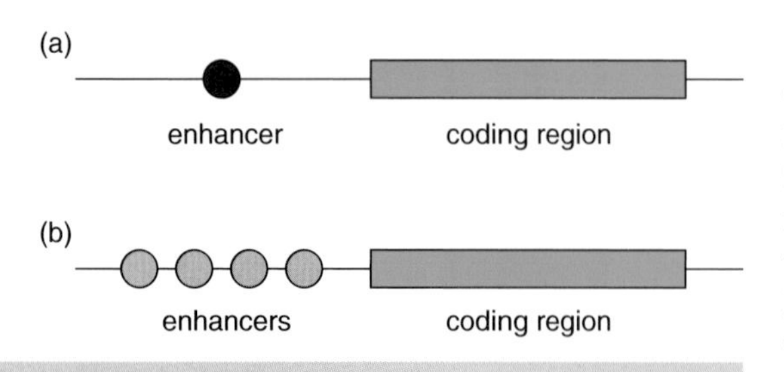

Figure 14.10 (a): Diagram of simplified gene structure with just a single enhancer. (b): The more realistic situation of a gene with multiple enhancers.

The probability of a mutation being selectively advantageous decreases with the number of its developmental effects. So, other things being equal, mutations in regulatory regions should be more often favoured by selection than those in coding regions. This is important, and may in a sense 'bias' evolution towards more frequent regulatory than coding changes; however, we need to be cautious in our interpretation of this point. First, the contrast is a quantitative rather than a qualitative one – we know for certain that evolution involves mutational changes in both coding and regulatory regions. Second, although the 'bias' referred to above is very real, it is not 'developmental bias' or 'developmental constraint' as used elsewhere in this book. Rather, it is a form of *selective* constraint. The importance of *developmental* constraint/bias in relation to the evolution of developmental genes will be discussed, in conjunction with the concept of gene co-option, in the following chapter.

SUGGESTIONS FOR FURTHER READING

The best student text on the subject of developmental genes and evolution is:

Carroll, S.B., Grenier, J.K. and Weatherbee, S.D. 2005 *From DNA to Diversity: Molecular Genetics and the Evolution of Animal Design*, second edition. Blackwell, Malden, MA.

For a more advanced treatment, see:

Wilkins, A.S. 2002 *The Evolution of Developmental Pathways*. Sinauer, Sunderland, MA.

Also very much worth reading is the evo-devo chapter at the end of the leading developmental biology text:

Gilbert, S.F. 2010 *Developmental Biology*, ninth edition. Sinauer, Sunderland, MA.

Earlier chapters in Gilbert's book also deal thoroughly with the roles of many developmental genes in particular taxa.

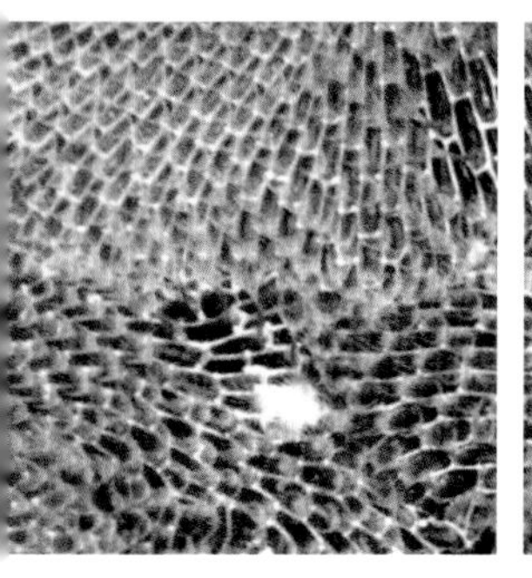

CHAPTER 15
Gene Co-option as an Evolutionary Mechanism

15.1 What is Gene Co-option?

The concept of gene **co-option** is one of the most important to have emerged from the field of evo-devo thus far. The basic idea is that:

a gene that has some initial well-established function in development becomes deployed, over the course of evolutionary time, to perform a new function – normally in addition to, rather than instead of, its original one.

One example that we will pursue in detail in this chapter involves the gene *distal-less* (*dll*). This has been co-opted in different ways in different **lineages**, often, but not always, in relation to the evolution of limbs.

Some authors use the term 'recruitment' instead of 'co-option', so it is worth noting that these are effectively synonymous in evo-devo. However, 'co-option' will be used throughout the present chapter (and indeed the book) in the interests of consistency.

The definition of co-option in evo-devo given above is in line with typical dictionary definitions relating to the general usage of the word. However, the definition only works if we know what is meant by '*new*'. The problem is that in all evolutionary changes in development, the role of at least one **developmental gene** is altered to some degree. That is, it becomes different. But how different does the role have to become in order to be considered a 'new' role? Is there a continuous spectrum of possible magnitudes of change? Or is there some clear line of demarcation between an altered role and a new one?

Luckily, the answer to the last version of the question is often 'yes'. Consider the gene *engrailed*, which we have encountered before in relation to animal segmentation. This is normally thought of as a **segment polarity** gene in arthropods, expressed in one transverse stripe per segment. As different lineages of arthropods have diverged, they have often changed in their number of segments. This means that the whole developmental-genetic machinery producing segments has changed, at least to the extent that its action has become iterated more or fewer times, so that more or fewer segments are produced. But this could hardly be thought of as a new role for the genes concerned, including *engrailed*. Rather, they have the same role; they produce a different morphological result only because of the different number of times that role is implemented. However, *engrailed* is also used in the development of the nervous system. That role is fundamentally different; we will examine it further in Section 15.2. So the first sort of change (number of segments) does not involve co-option; the second does.

Another example of the distinction between an altered role and a new one involves the **Hox genes**. Their role in determining **segment identity** in arthropods and vertebrates is well known. Changes in segment identity in the course of evolution are widespread, and are always accompanied by changes in Hox gene expression, as in the case of the centipede poison claw segment examined in Chapter 9. Yet even there, where a new type of segment has arisen, the role of the Hox genes is not itself 'new'; rather, the same sort of role – involving a defined antero-posterior expression domain in the trunk – produces a new developmental result, this time because of a new combination of Hox genes being expressed in a particular segment. In contrast, some aspects of the role of Hox genes (but not others[42]) in the development of **tetrapod** digits, which we also examined in Chapter 9, *are* new.

However, in both of these cases, involving *engrailed* and Hox genes, the role of the gene concerned is new only at one level. At another level it is unchanged. In both cases the gene continues to make a transcription factor in its co-opted role. The 'newness' emerges despite this. For example, the transcription factor may have evolved in its coding sequence so that it binds to new downstream genes. Alternatively, it may have evolved in one or more of its regulatory regions so that it becomes activated in new parts of the embryo and/or in new periods of development, as shown in Fig. 15.1, in which a gene making a transcription factor normally only found in the head is co-opted for a new role in the trunk.

Since co-option often involves expression in different parts of the embryo, this would seem to argue for the evolution of regulatory regions being more important. However,

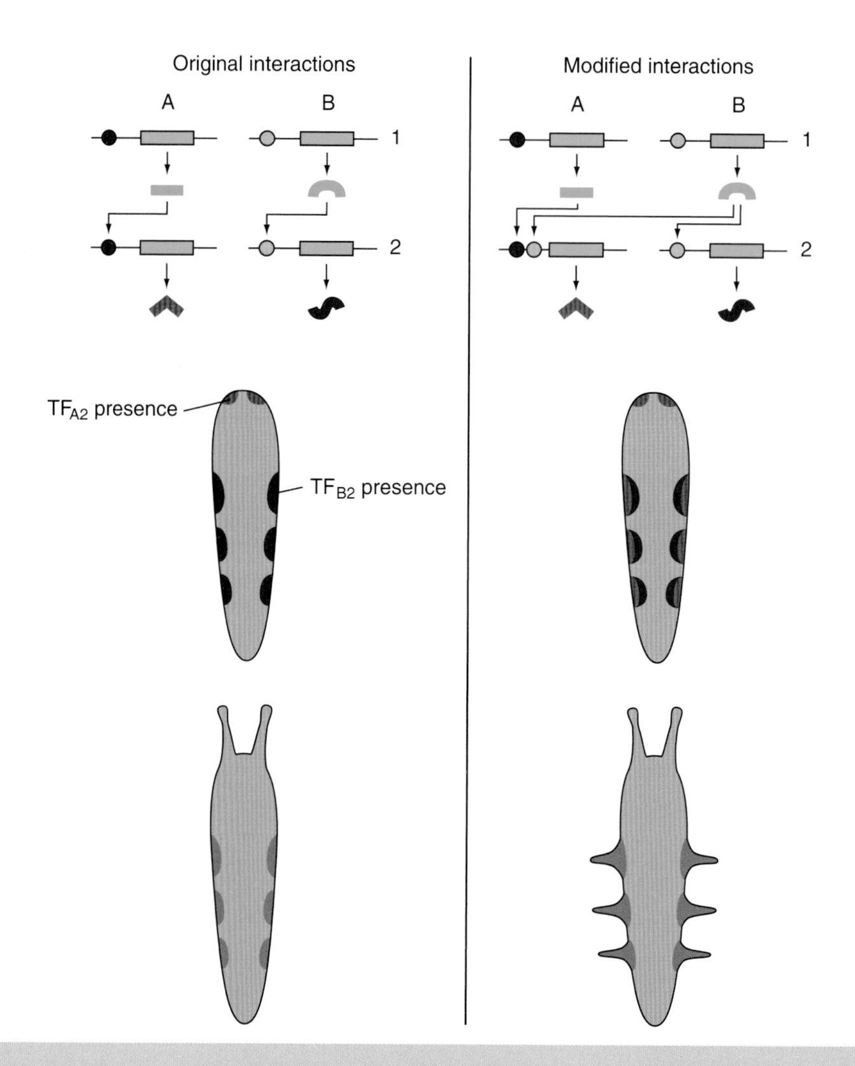

Figure 15.1 One possible form of gene co-option. Here, in the original situation (LHS) one cascade of interactions is operating in the head, and causing the outgrowth of antennae; another is operating in parts of the trunk and producing thickenings of the skin. In the modified situation (RHS), a mutation leading to the gene producing transcription factor A2 acquiring a regulatory region identical or very similar to that of the gene producing TF B2 results in trunk outgrowths in the form of rudimentary legs.

the co-option of gene A2 in the trunk (Fig. 15.1) could arise from a change in its regulatory region (as shown in the figure) *or* from a change in the coding region of upstream gene B1.

In some cases, the direction of co-option is clear. In other words, it is readily apparent which role was the original and which was the co-opted one. However, in other cases it is not at all clear which is which. We will revisit this issue in subsequent sections. A related issue is that there can be multiple co-options from role A to role B in

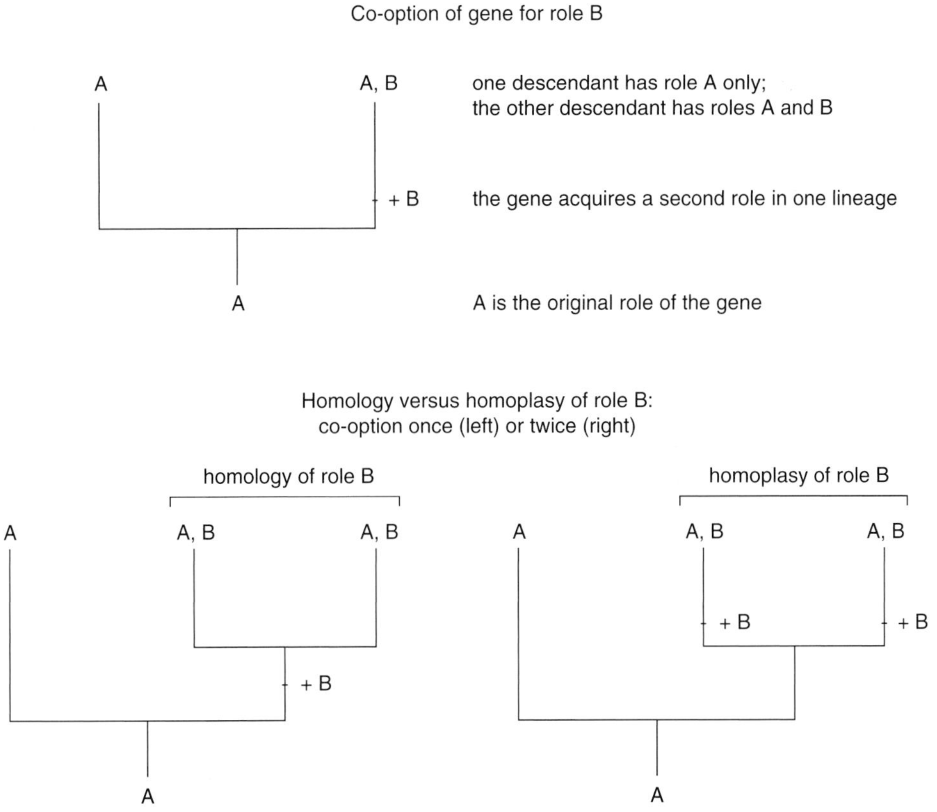

Figure 15.2 Top: Co-option of a gene product from role A to role B. Bottom: Two independent co-options for role B (RHS) contrasted with one such co-option followed by lineage divergence. This contrast connects with the issue of whether the B role is homoplastic (RHS) or homologous.

the sense that such co-options have occurred independently in two or more lineages. This may have occurred in relation to the origins of segments, eyes and limbs. See Fig. 15.2, which also relates co-option to **homology** and **homoplasy**.

15.2 Co-option in the Evolution of Segments and Eyes

Continuing the Story of *engrailed* and Segments

We saw in Chapter 11 that there are two possible interpretations of the fact that the arthropods and **chordates** have some similarities in the developmental genes underlying their segmentation. First, the LCA, in this case the **urbilaterian**, may have had at least a rudimentary form of segmentation, in which case the segments of arthropods and chordates are homologous. Second, the urbilaterian may have been completely unsegmented, meaning that arthropod and chordate segments were evolved independently, but that each of these independent origins was based on the co-option of some of the same genes, such as *notch*, *delta* and *engrailed*.

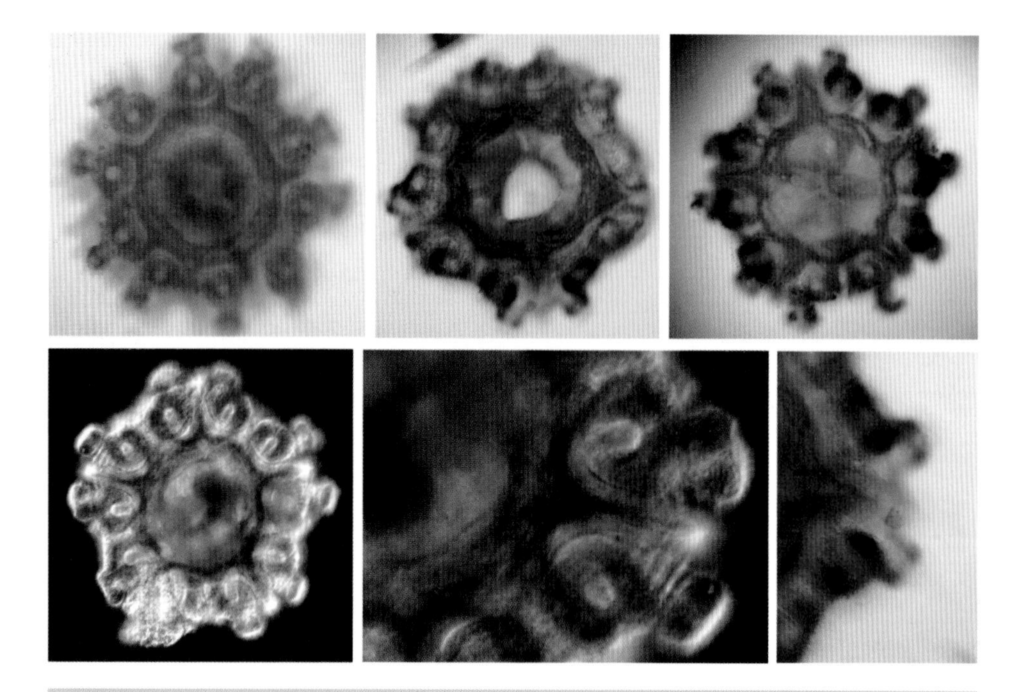

Figure 15.3 Expression of the gene *engrailed* in the nervous system (including the main circum-oral nerve ring, which shows up as a dark-blue circle in most of the pictures) of an echinoderm. The widespread nervous system expression of *engrailed* suggests that this may have been its original function in the animal kingdom, with its role as a segment polarity gene in arthropods being the result of its later co-option for that role. (From Byrne *et al.* 2005, *Dev. Genes Evol.*, **215**: 608–617.)

This issue can be examined further by looking at the **phylogenetic** range of genes such as *engrailed*, as follows. So far, *engrailed* homologues have been found in all bilaterian groups examined, including **nematodes**, molluscs and **echinoderms**. Two of these three groups, nematodes and echinoderms, show no signs of segmentation, while some members of the third, the molluscs, have a few segmentally reiterated structures (Fig. 11.4). In all three groups, *engrailed* is expressed in the developing nervous system (Fig. 15.3). In molluscs it is also involved in the reiterated structures and in shell formation[43].

A homologue of *engrailed* in the basally-branching non-bilaterian **cnidarians** is also known. Therefore, it seems likely that the original *engrailed* gene arose prior to the cnidarian-urbilaterian divergence, perhaps as a result of the duplication and divergence of a related **homeobox** gene. A reasonable hypothesis for evolution of *engrailed* function from that point on is as follows:

The original function may have been in the development of the nervous system – a role that continues in many bilaterian groups today. Then, *engrailed* was co-opted as a segmentation gene, either once in a segmented urbilaterian or, more likely,

Figure 15.4 Examples of the many different forms that eyes can take in the animal kingdom, including 'simple eyes', compound eyes, and 'camera-type' eyes. Top left: crustacean; top right: insect; bottom left: vertebrate; bottom right: mollusc. In the last case, the eyes are tiny black spots at the ends of the tentacles.

more than once in the stem lineages leading to each of the phyla characterised by a segmented **body plan**. In some restricted groups, it was co-opted for yet other functions, such as the role in mollusc shell formation. In addition, the gene duplicated in some lineages, including the one leading to *Drosophila*, in which *invected* is recognised to have arisen from *engrailed*. This duplication seems to have led to a combined role of the two genes in segment formation, as well as in the development of the hindgut; co-option to a completely new role does not seem to have happened (yet) with *invected*, although often co-option can indeed be a consequence of gene duplication and divergence.

Continuing the Story of *pax6* Genes and Eyes

Eyes, broadly defined to include all light-sensitive organs from 'simple eyes' to compound and camera-type eyes, are found in almost all groups of animals. Examples are shown in Fig. 15.4. The only exceptions seem to be the basal groups of Porifera (sponges) and **Placozoa** plus some apparently degenerate bilaterians, the 'mesozoans', which were originally thought to be **basal** but now appear not to be. A key gene in eye formation is a member of the Pax family, *pax6*, as we saw in the previous chapter. Here, we pick up this story where we left off there, and ask the specific question: has co-option of *pax6* genes (and others) been involved in the evolution of eyes?

Like *engrailed*, the Pax family of genes is found in a broad range of **taxa**. Family members are found in basal animals, including those without eyes, such as corals[44].

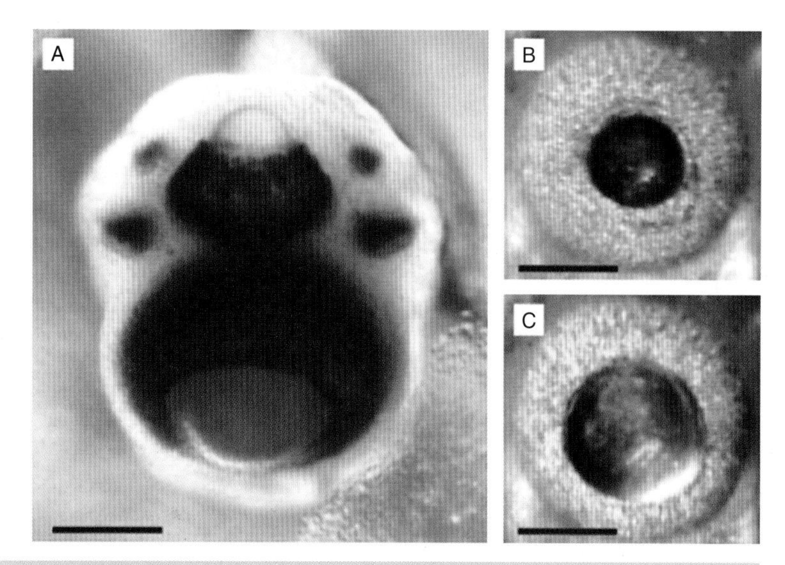

Figure 15.5 The surprisingly complex eyes (A, B, C) of the 'box jellies' (a class of cnidarians) that present something of a paradox because they appear to be capable of forming images, yet these animals have no brain to process images. (From Nilsson *et al.* 2005, *Nature*, **435:** 201–205.)

The initial function of *pax6* and its forerunners (such as *paxB*) was thus presumably not in eye development but rather in some other developmental process – though it is not as clear as in the case of *engrailed* what that process was. So, its role in eye formation is an example of co-option. But is this a case of a single co-option and homologous eyes throughout the animal kingdom, or a case of two or more independent co-options?

The answer to this question can be found in the Cnidaria – the phylum including sea anemones, corals, hydroids and jellyfish – and in particular, the cnidarian group Cubozoa ('box jellies'; Fig. 15.5). These have quite advanced, apparently image-forming eyes, which is odd since they have no brain to process images. Perhaps the main function of the eye is to detect movement in a way that does not rely on detailed images of the moving objects (potential prey). In any event, it is clear that the Cubozoa are very **derived** cnidarians. The most basal extant cnidarians, whether they are anthozoans or hydrozoans, have no eyes. Thus we have a situation where: the original animals had no eyes; the sponges and placozoans have no eyes; most cnidarians have no eyes; cubozoans do have eyes; and the vast majority of bilaterians have eyes. The most reasonable interpretation of this information is that a Pax-group gene has been co-opted at least twice to the role of eye formation.

Whether there have been *more than* two co-options is less clear. As with segments, the issue is whether the urbilaterian had eyes or not. The competing hypotheses of simple and complex urbilaterians are still being debated at the time of writing. At one extreme we have a simple animal such as an acoel flatworm with no segments or eyes. At the other, we have a much more complex animal, with both segments and eyes (and various other organs too, e.g. a heart). And, of course, intermediate forms are possible also.

There are three possibilities in relation to eyes. First, the urbilaterian was completely eyeless, and various lineages diverging from it independently 'invented' eyes. Second, the urbilaterian had some very simple light-sensitive pigment spots, which were elaborated into more complex eyes independently in several of the descendant lineages. Third, the urbilaterian had something more than that, perhaps akin to the ocelli observed in many of today's arthropods. In the first of these three scenarios there must have been more than two co-option events; in the second and third there need not have been more than two.

Figure 15.6 A small sample of the vast range of types of appendages that are found in the animal kingdom. Top left: the flight balancing organ (haltere) of a fly; top right: the wings of a bat; bottom left: the 'tentacles' of an octopus; bottom right: the 'arms' of a starfish (each of which bears many smaller appendages called tube-feet).

15.3 Appendage Evolution and Gene Co-option

There is some confusion in the literature as to what constitutes a 'limb'; hence the use of the broader term 'appendage' in this section heading. An appendage, in the biological sense, is a very broad term, defined by the *Concise Oxford English Dictionary* as 'a projecting part of an organism'. The dictionary definition of a limb, in contrast, is much more restricted: 'an arm, leg, or wing'. This usage is consistent with the idea of a vertebrate 'fin-to-limb transition', as discussed in Chapter 11, because the fins of a fish are generally not described as limbs, though they clearly are appendages. However, it is less consistent with the common usage of 'limb' (and even 'leg') to describe the paired appendages of aquatic crustaceans, which may be used for walking, swimming, feeding, or a mixture of these activities, depending on the group of crustaceans under consideration, such as krill, prawns, lobsters and crabs.

What all appendages have in common is that they arise from one (or both) of two processes: an altered plane of cell division so that part of the body wall grows outwards; or a process of cell migration that has the same effect. The only exceptions to this rule are 'appendages' that are non-cellular, such as sensory bristles in arthropods or mammalian hairs. These will not be considered further here.

Different kinds of animals have different kinds of appendages (see Fig. 15.6 for some examples). These include, in addition to the familiar fins and limbs: various kinds of feeding appendages, ranging from arthropod mandibles and maxillae to

Figure 15.7 A 'typical' cactus body form (top) and an interpretive drawing (bottom) showing the roots which, as in most plants, are long thin ramifying structures that could be described as appendages.

bryozoan lophophores; sensory appendages, such as the tentacles of snails; and a diversity of other kinds of appendages, for example, spider spinnerets and echinoderm tube-feet.

The term 'appendage' is less often used by botanists than by zoologists, despite the use of 'organism' in the dictionary definition. Some plants could be described as being composed largely of appendages, as in the case of the growth form of a tree or bush. In other cases, such as some cacti, there is indeed a main compact body as in an animal, with two main types of appendage – the non-cellular spikes above ground, and the roots below ground (Fig. 15.7).

As we now examine gene co-option in the evolution of appendages, we will focus on animals rather than plants, and on cellular appendages rather than non-cellular ones. We will focus in particular on the gene *distal-less* and its homologues. In some ways the story is similar to those of segments and eyes discussed in the previous section. However, in one important way the appendage story is clearer. It is conceivable (though perhaps unlikely) that the urbilaterian had simple eyes and proto-segments to which all extant bilaterian eyes and segments are homologous; but a comparably neat picture is not conceivable in relation to appendages. For example, the tube-feet of echinoderms are not considered to be homologous to any appendages in their **deuterostome** relatives the vertebrates, or indeed to any appendages of animals, extant or extinct, outside their own phylum.

The original function of *dll* in the earliest animals in which it occurred may have been to control **downstream** genes that themselves controlled aspects of the pattern of cell proliferation during growth, possibly unrelated to any kind of appendages. If so, then its use in the earliest animal appendages was itself a co-option, and one that has been repeated many times, as different appendage types have arisen independently on many occasions in different lineages. Certainly, *dll* expression in growing appendage tips is well known in many animal groups, and has been particularly thoroughly studied in *Drosophila* (Fig. 15.8). However, *dll* has been co-opted for additional roles, both in appendages and in the trunk. We will look at one example of each.

In fact, we have already seen one example of *dll* co-option in a particular type of appendage – the butterfly wing. We noted in Chapter 8 that *dll* expression was

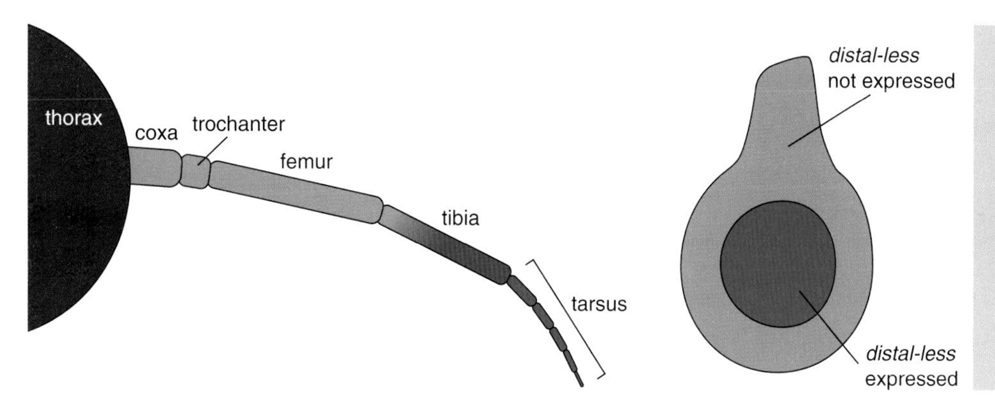

Figure 15.8 Expression of *distal-less* in a *Drosophila* leg disc (RHS) and, later, in the corresponding leg.

involved in determining the focal point of wing eyespots in the genus *Bicyclus*, and it may well also have a similar role in the development of eyespots in all species of butterflies that have them – though this is not yet certain, especially given that eyespots have arisen independently in different butterfly **clades**. This (repeated) co-option is clearly nested within the earlier co-option of using *dll* in the development of limbs.

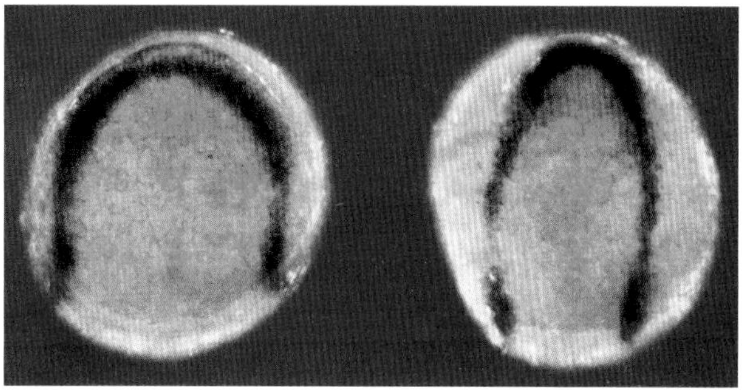

Figure 15.9 Expression of *distal-less* at or near the left- and right-hand borders of the neural plate in a zebrafish embryo – it is from these border regions that the neural crest cells arise. (Reproduced with permission from Akimenko *et al.* 1994 *J. Neurosci.*, **14**.)

Regarding co-option of *dll* in the trunk, there have been several such events in the vertebrates, associated with the expansion of the *dll*-related *dlx* family through gene duplications. Some of these have involved aspects of the development of the nervous system. For example, *dlx* genes are expressed at the borders between fore-, mid- and hind-brain; and at the lateral borders of the neural plate[45] in or near the regions from which neural crest[46] cells arise (Fig. 15.9).

However, although some of these co-options will probably turn out to be genuinely vertebrate-specific (as in the tri-partite brain), others have already been found outside of the vertebrates, and so cannot be specifically connected with the proliferation of the vertebrate Dlx gene family. Expression of *dll* at the borders of the developing neural plate, or its equivalent, has been found in several invertebrate groups. And these include not just invertebrate chordates, such as amphioxus, but also much more distantly related invertebrates such as **annelids**.

Perhaps some nervous system roles of *dll* are in fact pan-bilaterian; but even if this is so, they are still co-options, since *dll* is found outside the Bilateria, for example, in the Cnidaria, which lack a central nervous system and thus a neural plate equivalent.

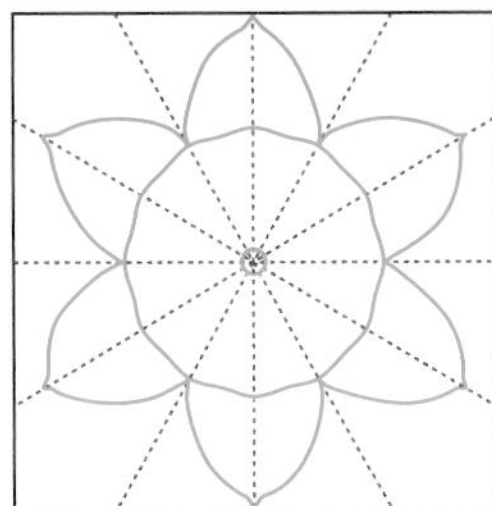
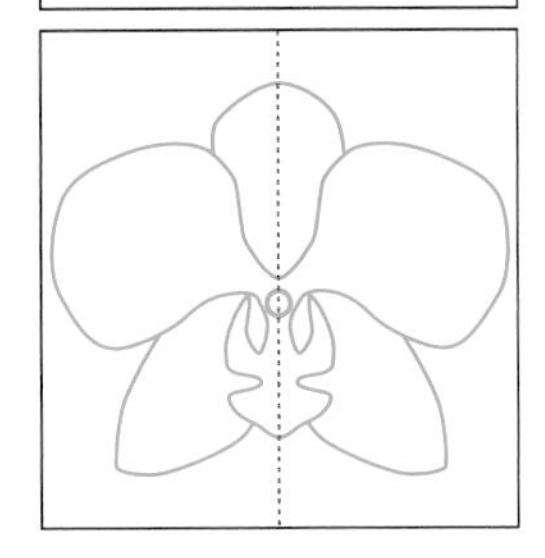

Figure 15.10 Two types of symmetry found in flowers: the radial type, with many planes of symmetry (actinomorphic) and the type that has only a single plane of symmetry (zygomorphic).

15.4 Co-option in the Evolution of Zygomorphic Flowers

In Chapter 7, we looked at the evolution of an unusual flower structure, the 'Chinese lantern', which is found in a single clade – the family Solanaceae. Here we will examine another type of flower structure that has arisen not just in one family but in many. This is the zygomorphic type of flower, in which the radial symmetry of the more basal actinomorphic flower type has been replaced by a single plane of symmetry (Fig. 15.10).

The phylogenetic distribution of the zygomorphic type of flower indicates that it has arisen more than 25 times. It is characteristic of some major families, including the Orchidaceae, the Fabaceae (the pea family) and the Veronicaceae (the family that includes *Antirrhinum*). Each of these families contains many thousands of species.

One common selective scenario for the evolution of zygomorphic flowers relates to their pollination by specialist insect pollinators rather than by more generalist ones or by the wind. Often the petals at the base of the flower are modified into a kind of 'landing platform' for the insects concerned.

A key gene involved in the development of flowers[47] is *cycloidea* (*cyc*), which encodes a **transcription factor** of the TCP family. This gene is widespread in the **angiosperms**. In some groups, including the Veronicaceae, duplication of the *cyc* gene has produced a related gene *dichotoma* (*dich*), which again produces a TCP-family transcription factor. In the 'model plant' *Arabidopsis*, flowers are actinomorphic; and there is a homologue of the *Antirrhinum cyc* gene, called *tcp1*. An obvious comparison to make then, is of the expression patterns of this gene in *Arabidopsis* and in a species with zygmorphic flowers. Such a comparison is shown in Fig. 15.11. Three points of interest emerge.

First, in both *Arabidopsis* and *Linaria* (a close relative of *Antirrhinum*) there is asymmetric expression in the young flowers. Expression is restricted to the adaxial part of the young flower meristems. A meristem is an area of tissue consisting of undifferentiated stem cells; 'adaxial' is a botanical term broadly equivalent to the zoological term 'dorsal'. Second, this asymmetric expression persists into later floral development in *Linaria* but not in *Arabidopsis*, suggesting a role of late adaxial expression in producing a zygomorphic flower. Third, in *Arabidopsis*, expression can be seen in vegetative shoot axillary meristems, suggesting an ancestral role of the *tcp1* gene in the development of meristems in general, not just those that lead to the production of flowers.

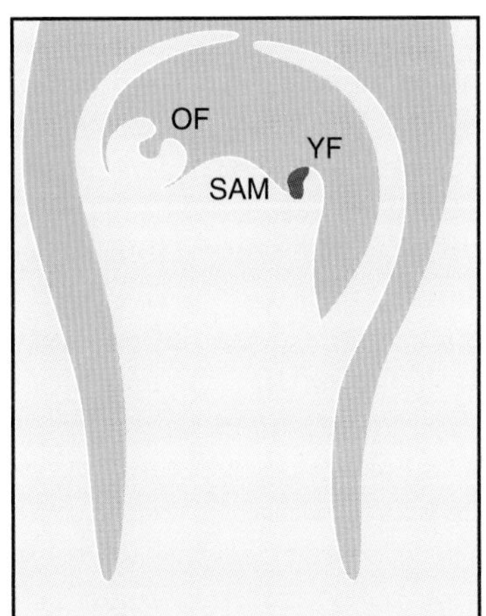

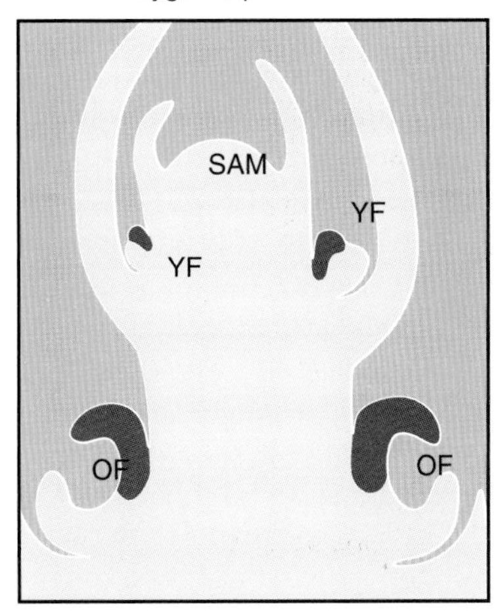

Figure 15.11 Expression (grey) of the gene *cycloidea* (*cyc*) or its homologue in cross-sections of developing plants with actinomorphic and zygomorphic flowers: respectively *Arabidopsis* and *Linaria*. SAM: shoot apical meristem; YF: young flower; OF: older flower (Diagram produced from a photomicrograph in Cubas, 2004, *BioEssays*, **26**: 1175–1184.)

The difference between zygomorphic and actinomorphic flowers is an example of **heterotopy** in evolution – the form of developmental repatterning that involves alteration in spatial positioning. Yet the difference in expression patterns in the floral meristems of *Arabidopsis* and *Linaria* noted in Fig. 15.12 is, rather, an example of **heterochrony** – a change in the *timing* of expression, with late expression being associated with production of a zygomorphic flower. This reinforces the message of Chapter 10 – the integrative nature of **developmental repatterning**.

The pioneering work on the developmental genetic basis of zygomorphy was carried out on *Antirrhinum* and its close relatives such as *Linaria* – i.e. members of the family Veronicaceae. More recent work has shown that in other families that have independently evolved zygomorphic flowers, altered expression of the *cyc* gene is also involved. So this would seem to be one of the clearest examples, along with *dll* in animal limbs, of repeated co-option of homologous genes for the development of **convergent**, non-homologous structures.

15.5 Evolution of the 'Genetic Toolkit'

We will now return from plants to animals to look at a concept that, while it may be equally applicable to these two kingdoms, has been mostly developed in relation to animals: the genetic toolkit for development.

The basic idea is this: the widespread existence of homologous developmental genes across many (sometimes all) animal phyla constitutes a general toolkit used for making animals. The toolkit has sufficient evolutionary inertia that it can be recognised in all taxa; yet it also evolves sufficiently, especially in terms of the roles of the genes included in it, to be responsible for the various kinds of developmental repatterning that occur in the various animal lineages. The toolkit itself can be expanded by duplication of its constituent genes; and this process relates to repatterning because new developmental roles often arise through the divergence of duplicated genes.

The animal genetic toolkit comprises the genes we have discussed thus far in this book, including the Hox genes, the Pax genes, many other transcription factor genes (*engrailed*, *distal-less*, etc.), and genes making the protein components of **signalling pathways** (e.g. *wingless*, *hedgehog* and *decapentaplegic* in *Drosophila* and their homologues in other animals). The toolkit is smallest in basal animals such as sponges, largest in vertebrates. However, the difference in the size of the toolkit, in terms of its constituent gene number, is less between simple and complex animals than was previously expected.

We will now look at the toolkit in two ways: positive and negative. On the positive side we will explore what has been learned about toolkit evolution and what this information tells us about the evolution of development. On the negative side, we will examine problems with the general concept of a genetic toolkit for development, which can be misleading in several ways.

One of the most important conclusions to have emerged from studies of the toolkit in various animals, and indeed in unicellular **outgroups**, is that the genes underlying major morphological transitions are often already present in the genome of the pre-transition organisms[48]. This in itself indicates that the genes were involved in some original role before contributing to a new role that became necessary only after the transition – and so constitutes yet more evidence for gene co-option.

Two types of major transition in early evolution were the (multiple) origin of multicellularity and the (single) origin of animal bilateral symmetry. In both cases, genes involved in the new feature were pre-existent. We have briefly examined examples earlier, but it is worth elaborating on them here.

Each of the several origins of multicellularity, including that which occurred in the stem animal, was based on cells that previously remained separate beginning to stick together. In other words, by definition a central element, perhaps *the* central element, of a multicellular body form is cell-cell adhesion. In **extant** animals there are several classes of adhesion molecule, notably the cadherins, **integrins** and lectins. Within each class there is considerable diversity of members, often associated with different tissue types – for example, different kinds of cadherin present in epithelial, neural and other cell/tissue types. Also, some proteins involved in cell-adhesion work by direct binding of one cell to another, as in the case of homophilic binding between cells with the same kind of cadherin projecting from their membranes. Others work by linking the cell to the extracellular matrix.

The genome of the unicellular creature *Monisiga brevicollis*, a member of the group **Choanoflagellata**, which is thought to be the **sister-group** of the animals, contains genes coding for members of all three of the above-mentioned groups of adhesion proteins[48] (Fig. 15.12). What does it use these for? More importantly, but a harder question to answer: what did the immediate animal ancestor use these proteins for, assuming that it, like *Monosiga*, possessed them?

There are several possible answers to these questions. Recall that many of the creatures that we think of as unicells are capable of forming filaments, sheets or mats of cells. Even **prokaryotic** 'unicells' can form impressive matted structures – such as

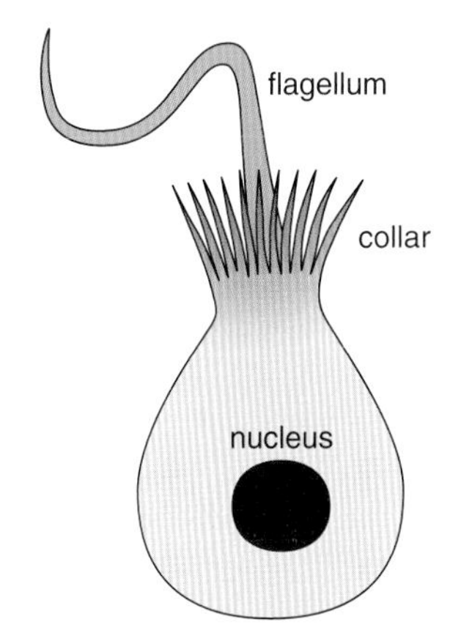

Some of the 'animal developmental toolkit' genes/proteins found in the unicellular choanoflagellate *Monosiga brevicollis*:

Gene/protein family	Protein function in animals
Cadherins	cell-cell adhesion proteins
Integrins	cell-cell movement proteins
Collagens	proteins characteristic of cartilage
Fibronectins	extracellular matrix proteins
Laminins	extracellular matrix proteins
Homeodomains	transcription factors
TCF proteins	transcription factors

Figure 15.12 Diagrammatic morphology of the unicellular choanoflagellate organism *Monosiga*, together with a list of some of its developmental toolkit genes/proteins, including genes coding for what we think of in animals as cell-cell adhesion proteins.

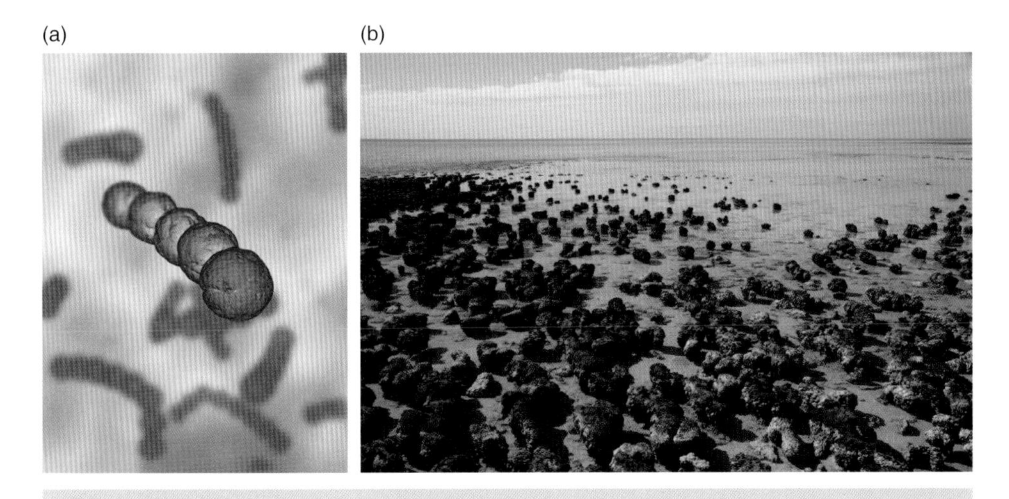

Figure 15.13 Although prokaryotic organisms are typically thought of as being unicellular, many of them are capable of forming multicellular structures such as those shown here, which can take the form of strings (left – streptococcal bacteria) and mats (right – cyanobacterial stromatolites) of cells. However, they never produce multicellular forms that contain organs in the way that many eukaryotic organisms do.

the stromatolites formed by cyanobacteria, fossilised examples of which provide some of the best evidence of early life-forms, with present-day examples sometimes reaching very large proportions (Fig. 15.13).

However, a unicellular creature that does not form aggregations (whether filaments, sheets or mats) may still need to be capable of at least transient adhesion to other cells. This may be necessary as a first stage in the consumption of smaller cells. It may also be necessary, in reproduction, as a first stage in the exchange of genetic material between the cells concerned.

So what initially looked like a problem is not a problem at all. No unicellular creatures exist in a perfectly isolated state. Thus genes making adhesion proteins were already part of their 'toolkit', not for development but rather for feeding, reproduction or other basic life functions. The co-option of these genes to roles in multicellular development is probably the 'easiest' way for genuine ('obligate') multicellularity to arise.

While multicellularity has clearly arisen on several occasions in evolution, bilateral symmetry of the body seems to have arisen just once in the evolution of animals. Can a similar scenario of co-option of pre-existing genes for new roles be imagined here? This is a more difficult question because, while the key feature of the transition from unicell to multicell is clearly cells sticking together, the key feature of the transition to bilateral symmetry is less clear. Not only that, but the pre-bilateral symmetry state is itself less clear. It was probably a form of radial symmetry, as exhibited by present-day cnidarians and ctenophores, but it might have been an essentially asymmetric body form, as exhibited by most sponges.

In either case, what are the key features that distinguish a bilaterally symmetrical body plan from its non-bilateral ancestor? Given the uncertainty associated with the nature of the urbilaterian, as discussed earlier, it is probably best not to look for the answer at this question in features such as segmentation, which apply to the 'complex urbilaterian' hypothesis but not to its 'simple' counterpart. Rather, we should look at features that apply to the urbilaterian, regardless of its degree of morphological complexity.

Taking this approach, the key features of the urbilaterian, and indeed of all bilaterally symmetrical animals, are the three main body axes: antero-posterior, dorso-ventral and left-right. None of these axes exist in the pre-bilateral animal body plans. A radially symmetric animal such as a jellyfish has only an oral-aboral axis. This probably does not correspond to any of the three body axes of a bilaterian. In the case of a sea anemone, we may think of it as a sort of A-P axis, with the 'head' bearing the tentacles as the anterior end, and the point of attachment to the substratum as the posterior end; and in the case of a jellyfish we may think of oral-aboral as a sort of D-V axis, because the aboral surface (dorsal) is normally at the top. But these are probably just mental conveniences rather than true axis homologies.

The genes that control axis development are now well known in bilaterians. They include the Hox genes (A-P axis), the *dorsal/dpp/sog* system (D-V axis) and a suite of genes that has not been discussed above, including *lefty* and *nodal* (L-R axis; Fig. 15.14). Do homologues of these genes exist in cnidarians? The answer is 'yes' for Hox genes, though the cnidarian complement of such genes is small, so duplication and divergence must have been involved to get to the higher number of Hox genes found in bilaterians. For the various other 'axis toolkit genes', the answer appears to be often 'yes' but sometimes 'no'. So in at least a few cases, expansion of the toolkit at the point of origin of the Bilateria may have involved the appearance of sufficiently divergent genes that they are not recognised as homologues of any known cnidarian genes.

Notice that throughout this account, and indeed earlier in the book, genes have often been described as homologues. This implies their divergence from a common ancestral gene. Because homology can be used to apply both to equivalent genes in e.g. animals belonging to different taxa (i.e. *engrailed* in *Drosophila* and amphioxus) or to

Genes/proteins involved in L-R asymmetry

Gene/protein	Protein type
Nodal	TGFβ signalling molecule
Lefty	TGFβ signalling molecule
Activin	TGFβ signalling molecule
Sonic hedgehog	hh-class signalling molecule
Pitx	homeodomain transcription factor

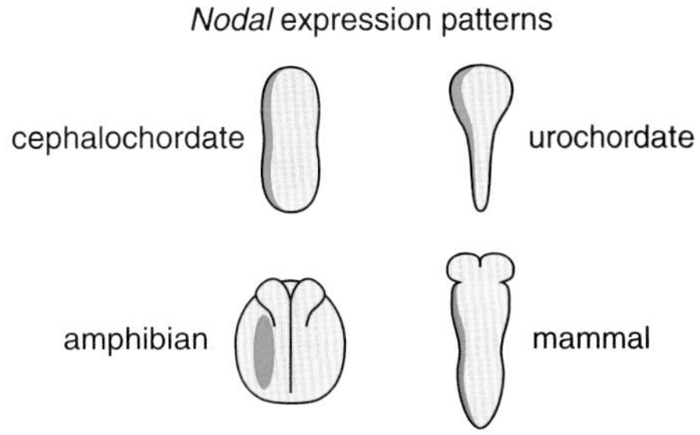

Nodal expression patterns

cephalochordate
urochordate
amphibian
mammal

Figure 15.14 The left-right axis of bilaterian animals used to be poorly understood, in terms of the genes involved, compared with the other two main body axes (A-P and D-V). However, many of the genes involved in establishing the L-R axis are now known (see list, left); and their expression patterns have been characterised in some bilaterians, such as the four chordates pictured (right).

related genes in the same animal (e.g. *engrailed* and *invected* in *Drosophila*), different names are increasingly often given to these two types of homology: orthology and paralogy, respectively (Fig. 15.15).

We turn now to criticisms of the 'genetic toolkit' concept. First, if development can indeed be thought of as being shaped by tools, then the main tools are proteins (mostly transcription factors and signalling proteins) rather than genes. The genes are more analogous to the machines that produce the tools, though of course this is only one way of looking at the constant interplay between genes and proteins. Second, should multicellular creatures, whether animals, plants or fungi, really be thought of as entities that are brought into being by the use of 'tools' on some, for want of a better phrase, 'raw material'? The answer to this question is a resounding 'no'. At the stage of the zygote there is little in the way of raw material, in the sense that there is only one cell. Also, many of the tools (in the sense of proteins) do not yet exist because the genes that make them have not yet been switched on. So, while the concept of a 'genetic toolkit' can be useful, it can also be problematic. It should thus be used with caution.

15.6 Co-option, Exaptation and Developmental Bias

It is now time to connect the gene-level concept of co-option with the organism-level concept of **exaptation**[9], and then to connect co-option with the idea of **developmental bias**. Further conceptual connections will follow later, especially in Chapter 19, which is devoted to elaborating such connections.

In fact, with some provisos, co-option and exaptation are the same thing. Apart from the difference in the level of biological organisation at which the two concepts

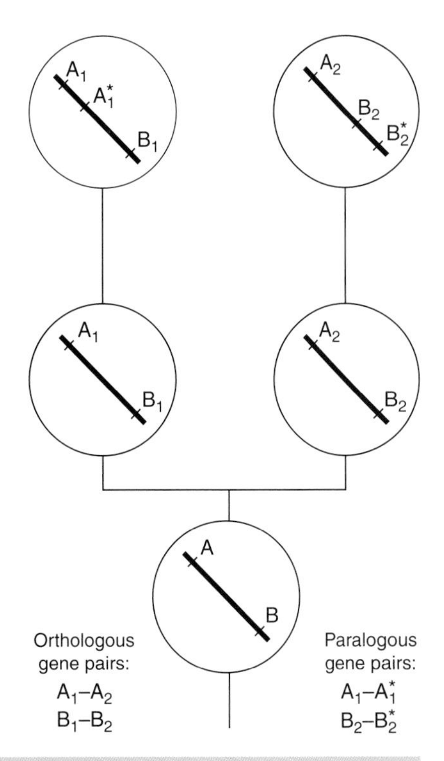

Figure 15.15 Diagram to illustrate two different forms of homology of genes: orthology and paralogy. The former refers to descendant genes of a common ancestral gene in different species (with the daughter genes arising via speciation and divergence); the latter refers to daughter genes of a common ancestral gene in the same species (in which case the daughter genes arise via duplication and divergence).

arose, and to which they are still usually applied, there is really just one other potential difference. It relates to whether the original function of the gene/structure is retained after the adoption of the new one.

In exaptation, the emphasis has often been on one selective pressure, related to a particular function, producing the initial stages of a new structure, with a different form of selection, related to a different function, being responsible for the later elaboration of that structure; for example, heat insulation and flight in relation to the origin and evolution of feathers. In contrast, the emphasis in co-option has usually been on acquiring a new developmental role through a new form of selection, without losing the original one; for example *engrailed* starting off as a gene involved in the development of the nervous system and ending up with an *additional* segment polarity role.

However, this difference is not a hard and fast one, for two reasons. First, in cases of exaptation, the original form of selection may remain alongside the new one, or even reappear at a later stage. Recent evolution of feather structure in most groups of birds may well be mostly due to selection for improved flight ability. But some might relate to insulation either as well as (arctic gulls) or instead of (penguins) selection for ability to fly. Second, in some cases of co-option, such as the co-option of *zen* and *ftz* for non-Hox roles (i.e. non-segment-identity roles) in *Drosophila*, the original Hox function appears to have been lost.

Now for a harder question: to what extent does the gene-level process of co-option constitute an example of developmental bias, as defined more broadly in Chapter 13 (some types of change being inherently more probable than others)? To answer this question, we need to go beyond the co-option of existing genes for new roles, as discussed so far, and examine the relative probability of such co-option versus that of *de novo* production of new genes for those new roles.

At this point it becomes apparent not only that gene co-option is a type of developmental bias but also that this bias is particularly strong. The production of *de novo* genes is not an easy matter. It requires either duplication followed by very significant divergence or some other process such as 'intron shuffling' (Fig. 15.16). In fact, in either case it can be argued that the new gene is not really new at all because it has been created from pre-existing components.

Indeed, this view of gene co-option as a form of developmental bias extends that latter concept from its original formulation. We first examined the idea of bias in the context of the different ways in which birds and mammals make longer necks – respectively by adding or elongating vertebrae. This comparison involved looking at the different morphological results of the (presumed) different structures of the two developmental systems, most genetic details of which remain unclear. Now we can also see developmental bias at the genetic level. The importance of bias in determining the

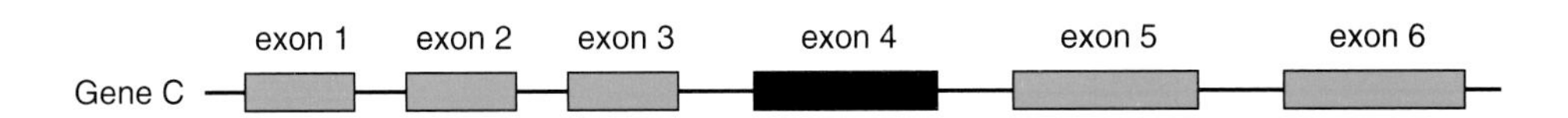

Figure 15.16 New genes can be created from old ones by a process known as 'exon shuffling', which takes place as shown.

ways in which developmental systems respond to selective pressures is now very apparent. Evolution is unquestionably an *interplay* between the biased availability of variants at the genetic/developmental level and the biased survival of variants that we call natural selection.

SUGGESTIONS FOR FURTHER READING

For general reviews of gene co-option in evolution, see:

True, J.R. and Carroll, S.B. 2002 Gene co-option in physiological and morphological evolution. *Ann. Rev. Cell Dev. Biol.* **18**: 53–80.

Sanetra, M., Begemann, G., Becker, M-B. and Meyer, A. 2005 Conservation and co-option in developmental programmes: the importance of homology relationships. *Frontiers in Zoology*, **2**: doi 10.1.1186/1742-9994-2-15.

For a review of the evolution of floral asymmetry, as discussed in Section 15.4, see:

Cubas, P. 2004 Floral zygomorphy, the recurring evolution of a successful trait. *BioEssays*, **26**: 1175–1184.

For a provocative review of gene co-option and 'paramorphism' in relation to the evolution of animal appendages, as discussed in Section 15.3, see:

Minelli, A. 2003 The origin and evolution of appendages. *Int. J. Dev. Biol.* **47**: 573–581.

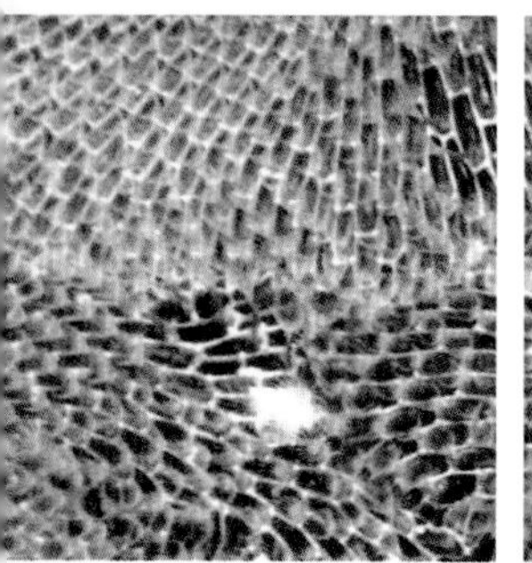
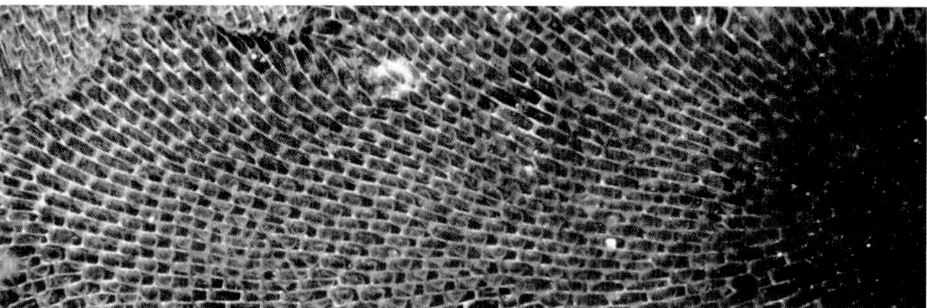

CHAPTER 16

Developmental Plasticity and Evolution

16.1 Types of Developmental Plasticity

Developmental **plasticity**, or phenotypic plasticity as it is sometimes called, involves the course of development being altered by one or more environmental factors – for example temperature, photoperiod or population density. In some cases, the alteration is extreme, such as the switch from wingless to winged aphids when population density is high. In other cases, the alteration is more subtle, such as the production of smaller adult *Drosophila* flies when there is less food available during the larval period. In between these discrete and continuous types of plasticity, there is another – meristic – where the number of some part of an organism alters. An example of this is the production of more segments in some centipedes as a result of higher temperatures during embryogenesis.

The structure of this chapter will reflect this particular classification of types of plasticity – discrete, meristic and continuous. But there are other ways to classify types of plasticity. The most important of these is whether plasticity is adaptive or not.

Some authors have even suggested restricting the use of 'plasticity' to cases which are adaptive, while others have urged a more inclusive usage. The latter approach will be adopted here for the pragmatic reason that it is often unclear whether a particular instance of plasticity is indeed adaptive.

This point can be exemplified by the relationship between body size and food supply in *Drosophila* fruit-flies[49] mentioned above (Fig. 16.1). The environmental variable here can equally well be measured either as population density (number of larvae using a given amount of food, as shown in the figure) or as amount of food available per larva. The only difference is that in the latter case the graph will show an upward rather than a downward trend.

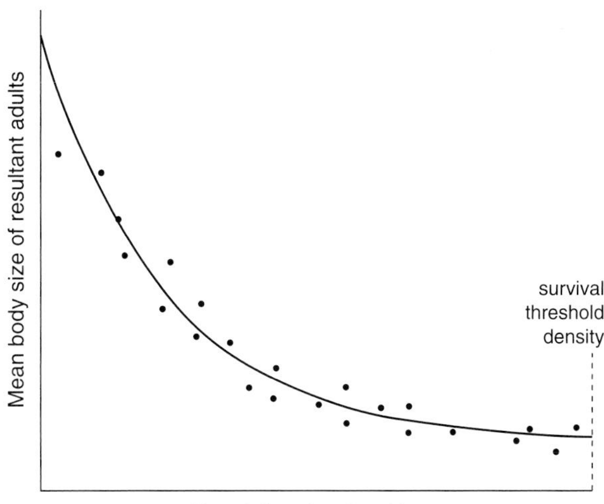

Figure 16.1 The relationship between larval population density and the body size of the resultant adults in laboratory populations of *Drosophila melanogaster*. As can be seen, the higher the density of a larval culture, the smaller the size of the adults that develop from the culture.

This relationship is known from a very wide range of species from many taxa. It is well known in fish (from fish farming practices) and also in crop plants. It may well be a near-universal phenomenon in the living world.

The most obvious interpretation of this type of plasticity is that it is not adaptive, but rather an inevitable consequence of the effect of a limited food supply on the developmental process. It is certainly true in *Drosophila* that smaller female flies produce fewer eggs and that they are thus less fit; and smaller male flies may also be less fit. So the plastic response to higher population density would seem to be *maladaptive*.

However, this is not necessarily true, because the production of gametes is only one component of **fitness**. Survival to adulthood is another key component of fitness, and the ability to pupate at a smaller size, and hence to produce a viable adult, even when food is scarce, may well be an adaptation. If an ancestral fly had no developmental flexibility in this respect, and simply died as a larva when there was insufficient food to reach its 'normal' size for pupation, a mutant possessing the flexibility to pupate at a smaller size and produce a viable (albeit not very fecund) adult would spread through the population – in other words, the new-found plasticity would be adaptive.

Even if this **adaptive scenario** happened in an early **holometabolous** insect (and we do not know if it did), the adaptive nature of the plasticity need not also be true in other taxa exhibiting the same pattern of plasticity. Where development is direct rather than via a **larva**, there is no equivalent of the 'metamorphosis barrier' that must be gone through to arrive at adulthood. Thus the *exact* adaptive scenario described above could not apply. However, becoming adult always involves, by definition, development of sexual maturity; and the flexibility to become reproductive at a smaller size in a direct developer might also be selectively advantageous.

In each of the following three sections, we will first take a broad perspective on the type of plasticity concerned – involving discrete, meristic and continuous developmental responses to environmental factors – and we will then explore one example in detail. In Section 16.5, we will briefly examine some of the genes involved in plasticity. In the final section of the chapter, we will look not at plasticity itself but rather at how patterns of plasticity may evolve, including the process of **genetic assimilation**. We will also ask the question: does the environment help to determine the direction of evolutionary change, not just by affecting the fitness relationships among phenotypes but also by affecting the phenotypes themselves?

16.2 Discrete Variants: Winged and Wingless Forms of Insects

When two or more discrete **phenotypes** are produced in a population in response to environmental cues, this is described as a **polyphenism**, in contrast to a **polymorphism**, where different phenotypes are determined genetically without any direct input from the environment. Many different environmental cues can produce a polyphenism, even in the same species. In some cases, the different forms are produced by environmental factors (e.g. photoperiod) that vary on a seasonal basis; these are described as seasonal polyphenisms. We noted in Chapter 8 that the African butterfly *Bicyclus anynana* has a pronounced seasonal polyphenism, with wet and dry season forms differing considerably in terms of their patterns of wing pigmentation. In other cases, polyphenisms are not seasonal, but rather are responses to environmental cues that need not be related to the time of year.

One important and widespread type of polyphenism in insects is the production of flying and flightless forms of the same species as a result of the influence of one or more environmental factors. Two of the best-known examples of this wing polyphenism are locusts and aphids[50] ('greenflies' or 'plant lice').

Figure 16.2 Part of a swarm of migratory locusts in South Africa.

In the desert locust, *Schistocerca gregaria*, which belongs to the order Orthoptera, the morphological difference between the two forms is subtle and complex. The non-migratory form is not wingless, but it does have smaller wings than its migratory counterpart. The difference is often measured as the ratio of forewing length to hind-femur length: this ratio is reduced in the non-migratory form. There are also differences in pigmentation. However, is this species the biggest difference is behavioural, with vast swarms of the migratory form occurring (Fig. 16.2).

Aphids belong to the order Hemiptera, and specifically to the

Figure 16.3 Some different species of aphids ('greenflies' – often a misnomer, as can be seen), illustrating a small part of the considerable diversity in this group of insects.

superfamily Aphidoidea. There are about 4500 species of aphids worldwide. Some have been extensively studied because of their importance as crop pests (as is the case with locusts), and an aphid genome project has recently been completed. Examples of different groups of aphids are shown in Fig. 16.3. Wing polyphenism is a feature of many aphid species; and in many cases this involves the flightless morph being completely wingless. This polyphenism has probably arisen independently in different aphid groups. In general, evolution has proceeded from a winged ancestral state to a polyphenic descendant state. That is, the always-winged condition is the primitive one. It is important to keep this in mind, because much work in this field focuses on the environmental cues that produce winged forms within polyphenic species, which may lead to an erroneous conclusion that the ancestral aphid was wingless. (The ancestral *insect* was almost certainly wingless, but that is an entirely different issue.)

Many aphids have complex life-histories, with both sexual and asexual (parthenogenetic) reproduction. Wing polyphenism is usually found in parthenogenetic females, but in some species may also be found in sexual females. In a small proportion of aphid species, males can be found with and without wings. Strangely, though, this appears to be a polymorphism rather than a polyphenism; and a particular gene (called *aphicarus*) causing the male polymorphism has been identified.

Factors that have been shown to influence the development of wings in one or more polyphenic species include the following:

- Host plant quality
- Population density
- Presence of competing aphid species
- Presence of predators
- Presence of ants ('aphid protectors')
- Abiotic factors, such as temperature and photoperiod.

These factors often interact with each other rather than acting independently. With regard to the biotic factors, the combination of high host plant quality, low population density, little threat from competitors and predators, and a significant presence of ants, will produce a very low frequency of winged females; while the opposite combination of factors will produce a high frequency.

The species that have been most studied are the pea aphid, *Acyrthosiphon pisum* and the bean aphid, *Aphis fabae*. Winged and wingless forms of one of these species are shown in Fig. 16.4. It is not yet clear, even in these best-studied species, how the environmental factors are perceived and the mechanism by which their perception is translated into an altered developmental pathway. Tactile stimuli provide one possible way in which high population density may be perceived, and the effect of **hormones** (such as ecdysone and juvenile hormone) on the development of wing *anlagen* may be part of the mechanism.

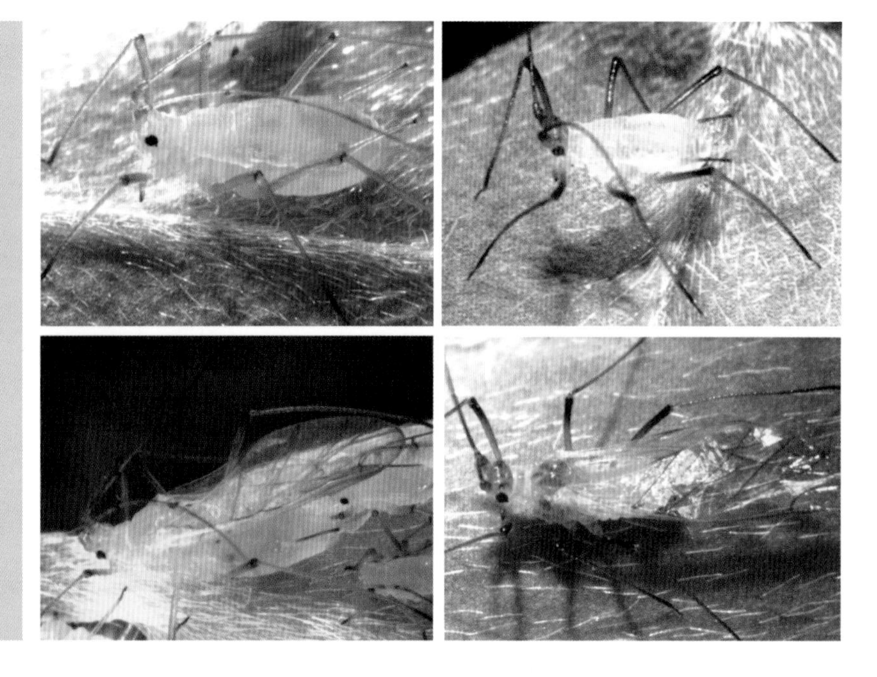

Figure 16.4 Winged and wingless forms of the pea aphid: the female polyphenism (left) and male genetic polymorphism (right). In both cases, discrete alternative wingless (top) or winged (bottom) morphs are produced (From Braendle *et al.* 2006, *Heredity*, **97**: 192–199.)

16.3 Meristic Variation: the Number of Segments in Centipedes

Although much work on the quantitative genetics of meristic variation has been carried out, notably on the number of bristles on various parts of the bodies of fruit-flies, comparatively little work has been carried out on plasticity involving meristic variation. Consequently, there is no special terminology in this area yet – equivalent, for example, to the 'polyphenisms' of the previous section or the 'reaction norms' of the next one. So we will proceed straight away to look at a particular example.

We have examined segmentation in centipedes in previous chapters. Here we will focus exclusively on the centipede order Geophilomorpha ('ground-loving forms'), as this is the only order to exhibit intraspecific variation in segment number. Although there are more than 1000 species of geophilomorphs, very few of them have been studied in detail and, of these, just one species stands out as being a 'model system': the coastal species *Strigamia maritima*[51] (Fig. 16.5).

The number of leg-bearing segments in this species varies from 43 to 53, with any individual population showing a more limited range than this – for example, 47 to 51. As we saw in Chapter 13, only the odd-numbered character states are present in any population, so the numbers 44, 46, etc. are absent. Part of the variation within a population is explicable in terms of sexual dimorphism – typically females have two segments more than males. However, in most if not all populations, there is also intra-sex variation. It has been shown that this has both genetic and environmental (temperature-related) components; here we will focus on the latter.

The first indication that segment number might be influenced by temperature came from a strong correlation between segment number and latitude[52] on the east coast of Britain (Fig. 16.6). This suggests a link with temperature because the other obvious latitude-correlated variable – photoperiod – is unlikely to have a major effect on an eyeless and largely subterranean animal. The second indication of a link came from observations showing that cohorts developing in unusually warm summers had higher segment numbers than other cohorts of the same population. It would take unusually strong selection to produce this result on the basis of an altered gene pool, so a plastic effect seems more likely. Confirmation of the hypothesised plastic effect of temperature on segment number came from an experiment in which *S. maritima* embryos were cultured at three different temperatures[53] (Fig. 16.7).

Figure 16.5 The coastal centipede *Strigamia maritima*. The specimen shown is a mother coiled around her brood in a natural population (on the east coast of Scotland). (Photograph courtesy of Ariel Chipman.)

While the adaptiveness of the wing polyphenism in aphids is not in any doubt, the effect of temperature on centipede segment number is another case where, like the effect of population density on body size in *Drosophila*, the adaptiveness or otherwise of the

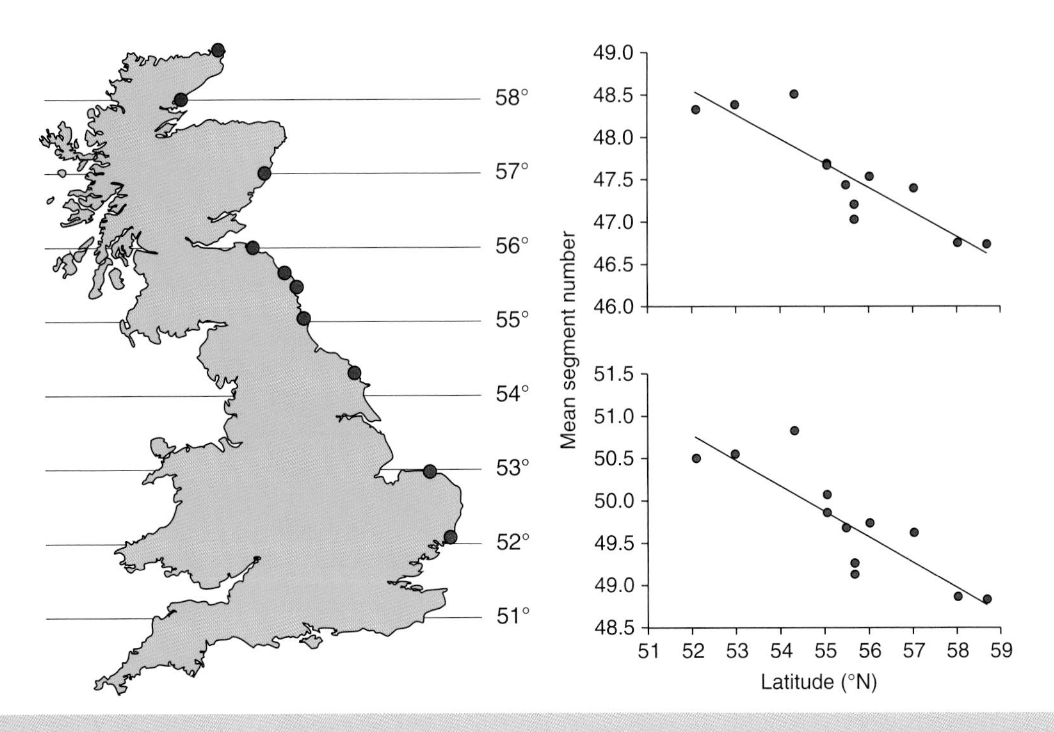

Figure 16.6 Latitudinal cline in segment number in *Strigamia maritima* on the East coast of Great Britain. The further north the population, the colder the average temperature in June (the month in which segments are formed in the embryos), and hence the smaller the number of segments. Top graph – males; bottom – females.

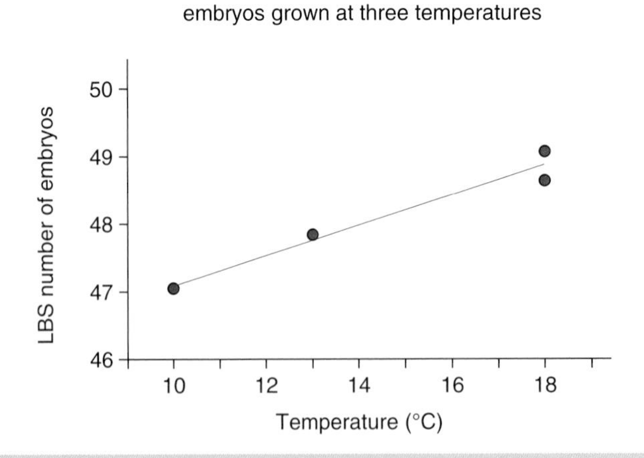

Figure 16.7 Results of an experiment in which batches of *Strigamia maritima* embryos were reared in the laboratory at different (constant) temperatures. As can be seen, the higher the rearing temperature, the greater the mean leg-bearing segment (LBS) number. (There are two points at 18°C because that treatment was repeated to test for consistency.) (Redrawn with permission, from Vedel *et al*. 2008, *Evolution & Development,* **10**.)

plasticity is unclear. This might be written off as a case of 'inevitable' rather than adaptive plasticity, were it not for the fact that there is also an interspecific pattern in the Geophilomorpha in which more northerly species typically have fewer segments than more southerly ones, thus suggesting that the intraspecific pattern may be acted on by selection in some way to produce its interspecific equivalent.

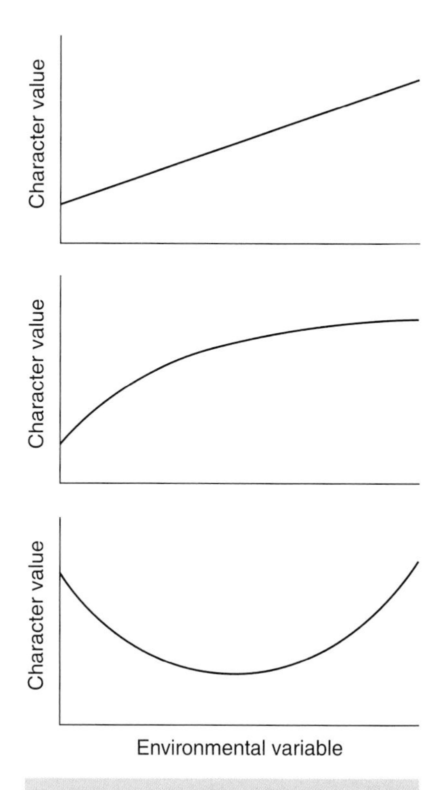

Figure 16.8 Three types of developmental reaction norm: linear, curved but still 'monotonic' (i.e. going in the same direction), and complex.

16.4 Continuous Variation: Plant Growth

The pattern of continuous variation in phenotype produced by variation in an environmental factor is called the **reaction norm**. Such patterns can be linear; curved but still monotonic; or complex (Fig. 16.8).

A measure that often yields a complex pattern of the type shown in the figure is **fluctuating asymmetry** (or FA). This is the term used to describe the small departures from perfect bilateral symmetry that characterise most **bilaterian** animals. FA can be measured by subtracting the size of the right-hand version of a character (e.g. the wing length of a fly or the tail length of a forked-tail bird such as a swallow) from the size of the left-hand one, or *vice versa*. Perfect bilateral symmetry thus gives a value of zero for FA; the greater the asymmetry the higher the FA value. Because extreme values of environmental factors such as temperature, in both directions, often represent 'stress' for the developmental system, the degree of FA tends to be highest at those values and lower at more central ones.

Overall, plants are more plastic than animals. This may be related to the fact that they cannot move as environmental conditions change in a particular place; thus selection may often have favoured increased plasticity. However, it is important to remember that not all plasticity is adaptive.

One example of plant plasticity that normally *is* adaptive is variation in growth rate or direction in response to light conditions. The tendency of many types of house-plant to grow towards light is well known. Also, plants living in shaded conditions, such as under a canopy of taller plants, may grow more, exhibiting 'shade-avoidance' plasticity. A field experiment conducted on the annual herb *Geranium carolinianum* showed that plants generally became more elongate, as measured, for example, by petiole length, when grown under more shaded conditions[54] (Fig. 16.9).

The plasticity of *Geranium* in response to light, and that of *Drosophila* in relation to population density, is depicted in the relevant figures (Figs 16.1 and 16.9) in terms of some size measure at a particular point in time. The measurements of *Geranium* were made on 'bolted' plants; those of *Drosophila* were made on adults. So what we are looking at, in such depictions, is the end result of plasticity in the developmental process. It would also be interesting, however, to look at plasticity *during* growth and development.

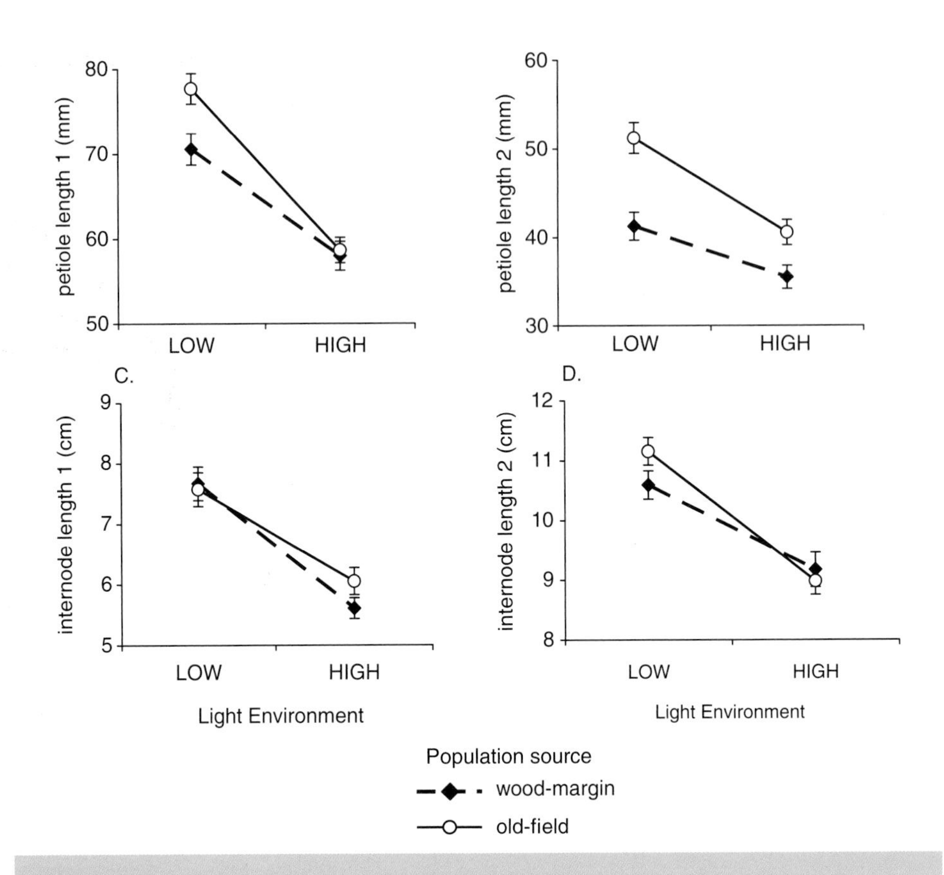

Figure 16.9 Results of a field experiment on a species of *Geranium*, showing that these plants have a more elongate growth form (measured in terms of petiole length and internode length) when exposed to shaded conditions. (Reproduced with permission from Bell and Galloway, 2008, *Am. J. Botany*, **95**.)

An example of this is shown in Fig. 16.10. Here we see the growth trajectories of *Eucalyptus* trees grown at different densities[55]. It is clear that higher density produces slower growth. This may be 'inevitable' rather than adaptive plasticity. However, it is also possible that the ability to grow slowly rather than not at all in dense conditions may have been selected for in the past, just as the flexibility to pupate at a smaller size in *Drosophila* may have been selected for, as we noted above.

16.5 Plasticity and Developmental Genes

One way to think about developmental plasticity is as an environmental influence 'breaking into' a cascade of gene interactions. Obvious questions to ask are when and how this 'breaking in' happens.

In the centipede segment number example (Section 16.4), shifts in rearing temperature part-way through embryogenesis have shown that the environmental effect occurs early rather than late. By the time when the first few trunk segments have formed, as visualised by stripes of expression of the segment-polarity gene *engrailed*, the embryo has lost the ability to respond to a change in temperature (Fig. 16.11). So, temperature is acting at some point earlier than this – though it is not yet known how much earlier. It is likely that the temperature effect is upstream of the start of the **segment-polarity** stage of the segmentation cascade.

One well-known example of plasticity not yet mentioned is the formation of castes (such as worker and queen) in social insects. Here, environmental cues such as food supply can cause the production of radically different **phenotypes** from the same **genome**. One study of a large number of genes in the social wasp *Polistes canadensis* showed that young females, workers and queens differed markedly in their levels of gene expression[56]. Different genes peaked in their expression in different castes (Fig. 16.12).

The plasticity underlying the production of social castes in hymenopterans is very complex – many developmental characters are affected. To make more headway into understanding the developmental-genetic basis of plasticity, it may be better to examine a relatively simple kind of plasticity, such as the varying amount of abdominal pigmentation in *Drosophila melanogaster* that is produced in response to developmental temperature[57]. A study of this system revealed that even an apparently simple form of phenotypic plasticity can have a complex genetic basis. However, some of the key players have been identified, including the **Hox gene** *abdominal-a* (Fig. 16.13).

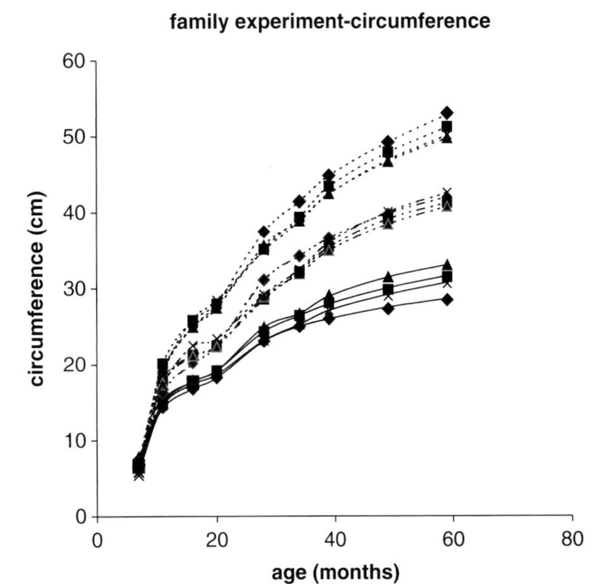

Figure 16.10 Developmental plasticity looked at in terms of growth rates rather than the sizes of adults. The growth rates shown are for *Eucalyptus* trees grown in the field at three different densities. Top curves – lowest density; bottom – highest density. Within each density, results are given separately for four genotypes; these are distinguished by different symbols – square, triangle, diamond and star. (Reproduced with permission from Bouvet *et al.* 2005, *Annals of Botany*, **96**.)

16.6 The Evolution of Patterns of Plasticity

This deceptively-simple heading covers an area that is complex and of the utmost importance. One way to look at the type of conventional evolutionary theory that involves genetically-fixed phenotypes is as a special case of more general evolutionary theory, in which the environment plays a role in the determination of phenotypes through altering the course of development. This more general theory includes both quantitative genetics and 'eco-devo' approaches. We can denote the magnitude of the environmental effect on

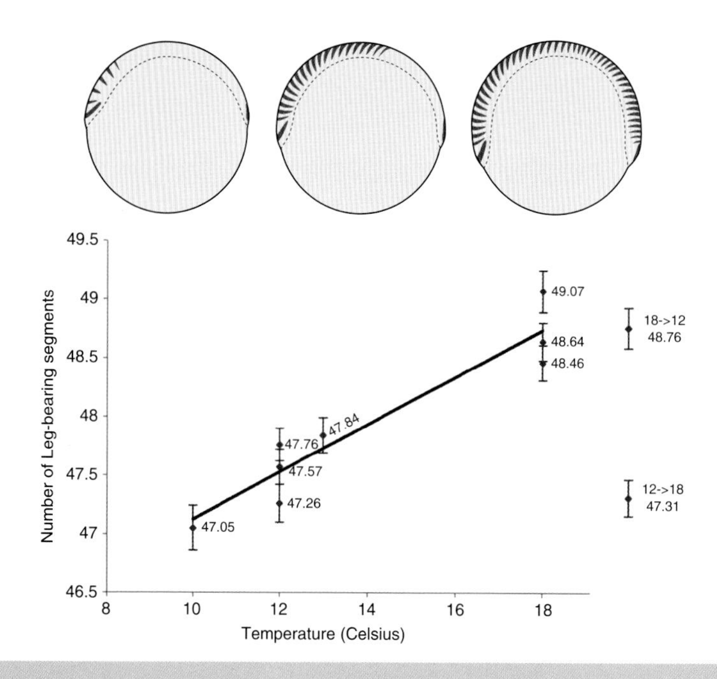

Figure 16.11 The centipede *Strigamia maritima* is capable of responding, in segment number, to a switch in the prevailing environmental temperature that occurs at an early stage of development, up to when just the first few segmental *engrailed* stripes are visible (top: left-hand embryo). Later, when more segments have been formed, there is no plasticity; at these late stages, the final number of segments is unresponsive to changes in temperature, even though segments are still forming. This was demonstrated by temperature-shift experiments (RHS of graph: broods shifted between 12°C and 18°C). (Graph reproduced with permission from Brena *et al.* 2010, *Evolution & Development*, **12**.)

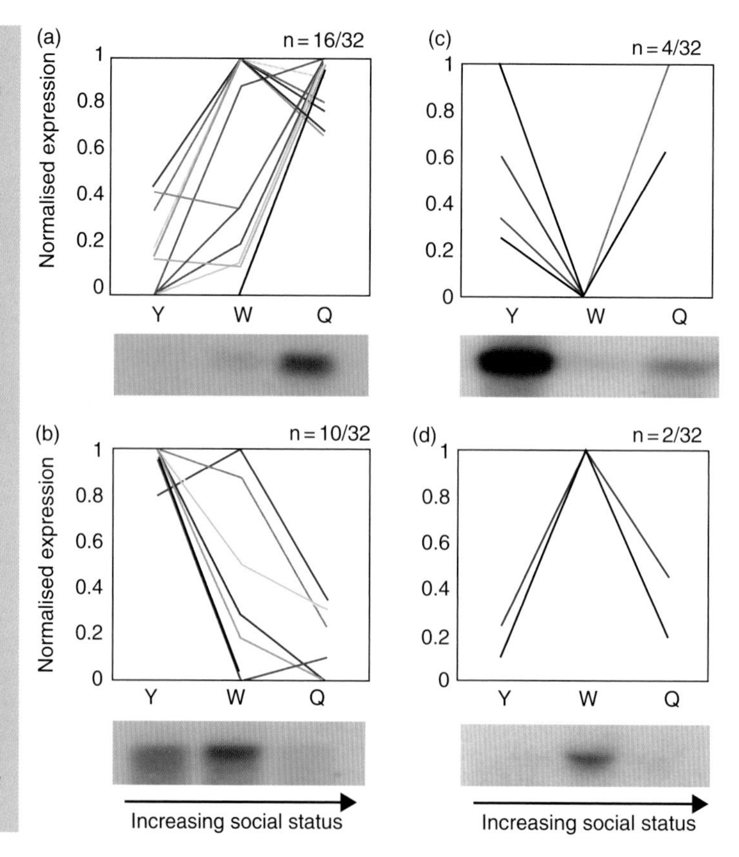

Figure 16.12 Different patterns of gene expression in different social castes in the wasp *Polistes canadensis*. Each line represents a different gene. Some genes vary in their expression in different ways among young females (Y), workers (W) and queens (Q). The simplest pattern of variation is that shown in panel d, in which two genes have a clear expression peak in workers. (Reproduced with permission from Sumner *et al.* 2006, *Proc. Roy. Soc. Lond. B.*, **273**, 19–26.)

development, and hence on the phenotype, as E, and arbitrarily allocate a scale from 0 to 1 for this magnitude of effect. It is then clear that fixed-phenotype evolutionary theory deals only with the special case of E = 0; that is, only one out of an almost infinite series of possibilities.

Patterns of plasticity can evolve in many ways. The classic early example is the evolutionary process called genetic assimilation, which we encountered briefly in Chapter 2 when dealing with the work of the geneticist C.H. Waddington. This is a good example with which to start; we will then examine other kinds of evolutionary process in which patterns of phenotypic plasticity become altered.

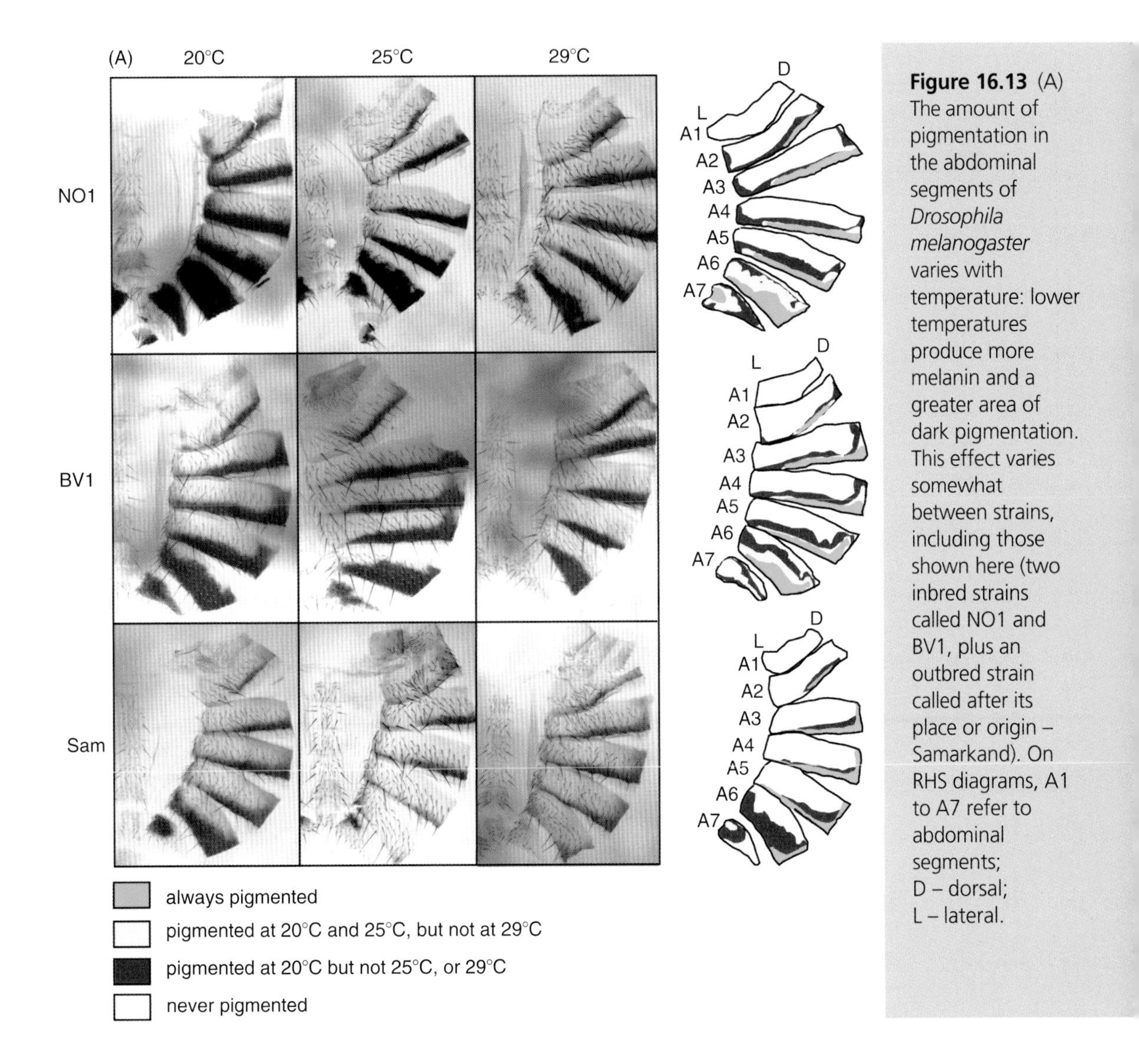

Figure 16.13 (A) The amount of pigmentation in the abdominal segments of *Drosophila melanogaster* varies with temperature: lower temperatures produce more melanin and a greater area of dark pigmentation. This effect varies somewhat between strains, including those shown here (two inbred strains called NO1 and BV1, plus an outbred strain called after its place or origin – Samarkand). On RHS diagrams, A1 to A7 refer to abdominal segments; D – dorsal; L – lateral.

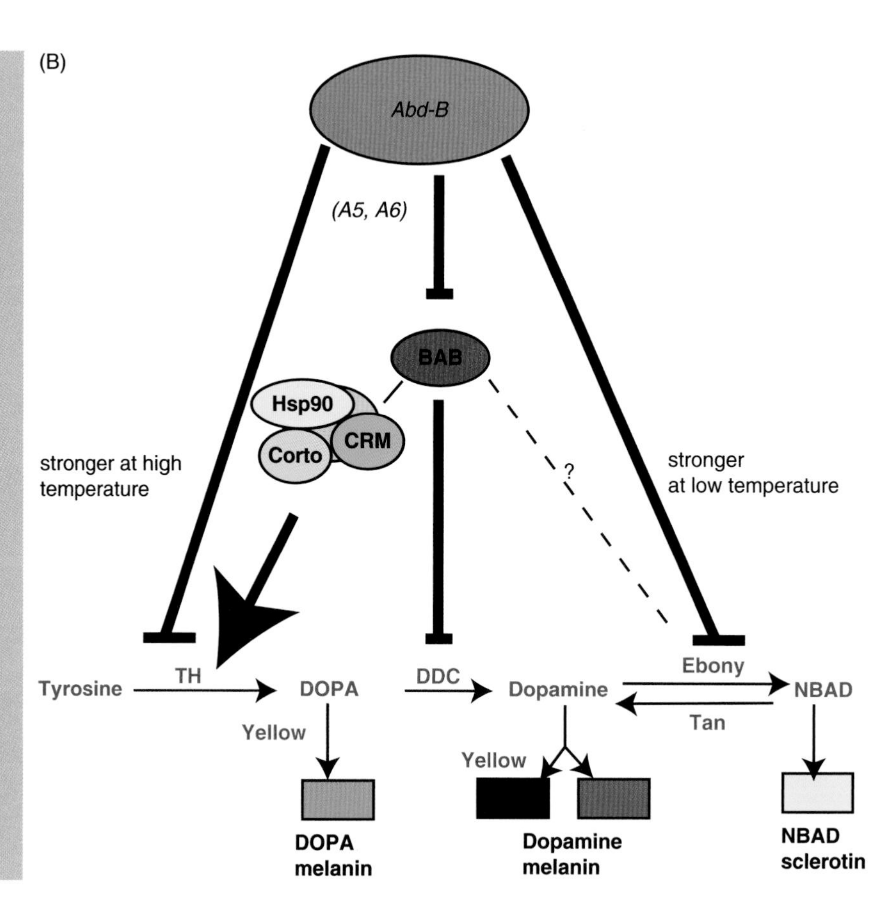

Figure 16.13 Contd (B) The genetic basis of this plasticity is complex; it includes a role for the Hox gene *abd-B*, (top of diagram; start of cascade) as well as several intermediates (many of whose names are given here in abbreviated form) in the activation of pigments such as melanin (base of diagram). (Reproduced with permission from Gibert *et al.* 2007, *PLoS Genetics*, **3**.)

Genetic Assimilation

Genetic assimilation is where a phenotype that initially arose in a population only in response to an environmental cue becomes 'constitutive' – that is, following evolution of the population, the phenotype concerned appears without the environmental cue. Waddington demonstrated this process by artificial selection experiments conducted in the laboratory. He used both the bithorax phenotype[58] (as we noted in Chapter 2) and another phenotype involving altered wing development – cross-veinless[59], in which flies lack a cross-vein connecting two of the main veins.

The experimental procedure was straightforward in both cases. The starting point is an outbred (and hence genetically variable) population of *Drosophila melanogaster*. This is exposed to an environmental stimulus that causes the altered phenotype to occur in a small proportion of individuals. Exposure to ether vapour at the egg stage does this for the bithorax phenotype, while heat shock at the pupal stage does it for the cross-veinless one.

In both cases, Waddington performed his artificial selection by breeding only from the flies in each generation that exhibited the altered phenotype. After many generations, flies appeared in the population in which the altered phenotype was produced without the environmental stimulus (Fig. 16.14). In these, the form of developmental repatterning leading to that phenotype had become genetically assimilated. The basis for this result is that **alleles** causing a greater tendency to produce the altered pattern of development were gradually accumulated, across many gene loci, by the selection process.

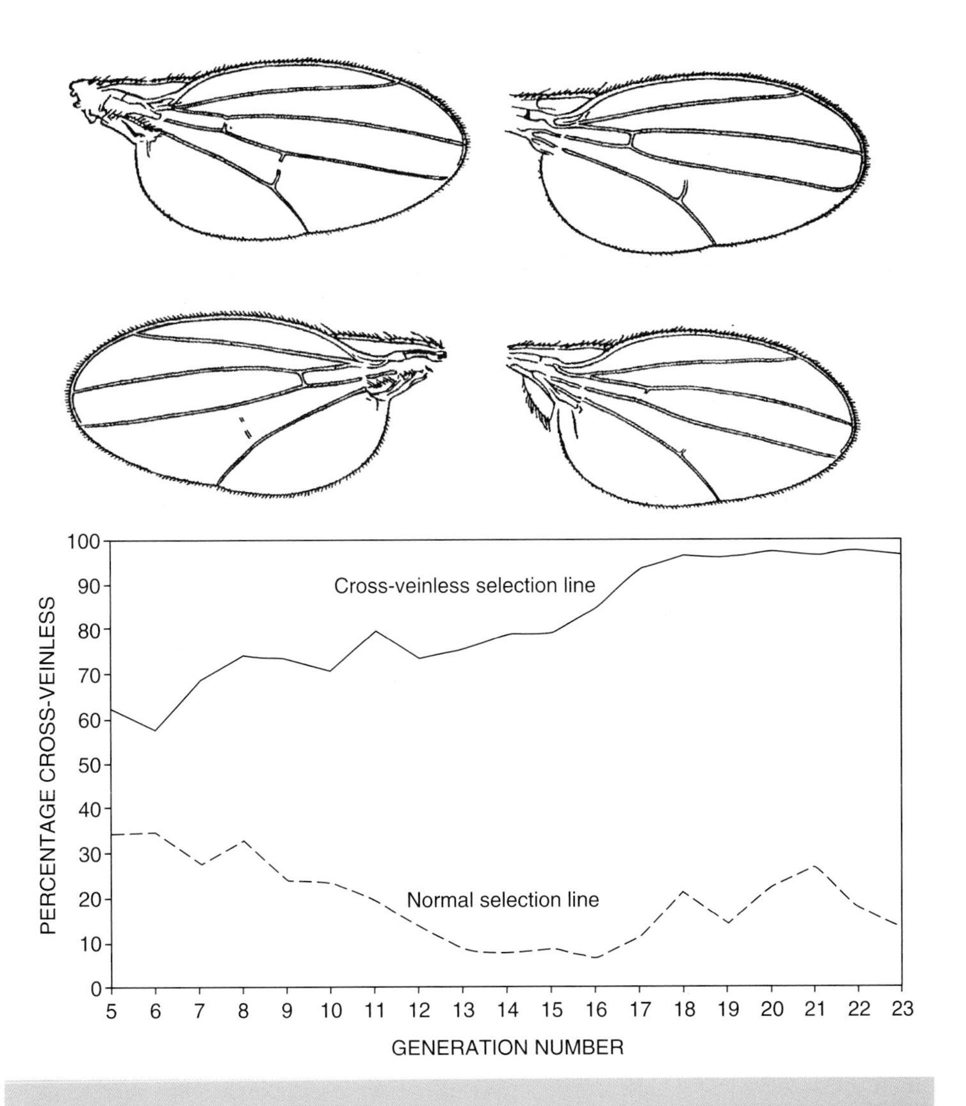

Figure 16.14 Top: Four degrees of reduction/loss of the cross-vein in a *Drosophila* wing. Bottom: The results of Waddington's selection experiments in *Drosophila melanogaster* in which cross-veinless phenotypes, which were initially produced only in response to a heat-shock (at pupation), became constitutive, i.e. appeared without the previously-necessary environmental cue. These began to appear in generation 14 of the 'up-selection' line in the experiment shown, at which stage the percentage cross-veinless *with* a heat-shock (as shown in the graph, bottom) had already risen considerably. (Reproduced with permission from Waddington, C.H. 1953, *Evolution*, **7**, 118–126.)

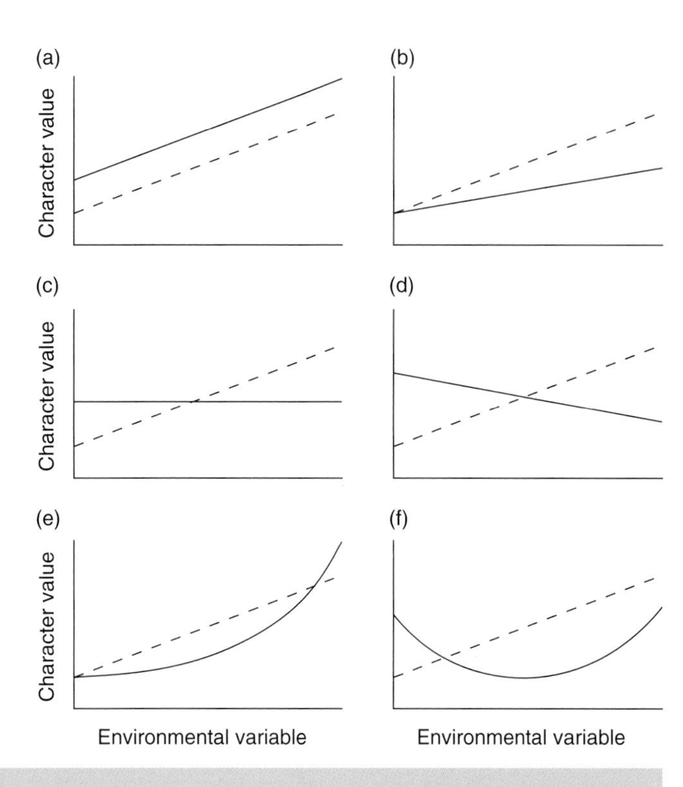

Figure 16.15 Six possible types of evolutionary change in an initially linear reaction norm with, in each case, the dashed line representing the original reaction norm, the solid line the altered one. (a) shift; (b) altered angle; (c) loss of plasticity; (d) altered direction; (e) from linear to curved; (f) from linear to complex.

Waddington knew that bithorax individuals are normally very unfit, and cross-veinless ones slightly so, in relation to 'wild-type' flies. But this was not important for his experiments, in which, because of the nature of the selective regime, those individuals were artificially made to be the fittest, since only they left surviving offspring. The purpose of the experiments was simply to demonstrate that the process of genetic assimilation could work. The extrapolation to nature then takes the form of proposing that genetic assimilation has occurred there too, as well as in the laboratory, but with other phenotypes that were advantageous under a particular regime of *natural* selection.

Evolution of Reaction Norms

In cases of genetic assimilation, as the phrase itself suggests, there is a reduction in the amount of phenotypic plasticity: the initially plastic response to an environmental stimulus is eventually produced as a non-plastic one. However, there is no rule saying that when patterns of plasticity evolve the amount of plasticity must decline. It may also increase, or remain the same in amount but alter in type. To investigate these possibilities we will now look at how the reaction norms that describe patterns of continuous rather than discrete plasticity can evolve. We will consider evolutionary changes from a starting point of linear reaction norms, because these represent the simplest situation (Fig. 16.15).

It is clear from the figure that there at least six ways in which an initially linear reaction norm of the type pictured can evolve:

1. it can shift upwards or downwards, while retaining the same slope;
2. it can alter in slope, becoming shallower or steeper;
3. it can become horizontal, in which case the plasticity has disappeared, in effect a special case of (2);
4. it can reverse direction while remaining linear;
5. it can become curved but still monotonic; or
6. it can become complex.

In any particular context, some of these may seem more probable than others. For example, in the context of the elongation of plant petioles as a form of shade-avoidance plasticity, where we saw linear declining patterns of petiole length with light intensity in Section 16.4, different light-regime environments might produce shifted, shallower, steeper or curved plastic responses, but it is hard to see why the response should reverse.

Genetic Accommodation

This process, which was given its name by the American evolutionary biologist Mary Jane West-Eberhard, involves selection refining a new phenotype that is produced *either* through a plastic response to an environmental stimulus *or* simply by a gene **mutation**. We must assume that the new phenotype is at least selectively **neutral** or mildly advantageous, for otherwise it is likely to be removed from the population by negative selection long before it can be refined. So we should not expect genetic accommodation to work, for example, on significantly unfit mutants such as bithorax *Drosophila*. However, one instance in which it may have been important is in the evolution of **chirality** in **gastropods**, which we examined in Chapter 7.

Stephen Jay Gould once noted that, in the West Indian snail *Cerion*, which is normally dextral, the rare sinistral form, when it occurred, was less stable in shape than its dextral counterpart[60]. One interpretation of this observation is that there are many so-called 'modifier genes' that act on the phenotype produced by the 'main' gene (the one for chirality) and that these become **coadapted** with the prevailing allele of the main gene so as to stabilise its effects. Following a switch to the alternative allele of the main gene, there is a period of instability (which is what Gould noted) but, eventually, the modifier genes become re-coadapted to the new version of the main gene. This has presumably happened in cases of entire sinistral families, such as the Clausiliidae, which show no greater instability of shell shape than do dextral families.

Experimental Evolution of a Polyphenism

Genetic assimilation can be thought of as evolution from a polyphenism to a stable phenotype. An experimental demonstration of the opposite process – evolution from a stable phenotype to a polyphenism – has recently been demonstrated in the laboratory of the American biologist Fred Nijhout, again (as in Waddington's work) using artificial selection[61].

The experiments were conducted on a species of lepidopteran, the tobacco hornworm *Manduca sexta*. The common name of this species is derived from the fact that the **larva** (not a worm, of course) feeds on tobacco plants. The larva is bright green, and this is a more-or-less fixed colour – it does not vary markedly from one natural environment to another. However, a closely-related species, *Manduca quinquemaculata*, has polyphenic larvae. The polyphenism is in response to temperature: larvae are black at lower temperatures, green at higher ones. The rationale underlying the experiment to be described was that, given the polyphenism in its close relative, it might be possible to make *Manduca sexta* polyphenic also, using an appropriate form of artificial selection.

The selection was imposed not on the **wild-type** green *M. sexta* larvae, but rather on the larvae of a laboratory strain fixed for the *black* mutant, which again is not polyphenic but rather, as its name implies, is characterised by black larvae in an environment-independent manner, at least if the environmental conditions are not extreme.

However, heat shocks (at 42°C) applied to the fourth-**instar** larvae of the *black* strain produced fifth-instar larvae with a range of pigmentations from the black characteristic of that strain in unshocked conditions to the green of the wild-type larvae (Fig. 16.16). The exposure of this variation through heat-shock allowed artificial selection to be conducted in both directions – towards the newly-exposed green (polyphenic line) and towards the 'normal', for that strain, black (monophenic line). The results are shown in Fig. 16.17. Clearly, selection is rapidly effective in both directions.

The reaction norms of the selected lines, and also of the unselected control line, were compared after 13 generations of selection (Fig. 16.18). It can be seen that in the polyphenic line the pattern of plasticity has been significantly altered in two ways. First, at any particular temperature the mean colour score is higher (more green) than

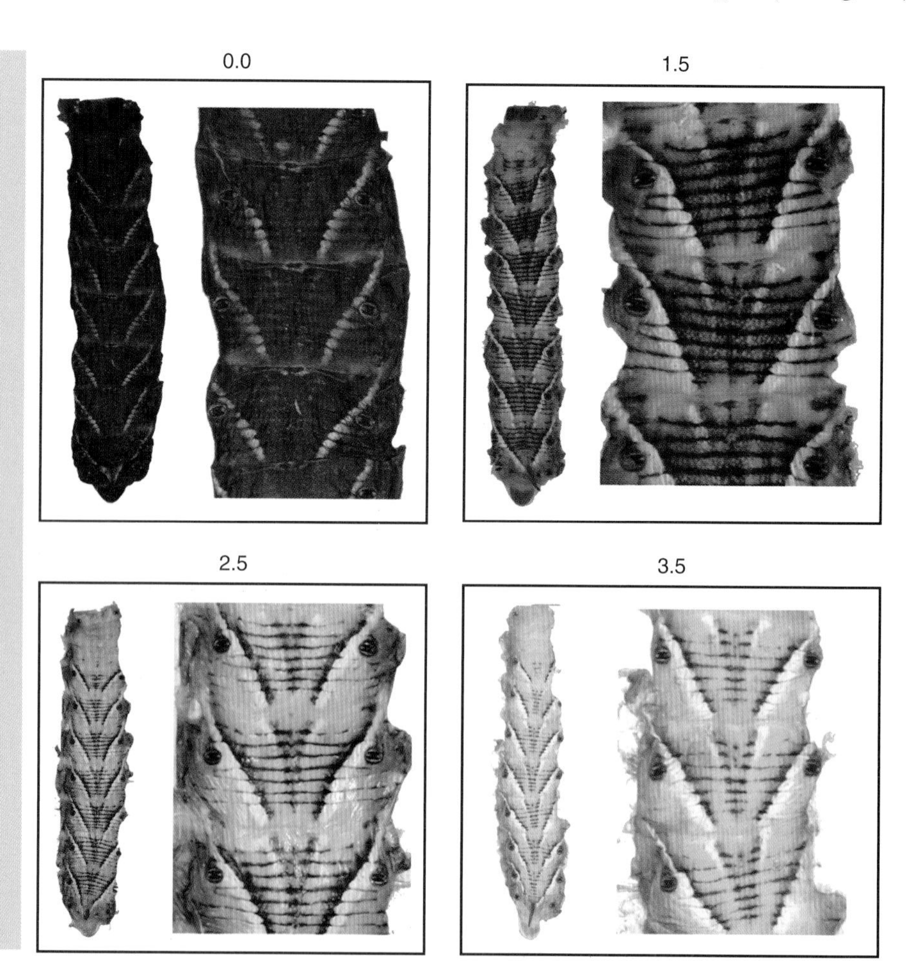

Figure 16.16 The range of colouration, from black to green, shown by heat-shocked larvae of the 'black' mutant of the tobacco hornworm, *Manduca sexta*. Note the colour scores from 0 to 3.5. The normal (i.e. non-mutant) larvae are green and have a score of 4. (Reproduced with permission from Suzuki and Nijhout, 2006 *Science*, **311**: 650–652.)

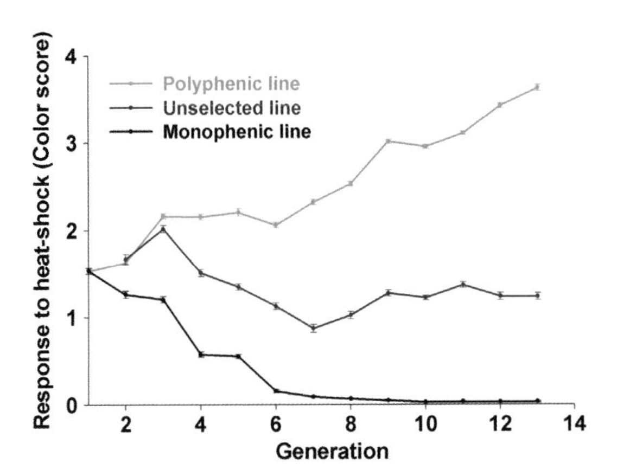

Figure 16.17 The results of a selection experiment in which the character selected was the colour of the larvae of the black mutant following heat-shock. In the 'up-selected' (or 'polyphenic') line, the degree of greenness produced in response to heat-shock increases; in the down-selected (or 'monophenic') line it decreases (to zero); and in the control (unselected) line it remains approximately constant. (Reproduced with permission from Suzuki and Nijhout, 2006, *Science*, **311**: 650–652.)

in the control or monophenic lines. Second, the pattern of change in colour between temperatures, i.e. the reaction norm, has become more abrupt or switch-like. The latter effect can be regarded as genetic accommodation if we imagine an adaptive scenario in which a pronounced polyphenism is advantageous.

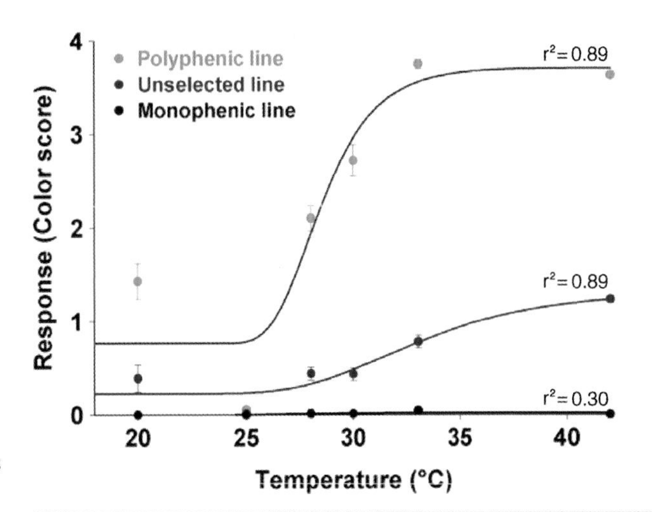

Figure 16.18 Reaction norms of the three selection/control lines referred to in Fig. 16.17, tested at generation 13 of the experiment. The up-selected line has become markedly polyphenic. The r^2 values indicate the proportion of the variation explained (e.g. 0.89 indicates 89% explained). (Reproduced with permission from Suzuki and Nijhout, 2006, *Science*, **311**: 650–652.)

Evolution of 'Co-development'

An interesting twist to the usual story of developmental plasticity, in which the environmental agent inducing the plastic response is external (e.g. light, temperature), is found in cases where the environmental agent is internal. This occurs in many symbioses. An interesting example involves the symbiotic bacteria that live in the mammalian gut. It has been shown that some of these are necessary for the normal development of the capillary network of the gut **epithelium**. Observation of gut development alone is not sufficient to reveal that this is actually a case of a plastic response to an environmental agent, because usually the gut bacteria are present. It takes experiments in which their absence is engineered[62] to reveal that the control of some aspects of gut development has evolved from a (presumed) ancestral state of being endogenous, to a **derived** state in which this control has been 'outsourced' to a symbiotic organism.

Plasticity and the Direction of Evolution

In Chapter 13, we saw that the direction of evolutionary change may be determined not solely by natural selection but rather by the interaction between selection and **developmental bias**. This message was reinforced, in Chapter 15, with the message that gene **co-option** may constitute a form of bias. Now, we ask the question: what do studies of the evolution of plasticity tell us about the determination of evolution's direction?

In the case of genetic assimilation, the direction of evolution is due to a combination of three things: a particular environmental change, the nature of the developmental repatterning produced by that change (which may be subject to developmental bias), and the fitness consequences of this repatterning. Thus once again evolution's direction is set by a combination of factors, but this time a three-way combination.

The evolution of developmental reaction norms would at first sight seem to be simple. Given a particular reaction norm as a starting point, it can be modified by selection in the various ways that were pictured in Fig. 16.15. But this pre-supposes that the reaction norm exists. In some cases, it may indeed exist as a result of 'inevitable plasticity', such as the negative correlation between body size and population density that we have seen applies to a wide range of organisms. In this case, selection may be able to shape 'inevitable plasticity' into adaptive plasticity, as discussed previously in relation to the potential fitness increase for a holometabolous insect of being able to pupate and become adult at a smaller size. The inevitable-to-adaptive plasticity transition falls into the category of genetic accommodation.

What if there is originally no reaction norm? In other words, what if there is no plasticity in response to variation in an environmental factor? This is like starting from the horizontal line that appeared in Fig. 16.15 as an *end-result* of one kind of selection.

For example, what if an early plant species shows no plasticity of growth in relation to light, no shade-avoidance response? Clearly, such a response would often be adaptive, but how can it be selected for if it does not exist in the first place? This is like the situation of a peppered moth population in a newly soot-polluted environment 'waiting for' mutation. Just as in the special case of the evolution of genetically-fixed phenotypes, evolution of plastic phenotypes is ultimately dependent on the introduction of variants through mutation and developmental repatterning.

This line of argument takes us again to the interesting idea of a connection between developmental bias and plasticity. As mutations occur that produce plasticity in development, where previously there was none, they may be selected for. This is the opposite of genetic assimilation, and may be equally or more frequent – we still lack the evidence to say which process is likely to be more common. But, while mutations themselves are random in many respects, developmental repatterning is not. The initial plastic variants made available to selection may reflect the relative 'ease' of making the developmental process plastic in different ways. So the determination of evolution's direction may again, as in the case of genetic assimilation, be a three-way interactive process.

SUGGESTIONS FOR FURTHER READING

There have been three books published in the last 15 years that have a strong focus on the evolutionary importance of plasticity:

Schlichting, C.D. and Pigliucci, M. 1997 *Phenotypic Evolution: A Reaction Norm Perspective*. Sinauer, Sunderland, MA.

West-Eberhard, M.J. 2003 *Developmental Plasticity and Evolution*. Oxford University Press, Oxford.

Gilbert, S.F. and Epel, D. 2009 *Ecological Developmental Biology: Integrating Epigenetics, Medicine and Evolution*. Sinauer, Sunderland, MA.

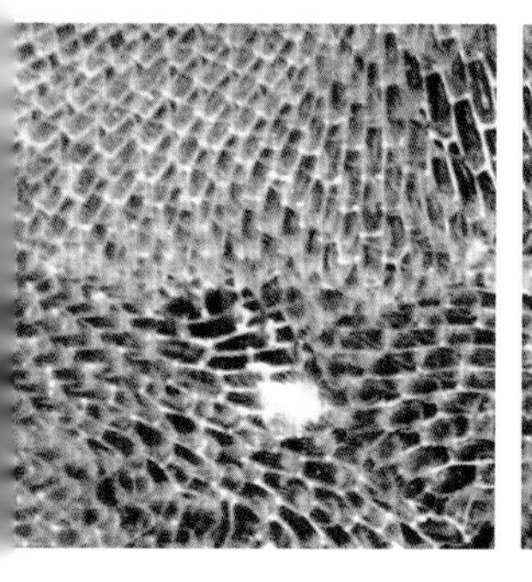
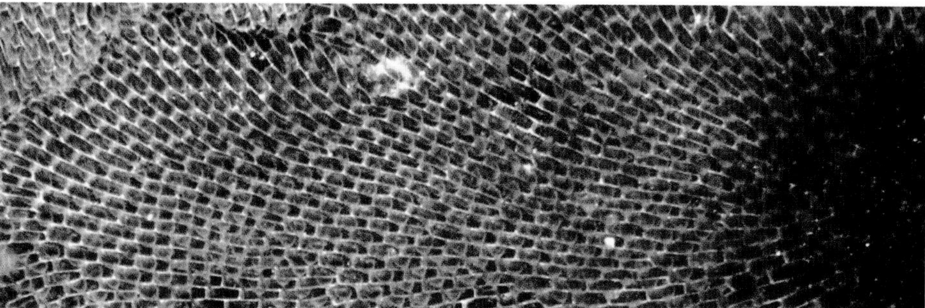

CHAPTER 17
The Origin of Species, Novelties and Body Plans

17.1 Is Evolution Scale-dependent?

There has been much discussion about the extent to which evolution is 'scale-dependent'. Recent publications in this area include those giving opposing answers to the question by Armand Leroi[63] and Douglas Erwin[64]. But what, exactly, does the phrase 'scale dependence' mean?

The key issue here is the extent to which large-scale, long-term evolution can be explained by the accumulation of the kinds of small-scale, short-term processes that we are able to observe in natural populations, including such cases as industrial **melanism** in the moth *Biston betularia*, which was discussed in Chapter 4. This involved change in a particular gene (along with modifier genes) in particular populations of this species. How easy is it to extrapolate from this sort of **micro-evolutionary** (intraspecific) evolutionary change to trans-specific changes?

Evolutionary processes that generate new species, and in the longer-term **novelties** and **body plans**, have been lumped together by many authors as 'macro-evolution'. However, it is better to split them into two groups, as suggested by G.G. Simpson in the mid-20th century[8].

Simpson used **macro-evolution** to describe changes that led to the proliferation of different species within a genus or family. In contrast, he used **mega-evolution** for changes that were larger again than macro-evolutionary ones. Although there is no clear line of distinction between these, they are nevertheless useful categories. After all, macro-evolution often produces a small **clade** of species that are developmentally barely different from each other; that is, the amount of **developmental repatterning** that has taken place during their proliferation is little more than can be found taking place in micro-evolution. In contrast, the origin of novelties and of new body plans necessarily involves major developmental repatterning of a sort that is never (or almost never) found in micro-evolutionary change. Note that much hangs on which of these it is – 'never' or 'almost never' – because in one case mega-evolution is not explicable in terms of micro-scale processes that have been gradually accumulated over long periods of time, while in the other it is explicable as something that is composed of many micro-scale changes, including some highly unusual ones. We will return to this issue in Section 17.3 and beyond, after first dealing with Darwin's 'origin of species', or **speciation** as it is usually called nowadays.

17.2 Speciation

Species are the only 'real' categories in the taxonomic hierarchy. We have no clear definition of what constitutes, say, a genus or an order; but we have a very clear definition of a 'biological species' – based on ability to interbreed and to produce viable, fertile offspring. Thus two animals or plants belong to the same species if they can (and do) interbreed under natural conditions; whereas they belong to different species if they cannot.

Although this is a very clear criterion, there are often practical difficulties in applying it. First, where the occurrence or lack of interbreeding can be directly studied, we often find that individuals of a particular pair of closely-related species usually do not interbreed, but a few of them do. We will look at an example of this shortly. This problem occurs in both animals and plants, though it is more widespread in the latter. Second, there are many cases where the possible interbreeding between two presumed sister-species cannot be directly studied. While this is true of many **extant** species, the problem is at its worst for palaeospecies. Consequently, palaeontologists define these on morphological grounds, using small differences in structure that are thought to (but might not) indicate a lack of interbreeding between the presumed separate species. In many taxonomic groups, extant species are also defined this way in practice, on the same basis – that consistent morphological differences are thought to indicate a lack of interbreeding. Most taxonomic keys used for the identification of specimens found in the wild work on the basis of observations on the *structure* of the specimens concerned. Notwithstanding these difficulties, species are still the most real of the taxonomic categories.

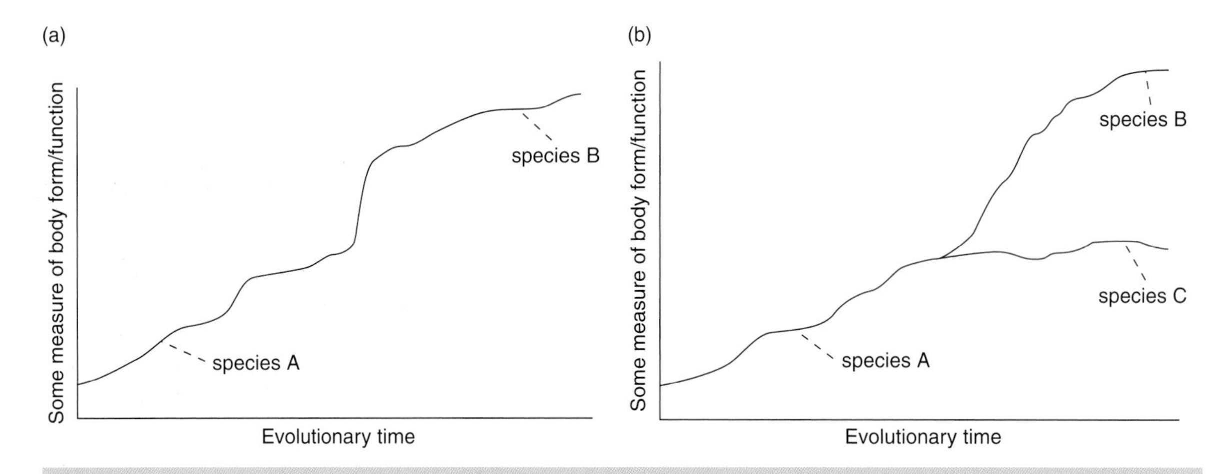

Figure 17.1 The two types of speciation: Left: Anagenesis, in which members of a lineage become sufficiently different over time that, even without splitting of the lineage, late members would be unable to interbreed with early ones (likely; but clearly untestable). Right: Cladogenesis, in which a lineage splits into two (or more) daughter species. This can be tested because a few species can be caught in the middle of this process and partial reproductive isolation observed.

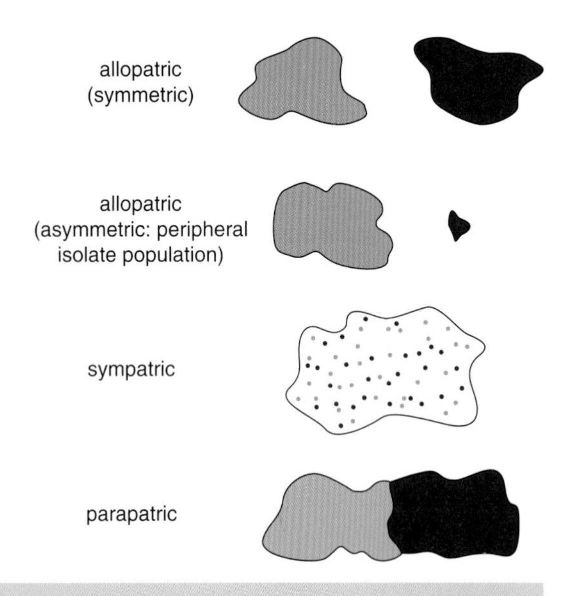

Figure 17.2 Geographical types of cladogenetic speciation. The main three are allopatric (different places), sympatric (same place) and parapatric (the intermediate situation of abutting species ranges). Allopatric, which can be divided into symmetric and asymmetric, is thought to be commonest and sympatric rarest on the basis that restricted gene flow favours speciation.

It is entirely possible that, if we were able to trace an evolutionary **lineage** through time, we would find that, even without any splitting of the lineage, there are sufficient accumulated changes that the organisms belonging to the lineage at times t_1 and t_2, separated by, say, a million generations, would not be able to interbreed with each other (Fig. 17.1a). This kind of speciation is called anagenesis. However, the alternative kind of speciation, the one that is associated with lineage splitting events and is called cladogenesis (Fig. 17.1b), is more accessible to practical studies. In cladogenesis, one or more **daughter species** branch off from a parental species, which itself continues to exist, albeit somewhat altered, given the passage of time. In cases where the 'parental' species has been sufficiently altered, then this label is no longer appropriate and we should simply talk of two (or more) distinct daughter species.

One way to classify types of cladogenesis is by the geography of the speciating populations (Fig. 17.2). Speciation is thought to occur most readily where there is geographical separation between the diverging populations – allopatric speciation. This is why islands seem to be such fertile places for the birth of new species. But geographical separation is not a necessary pre-requisite for speciation to occur. Species can also be formed by the divergence of populations whose distributions abut each other – parapatric speciation – and in some cases even if the distributions overlap – sympatric speciation.

It is easy to see why speciation can occur in allopatry. If natural selection favours different character values in different places because of different prevailing ecological conditions, the populations may diverge to the point where interbreeding is no longer possible. Because of the geographical separation, the action of selection is not diluted or negated by significant gene flow between the populations concerned. It is, however, much harder to see how such selection can be effective in cases where the populations are inhabiting the same area – i.e. the sympatric case. Often, the answer is that the species form suddenly by a process known as allopolyploidy (Fig.17.3), where a **polyploid** hybrid forms by the union of germ cells of the two parental species, yielding offspring that can breed with each other but with neither of the parent species. The marsh-grass *Spartina townsendii* (Fig. 17.3) is an example of a species formed in this way.

Figure 17.3 Sympatric speciation can occur through allopolyploidy – the process in which two parental species interbreed and their chromosomal complements are effectively added together. This is a much more rapid form of speciation than any other. It is common in plants, with a well-known example being the marsh-grass *Spartina townsendii* (pictured here in its typical habitat), which is a hybrid between two parental species of *Spartina*. It is less common, but still possible, in animals.

We will now look at the more common type of speciation that occurs allopatrically and by the accumulation of many small genetic changes. We will approach this type of speciation from both 'ends'. In other words, we will look at one example where it is just beginning and another in which it is almost complete. These examples involve flies and snails, respectively.

Drosophila pseudoobscura

This species of fruit-fly is distributed across a wide area of Western North America, but also has an outlying population in Colombia (Fig. 17.4). Such outlying populations are often described as peripheral isolates, and when they undergo speciation this is referred to as 'allopatric speciation by peripheral isolate' or, according to some authors, 'peripatric speciation'. Whichever term is used, this is clearly a sub-category of allopatric speciation; the other subcategory would be the more geographically symmetrical process that occurs when, for example, the populations of a holarctic species on either side of the Atlantic diverge to the point where they become separate species.

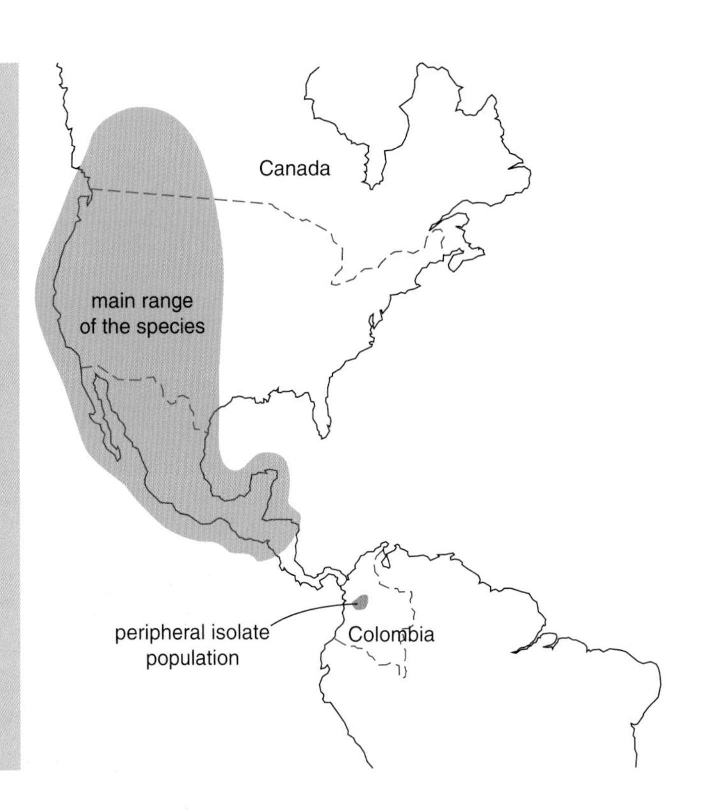

Figure 17.4 The geographical distribution of the fruit-fly *Drosophila pseudoobscura*. Note the outlying population in Colombia, which is separated from the main range of the species by a large distance, rendering gene flow negligible.

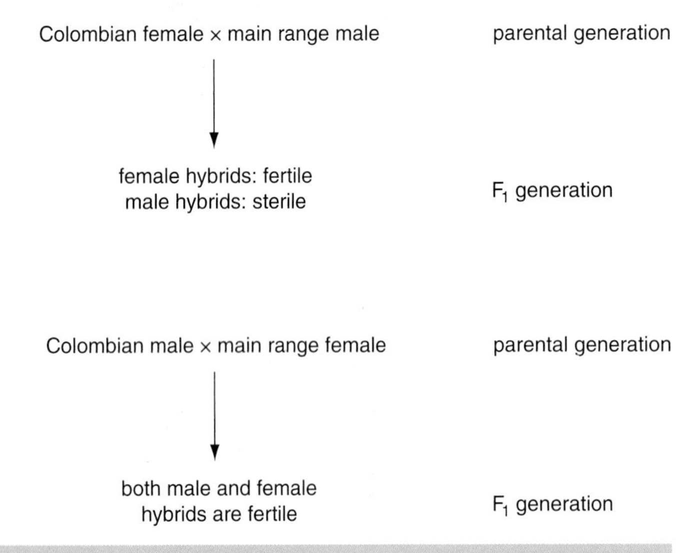

Figure 17.5 Results of breeding experiments involving 'main range' and Colombian *D. pseudoobscura*. Note that there is partial reproductive isolation in the form of male hybrid sterility when the cross is done in one direction (but not in the other).

It was discovered in the 1970s that the Colombian population of *D. pseudoobscura* had become partially reproductively isolated from its parent species[65]. Attempts to interbreed Colombian and North American flies gave the result shown in Fig. 17.5. As can be seen, a quarter of the offspring were sterile. These were all males, and since males are the heterogametic sex in flies, as in mammals, this is an example of Haldane's rule, which states that in cases of incipient reproductive isolation it is the heterogametic sex that suffers problems first.

This is an appropriate point at which to introduce the concept of 'reproductive isolating mechanisms'. If members of two populations have difficulty in interbreeding or, later in their divergence, are entirely unable to interbreed, there are several possible reasons for this. The primary division is into pre- and post-zygotic mechanisms (Fig. 17.6). The former, such as differences in

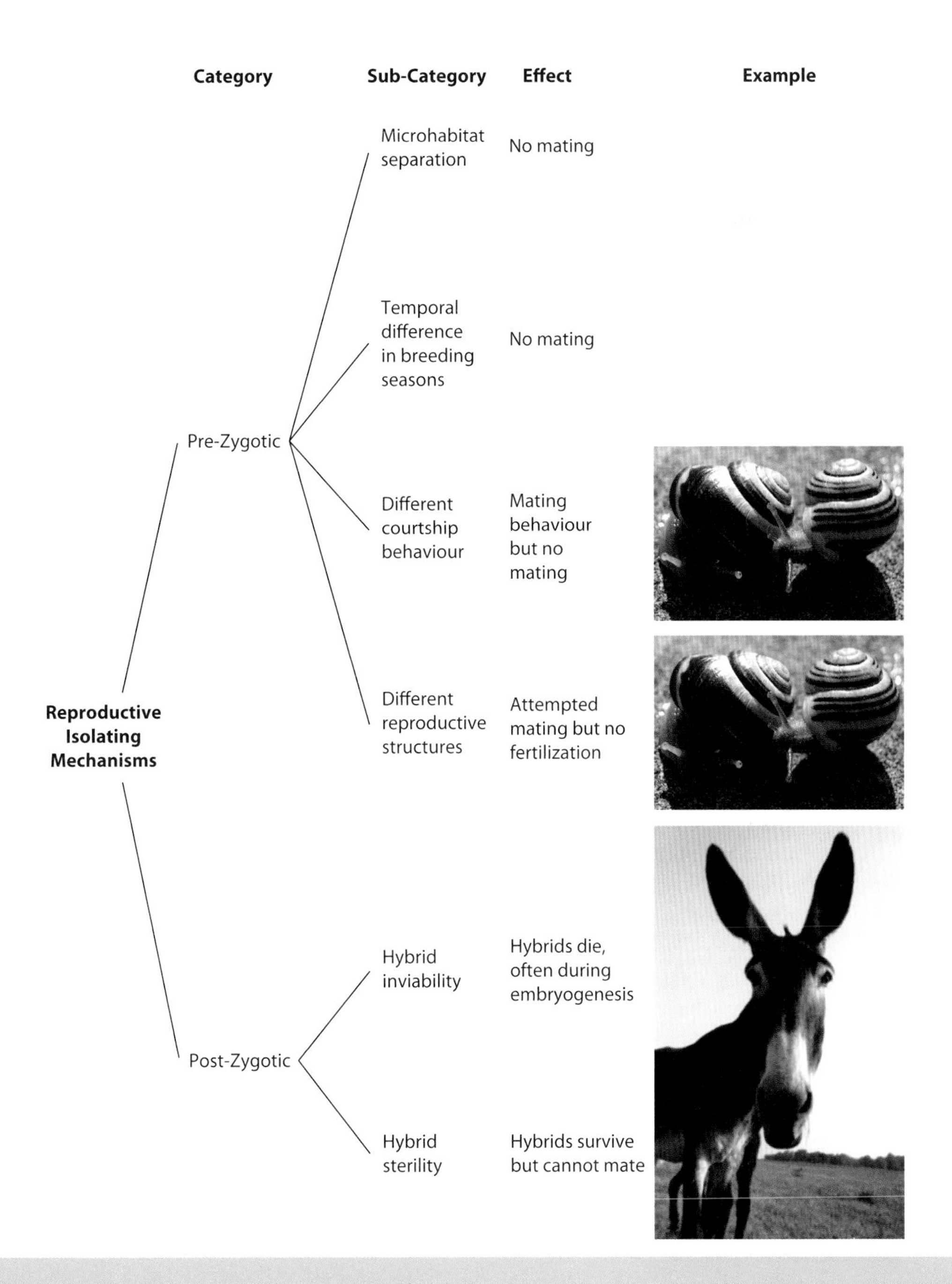

Figure 17.6 Types of reproductive isolating mechanism. The main distinction is between pre- and post-zygotic; but there are sub-types within each of these main two categories. Note the two examples given in pictorial form. The fact that one of them (snails) appears twice indicates that two (or more) forms of isolating mechanism can (and usually do) act to maintain the separation of recently-separated species. The other example is the well-known one of the sterile mule – hybrid offspring of a mating between a horse and a donkey.

courtship behaviour in animals, preclude the formation of an inter-type zygote. The latter act to prevent such a zygote, if it forms, from developing into a fertile adult. In *D. pseudoobscura* there may well be pre-zygotic mechanisms at work in the wild, but these are not yet known, as the attempts at interbreeding were conducted in the lab. So what this example gives evidence for is the start of post-zygotic isolation.

It cannot be emphasised too strongly that there is nothing purposive about the *start* of speciation – it is an entirely accidental process. When populations diverge past a certain point, they become partially reproductively isolated as a side-effect of the genetic differences they have accumulated as a result of the divergent selective pressures that have adapted each to its local environmental conditions. However, there may be a more purposive element to the *completion* of speciation.

The rationale for this is as follows: In nature, populations that have been allopatric (and thus out of genetic contact) for many generations may expand their ranges and come to overlap geographically again. This is referred to as neo-sympatry. When this occurs, there may be selection against individuals of each incipient species that breed with individuals of the other one, because such individuals are less fit than others due to the problem of hybrid inviability or sterility. So, given genetic variation in some character – say courtship behaviour – selection will favour variants that tend to engage in same-type rather than cross-type matings. This selective process, which may be responsible for completing speciation in many cases, has been referred to by three names: the Wallace effect (in honour of Alfred Russel Wallace); reproductive character displacement; or, most briefly, reinforcement.

Land Snails of the Genus *Cepaea*

While *Drosophila* is a highly **speciose** genus, with more than 2000 species, *Cepaea* is at the other end of the spectrum. It is a genus with just four species: *C. nemoralis, C. hortensis, C. vindobonensis* and *C. sylvatica*. All are European species, found in many continental European countries, but just two of them (*C. nemoralis* and *C. hortensis*) are also found in the British Isles. These two are thought to be sister-species (Fig. 17.7). We looked at one of them, *C. nemoralis*, in Chapter 9 in relation to its shell colour **polymorphism**.

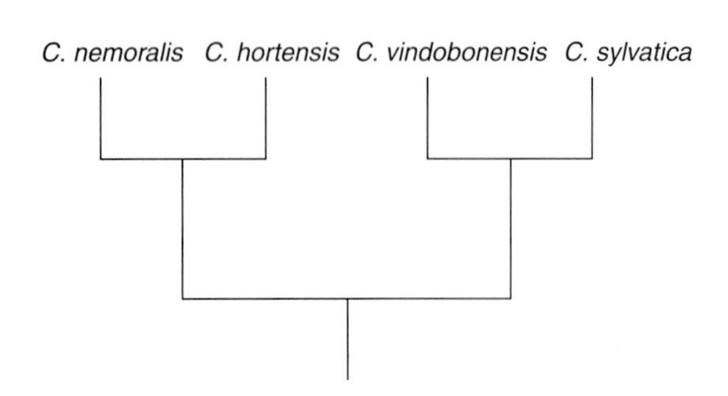

Figure 17.7 Cladogram showing the relationships between the four species of the land-snail genus *Cepaea*.

C. nemoralis and *C. hortensis* are 'good' biological species. This rather strange term is used to describe sister-species that are thought to rarely if ever interbreed – in other words, they are 'well-behaved' in relation to the definition of biological species, and are thus good examples of it. However, little in evolutionary biology is neat and tidy. A study of very large numbers of individuals of these two species revealed that there was a low level of hybridisation between them (almost certainly producing only sterile hybrids). The evidence for this came from a study of reproductive structures called darts, which are a feature of *Cepaea* and of the family Helicidae more generally (Fig. 17.8). These calcareous 'missiles' are fired at potential mates as part of the courtship behaviour of the snails (which are cross-fertilising hermaphrodites). As can be seen, the darts of the two species have different fin structures, and the darts of hybrids can be recognised because they are intermediate in structure.

The level of hybrids has been estimated as less than 1 in 1000. Why is it so low? Some of the various isolating mechanisms that were listed in Fig. 17.6 can be ruled out, while others cannot. Individuals of the two species often occur sympatrically, and indeed in the same microhabitats, such as on the stems of the umbelliferous plant called hogweed or underneath the leaves of dandelions. They have the same mating season (May) and are active at the same time of day (early morning while the plants are still moist). Pairs of snails attempting to mate can be readily observed in dense populations, including, occasionally, pairs in which one member belongs to one species and one to another. So the separation of the species is not maintained by spatial or temporal separation, but by a mixture of courtship differences, reproductive structure differences and, probably, post-zygotic incompatibility.

We should now add some detail to the inter-species difference. It is often impossible to distinguish individuals of the two species with certainty as juveniles. However, as adults, there is the difference in dart structure already mentioned. There is also a (statistical) difference in size – *C. nemoralis* is the bigger of the two, though the size distributions overlap. There is also an almost-consistent difference in the pigmentation of the lip of the shell – that of *C. nemoralis* is almost always darkly pigmented, while that of *C. hortensis* is almost always white or off-white. The rest of the shell pigmentation is highly polymorphic in both species with many parallel variants.

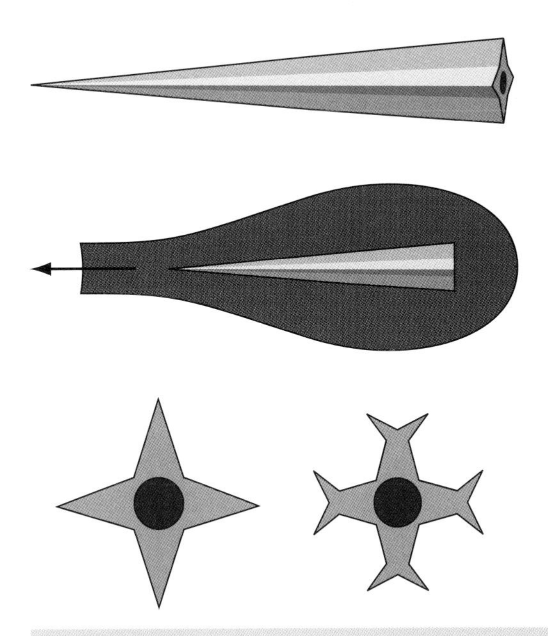

Figure 17.8 Diagram of a reproductive dart (top) that is 'fired' by one snail (from a muscular dart sac – centre) at a potential mate in species of the genus *Cepaea* and of other genera within the family Helicidae. In the case of the sister-species *C. nemoralis* and *C. hortensis*, these darts differ in their exact structure, as can be seen in cross-sections of the two (bottom). Rare hybrids can be recognised by having an intermediate dart structure, e.g. two fins 'simple' and the other two bifurcated.

Broadening Out

This chapter is part of the 'direction of evolution' section of the book. So, the question arises: in what way (if at all) is speciation linked to the determination of evolution's direction?

The most obvious answer is this: since speciation is a largely accidental process that effectively preserves the results of selection for local adaptation, it is a way of translating the directional effect of such selection from the intraspecific to the interspecific realm. However, it may also have a component of **developmental bias**, as follows:

In some cases, the genes contributing to the speciation process have been analysed. These include cases of entirely natural speciations and those cases where human artificial selection has produced what is effectively a new 'artificial species' from an old

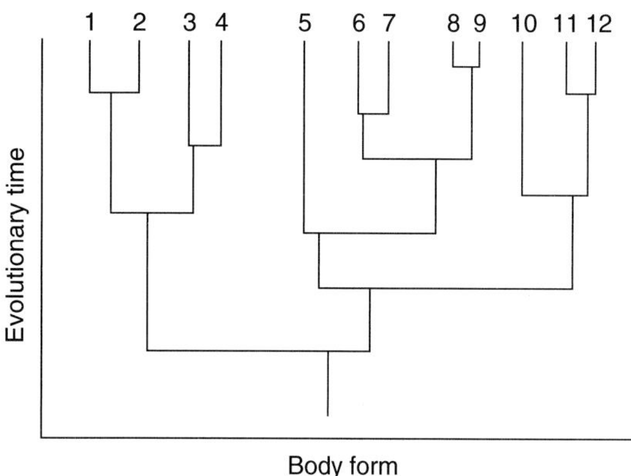

Figure 17.9 Diagrammatic representation of gradual and punctuational patterns of evolutionary change. The former includes a mixture of anagenesis and cladogenesis; the latter includes only cladogenesis, because in this pattern all evolutionary change in organismic form is associated with lineage splitting. Most biologists now believe that evolutionary pattern at the species level is intermediate between these extreme views.

natural one, an example of this being the production of maize from teosinte. It is usually found, in both cases, that the genetic basis is intermediate between two extremes: single **mutations** of large developmental effect (**macromutations**) and hundreds of genetic differences each of which has a tiny effect (**polygenes**). A typical speciation event seems to be underlain by perhaps a few tens of genes, most of them with very small effects but a few of them with larger effects (**mesomutations**). The exact form that speciation takes may be partly determined by which mesomutations happen to be available in the population at the time.

Finally, in addition to the standard Darwinian selection that causes the divergence of populations within a species, and may complete the speciation process via the 'Wallace effect' as discussed above, it has been suggested that there is a higher-level version of selection going on in evolution too. This is based on the differential reproductive success of species rather than of genetically different individual organisms. It has been given the name species selection[66] and it has been linked (though the linkage is not necessary) with the concept of **punctuated equilibrium**[67].

There was an intense debate, in the 1970s and 1980s, about whether developmental processes and their resultant adult forms change gradually in evolution, or whether they remain virtually the same throughout most a species' life-span and change abruptly only at the points in time when cladogeneses occur. The latter pattern was dubbed 'punctuated equilibrium' by Niles Eldredge and Stephen Jay Gould in 1972 (Fig. 17.9).

It now seems clear that in some **clades** the pattern is quite gradual, in others more punctuational, with a range of intermediates from one to the other. It is also clear that if speciation occurs via a peripheral isolate population, which then later extends its range and moves into neo-sympatry with its parent species, a series of fossils from a particular area may be revealing not the tempo of evolutionary divergence (which probably occurred elsewhere) but rather the Wallace effect or the ecological migration of an already-formed species to a new area. In other words, the entire punctuated equilibrium debate now looks like the proverbial storm in a teacup.

However, Stephen Stanley's idea of species selection is not so readily written off. There is a persuasive logic to his argument that transcends its origin against the backcloth of punctuated equilibrium. The argument is as follows. If we consider the origin of a species as being its birth and the extinction of a species as its death, we can easily imagine a higher-level analogue of Darwinian individual selection: those parental species that give rise to more surviving daughter species (either because they persist for longer or because they spin off more daughter species per unit time) will in a sense be 'fitter' than others. The fauna or flora will, over evolutionary time, come to be composed of a greater number of species of this 'fitter' kind.

Species selection cannot determine the direction of evolutionary change within any one species, but it may well help to determine the direction of evolution at a higher level, especially given that more speciose clades will, other things being equal, have lower extinction probabilities than less speciose ones. Such 'other things' include the geographical ranges of the species concerned. So, over long periods of evolutionary time, the array of creatures that is found may be partly a consequence of the Darwinian selection (and developmental bias) that brought each species into existence; but also partly a consequence of the process of species selection acting among the species so formed.

17.3 The Origin of Novelties

As we saw in relation to the pair of snail sister-species, *Cepaea nemoralis* and *C. hortensis*, in Section 17.2, it is often the case that speciation does not produce radical developmental repatterning. The slight difference in shell size (with accompanying change in body size) and the slight difference in the shape of the reproductive dart represent very minor forms of repatterning indeed. Hence it is difficult to analyse the developmental differences between the recently-separated species. Sometimes they can be analysed genetically, as noted above, in terms of the relative sizes of the contributions of different genes to the developmental repatterning. This is referred to as analysing the quantitative trait loci (Chapter 12) that contribute to a character (or trait) such as shell size. But even when this is possible, it does not reveal exactly how the **developmental trajectory** has been changed during speciation.

Analysing the nature of developmental repatterning is easier when its amount is greater; and this is true in the case of the origin of novelties. Although these are harder to define than species, there are some particular examples that virtually all biologists accept as qualifying for novelty status. One of the most celebrated of these is the turtle's shell, the story of which we will now pick up from where we left it at the end of Chapter 13.

Turtles possess a feature – the shell – that is shared with no other vertebrate. How did this novelty arise in terms of developmental repatterning? It is not yet possible to give a complete answer to this question, but some aspects of the repatterning involved are now understood[28].

The point of origin of the chelonian clade from a more typical early **amniote** is not certain. But this does not matter because, whichever group chelonians arose from, they were almost certainly shell-less. So, by comparing the development of turtle embryos with that of 'typical tetrapods', we can get at the nature of the repatterning that took place to produce this evolutionary novelty. Before looking at chelonian development, we will look at its end result – the morphology of the adult. In particular, we need to establish exactly how this is different from most other vertebrates.

The chelonian shell consists of a dorsal carapace (Fig. 17.10) and a ventral plastron. The vertebrae are fused into the carapace, as are the ribs. Consequently, the pectoral

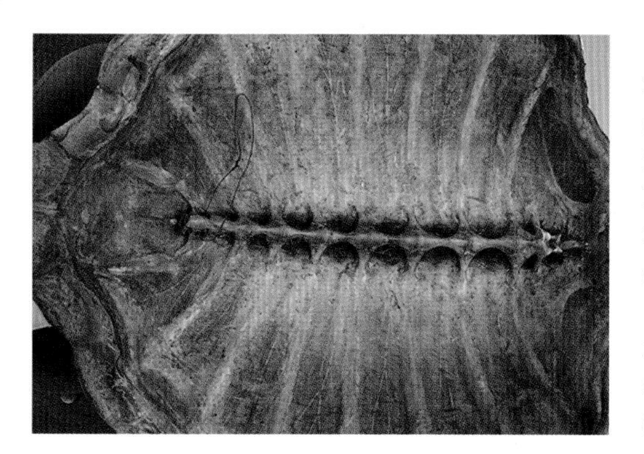

Figure 17.10 A turtle carapace (the dorsal part of the shell) seen from underneath. Note the way the ribs fuse with the shell. This means that the scapula (shoulder blade) that is normally exterior to the ribcage in tetrapods (including ourselves) is interior.

girdle, which is outside the ribcage in all other vertebrates, is inside it in chelonians. Even without considering the situation in more detail, it is immediately clear that this is no 'routine' form of evolutionary change in development. It is hard, but perhaps not impossible (see below) to imagine how it could have proceeded by a series of small steps; however, it is very unlikely to have arisen by a single 'macromutation'. As we have already noted, such mutations are normally characterised by major **fitness** decreases and so will be removed from populations, not spread through them, by selection.

The turtle shell is made up of more than 50 bones, fused together. These bones are not found in any vertebrate outside the clade Chelonia. The bones of the carapace are a mixture of endochondral and dermal – that is, formed from cartilage and skin tissue, respectively. The bones of the plastron seem to be exclusively dermal. The formation of the shell, along with the atypical development of the ribs – fusing dorsally with the carapace rather than bending round ventrally and meeting at the sternum – is initiated by an embryonic structure called the carapacial ridge.

This ridge is an outgrowth of thickened epidermis and underlying **mesodermal** tissue that forms initially in the middle region of the body (in terms of the antero-posterior axis), subsequently extending in both anterior and posterior directions. In terms of the dorso-ventral axis, the carapacial ridge lies above (i.e. dorsal to) the limb buds. The ridge has been shown to 'ensnare' the ribs and cause their growth to deflect dorsally compared with the usual vertebrate situation. If, in turtle **embryogenesis**, the scapula arises anteriorly to the rib cage, in contrast to the situation in other **amniotes**, then during development some of the anterior ribs may overgrow it, as a result of their attraction to signals from the carapacial ridge (Fig. 17.11).

But what genes and proteins are involved in the formation of the ridge and its 'instructions' to the ribs? Initial studies have shown that a common growth factor, Fibroblast Growth Factor 10 (FGF-10) plays a role in one or both of these. The gene producing this growth factor is expressed in the mesoderm of the carapacial ridge. Further studies on other genes/proteins involved in the development of the unique chelonian design have been conducted recently. These have revealed the involvement of several genes known to be involved in the development of the vertebrate limb bud; these may have been **co-opted**, in chelonians, for a new role in the carapacial ridge.

Now is an appropriate point at which to return to the question raised at the start of this chapter – is evolution scale-dependent? Studies of the proliferation of species within genera and families often suggest that macro-evolution is simply accumulated micro-evolution, with the addition of reproductive isolation. However, studies on the origin of novelties (in the realm of mega-evolution, and in the turtle case associated with the origin of an order) suggest that some things are not just accumulated micro-evolution.

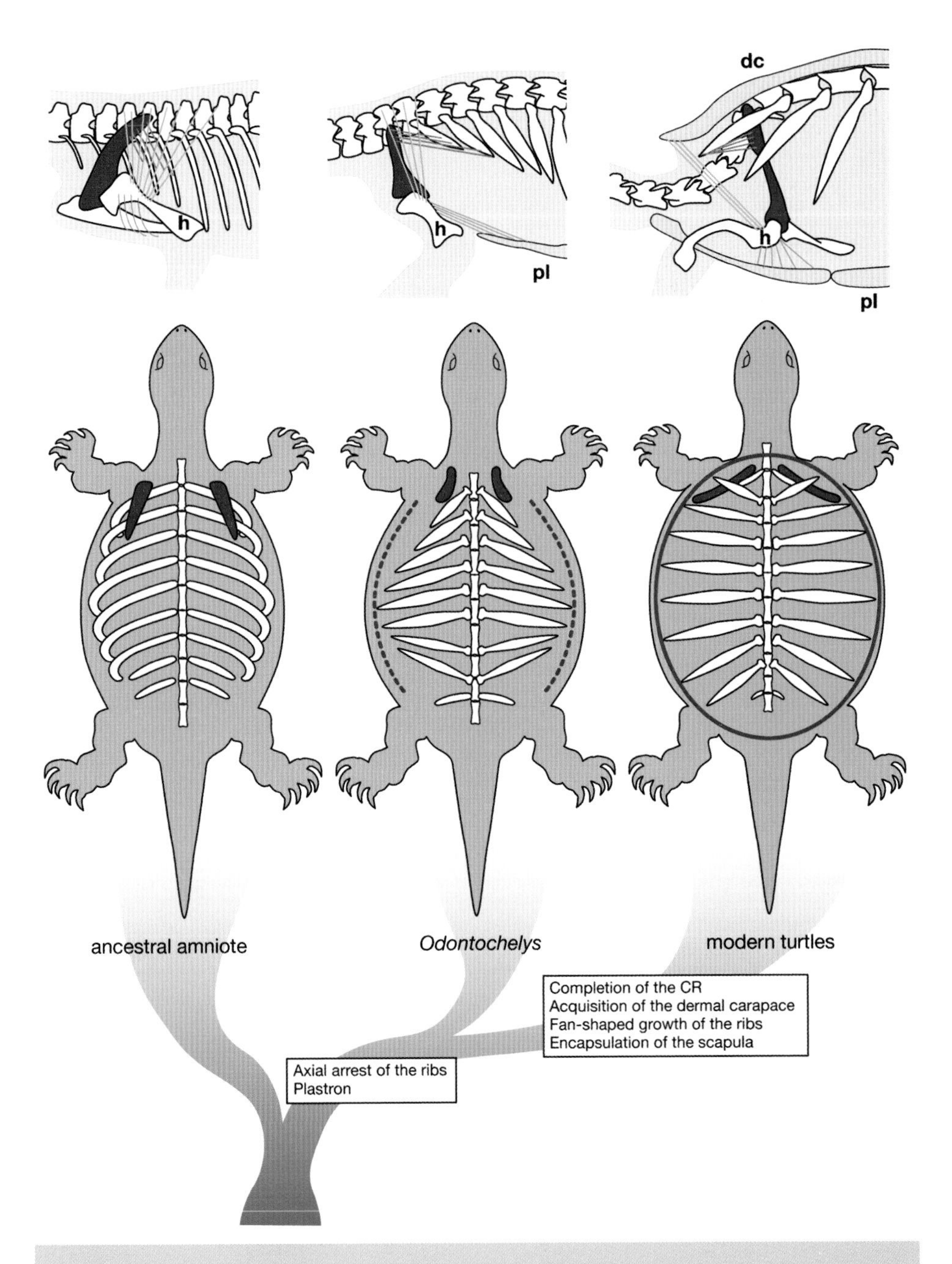

Figure 17.11 A possible scenario for the evolution of the turtle shell, with its internalised scapula. Top: proposed changes in the connections between muscles and bones. Bottom: schematic phylogeny in which the fossil turtle *Odontochelys* represents a kind of intermediate stage between the 'typical tetrapod' (left) and the chelonian (right) rib/pectoral girdle arrangements. dc – dermal carapace; pl – plastron; h – humerus; CR – Carapacial ridge. (Reproduced with permission from Nagashima *et al.* 2009 *Science*, **325**.)

Our conclusions, then, are that the origin of a novelty is at least sometimes not explicable through accumulated typical micro-evolutionary processes; and therefore that the relative frequencies with which evolution moves in one novel direction rather than a different one may be explicable in terms of developmental bias as well as in terms of selection.

17.4 Body Plans I: Overview

Body plans are no easier to define than novelties. Yet they are easy to recognise, as the widespread usage of the term 'body plan' in the evolutionary literature shows. In the case of a turtle, it is equally clear that it is a member of the order Chelonia and of the sub-phylum Vertebrata.

Note the taxonomic rank of 'sub-phylum' above. Two things usually characterise body plans. The first is that they relate to high ranks, normally phyla or sub-phyla. Within a taxon of this rank, the key aspects of the body plan are usually retained in all members of the taxon, as in the vertebrate skeleton, though in some cases there can be a few secondary losses, as we saw earlier in relation to **annelid** segmentation. The second thing about body plans, as these examples of skeleton and segmentation suggest, is that they are deeply-embedded features of the architecture of the animals concerned, usually with early embryological origins. However, in cases of complex life histories, it may be necessary to separate **larval** and adult body plans. In such cases the latter may have a later developmental origin, for example, at **metamorphosis** in **holometabolous** insects.

As well as arising early in development, it appears that most animal body plans arose early in evolution. All the animal **phyla** that readily leave fossils, including arthropods, molluscs, chordates, brachiopods and many others, are represented in the fauna of the Cambrian, while none of them are found, with certainty, before the base of that period (543 MYA). This fact has led to the assumption that those phyla that do not readily fossilise were also present in the Cambrian, and thus to the idea of a **Cambrian explosion** of animal body plans. This idea, and criticisms of it, will be discussed in Section 17.6.

You may have noticed that 'animal' rather than 'organismic' has been used above in our introductory discussion of body plans. This is because it is less clear whether 'body plan' is a useful term in describing the evolution of the various major groups of plants and their defining features. The **angiosperms**, as a high-ranking taxon in the plant kingdom, are most obviously distinguished from other plants by their possession of flowers; but these structures do not arise, in developmental terms, until a relatively late stage of **ontogeny**. Also, the group itself did not arise, as we saw previously, until relatively late in evolution – perhaps c. 200 MYA as opposed to 500+ MYA. So there is a fascinating question here: does this represent a real difference between animal and plant evolution; and if so, what is its basis?

This question cannot yet be answered, and so the utility of 'body plan' in the plant kingdom must remain in doubt for the moment. Consequently, the examples given in Sections 17.5 and 17.6 are animal ones.

17.5 Body Plans II: the Origin of the Vertebrates

We will begin here by looking at how both the chordates and the vertebrates arose within the context of the evolution of the **deuterostomes**, building on the outline picture of deuterostome **phylogeny** given in Chapter 7. A central feature of the chordate body plan is of course the possession of a **notochord**, after which structure the phylum is named; likewise, a central feature of the vertebrate body plan is the possession of a vertebral column, plus associated skeletal elements (Fig. 17.12).

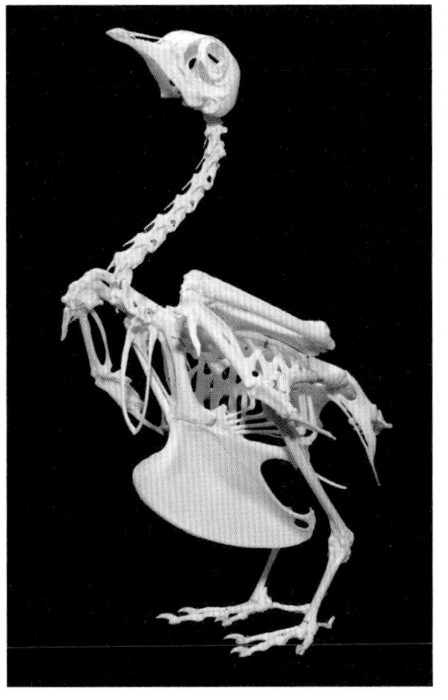

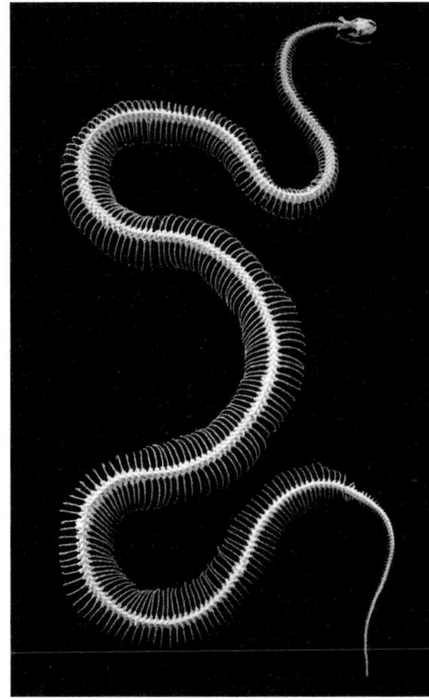

Figure 17.12 Three vertebrate skeletons which, as can be seen, are all variations on a theme. This skeletal theme is the central feature of the vertebrate body plan. No vertebrates lack such skeletons (though hagfish have very unusual ones); while no non-vertebrate groups possess them. The examples shown are a bird, a snake and a dinosaur.

Because we will be concentrating here on the vertebrate body plan, we can deal with the notochord quite quickly. This stiff rod, which provides support for the body in non-vertebrate chordates such as amphioxus, is retained in vertebrate embryos and functions as a signalling centre, secreting the Sonic Hedgehog protein that induces certain developmental fates in surrounding tissues, as we noted in Chapter 10. However, as the development of a vertebrate proceeds, the notochord 'disappears', though some of it may remain in the inter-vertebral discs. For all vertebrates, the central supporting skeletal structure, from an early developmental stage, is the vertebral column. All vertebrates also possess a skull and ribs. The degree to which there is an 'appendicular skeleton' too, that is bones (or in some cases cartilage) that supports fins or limbs, varies. In a few groups, such as the more **derived** snakes (e.g. vipers), there is no appendicular skeleton.

So, to investigate the origin of the vertebrate body plan in the fossil record, what we need to look for are the earliest fossil remains of vertebrae, skulls, and perhaps also fin-supporting bones. Recall that all the earliest vertebrates were marine, so we would not expect to find limb bones at early stages of vertebrate evolution.

At present, the earliest known vertebrate fossils are those found at the now-famous Chengjiang locality in China. These are early Cambrian, being approximately 530 MY old. Two genera have been described are *Haikouichthys* and *Myllokunmingia* (Fig. 17.13), though it is possible that the specimens involved might be conspecific. These fossil specimens resemble present-day jawless fish, such as lampreys and hagfish.

Even if these remain the oldest known fossil vertebrates, as further palaeontological research reveals more and more fossils (and there have already been claims, though not yet widely supported, of older jawless fish), we must consider an additional problem. Whenever the vertebrate lineage split from its sister-lineage, regardless of whether that was a **cephalochordate** or **urochordate** one, it may not at that time have been characterised by even the rudimentary-looking vertebrae that are seen in, for example, *Haikouichthys*. In other words, although we associate the main vertebrate body-plan features, such as a skull and vertebral column, with the origin of the vertebrate lineage, it is possible, indeed likely, that the two events did not occur simultaneously. Recall that a lineage split, such as occurs in a standard cladogenetic speciation event, can occur very quickly. The assembly, through developmental repatterning, of a vertebrate skeleton is likely to take much longer. So there may, ironically, have been proto-vertebrates that lacked vertebrae (or other skeletal elements) and so left no fossil remains. This kind of problem may never be completely overcome in our search for the origin of the vertebrate body plan, or indeed the body plans of other major animal groups.

17.6 Body Plans III: the 'Cambrian Explosion'

In the early 1990s, most evolutionary biologists were convinced that the Cambrian explosion was a very real event, and that it took place in a very short period of geological time, perhaps about 10 MY, from around 540 to 530 MYA. This view was based on assessment of the available fossil evidence, which seemed to show that most

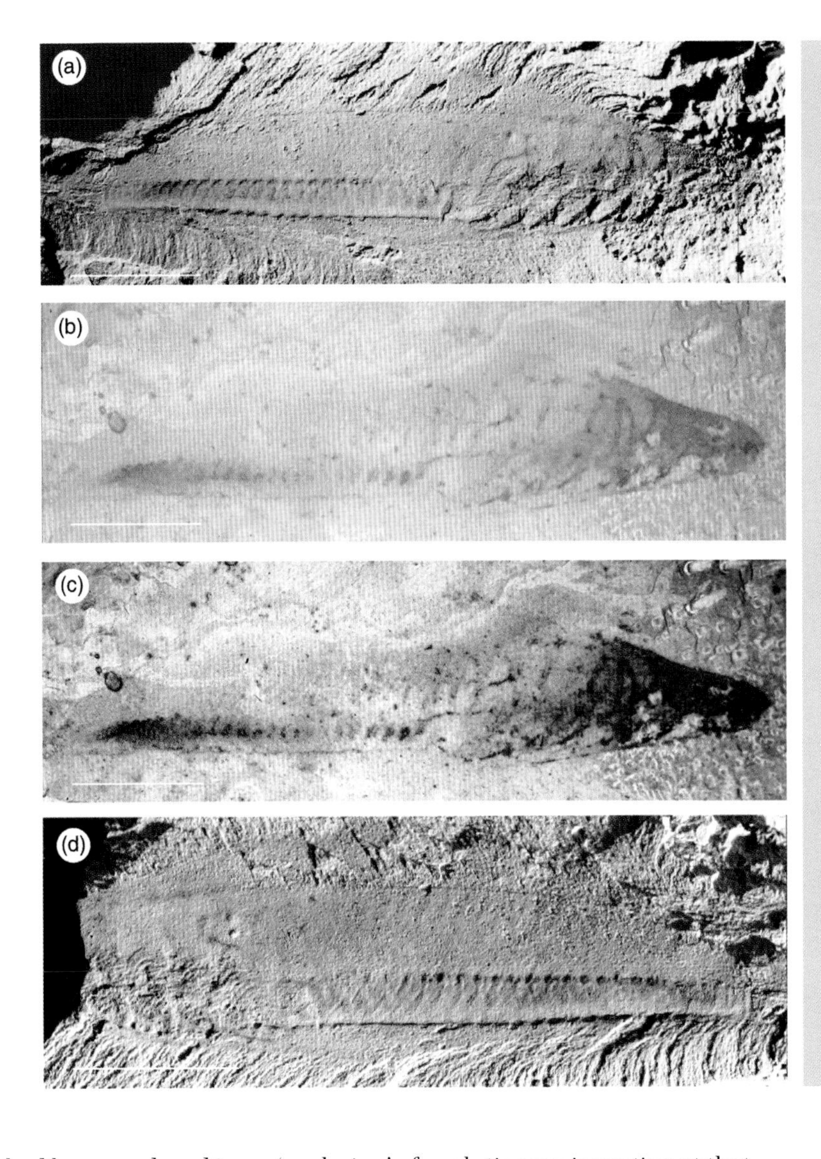

Figure 17.13 One of the earliest known vertebrate fossils, *Myllokunmingia*, (from a site in China). (a) and (d) – Part and Counterpart; (b) – the 'part' viewed with polarised light; and (c) – with computer enhancement. (Reproduced with permission from Xian-guang *et al.* 2002, *Proc. Roy. Soc. Lond.* B, **269**.)

animal phyla had been produced in an 'explosion' of evolutionary invention at that time. However, in the mid- and late 1990s, a series of papers appeared in which the times of origin of the major animal groups were estimated using molecular methods[68]. These methods were based on the assumption of an approximately constant rate of molecular evolution at the level of DNA and proteins – the so-called molecular clock.

The *modus operandi*, in simplified form, was as follows. Take two present-day species, say a crocodile and a bird, whose divergence time has been well established from the fossil record. Then take a gene that occurs in both of these animals and sequence it. The percentage sequence difference in the two orthologues can be divided by the time since their divergence to give a 'sequence difference per unit time' estimate. Then, to determine the time of divergence of two more distantly related species, say a

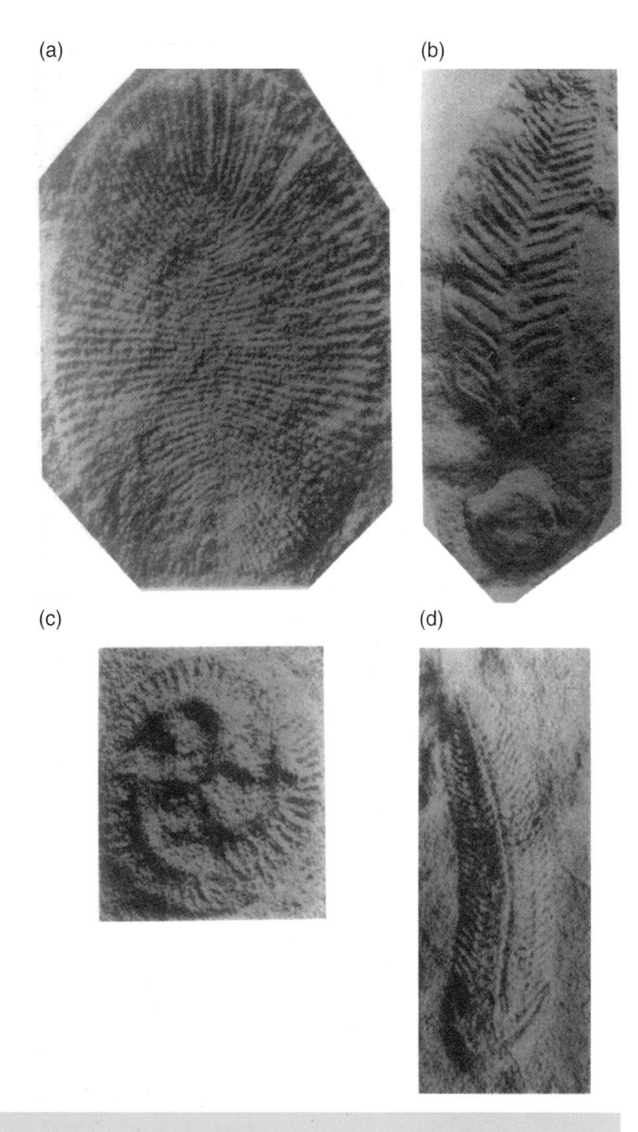

Figure 17.14 A selection of fossils of Vendian age (c. 600 MYA), whose interpretation is contentious. Collectively they are often referred to as the Vendobionta (simply meaning life-forms from the Vendian Period). In contrast to the fossil shown in Fig. 17.15, the ones shown here are probably adults. (From Glaessner, 1984, *The Dawn of Animal Life*, Cambridge University Press.)

protostome and a **deuterostome** (e.g. a fruit-fly and a mouse), repeat the sequence analysis, on orthologues of the same gene (because different genes are known to evolve at different rates). This sequence difference will be greater than the first one. If there really is a molecular clock, even a 'sloppy clock' as it has sometimes been called, the divergence time of protostome and deuterostome lineages can be estimated.

The molecular studies of the 1990s produced widely varying estimates of the divergence times for protostomes and deuterostomes, but most were vastly older than the estimates that had been previously made from study of the fossil record. Instead of a divergence time of about 540 MYA, which would be consistent with a genuine Cambrian explosion, some of the molecular studies produced estimates of more than double that – around 1200 MYA. These estimates presented biologists with a major dilemma – should they believe the fossils or the DNA sequences?

Although this debate is far from over, the balance of opinion is shifting back to the earlier, fossil-based view. It seems likely that the molecular clock hypothesis suffers from a fundamental flaw. Although the rate of evolution of genes and proteins may have been quasi-clocklike in the later stages of animal evolution, it was probably not so in the earliest stages. The reason for this is that when a new gene is formed, for example by duplication of an 'old' one followed by divergence and the adoption of a new role, it is likely to be under intense selection to improve its ability in that new role. After a period of rapid evolution, its sequence is likely to change more slowly. Associated with this slow-down, it is reasonable to expect that the ratio of selected to **neutral** sequence changes will decline.

This flaw in the molecular approach, combined with the continuing lack of animal fossils from the period before about 650 MYA (pre-**Vendian**), together make a strong case that there was indeed an explosion of animal life. But was it really a Cambrian explosion, and hence later than 543 MYA, or was it a somewhat earlier Vendian one?

The Vendian period contains a diverse array of macroscopic fossils variously referred to as the Ediacaran fauna (after the Ediacaran Hills in southern Australia where they were first discovered), the Vendozoa[69] or the Vendobionta. Note that the first two of these terms imply that the creatures concerned were animals, while the third does not. Some of the Vendian fossils are shown in Fig. 17.14. Interpretations include:

a) they are representatives of existing animal phyla;
b) they collectively constitute a diverse clade of animals not belonging to existing phyla – a clade that died out around the base of the Cambrian;
c) they constitute a diverse clade of multicellular creatures not belonging to the animal kingdom (or to any other kingdom with extant members);
d) they were a strange group of marine lichens; and
e) they were a mixture of two or more of the above.

There is not yet agreement on which of these hypotheses is correct, though there is little support for the 'marine lichens' one. The debate has been complicated by the discovery of cellular-level Vendian fossils, which have been interpreted both as animal embryos[70] and as groups of bacterial cells[71] (Fig. 17.15).

If there was indeed an 'explosion' of animal body plans at an early stage in animal evolution, whether in the Vendian or the Cambrian, what were the reasons for it? Two hypotheses have been proposed, effectively internal and external ones. According to the former, major developmental repatterning was easier in early animal evolution, because developmental systems had not yet become deeply **coadapted** and resistant to change. According to the latter, early animal evolution was characterised by an ecological 'runaway' process, in which new body plans proliferated during a phase of comparatively empty ecospace and relative freedom from competition. Of course, these two hypotheses are not mutually exclusive.

Finally, we return to the question of what determines the direction of evolution – this time in the context of the origin of body plans. There seems little doubt that natural selection was again a key player. But, as in the origin of novelties such as the chelonian shell, it seems likely that body plan origins involved the availability of unusual variants to natural selection, and that therefore developmental bias played a role too. If so, then it is again a case of the direction of evolution being determined jointly by *structured* (as opposed to amorphous or universal) variation as wall as the dynamics of natural selection.

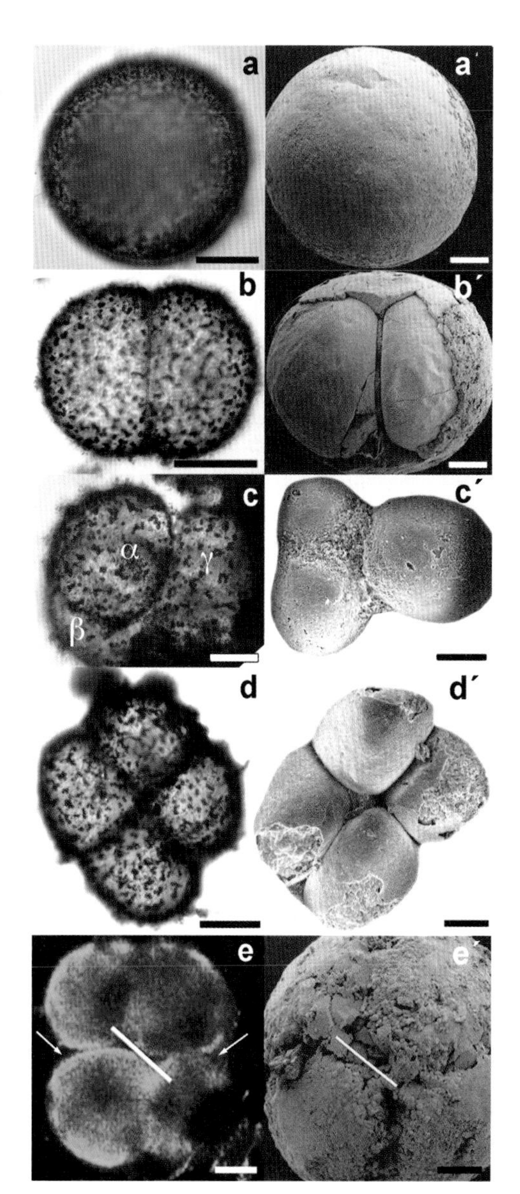

Figure 17.15 Cellular fossils of Vendian age (right, a′ to e′), have been interpreted by some as animal embryos. However, others interpret them as giant sulphur bacterial cells, because of their similarity to extant such cells (a–e). Scale-bars – 150 μm. Diagonal lines – offset positioning of cells. Arrows – thin sheath surrounding cell clusters. (Reproduced with permission from Bailey *et al.* 2007, *Nature*, **445**.)

SUGGESTIONS FOR FURTHER READING

For an account of speciation focused on Darwin's finches, see the book that was also recommended at the end of Chapter 12:

Grant, P.R. and Grant, B.R. 2008 *How and Why Species Multiply: The Radiation of Darwin's Finches*. Princeton University Press, Princeton, NJ.

For a general discussion of the origin of animal body plans, see:

Arthur, W. 1997 *The Origin of Animal Body Plans: A Study in Evolutionary Developmental Biology*. Cambridge University Press, Cambridge.

For two very different accounts of the Cambrian explosion, see:

Gould, S.J. 1990 *Wonderful Life: The Burgess Shale and the Nature of History*. Norton, New York.

Conway Morris, S. 1999 *The Crucible of Creation: The Burgess Shale and the Rise of Animals*. Oxford University Press, Oxford.

CHAPTER 18 The Evolution of Complexity

18.1 Defining Complexity

One of the most striking results of evolution is the production of complex organisms from simple ones. The actual evolutionary process that has occurred on Earth over the last four billion years is very different from another evolutionary process which, over a similar period of time, produces only a wide range of bacteria. This latter type of evolutionary process may be merely hypothetical, or it may actually have happened on a hitherto undiscovered planet in some distant solar system.

However, considerable caution is needed in dealing with the evolution of complexity, for two reasons. First, although we all have intuitive ideas about what complexity means, it is not so easy to define. There are at least 20 different definitions of it, and probably many more than that. We need to settle upon one of these. Second, under the old, discredited view of a *scala naturae*, or 'natural scale', evolution was seen as having an inexorable tendency to produce ever-more complex creatures. But we now know that there is no such 'law' – rather, complexity can increase, decrease or stay the same in a **lineage** over evolutionary time. This section deals with the definition of complexity; the next one deals with the lack of a law of complexity evolution.

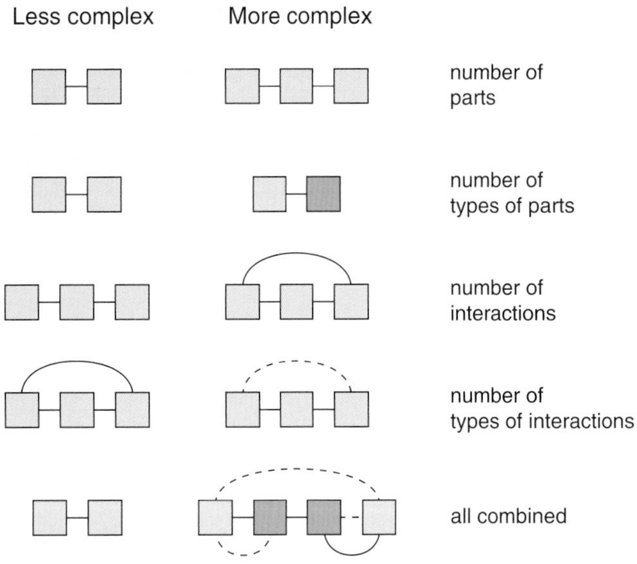

Less complex More complex

number of parts

number of types of parts

number of interactions

number of types of interactions

all combined

Figure 18.1 The complexity of an organism can be defined in terms of the number of parts, and the number of types of parts, of which it is made up. To this can be added the number of interactions, and of types of interactions, among those parts.

So, how should we define the complexity of an organism? One of the simplest definitions is 'the number of component parts'. This is a good first step. Multicellular organisms are usually perceived as being more complex than unicellular ones. However, 'parts' need not be cells. They might be smaller (e.g. genes) or larger (e.g. body segments). Is a centipede with 191 trunk segments more complex than a fly with only 12? Probably not. The problem with this comparison is that although the trunk of a centipede is made up of many segments, most of these are not very different from each other; this is sometimes referred to as homonomous segmentation. We can begin to refine our definition of complexity to take account of this problem: complexity can be defined as 'the number of different *types* of component parts'.

Now we encounter another problem. If 'parts' refers to physical entities such as genes, cells or body segments, it is also important to include in our definition of complexity the extent to which parts interact. Both a greater number of interactions, and a greater number of *types* of interaction among a fixed number of parts, should be considered to confer greater complexity. Thus we end up defining the level of complexity of an organism as 'the number of different types of parts and the number and types of their interactions'. This definition is illustrated diagrammatically in Fig. 18.1.

A final complication concerns the life-history dimension of an organism. The discussion so far has been implicitly based on adults. When the term 'fly' is used, few people picture the fly as the egg that it starts off as. In direct developers such as mammals, the developmental process produces an increase in complexity over developmental time that parallels the increase in complexity over evolutionary time, characterising the lineage that led from unicellular ancestors of the distant past to the mammals of today. This is a relatively simple situation.

In indirect developers though, it is not so clear what happens to the level of complexity of the organism as it progresses through its life-history. Indeed, given the variety of modes of indirect development, there can be no general answer to this question. In a butterfly with a **larva** and an adult, complexity increases twice – once in the egg, producing a larva, and once in the chrysalis, producing the adult. In contrast, during the larval growth phase complexity changes little, and during the adult lifespan it remains approximately constant. But in some parasitic worms of the phylum Platyhelminthes, the number of stages in the life-history is greater – as many as eight in some species (Fig. 18.2). Here the pattern of change in complexity through the life-history may increase (and decrease) several times. Also, the origin of indirect from direct development in a lineage can itself be considered as an evolutionary increase in complexity: such increases (and equivalent decreases) will be discussed in Section 18.4.

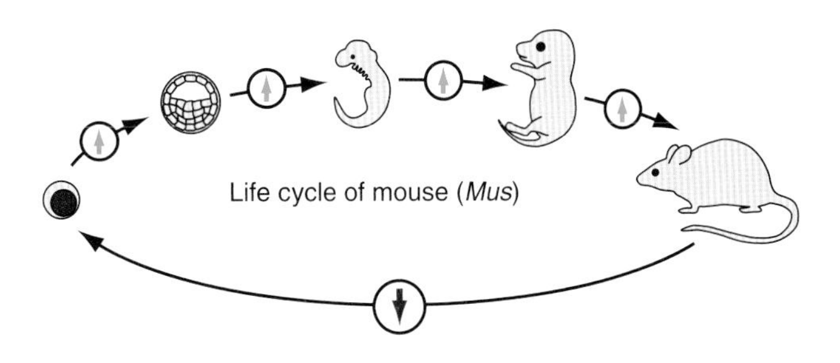

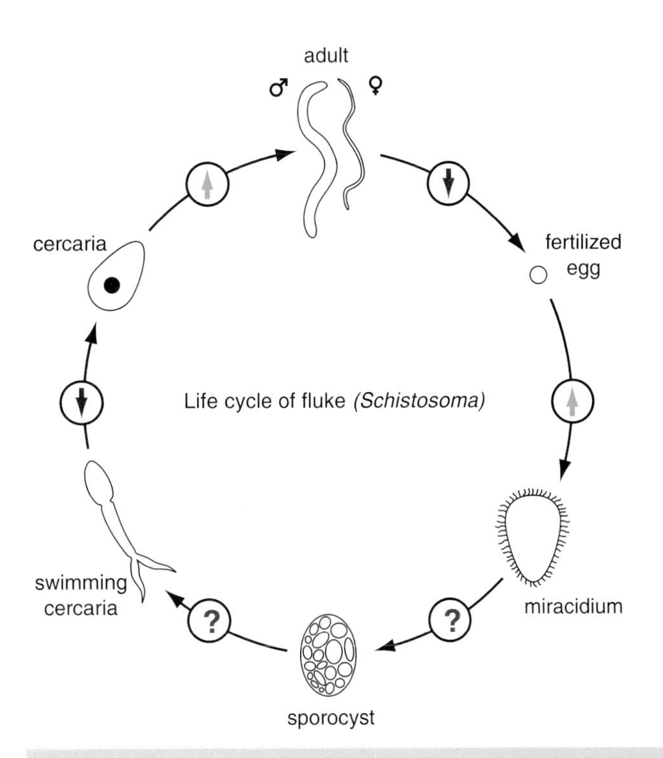

Figure 18.2 While in some organisms (e.g. mammals) there is direct development towards an adult that is clearly the most complex stage of the life-cycle, other organisms (e.g. parasitic Platyhelminthes) have indirect development, meaning that the life-cycle itself has become more complex and that the complexity of its constituent stages can vary considerably.

18.2 The Lack of a 'Law of Increasing Complexity'

Some early depictions of overall evolution saw it as a broad upward thrust driving inexorably from simple beginnings to a pinnacle of complexity represented by mammals and, at the very top, humans[72]. Other taxa were seen merely as side-branches (Fig. 18.3a). In contrast, most recent depictions of evolution, in terms of an overall tree encompassing either all life forms or all animals, show evolution to be a diversifying process leading in many directions, one of which, 'accidentally', produced humans – the only animals capable of thinking, writing and reading about the process through which they arose (Fig. 18.3b).

This shift, from an old *scala naturae* approach to a modern 'diversification' one, could be compared with the shift from an Earth-centred to a Sun-centred view of the solar system. In both cases 'we' (our planet; our species) have been dislodged from the centre/pinnacle to an unremarkable, 'ordinary-rather-than-special', position. Of course, 'we' are still unique, both in the sense of being the only life-bearing planet in our solar system and being the only animal species capable of abstract thought (as far as we know). But the lesson that has been learned from making the transition from the old to the new view of evolution is that the evolutionary process has no foresight. Rather, it is, as captured in the title of the book *The Blind Watchmaker* by English biologist Richard Dawkins[73], 'blind' to whatever may happen in the future. So an upward trend in complexity, with some side branches, is a very inappropriate way of picturing it. Some lineages do not just go off to the side: in complexity terms, some of them go significantly downwards.

As lineages gradually separate through the process of multiple **speciation** events (Chapter 17), each may do one of three things in terms of complexity: increase,

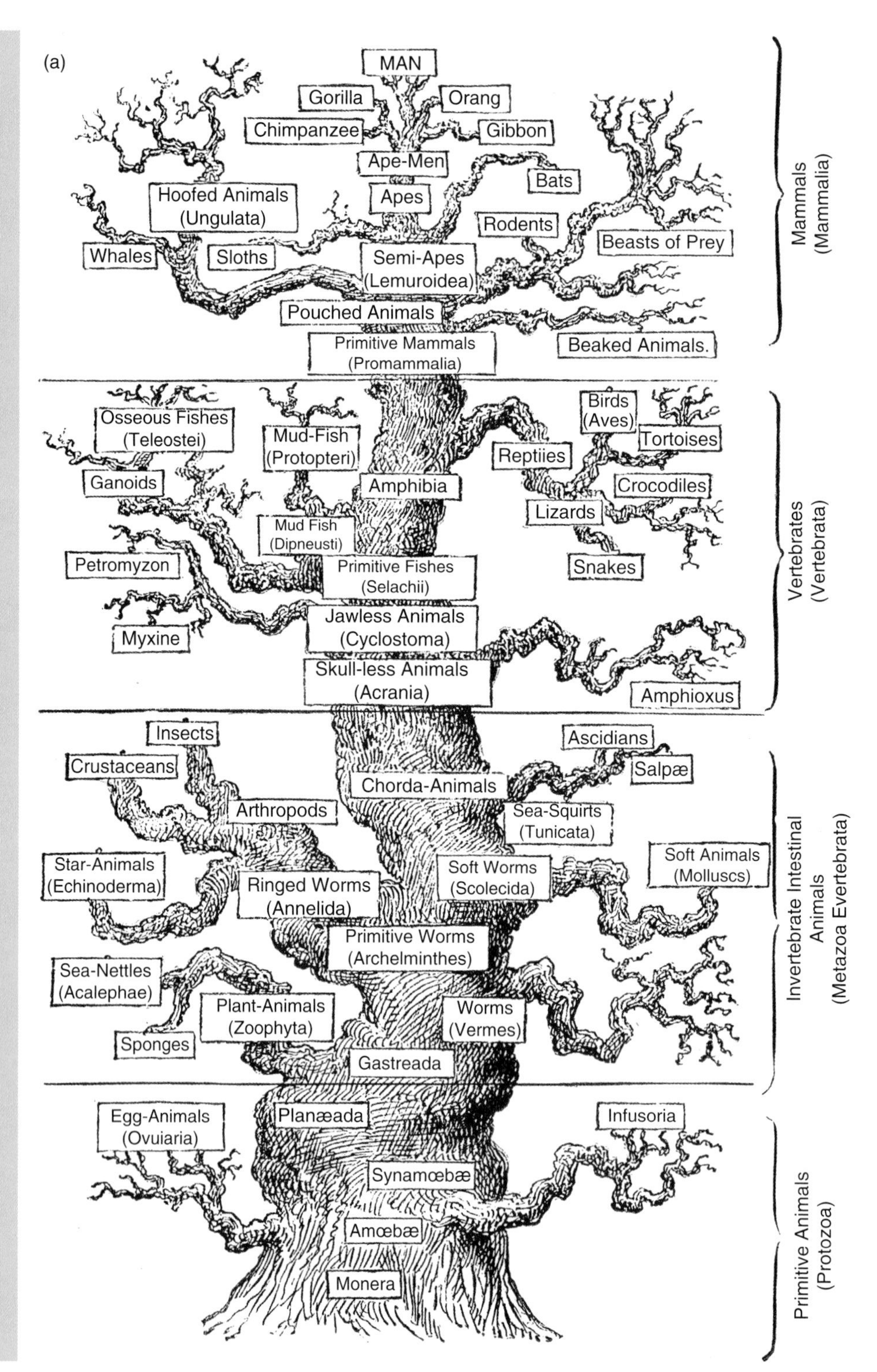

Figure 18.3 (a) An early view of the tree of life, produced by Ernst Haeckel, showing an inexorable central upward thrust towards humans, with other types of organisms as mere side-branches. (*Continued on next page*)

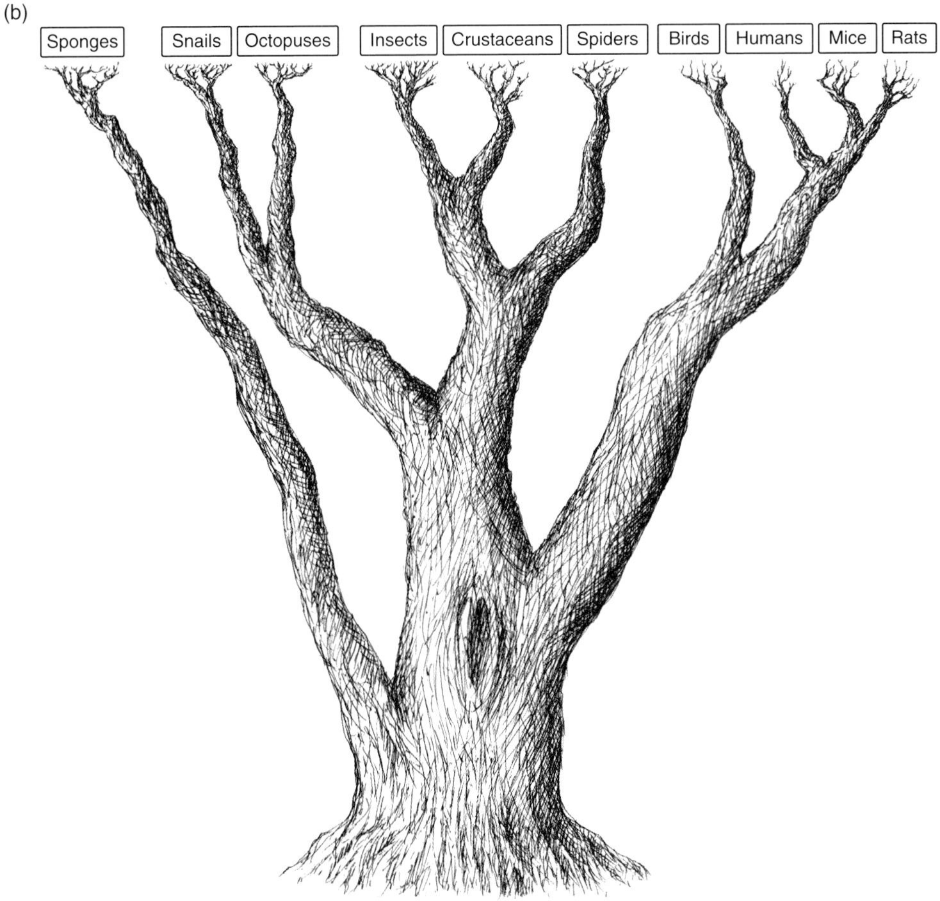

Figure 18.3 Contd (b) The current view, also shown as an actual tree to allow direct comparison that is not obscured by different forms of presentation, in which lineages diversify in many directions, one of which, by chance, led to humans.

decrease or stay the same. There is not yet any available quantitative estimate of the relative frequencies of these three possibilities. However, it seems likely that staying the same is the commonest, decreasing in complexity the rarest, and increasing in complexity intermediate between the other two. But even such a qualitative ranking of the three is risky, and it may depend on which measure of complexity is used, among other things.

For this section and the next, we will use number of cells, and number of cell-types, as our measures of complexity. This choice will enable comparisons to be made across almost all organisms, the only exceptions being those that are not truly cellular, including the viruses and the **syncytial** slime moulds. An alternative choice of the number of types of some other component part, such as body segments, would exclude many more taxa from possible consideration.

To illustrate the fact that evolution can produce decreases in complexity over time, we will look at some strange animals called mesozoans, whose tiny bodies are made up

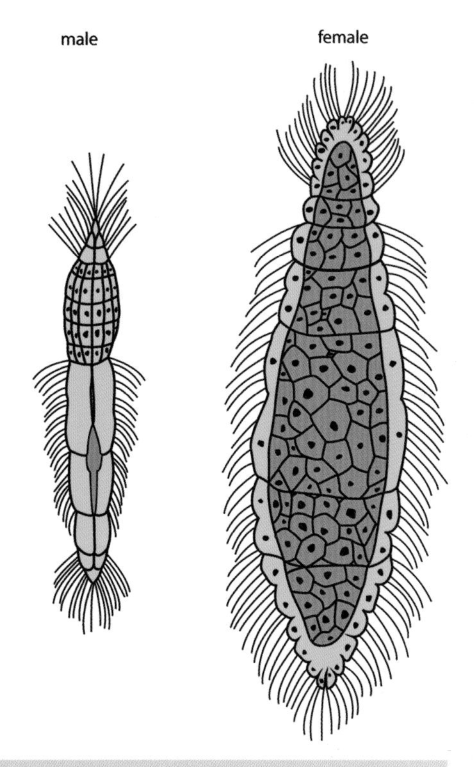

Figure 18.4 A 'mesozoan'. In fact, the previously recognised group called the Mesozoa has now been abandoned because it consists of two sub-groups that are not closely related. The 'mesozoan' shown here belongs to one of these (the Orthonectida). A representative of the other (Dicyemida) is shown in Fig. 18.6.

of very few cells and also of very few types of cells (Fig. 18.4). But before discussing these fascinating creatures in detail, we should pause to discuss the potentially misleading meaning of their name, and its relationship to the names of other groups of 'animals' – protozoans, metazoans and eumetazoans.

The obsolete taxon Protozoa means 'first animals'. It is now clear that these single-celled creatures are of many fundamentally different kinds (Fig. 18.5) and that 'Protozoa' is a **polyphyletic**, and hence invalid, group. These creatures are no longer considered to be animals at all. In contrast, the taxon Metazoa, which is synonymous with the modern usage of Animalia, is still often used. This is fine in the sense that animals, or metazoans, are **monophyletic**; but Metazoa is clearly a redundant term, and it seems simpler to talk about animals than about metazoans. Why use jargon when it is not necessary?

Having disposed of Protozoa and Metazoa, we can in fact get rid of the other two terms as well. Eumetazoa ('true animals') has been used to refer to all animals other than sponges. However, given the current uncertainty surrounding the exact structure of the base of the animal tree, such a grouping seems both unnecessary and unwise. And Mesozoa, meaning 'middle animals', in the sense of being intermediate between Protozoa and Metazoa, is inappropriate for two reasons.

First, it now seems likely (though it is not yet certain) that the Mesozoa are secondarily simplified: they are not 'living fossils' of the first multicellular animals, but rather are the degenerate descendants of more complex ancestors, probably some lophotrochozoan group, with the most likely one being the Platyhelminthes. Second, they are composed of two sub-groups that may not be particularly closely related. In other words, Mesozoa, like Protozoa, is probably a polyphyletic group.

The two groups that make up the Mesozoa are both parasitic, and this may well be the clue to their significantly simplified bodies. Since the two are probably not **sister-groups**, we will deal with them separately.

First, the Dicyemida[74] (also known as Rhombozoa). These are parasites of cephalopod molluscs (Fig. 18.6). Their bodies, when adult, are made up of between 8 and 40 cells belonging to about 6 cell types. They live in the renal sac (or 'kidney') of their host, which is fluid-filled. They attach to the wall of this sac via the calotte, a structure at the anterior end. Dicyemids have no organs as such at any stage of their life-cycles.

Second, the Orthonectida[75]. These are also parasites of marine animals, but are found in a wider variety of hosts. Compared with the restriction of dicyemids to a single class of a single **phylum** of host, orthonectids parasitise species belonging to at least four phyla: nemerteans, platyhelminthes, molluscs and echinoderms. Again, the body form is very simple. A typical adult orthonectid may have more cells than a dicyemid, but these belong only to two broad types – outer ciliated cells and inner reproductive cells. Again there are no organs.

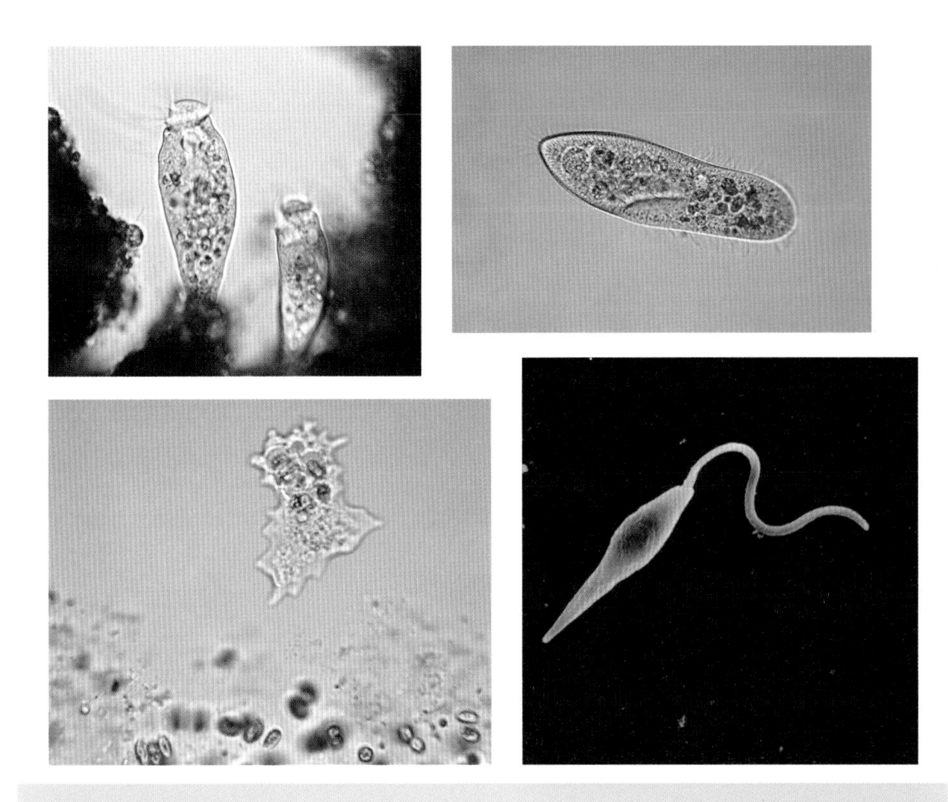

Figure 18.5 A few of the many kinds of 'Protozoa'. This group, like the 'Mesozoa', has now been abandoned. Protozoans are in fact diverse kinds of unicells and, since one of these, perhaps a choanoflagellate, gave rise to the animals, 'Protozoa' is certainly a paraphyletic group (and it is probably polyphyletic also).

The adoption of a parasitic life-style has often been linked with evolutionary decreases in the size of the adult and in its structural complexity. Certainly these things can often be linked, and the two 'mesozoan' groups may provide a good example of this linkage. However, care must be taken here in assuming general evolutionary rules where none exist.

It is clear that an organism living inside another may have access to a ready food supply that renders obsolete many food-acquiring and food-processing structures. Hence a decrease in complexity, at least of those structures, might be expected. Also, parasites generally need to be much smaller than their hosts. But there are also counter-arguments. Parasites often have very complex life-cycles, relating to the fact that while living within another organism can make food-acquisition easier, it may also make reproduction and dispersal harder.

It is important too, when considering the possible reduction in adult body size (and associated complexity of the adult) to make the right comparison. The dicyemids and orthonectids are all parasites, so no comparison can be made with free-living species within these groups. In nematodes and crustaceans, taxa that

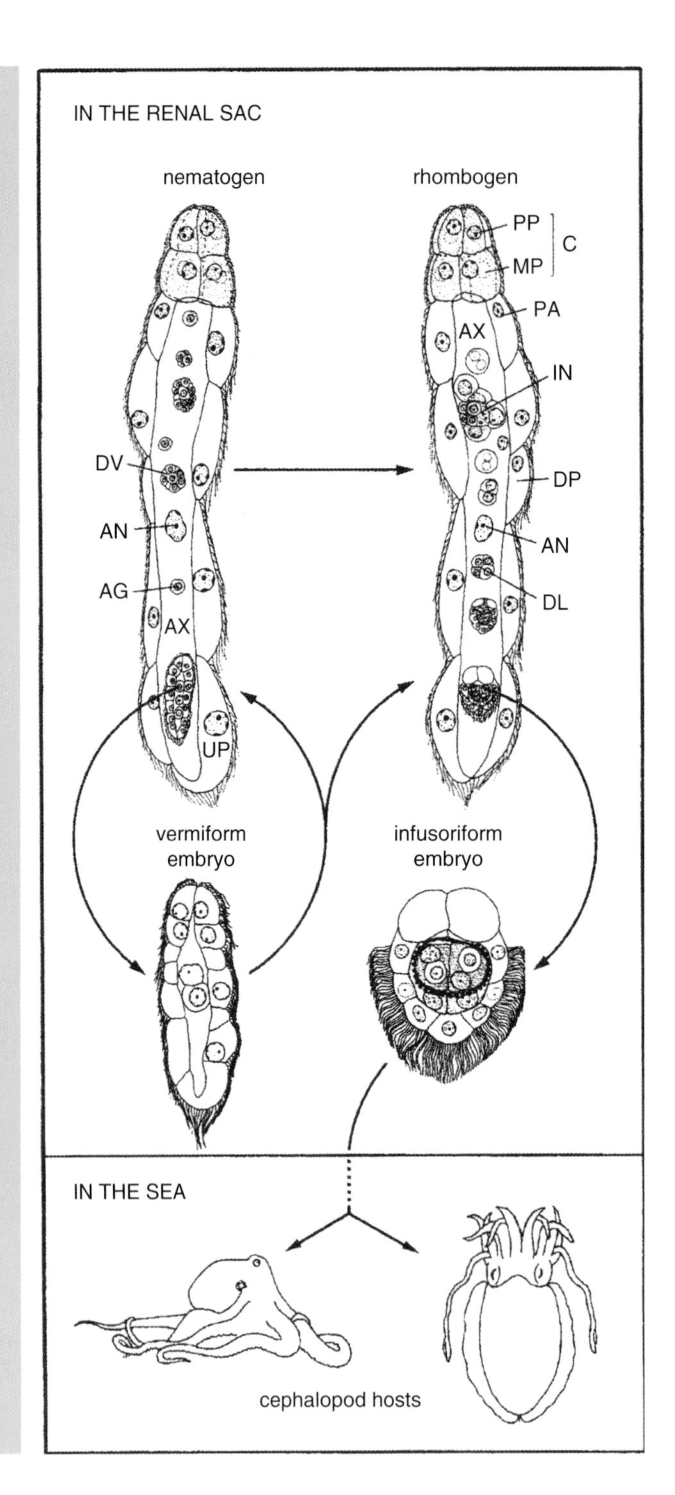

Figure 18.6 The life-cycle of a dicyemid, a parasite of cephalopod molluscs. AG, agamete; AN, axial cell nucleus; AX, axial cell; C, calotte; DI, developing infusiform embryo; DP, diapolar cell; DV, developing vermiform embryo; IN, infusorigen; MP, metapolar cell; PA, parapolar cell; PP, propolar cell; UP, uropolar cell. (From Furuya and Tsuneki, 2003, *Zool. Sci.*, **20**: 519–532.)

contain both parasitic and free-living forms, the available evidence suggests that species parasitic on other invertebrates are often not much different in body size from their free-living relatives, while those that parasitise vertebrates are significantly larger[76] (Fig. 18.7). Also, the relationship between size and complexity is a statistical rather than an absolute one.

Our conclusion, then, is four-fold. First, complexity decrease occurs in evolution; clear examples can be given of this. Second, there are many cases, not just a few. Third, parasitism can be a cause of such decreases. But, fourth, it does not *always* cause complexity decrease; and some other factors do. Examples include the loss of limbs (and hence a reduction in the number of types of parts) in snakes, associated with the adoption of a different form of movement; and the loss of eyes in many subterranean animals. To put it another way, parasitism is neither a necessary nor a sufficient condition for decreased complexity of adult structure in evolution.

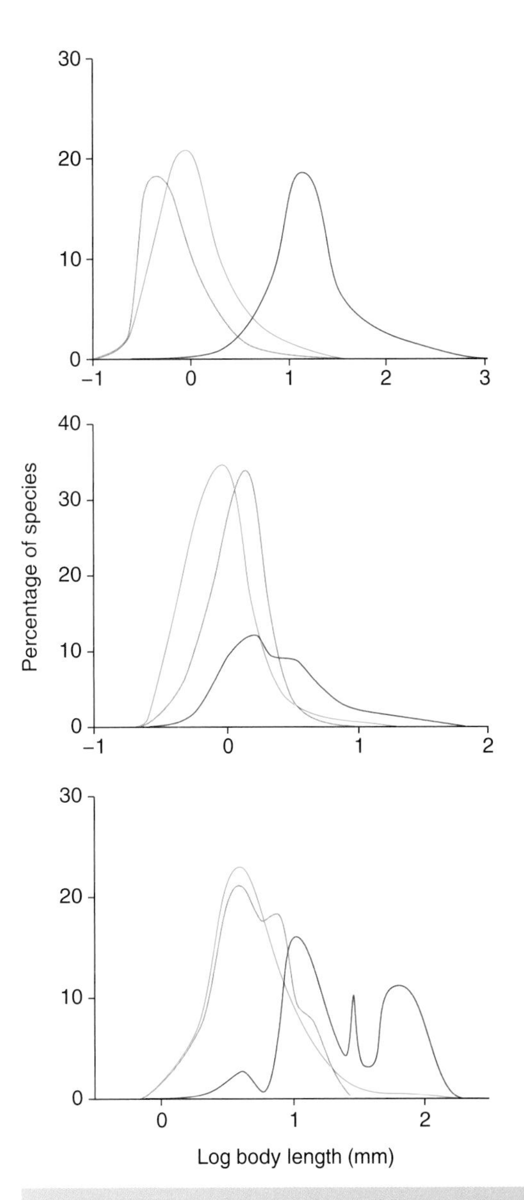

Figure 18.7 There are both parasitic and free-living forms of nematode worms (top), and of the crustacean groups Copepoda (centre) and Isopoda (bottom). In all three of these groups, the species parasitic on other invertebrates (orange) have roughly the same body size as their free-living counterparts (blue), whereas the species that parasitise vertebrates (red) are consistently larger (courtesy of Robert Poulin).

18.3 Increases in the Complexity of Adults

The preceding section fulfilled a very great need: the need to avoid inventing spurious evolutionary laws – in this case a 'law of increasing complexity'. As we have now seen very clearly, there is no such law. Indeed, not only does evolution not *always* produce increases in complexity, it may not even produce them *often*. But when it does produce them, they can, especially in the long term, be spectacular. They are significant features of the overall evolutionary process and deserve special attention.

The best way to consider increases in complexity is to think not in terms of **taxa**, as some authors have done[77] (Fig. 18.8) but rather in terms of lineages (Fig. 18.9). In Fig. 18.9a, we see the contrast between the pattern of complexity change in two lineages over the period of evolutionary time from the dawn of animals to the present day: the lineage leading to an extant orthonectid species and the lineage leading to humans. As can be seen, the patterns of change are very different – one goes up and then back down; the other goes upward all the way, though possibly towards a plateau. The lineages in the picture are greatly simplified (in reality they will exhibit much 'wobble'), but the contrast between them serves to

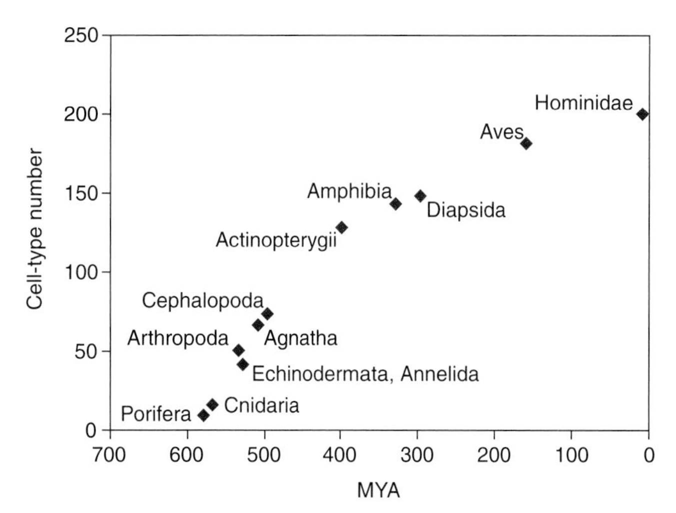

Figure 18.8 An apparent relationship between recentness of evolutionary origin of a taxon and its level of complexity: later arising groups appear here to be more complex. However, this picture conveys a false message because the level (or 'rank') of taxon varies dramatically. If the family Hominidae were replaced with its phylum (Chordata), then its origin would shift dramatically to the left (from near zero MYA to >500 MYA). Conversely, if Arthropoda were replaced with a constituent family, such as Drosophilidae, its origin would shift significantly to the right. (Reproduced with permission from Valentine, *et al.* 1994, *Paleobiology*, 20.)

make the point: different lineages can exhibit very different patterns of change in complexity over the same period of evolutionary time.

Because a single evolutionary lineage is a series of species joined end-to-end, with offshoot lineages being ignored, the level of complexity depicted for a lineage at any point in its history consists of a single value rather than an average. This should be true providing we consistently use the same life-stage (usually the adult), and providing also that the picture is not complicated by alternative sexes, morphs, castes, or other sub-groups within the species having significantly different levels of complexity.

If we wish to get some idea of how minimum, average and maximum complexity have changed over the course of animal evolution, we need to envisage a picture in which the individual histories of complexity change in *all* lineages of animals are included, rather than just two representative lineages. This second approach is shown in Fig. 18.9b. Although it is more of a 'thought experiment' than a plotting of actual data (most of which is unavailable), it yields three robust

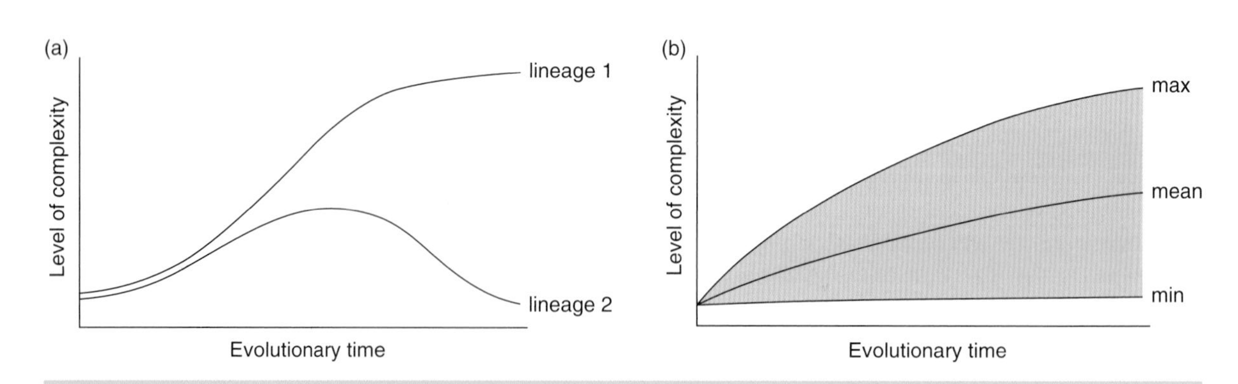

Figure 18.9 A lineage-based (rather than taxon-based) view of animal complexity. (a) Comparison of two lineages (e.g. human and orthonectid) taking very different evolutionary trajectories in terms of complexity change. (b) Patterns of change in maximum, minimum and mean complexity over evolutionary time.

conclusions – one each for minimum, average and maximum complexity – that are almost certain to remain intact as more data become available.

Minimum complexity has stayed approximately constant. If we include the choanoflagellate 'protozoans', as the most likely sister-group to the animal kingdom, then the minimum complexity of the combined animal/choanoflagellate clade has remained at a single cell-type since animals began, because many species of unicellular choanoflagellates are still alive today. Alternatively, if we exclude these, and consider only the animals (or 'metazoans') themselves, then minimum complexity has remained at the level of 'a few cells' and 'a few cell types'.

Maximum complexity, on the other hand, has risen dramatically, to trillions of cells, belonging to 200+ types. And average complexity has also risen, as it must do when the minimum remains the same and the maximum increases many-fold. Does this increase in average complexity constitute a general evolutionary law, despite the fact that no such law exists for individual lineages? This is currently an open question, or a hypothesis if you prefer to call it that. One way to begin to test it is to examine an independent data-set and to see if it exhibits the same trend.

The plant kingdom provides just such a data-set. Plants had unicellular ancestors, just as animals did – though in a different group than the flagellate protozoans. Plant evolution, like that of animals, has taken place over several hundred million years. Some individual lineages have massively increased in complexity over that time, such as one leading to a present-day **angiosperm** species, whether a herbaceous one, such as the 'model plant' *Arabidopsis thaliana*, or a woody one, such as the oak tree *Quercus robur*. Other lineages have remained very simple; and others have exhibited a variety of patterns of change, doubtless including equivalents of the 'mesozoa' in which complexity has first risen and then fallen. The composite of all these different patterns characterising different plant lineages is probably the same as that shown in Fig. 18.9b for animals. Indeed, since we have done equivalent 'thought experiments' in both kingdoms, Fig. 18.9b could equally well have been produced first by considering plants and then reproduced by considering animals – the opposite approach to the one we actually used – each approach is equally valid.

In both kingdoms, it is possible to discern, albeit very broadly, the nature of the body forms of intermediate complexity that those lineages reaching the greatest complexity passed through. This can be done by reference to outline **cladograms** of animal and plant evolution, onto which have been mapped the additional components that have arisen at or between the various lineage-splitting events (Fig. 18.10). Note that here the 'components' are at a higher level than cells. Note also the similarity to the main thrust of Chapter 11 ('Mapping Repatterning to Trees'). The current exercise is a special case of that mapping: here we are restricting our approach to those forms of repatterning that result in the production of new types of component parts.

As noted in Chapter 1, there are not just two kingdoms that include diverse multicellular forms; there are several. However, the others (i.e other than the plant and animal kingdoms) have achieved less impressive levels of maximum complexity. These include the fungi and the brown algae. Do these conform to the pattern shown in Fig. 18.9b? It is too early to say, for two very different reasons. First, less work has been done on these taxa. Second, given that evolution has not stopped, and that these groups *may* become more complex over time, they might show the animal/plant pattern in the future much more clearly than they do now. This point

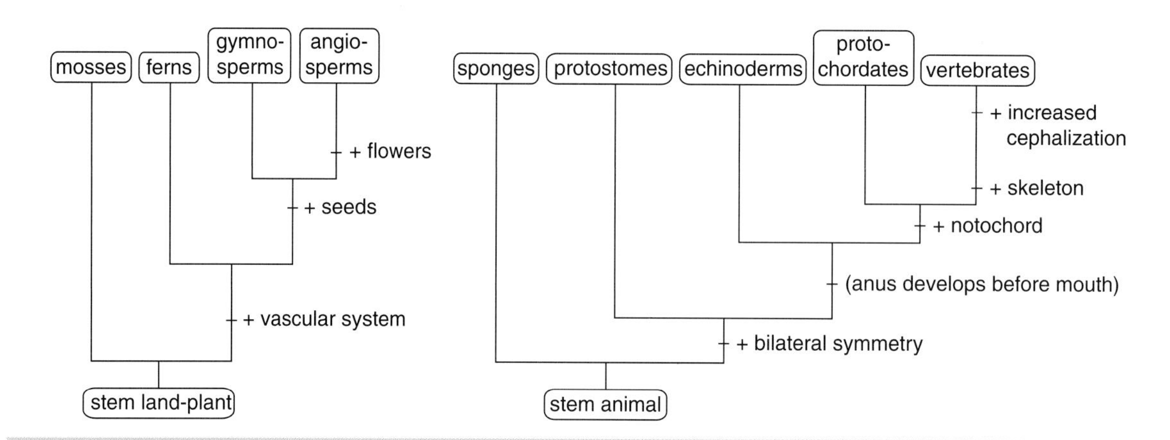

Figure 18.10 Outline cladograms of the plant and animal kingdoms, focusing in each case on the pattern of acquisition of additional features that led to increased complexity – specifically in vertebrates and angiosperms.

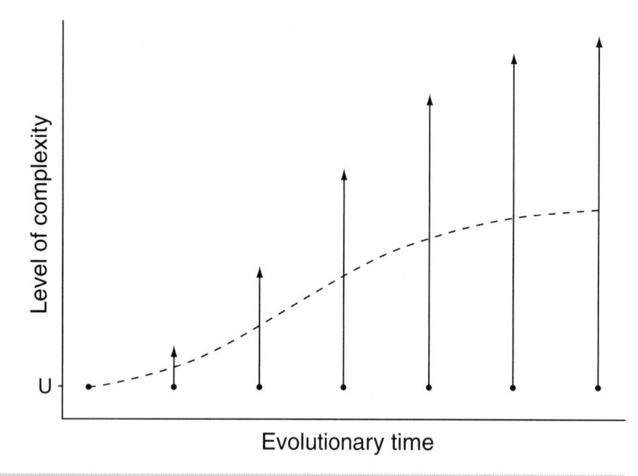

Figure 18.11 A developmentally explicit version of the pattern of complexity change in a lineage whose adults have consistently increased in complexity over evolutionary time (such as one of the two shown in Fig. 18.9a). U – unicellular stage.

serves to remind us that we are always judging the evolutionary process at a particular point in its history – the point at which we ourselves have arrived on the scene.

Finally, we return to the developmental (as opposed to the evolutionary) dimension of organismic complexity. Even for the relatively straightforward case of direct-developing animals in a lineage that has increased in complexity over evolutionary time, the picture shown in Fig. 18.9a is too simple, because it relates only to the complexity of the adult. An equivalent picture, taking into account the number of cell types present at all stages of the life-cycle, is shown in Fig. 18.11. Also, if we trace the lineage of a direct-developer such as a human back over a long enough period of time, we may find that not all the ancestors were also direct developers. This leads into the subject of the next section – evolutionary switches between direct and indirect development.

18.4 Changes in the Complexity of Life-histories

Just as there is no law of increasing complexity of adult body form in evolution, so too there is no law of increasing complexity of life-cycle. It is clear that direct development can evolve into its indirect equivalent (with one or more larval stages) and *vice versa*.

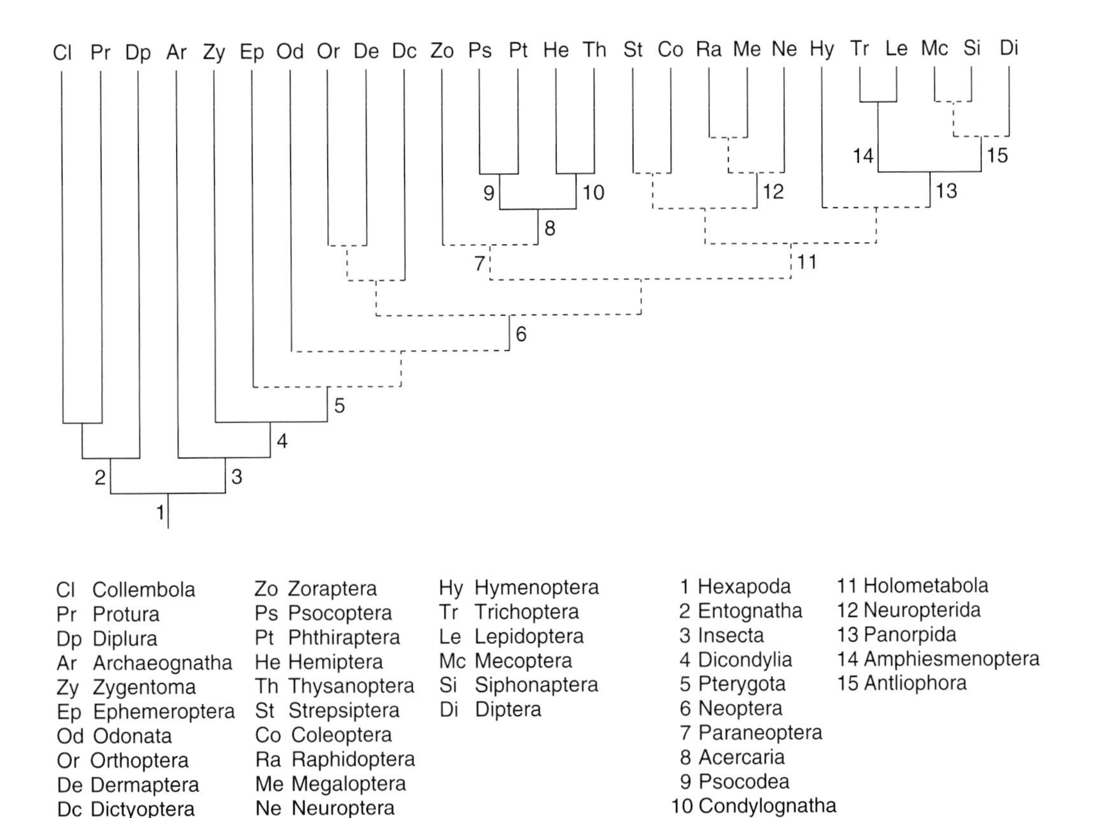

Figure 18.12 An outline phylogeny of most of the many orders of insects. Note that some groups are still difficult to place with certainty (dashed lines). Also note that holometabolous development (node 11), with complete metamorphosis between larva and adult, is a derived rather than primitive feature. (Modified from Kristensen, 1981, Ann. *Rev. Entomol.*, **26**: 135–157.)

We will approach this interesting topic by looking at three very different groups of animals, characterised by very different kinds of development: insects, snails and frogs.

The insects comprise about 80 per cent of all known animal species. They are grouped into about 30 orders, one of which, the Coleoptera (beetles) has about the same number of species as the entire plant kingdom. A **phylogeny** of the insect orders[78] is shown in Fig. 18.12. While this will no doubt change somewhat as more data on particular groups emerge and as more phylogenetic analyses are carried out, certain features are likely to be robust, including the one of most interest to us here – the phylogenetic position of the group known as the **Holometabola**. This is the group that includes those insect orders (e.g. Lepidoptera, Diptera) in which development is indirect in the sense that there is complete **metamorphosis**, via a pupal stage, from a larva to a (very different) adult. In other, **hemimetabolous**, insects, such as the Orthoptera (locusts, crickets and grasshoppers), development consists of a series of nymphal stages that grow gradually more like the adult as the series progresses.

It is clear from Fig. 18.12 that holometabolous development, with its characteristic larval stage that is so unlike the adult, is a derived condition within the insects; hemimetabolous development, which is similar to 'typical' direct development but with the process being punctuated by a series of moulting events, is the ancestral situation. So the kind of evolutionary change that occurred at the point of origin of the Holometabola was direct-to-indirect (or D-to-I). However, given the wide acceptance of the **Pancrustacea** hypothesis, in which insects are either the sister-group of the Crustacea, or even a **clade** within this latter group, the origin of insects was probably associated with a transition of the opposite kind (I-to-D). This is because most crustaceans are marine, and most marine species of crustaceans have indirect development, albeit of a very different kind to that of the Holometabola – for example, many have a **nauplius larva**, as we saw in Chapter 8.

In fact, whether development is direct or not in a particular animal species is much influenced by whether that species is marine, freshwater or terrestrial. Not only did the first insects make the I-to-D transition, but so too did the terrestrial woodlice (crustacean order Isopoda), whose marine ancestors very likely had a dispersive larval stage, as do their **extant** equivalents.

This habitat link to type of development is also seen in the molluscan class Gastropoda (snails and slugs). The original gastropods were marine animals: fossil gastropods are known from the Cambrian (>500 MYA), when a terrestrial fauna was yet to appear. Their extant equivalents – sea snails such as the periwinkle *Littorina* – have dispersive **trochophore larvae** (Fig. 18.13). Indeed, as we saw earlier (Chapter 11), one of the three main branches of the Bilateria – the Lophotrochozoa, which includes the molluscs and annelids – is named after two features, one of which is the trochophore larva that characterises the development of so many members of that group.

In contrast to marine snails, freshwater and land snails typically have direct development, so the switch in this case is I-to-D. For example, freshwater snails of the genus *Lymnaea* lay egg-masses on the undersides of the leaves of aquatic plants, and these hatch directly into miniature snails (Fig. 18.13) – there is no larval stage. The same is true of land snails of the genus *Cepaea*, which lay their eggs in the soil. Again, these hatch directly into miniature versions of the adult – there is no larva.

Frogs and toads (amphibian order **Anura**) are typically indirect developers, with the familiar freshwater tadpole stage usually intervening, in the life-cycle, between egg and adult. However, some groups of frogs have undergone a switch from I-to-D, and the eggs hatch directly into tiny frogs – there is no tadpole stage (Fig. 18.14). This switch has clearly happened many times, as direct-developing frogs are found in several different anuran families. In many cases it is associated with a terrestrial life-style, in which case the eggs may be carried on the backs of their mothers or laid in damp vegetation. The hatching froglets are near-perfect miniature frogs in some species; in others they have tails that are quickly lost. However, some groups of frogs that have remained aquatic also have direct-developing member species. So the link between an aquatic ecology and a larval stage in the life-cycle is, like so many links in evolution, a statistical rather than an absolute one.

Figure 18.13 Most gastropods have a larval stage known as a trochophore. However, groups which left the sea and became terrestrial (or freshwater) have lost this dispersive larval stage. Left – the marine *Littorina*, which has a trochophore larva. Right – the freshwater *Lymnaea*, which has direct-developing eggs. (Trochophore picture from Jackson *et al*. 2007, *BMC Evolutionary Biology*, **7**: 160.)

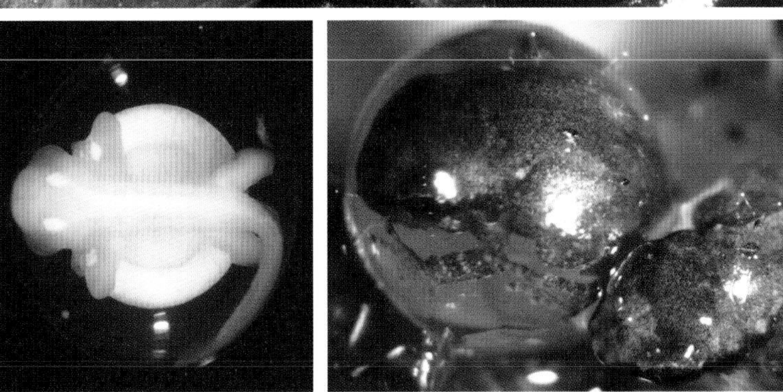

Figure 18.14 An example of a direct-developing frog, whose life-cycle has no tadpole stage. Such an evolutionary switch from indirect to direct development has occurred independently (i.e. convergently) in different groups of frogs (photographs courtesy of Gerhard Schlosser.)

18.5 Complexity at the Molecular Level

It seems reasonable to expect that complexity of adult bodies and of life-cycles would be associated with, and indeed caused by, underlying complexity at the molecular level. This is because to make the additional organs or life-stages that go with increased complexity at the morphological level, it is necessary to have the additional cell types of which those organs/stages are made (some of which will be unique in each case); and different cell-types are usually characterised by different combinations of expressed genes. So complexity at morphological and genetic/molecular levels should correspond.

However, there are many nuances in this correspondence. For example, the most complex organisms (as measured by the number of different body parts) do not necessarily have the most complex **genomes** (as measured by the number of different genes). It is true that **eukaryote** genome sizes are often in the range 10,000 to 30,000, which puts them well above the typical number of genes in a prokaryote genome (often in the range 1000 to 5000). But comparisons among different eukaryotes do not yield an equally clear picture. The numbers of genes in a nematode with less than 1000 cells and a human with trillions of cells are not so very different.

What is the resolution of this paradox? Indeed, is it a paradox at all? Perhaps not, for the following reason. As noted above, a cell-type is characterised by a particular combination of *expressed* genes. So, in order to have many different cell types it is necessary to have many different such combinations. However, even with a small genome size (for a multicellular eukaryote) of 10,000 genes, the number of possible combinations of expressed genes is almost infinite.

This line of reasoning takes us back to our discussion of how to define complexity in Section 18.1. We saw there that ideally a measure of complexity should include not just the number of different types of component parts but also the number of interactions between them. For pragmatic reasons, we have temporarily lost sight of these interactions: it is easier to count the number of cell types in an animal than to count (or even identify) all the interactions involved. But for a gene to go from silent to expressed (or from off to on, if you prefer) it must receive some signal, such as those examined in Chapter 3. Signals between genes, and between cells, constitute interactions. Perhaps this is where we should look to find the molecular-level explanation for changes in morphological complexity?

Genes and their products are regulated at several levels: transcription into pre-mRNA; **intron** splicing, yielding mature mRNA; translation into protein; and post-translational modification of the protein concerned. All of these may be important in generating the number of different cell types in a particular species, and hence its level of complexity.

One particular possibility that has aroused much interest recently[79] is that the complexity of an organism may be associated with the complexity of its set of micro-RNAs (miRNAs), small molecules (between 20 and 25 nucleotides long) that are now recognised to play a key role in the regulation of gene expression. The number of families of miRNAs has increased in most or all animal groups over time (Fig. 18.15). Note, however, that although a major expansion of the number of miRNA families occurred at the origin of the **Bilateria**, only a very small expansion in family number occurred at the origin of the chordates.

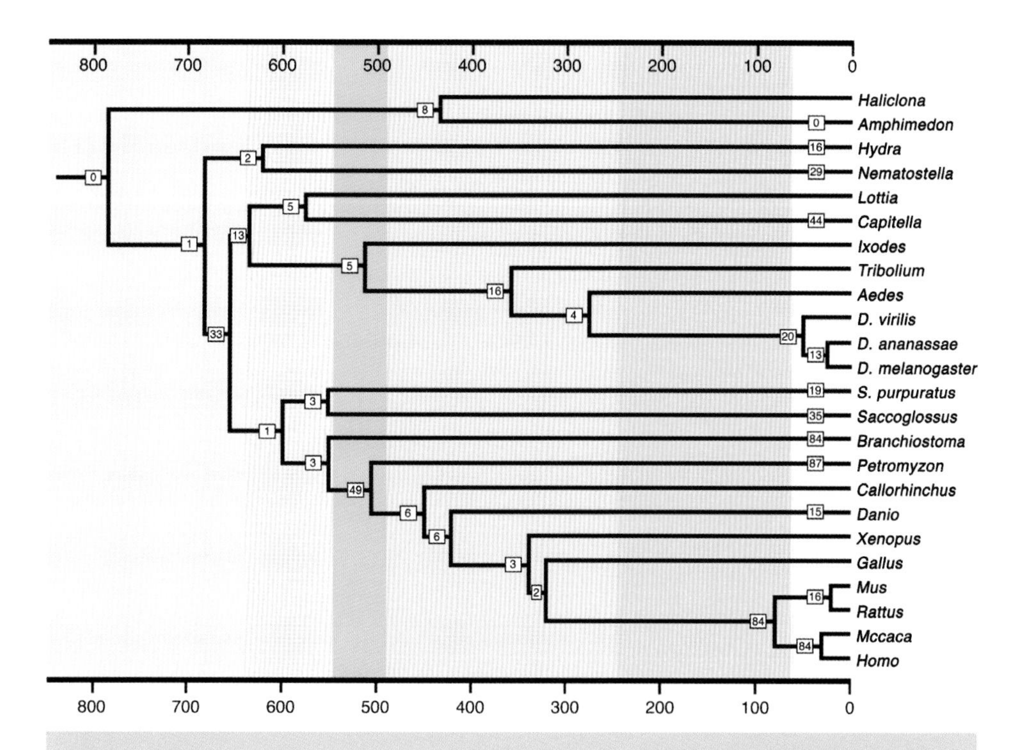

Figure 18.15 A mapping of the number of gains of families of micro-RNA onto some of the main branches of the animal kingdom. Note the major increase in the number of these families at the point of origin of the Bilateria. The main animal groups are delineated as follows: Sponges: *Haliclona* and *Amphimedon*; Cnidarians: *Hydra* and *Nematostella*; Bilaterian animals: the rest. The Bilateria include the protosomes (*Lottia* to *Drosophila melanogaster*) and the deuterostomes (*Strongylocentrotus purpuratus* to *Homo*). (From Peterson et al. 2009, *BioEssays*, **31**: 736–747.)

The problem in trying to explain the level of complexity of an animal's body in terms of counting its number of miRNA families may be the same problem that we encountered in trying to explain complexity in terms of counting its number of genes. In both cases, a simple count of the number of components is too simple. What matters is the pattern of interactions. The ways in which miRNAs help to regulate gene expression are only beginning to be known. Some types of miRNA may end up regulating far more genes than others. Until we have a more complete understanding of the regulation of gene expression, both by miRNAs and by other factors, the molecular basis of increases in morphological complexity must remain a matter of conjecture.

Despite this, it is possible to end on a positive rather than negative note. Complexity increase in both the animal and plant kingdoms is explicable largely in terms of the process often called **duplication and divergence**, though a better (but less alliterative) name for it would be 'replication and diversification'. The basic idea is that, when you have more than you need of something (be it a segment, a pair of limbs, a leaf, a cell-type, a gene[80], or a particular kind of miRNA), the role of natural selection, which in other contexts is often negative so that existing functions are preserved, can become more positive in that the 'extra' copies can diversify in function to do new things.

The importance of the process of replication-and-diversification in enabling complexity to increase *in some lineages* is almost certainly true; it is just that we do not yet know where best to apply it. To put it another way, we do not know the relative importance of the process at different levels, especially molecular ones, such as genes versus miRNAs versus types of regulatory interactions.

SUGGESTIONS FOR FURTHER READING

For a critical review of the idea of evolutionary increase in animal complexity, see:

McShea, D.W. 1996 Metazoan complexity and evolution: is there a trend? *Evolution*, **50**: 477–492.

For a discussion of the evolution of complexity with a particular focus on the link between complexity and body size, see:

Bonner, J.T. 1988 *The Evolution of Complexity by Means of Natural Selection*. Princeton University Press, Princeton, NJ.

For a 'popular science' approach to the evolution of complexity, including a refutation of the notion that organismic complexity is too great to be accounted for by natural processes (and so must be accounted for instead by 'intelligent design'), see:

Arthur, W. 2006 *Creatures of Accident: The Rise of the Animal Kingdom*. Hill & Wang, New York.

PART IV

Conclusions

'The aim of science is to seek the simplest explanations of complex facts.'
Alfred North Whitehead, *The Concept of Nature*, 1920

We are nearing the end of our journey through the ways in which the development of animals and plants evolves, and the impact that increased understanding of these, in the era of evo-devo, has had on evolutionary theory in general. The only things that now remain, forming the foci of the last two chapters, are as follows:

Chapter 19: Many individual chapters have dealt with particular concepts in the evolution of development – for example, heterochrony in Chapter 7, adaptation in Chapter 12 and gene co-option in Chapter 15. But ultimately we need to be able to see the connections between all these different concepts. That is the focus of this chapter, which also includes connections to other concepts not covered elsewhere in the book.

Chapter 20: Here, some important areas for future research are considered, with a focus on the short-term and mid-term future. Key messages include: the use of new techniques; the need for evo-devo to spread out to include function as well as form (i.e. to include physiological as well as morphological evolution); the need to spread out phylogenetically, including, as a key part of that, much-improved coverage of the evolution of multicellular creatures outside the animal and plant kingdoms; and the desirability of placing the whole endeavour of understanding the evolution of development in a philosophical context.

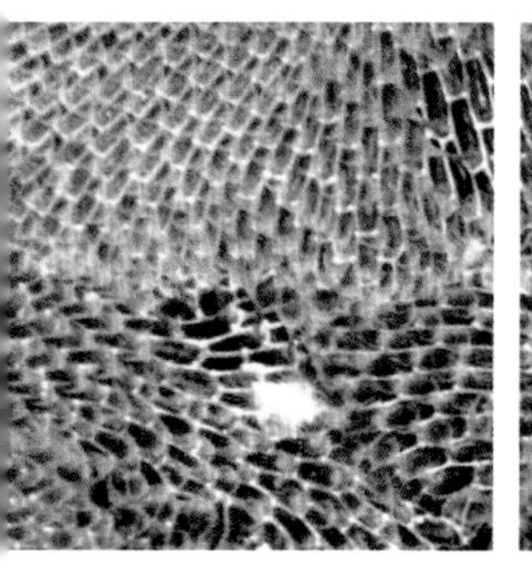
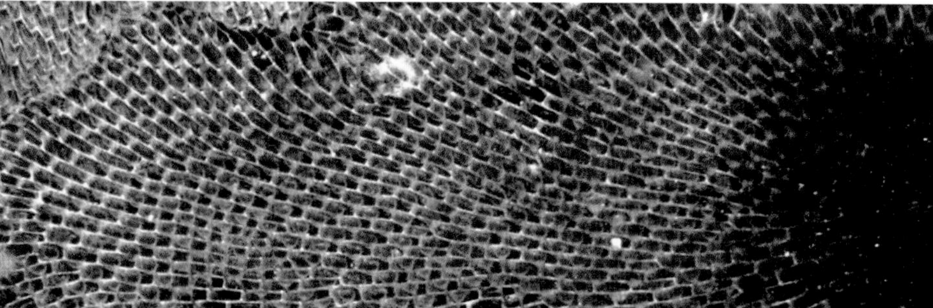

CHAPTER 19
Key Concepts and Connections

19.1 Introduction: From Original Idea to Mature Scientific Discipline

A new branch of science tends to follow a certain sequence. It starts from a new realisation, a new insight, a new method, a new result or, usually, from a combination of these. It then expands its data-set, using the new technique(s), usually in conjunction with older ones. New concepts emerge at various points in this process, but often in a rather disjointed way. The new concepts gradually link up, giving a coherent conceptual structure to the new discipline. The concepts often shift from being qualitative to quantitative. More data are acquired, which can test the validity of the new concepts (since these are originally just hypotheses). If the concepts are confirmed rather than rejected, the new field is new no more – it has matured and, if you like, 'solidified'.

Of course, this pattern is not rigid and, looking across different scientific disciplines, we can see many variations of it. Darwin's insight into the potential power and generality of natural selection as an evolutionary mechanism was united with Mendelian genetics (stemming from a different insight: into how inheritance operates in families). Some evidence was acquired for the action of natural selection in the wild.

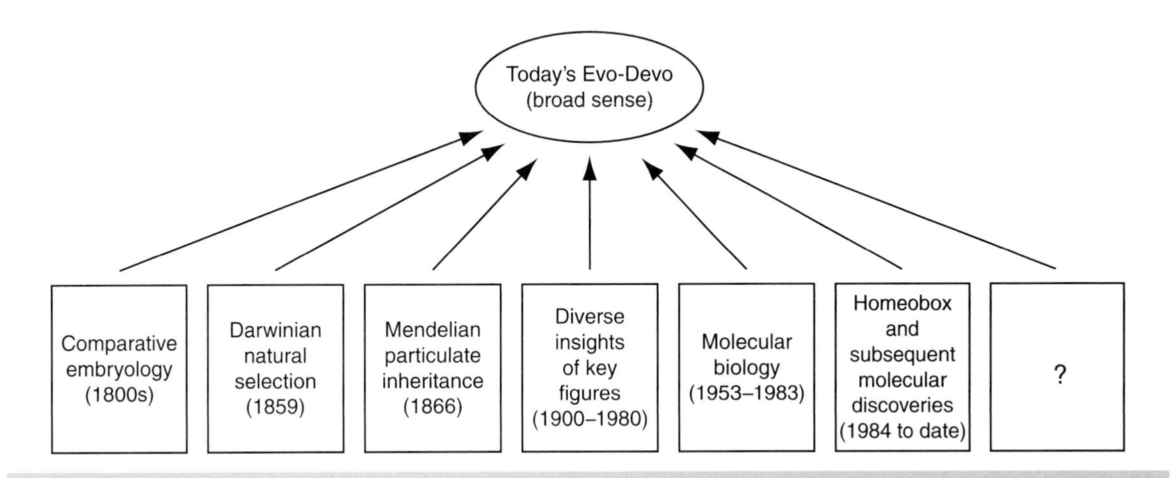

Figure 19.1 Some of the main foundations on which the modern science of evo-devo has been able to build. The extra box with the question-mark is there to make readers ask themselves whether there are any significant foundations of evo-devo, apart from those given in the other six boxes.

Then the combined Darwinian/Mendelian approach was quantified, producing population genetics theory. Much more evidence for selection in natural populations was built up, and the theory gradually solidified into the **modern synthesis** of the mid-20th century. So, here we see a combination of two quite separate insights at the outset rather than just one.

What of our new, developmental approach to evolution, or 'evo-devo' in the broader usage of that phrase? How has it progressed from its birth in the early 1980s to its current, more established (but still not mature) form?

This question is more difficult to answer, for the following reason. Prior to Darwin and Mendel, there was little to build on in terms of mechanisms of evolution or inheritance. Previously-conceived mechanisms, such as Lamarck's 'inheritance of acquired characters' were widely regarded as wrong. In contrast, evo-devo has had much to build on, including not only Darwinian natural selection and Mendelian genetics but also a wide range of other endeavours, ranging from the comparative embryology of von Baer and Haeckel dating from the 1800s to the whole field of molecular genetics that erupted out of the famous Watson-Crick paper[1] on the structure of DNA in 1953 (Fig. 19.1).

The current chapter is organised in a way that takes account of this different situation, in which a new discipline (evo-devo) is being built on top of many 'old' (i.e. longer-established) ones in the same general area. Thus, in Section 19.3, we will look both at the concepts that were established before 1980 as part of general evolutionary theory and also at those newer concepts that have stemmed directly or indirectly from evo-devo studies. We will examine:

a) the extent to which the earlier concepts constitute a coherent body of theory;

b) the nature of the major gap in classical theory;

c) the extent to which the newer evo-devo concepts constitute a coherent body of theory; and

d) the extent to which the new concepts are compatible with the older ones, and hence the extent to which the integration of the two groups of concepts forms a coherent, combined theory.

Before such examinations though, it is important to recap the main points that have arisen in the preceding 18 chapters about evo-devo in particular and about evolutionary biology more generally. After all, much has been discussed and our memories are imperfect. So Section 19.2 constitutes a reminder of the book's main points, in the form of a series of brief statements, grouped so as to correspond to the book's four Parts, and hence in chronological sequence of being presented.

19.2 A List of the Book's Main Points, and the Emergence of Key Concepts

Here, then, is a recap of the book's main points. Because we use particular words or phrases to denote important concepts, in the following list such words and phrases are highlighted in 'bold blue'. The blue helps to distinguish them from Glossary entries, which are suspended during this list. These words, and the concepts they represent, will form key issues for discussion in Section 19.3, where concepts will be cross-related. Also, the identification of key words and phrases will help to maintain continuity between the 'list approach' of the present section and the more complex 'integrative approach' of the next.

* * *

Part I

- Understanding the evolution of multicellular creatures requires a developmental approach.
- Evolution of adult form can only be achieved by **developmental repatterning**.
- The same is true of the evolution of juvenile and larval forms.
- Two-dimensional evolutionary trees are shorthand versions of **three-dimensional trees**.
- Nineteenth-century comparative embryology provides a foundation for evo-devo.
- Both **recapitulation** and **von Baerian divergence** occur, but neither are 'laws'.
- The period from 1900 to about 1980 saw a scattering of diverse approaches to the evolution of development.

- Evo-devo was born in the period from 1979 to 1984 – from **spandrels** to **homeobox**.
- The phrase **'evo-devo'** can be interpreted in broad and narrow ways.
- It is necessary to search for general principles among many, diverse 'facts'.
- Development can be analysed in terms of the **behaviour of cells.**
- Cells, and populations of cells, respond to many types of **signals**.
- **Gene expression**, signalling and cell behaviour produce a **developmental network**.
- Such networks become changed in evolution: this is developmental repatterning.
- All evolution takes place in the context of natural **populations** inhabiting particular environments.
- Repatterning affects individuals; **natural selection** affects populations.

Part II

- Genes, including developmental ones, get altered accidentally by **mutation**.
- This sometimes leads to developmental repatterning of mutant individuals.
- In cases where repatterning leads to enhanced **fitness**, it will spread by selection.
- There is no one pre-eminent form of repatterning.
- Rather, four types of repatterning can occur, all of them important in evolution.
- The four types of repatterning involve changes in time, place, amount or type.
- These are, respectively, **heterochrony**, **heterotopy**, **heterometry** and **heterotypy**.
- The type of repatterning occurring at molecular and higher levels may be different.
- Most cases of repatterning involve a mixture of two or more of the four types.
- Instances of repatterning can be mapped to evolutionary trees.
- This provides a basis for understanding **homology** and **homoplasy** of development.

Part III

- Natural selection is a major determinant of evolution's **direction**.
- Sometimes, selection is associated with fitness in particular **environments**.
- In other cases, however, selection is associated with internal **coadaptation**.
- There is a spectrum from the first of these scenarios to the second.
- A good way to picture this spectrum is the **trans-environment fitness profile**.
- Another good way to picture selection is the **adaptive** (or fitness) **landscape**.
- The selective advantage conferred by a new character state may change over time.

- When this happens, the initial **adaptation** gives way to later **exaptation**.
- Evolution's direction is not determined by selection alone.
- It is also determined by the **structure of development** and of **developmental variation**.
- This structure can be referred to as **developmental bias** or **constraint**.
- The former is a better, more inclusive term for structured developmental variation.
- Bias can then be divided into more and less probable variants: **drive** and constraint.
- In **micro-evolution**, developmental bias is related to **character correlation**.
- Bias can be thought of as the mechanistic basis for correlation of characters.
- Bias may be especially important in cases of the origin of **evolutionary novelties**.
- Ultimately, developmental bias can be traced to the genetic level.
- Some forms of evolutionary change of developmental genes are **'easier'** than others.
- **Co-option** of an existing gene for a new function is easier than producing a new gene.
- Changes in **regulatory regions** may be easier than changes in **coding regions**.
- Ease of evolution can be called **evolvability**, though care is needed in defining this.
- Developmental trajectories are often determined partly by the environment.
- Environmental variation can produce discrete, meristic or continuous responses.
- In all these cases, the phenomenon can be described as **plasticity**.
- Plastic variation is not inherited but a pattern of plastic response may be.
- Thus the environment can play a role in determining the structure of variation.
- The process of **speciation** (the 'origin of species') is reasonably well understood.
- Higher levels of evolution – origins of novelties and **body plans** – are not.
- That is, **mega-evolution** may involve more than accumulated microevolution.
- In other words, evolution may not be completely **scale-independent**.
- The jury is still out on this question, which is of the utmost importance.
- There is no law of increasing complexity in evolution.
- Rather, some lineages increase in **complexity**, others decrease, and others 'stand still'.
- However, **increases in complexity** are worthy of special attention.
- The link between molecular and morphological complexity is not yet clear.

* * *

As can be seen, each key word or phrase that has been deemed important enough to emphasise in bold has been emphasised in that way only on its first usage – otherwise the amount of bold text would become confusing rather than helpful. If we count the number of bold words/phrases in the list, we find about 50 of them. While this large number may seem daunting, it will reduce greatly if we group related concepts into a much smaller number of categories, as we shortly will.

Note, however, that this list is a recap of the book, and the book itself (like most others) is not an exhaustive account of its field. So before we begin to condense concepts into groups, we should ask the question: are there some additional concepts that should be included in our quest? A good way to answer this question is to look at the contents list of the book *Keywords and Concepts in Evolutionary Developmental Biology* (see Suggestions for Further Reading and Fig. 19.2). This book is a multi-authored collection of brief accounts of key concepts, and so is effectively an encyclopedia on the subject. It has about 50 entries, many of which are the same as, or are closely related to, the concepts given in bold in the above list. However, some are additional to the above. We will ignore most of the additional ones either because they are very general – such as 'Animal Phyla', or because they are concepts to do with development itself rather than its evolution, or because they relate to specific genes or specific body parts, which have not been included in our list anyhow.

Figure 19.2 Picture of the important edited collection of works entitled *Keywords and Concepts in Evolutionary Developmental Biology*, published in 2003 (courtesy of Harvard University Press.)

Ignoring those still leaves two important evo-devo concepts, **modularity** and the **phylotypic stage**, which have escaped detailed attention in the preceding 18 chapters. These two are in fact related, and a digression is now appropriate to explain them.

In Chapter 2, we noted von Baer's 'laws', which stated that a comparison of the developmental trajectories of two animals from different taxa (e.g. families, orders or classes) often reveals a pattern of early similarity giving way to later differences – this was pictured in Fig. 2.1. However, it was pointed out in the 1980s that the very earliest developmental stages are often rather different, and that comparisons of the developmental trajectories of different animals often yield a variant of the von Baerian pattern: the point of maximum similarity, or phylotypic stage, occurs near, but not at, the beginning of the developmental process[2] (Fig. 19.3). Examples of this stage are the insect germ band and the vertebrate pharyngula. The term 'phylotypic stage' was first used in relation to the insect germ band by the German biologist Klaus Sander[3].

Because the pattern of similarity and difference in the developmental trajectories of two animals is frequently as shown in Fig. 19.3, rather than as stated by von Baer, it is often now referred to as the developmental egg-timer or hour-glass. These are

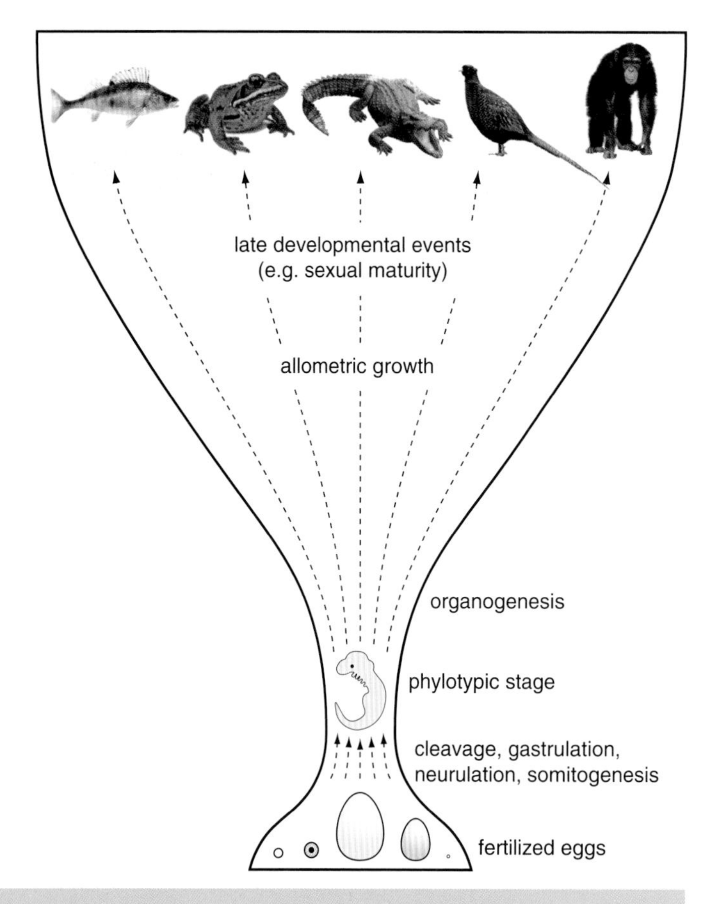

Figure 19.3 The developmental 'egg-timer' or 'hour-glass' model. This depicts the relationship often found when comparing the degree of similarity of the developmental trajectories of two or more animals. It is a modification of the pattern proposed by von Baer in 1828 (developmental divergence: early similarity giving way to later diversity) to take account of the fact that the very earliest stages are often quite different, so the point of maximal similarity is near, but not at, the start of development.

memorable phrases, but ones that may serve to mislead. An egg-timer or hourglass is symmetrical, with its point of constriction at the centre, whereas its developmental equivalent is highly asymmetrical, with the point of 'constriction' (i.e. maximum similarity) much closer to the start of development than to the end. Perhaps 'developmental wineglass' would be a better term.

There is a relationship between the phylotypic stage and the concept of modularity[4] in development. Modularity is one of those nice terms that are largely self-explanatory (Fig. 19.4). In the development of an animal or plant, we can often recognise quasi-autonomous modules, such as the **primordia** of segments, limbs, leaves or flowers. In evolution, it may be possible for these modules to change without disturbing the development of other parts of the body. Thus a highly modularised developmental system may be more readily 'evolvable' than a less modularised one.

But any particular developmental system may vary in its degree of modularity from start to finish. This is where modularity connects with the phylotypic stage. Consider, for example, the comparison of the developmental trajectories of a mammal and a bird. The earliest stages differ considerably because of their very different eggs. The latest stages also differ considerably, partly because of the fact that each is composed of modules (e.g. limb buds) that can to some extent evolve independently of the rest of the body, as noted above. But in between, at the phylotypic stage, the initial egg environment has less effect than before. Also, the embryo is developing as an integrated whole, before modules appear, and hence is more resistant to selectively-driven change, because a change that is helpful in one place may hinder in another, thus rendering an overall fitness increase harder to achieve.

So, the addition of modularity and the phylotypic stage (with its associated egg-timer pattern) completes our earlier list of key concepts in evo-devo and evolutionary biology more generally. We must now try to convert a long list into a small number of key concept groups. That is the purpose of the next section, whose four subsections correspond to the four questions identified towards the end of Section 18.1, all of which relate to interconnections. It is by looking at patterns of strong versus weak interconnections that we can see groups of concepts emerging, and can glimpse the overall structure of a new, combined body of evolutionary theory.

To anticipate the result of this exercise, it is worth watching out for three things: (i) the remaining centrality of natural selection at the population level; (ii) the emerging centrality of developmental repatterning (and its four sub-categories) at the individual level; and (iii) the complementarity of these two concepts, but also the possibility of tension between them.

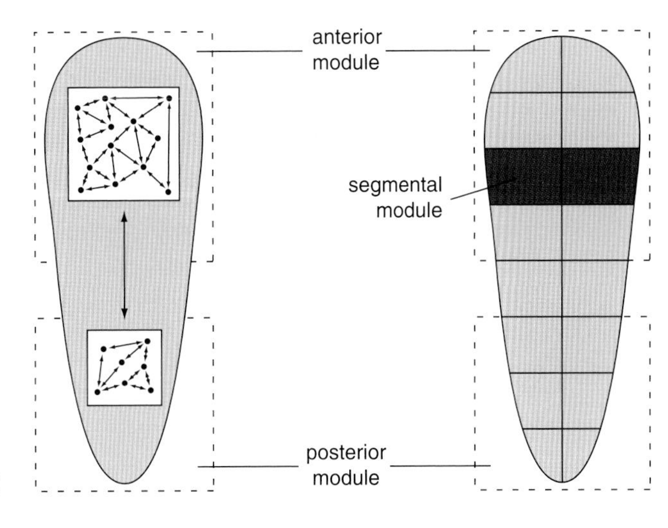

Figure 19.4 Diagrammatic representation of the idea of developmental 'modules' – parts of the embryo within each of which the density of interactions is high while between them it is low. Also, a contrast in the degree of modularity between two embryos: two modules (LHS) versus many (RHS). The bottom diagrams are cross-sections. DLM – dorsal left module; DRM – dorsal right module; VLM – ventral left module; VRM – ventral right module.

19.3 How do They Inter-Connect?

Connections among Classical Evolutionary Concepts

This is the easy bit; well, in relative terms anyhow. Classical evolutionary biology tends to be divided into two domains – the study of *pattern* and *process*. Views differ as to which of these it is logical to treat first. Here we will follow the cladist approach of 'pattern before process', though it may not affect the outcome much (or at all) if the alternative approach were taken.

Darwin's main success, in addition to proposing an important mechanism that drives evolutionary change, was in gaining the acceptance of most of the scientific community of the late 19th century for the *fact* of evolution – or 'descent with modification' as he called it. Given that the living world is now interpreted in this way by the *entire* scientific community (in my view 'creation scientists' are not a part of this community), the relationships between its constituent species can be represented in the form of an evolutionary tree or **phylogeny**, of which a **cladogram** is an abstract version – omitting, for example, actual time units and retaining just a time *sequence* of **lineage**-splitting events.

Given the pattern of origins of, and changes in, particular characters on the tree, we can describe their relationships as **homologous**[5] or **homoplastic**, as noted in Chapter 11. For example, it is accepted that the legs of mice and dinosaurs are homologous, whereas the legs of mice and the limbs of starfish (whether the main 'arms' or the smaller tube-feet on the undersides of these arms) are homoplastic. Equally, the tails of the zebrafish and the trout are homologous, whereas the relationship between either of these and the whale's tale is homoplastic. Homology results from the divergence of two or more lineages from a common ancestor that possessed the character under consideration. Homoplasy, in contrast, results from the **convergence** of two or more lineages towards a similar morphological feature that was not shared by their LCA.

Speciation can be thought of as a link between pattern and process, because it represents both the splitting of branches on an evolutionary tree (pattern) and also the end point of the micro-evolutionary process of selectively driven divergence of (usually) geographically discrete populations of a parental species. This process started with **natural selection** acting on variation, ultimately produced by **mutation**, in the standard way: enhanced fitness of particular variants in particular environments producing local **adaptation**. The variation acted upon by selection may be structured in various ways, including the correlation of characters, in which case selection on one character will produce responses not just in that one but in others too.

Of course, there are many simplifications in the above picture. Trees are inappropriate to describe situations in which two parental species combine to give a hybrid **daughter species** by allopolyploidy (reticulate evolution); and they are also insufficient to capture the process of horizontal gene transfer. Populations diverge not just through natural selection but through random processes too – the founder effect, **genetic drift** and population bottlenecks. Sometimes, natural selection acts to maintain the *status quo* (stabilising selection) rather than acting as an agent of change. And so on. But none of these complications threatens the coherence of the overall combined pattern-and-process picture of conventional evolutionary theory. This body of theory is not wrong; but it is seriously incomplete – this is where evo-devo comes in.

The Major Gap in Classical Theory

On the process (as opposed to pattern) side, classical evolutionary theory has an obvious gap that is acknowledged by all. It deals with the interplay between mutation and selection, thereby connecting the level of the gene with the level of the population, apparently jumping over the intermediate level – that of the individual organism.

But this is too extreme a criticism. After all, many classic evolutionary studies, whether on the pigmentation patterns of moths, the beak shapes of birds, or other characters in other taxa, do include reference to the individual's **phenotype** – indeed it is the phenotype that natural selection acts on, and that relates to local adaptation. So we need to refine and restrict our criticism of classical theory. It deals with phenotypes, but usually those of adults, and usually without much reference to their fourth, developmental, dimension. Thus the key process that changes genes – mutation – is included in the theory, as is the key process that changes populations – selection; but the key process that changes developing individuals – repatterning – is absent.

Evo-devo is all about developmental repatterning in evolution. If any of the 'key concepts' of evo-devo is logically prior to the rest, this is it. So we need to ask – do the various strands of evo-devo connect up to form a coherent theory of developmental repatterning? And if so, how does this intermesh with the classical theory – does it neatly 'plug the gap' or is the relationship between the two bodies of theory more complicated?

Connections of Evo-Devo Concepts with Each Other

The following discussion of 'developmental repatterning theory' is linked to Fig. 19.5, which shows how various concepts relate to each other. We will discuss the various aspects of repatterning in the top-down order in which they are arranged in the figure.

Causes

These can be both genetic and environmental. The genetic origin of developmental repatterning lies ultimately in mutation. However, given that offspring are not identical to their parents, much of the short-term repatterning that is observed in a population results from reshuffling of the **standing variation** that is an accumulated product of many past mutations. This reshuffling includes both the 'independent assortment' of different chromosomes and intra-chromosome recombination caused by the process of 'crossing-over'.

Environmentally-induced repatterning of development varies in its importance among taxa but occurs to some degree in all of them. Examples that we have covered include the production of winged and wingless forms of an insect species in response to variation in population density and the alteration of patterns of growth in a plant species in response to variation in the level of shade. Although such plastic responses are not heritable, a particular pattern of response may be – this is the developmental **reaction norm** or DRN. Various kinds of DRN evolution are known, including cases of increased genetic control (**genetic assimilation**).

Levels

Developmental repatterning can be observed at the molecular/cellular level or at the level of the overall **developmental trajectory**, including the adult phenotype and the phenotypes of all other stages of the life-cycle. Ultimately, changes at the latter level result from changes at the former one. Evo-devo has given rise to important new concepts relating to repatterning at the molecular level, notably gene **co-option**. It has also served to re-emphasise the importance of some older molecular concepts, such as gene **duplication and divergence**.

Types

Developmental entities or processes can be changed in time, space, amount or type. These are called, respectively, **heterochrony, heterotopy, heterometry** and **heterotypy**. These four types can be observed at the different levels discussed above; but are perhaps easiest to picture at the level of a single **developmental gene**, such as one that makes a **transcription factor**. In this case, they relate to the timing, spatial positioning, amount and type of its product. The first three of

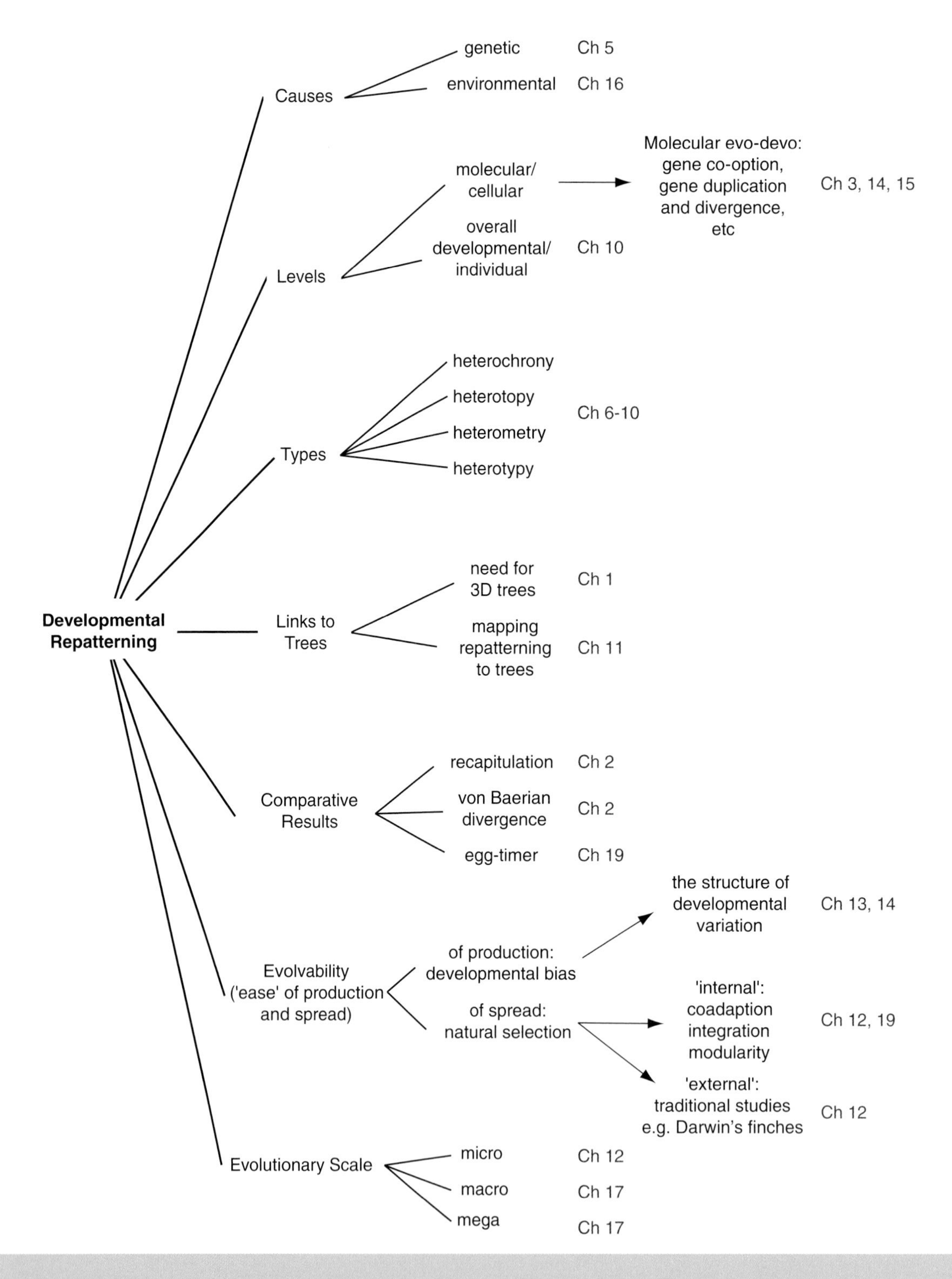

Figure 19.5 Developmental repatterning: outline of the relationships between this key concept and a wide range of others. Links to chapters are also given.

these are likely to arise from mutational changes in the regulatory regions of the gene, while the last – heterotypy – arises from changes in the coding region.

In practice, most instances of repatterning are complex – they consist of a combination of two or more of the four fundamental types. This is true at any one level, but it is even more apparent when different levels from molecular to morphological are considered, because any one type of change at the former level can cause a different type of change at the latter.

Links to Trees

There are two rather different connections between evo-devo and evolutionary trees. The first is that taking a developmental approach to evolution results in a realisation that the two-dimensional trees normally used are 'shorthand' trees. Trees should have an additional developmental dimension, so they should be three-dimensional, as was pictured in Fig. 1.3.

The second link between evo-devo and trees is the mapping of repatterning to trees; this reveals whether particular developmental features are related in a homologous or homoplastic way among particular **taxa**. Such mapping is of course not new, but the emphasis is now shifted away from the adult phenotype to the underlying developmental processes.

Comparative results

Given a long enough period of evolutionary time, say tens of millions of years, different lineages can diverge enough from their LCA that comparisons of the developmental trajectories of these lineages reveals certain types of pattern. These include the **von Baerian** pattern of developmental divergence, the variant on that called the egg-timer pattern (Fig. 19.1), and the Haeckelian pattern called **recapitulation**. The last of these is most likely to occur when one of the lineages being compared has undergone much complexification while the other has remained simple. However, neither these nor any other comparative patterns have the status of a law, as used to be thought. Evolution is a messy process; its results, in terms of long-term developmental repatterning, are many and varied.

Evolvability

This phrase relates to the relative 'ease' of different kinds of developmental repatterning, in terms of both the production of the relevant variants (at the individual level) and their spread (at the population level). The key concepts are very different for these two levels.

At the level of initial production, the most crucial concept is that of **developmental bias**. This refers to the fact that both the developmental process of an individual and the variation in that process in the population are *structured* rather than amorphous. This structure makes repatterning in some directions easier than others. In other words, in statistical terms, evolution is driven in some directions and constrained from going in other directions by the availability (or lack of it) of the appropriate developmental variation.

If variation is available in several directions, then the main force determining which way evolution goes, in a particular lineage, is natural selection. However, while this may be of the usually-envisaged sort, involving adaptation to local environmental conditions, it may also be of another sort that is related to internal **coadaptation**. In fact, there is a range of possibilities between these two extremes. This range is represented by the **trans-environment fitness profile** that we examined in Chapter 12. At the population level, stabilising selection can be thought of as countering evolvability, just as **developmental constraint** can be thought of as having that role at the individual level.

Evolutionary Scale

Evolution is sometimes divided into three broad categories – micro, macro and mega[6]. These refer to: (i) intraspecific changes; (ii) changes occurring in association with speciation and the proliferation of closely-related species within a genus or family; and (iii) changes occurring only very rarely in evolution, producing **novelties** and **body plans**; these are often associated with the appearance of new high-level taxa, such as orders, classes and **phyla**.

It is generally accepted that **macro-evolution** is explicable in terms of accumulated **micro-evolutionary** divergence and reproductive isolation. However, it is still unclear whether the **mega-evolutionary** changes resulting in the origin of novel body forms are explicable simply in terms of a similar 'accumulation effect' operating over an even longer time-scale. Another possibility is that mega-evolution involves highly unusual kinds of developmental variation that are normally absent from populations on a short-term or mid-term basis. Put another way, we do not yet know if evolution is a 'scale-independent' process. This is one of the most important outstanding questions for evolutionary theory as a whole.

* * *

It is clear from the above account, i.e. "Connections of evo-devo concepts with each other", that there is now a coherent body of theory related to how evolution occurs at the individual level. But how does this relate to classical, population-based evolutionary theory? This question has been partly answered 'in passing' already; but it becomes the main focus of attention in the next section.

The Relationship between Evo-Devo and Classical Evolutionary Theory

This relationship has two parts: definite complementarity and possible tension. We will deal with them in that order.

The complementarity is readily apparent. There was a clear need for the major gap in classical evolutionary theory to be filled: that is, the gap between changes at the gene level (caused by mutation) and changes at the population level (caused by natural selection, along with stochastic processes such as genetic drift). Because evo-devo, in the broad interpretation of that term, covers all levels from gene expression through cell-cell signalling pathways and larger-scale developmental processes up to the level of the individual and the life-cycle, it neatly plugs the gap in classical theory. Indeed, it can be argued that there is no need to see the two as separate any more.

But what of the possible tension between them? This is focused, as we have already seen, on two inter-related issues. These are: (i) the extent to which developmental bias can help to determine the direction of evolution; and (ii) the question of whether the origins of novelties and body plans require special explanations, over and above accumulated micro-evolutionary changes in development.

The question of the importance of developmental bias can be addressed on various evolutionary scales. At the micro-evolutionary scale, it can be addressed through artificial selection experiments, as we saw in Chapter 13 in relation to the case study[7] of the butterfly *Bicyclus anynana*. Interestingly, this study showed that bias is important in affecting the result of selection in some cases but not others, thus emphasising the open nature of this question.

The importance of developmental bias in the origin of novelties and body plans may well be much greater than its importance in micro-evolution, whatever that latter importance turns out to be. However, the origins of novelties and body plans are less amenable than micro-evolutionary processes to an experimental approach. Progress on this front therefore depends on a *comparative* approach instead. For example, consider the comparison of the relative frequencies of origin of shelled (rare) and legless (common) tetrapods that we discussed in Chapter 13. The striking contrast in the frequency of origins of the two body forms may be explicable:

1 in terms of the relative frequency with which each is selectively favoured;
2 in terms of the frequency with which the appropriate developmental variation is available; or
3 in terms of an equal combination of the two.

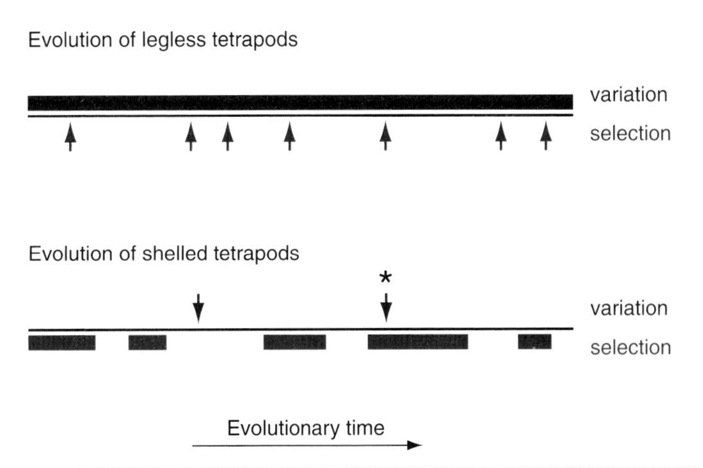

Figure 19.6 What is the reason for the very different relative frequencies of origin of *legless* and *shelled* tetrapods? This diagram represents a hypothesis in which the difference is due in large measure to differences in the availability of certain kinds of variation rather than to differences in the frequency of these two types of body form being adaptive. Bars – continuity of occurrence (of variation or selection); arrows – fleeting occurrence. Asterisk on lower picture indicates the only co-occurrence of variation and selection, and hence the only time that an evolutionary change toward the shelled condition will occur.

Shown in Fig. 19.6 is a diagram of a hypothesis mid-way between **1** and **2** – both availability of variation and natural selection are involved, but the former is more important than the latter in explaining the rarity of origins of shelled tetrapods. As with the possible role of developmental bias in micro-evolution, its possibly *greater* role in the origin of novelties and body plans remains an open, and fascinating, question.

SUGGESTIONS FOR FURTHER READING

An important compendium of key concepts in evo-devo and its overlap with general evolutionary theory is, as mentioned in Section 19.2:

Hall, B.K. and Olson, W.M. (eds) 2003 *Keywords and Concepts in Evolutionary Developmental Biology*. Harvard University Press, Cambridge, MA.

Another important compendium, this time placing evo-devo in a historical context, is:

Laubichler, M.D. and Maienschein, J. (eds) 2007 *From Embryology to Evo-Devo: A History of Developmental Evolution*. MIT Press, Cambridge, MA.

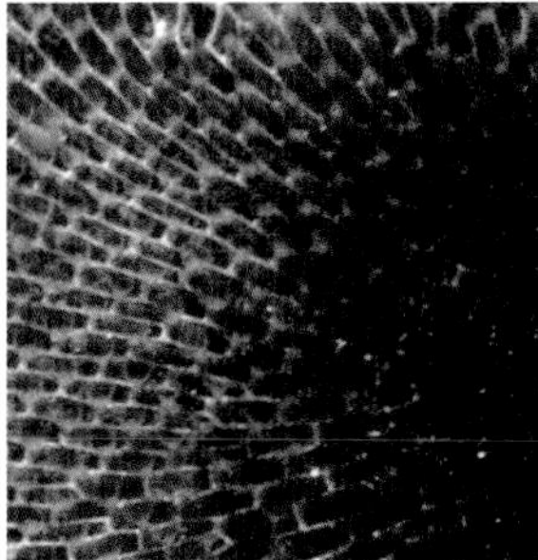

CHAPTER 20
Prospects

20.1 Introduction: From the Present into the Future

The previous chapter outlined the current state of play in evo-devo and its relationship with more general evolutionary theory. This one looks ahead into the future and asks the question: what can we expect in terms of progress over the next couple of decades? What are the key questions that we should be addressing? What are going to be the main research areas, and what approaches and techniques will they employ?

This final chapter is deliberately brief and focused on the short-term rather than long-term future. No-one can accurately predict the future of a branch of science; and the further forward we try to predict, the less accurate will our forecasts be.

20.2 Molecular Evo-Devo

The next two decades are likely to witness a continuation of the flow of new data in this area that started in the 1980s. This is where new 'facts' are emerging at a high rate. It can be compared with the flow of new facts in comparative embryology in the 19th century. Then, the *morphology* of the sequence of embryonic stages in animals belonging to several taxa was the focus of attention. Now, the *molecular basis* of

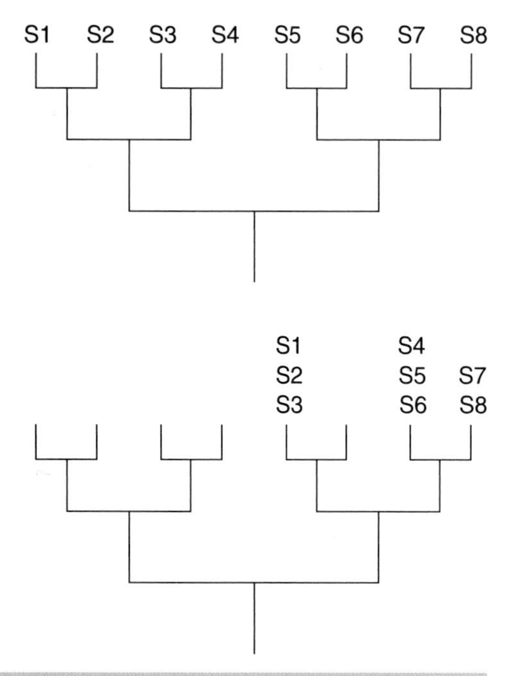

Figure 20.1 Good (top) and bad (bottom) patterns of taxon sampling. In the former, the points at which we have samples (S) are evenly spaced out across the tree, enabling conclusions to be drawn about how phylogenetically general something is (e.g. a particular developmental mechanism). In the latter, the sampling points are too clustered, leaving many branches of the tree unsampled.

development has become the main focus; but the comparative approach remains, and indeed is widening in its scope.

The way in which studies of the molecular basis of development are distributed on the overall tree of life, or some particular part of it, is referred to as **taxon** sampling. The better the taxon sampling, the more complete will be our understanding of how **developmental repatterning** has occurred in evolution. Figure 20.1 illustrates extreme examples of 'good' and 'bad' taxon sampling. Figure 20.2 shows the patterns of taxon sampling involved in (i) comparison of the current main model organisms only and (ii) ideal comparisons for the future.

It can be seen that one of our future goals is the inclusion of those animal **phyla** about which almost nothing is currently known. The same general goal applies to plants: comparative studies need to be extended to poorly-known high-level taxa, especially outside the **angiosperms**.

As an aside, it is worth noting that there is an interesting difference between the structures of the animal and plant kingdoms. Vertebrates are the most complex animals (using the criteria of Chapter 18, especially number of cell types); and angiosperms are the most complex plants. Although both are large **clades** with many thousands of species, they represent very different fractions of the total number of **extant** species in their respective kingdoms. Vertebrates, with about 50,000 known species, represent less than 5 per cent of the animal kingdom, with its million-plus known species overall. Angiosperms, with more than 250,000 species, make up more than 80 per cent of the plant kingdom (Fig. 20.3). Why this should be so is unclear. It may simply be an indication of the arbitrary, as opposed to goal-directed, nature of evolution.

There is another feature of 'good' taxon sampling that should be noted in relation to Fig. 20.1. As well as *extensive* sampling, to include the under-studied higher taxa, it is also important to do some *intensive* sampling. This has two components. One is to sample many individuals within a species to assess the amount and type of molecular intraspecific variation in **developmental genes**, since this is what selection works on. The other is to determine whether closely-related species (say in the same genus, family or order) have similar molecular-developmental processes, as might be expected. There have already been some surprises in this area. One we noted earlier is that the use of the protein Bicoid as an anterior determinant in the Drosophilidae turns out to be shared only by a few other families of flies. In contrast, all arthropods use the Engrailed protein as part of their system of determining the antero-posterior polarity of their segments.

This particular pair of contrasting findings is interesting, because there was initially an expectation that it would be the earliest-acting genes and proteins in development that would be most evolutionarily conserved. Clearly this is not so, at least in general – though

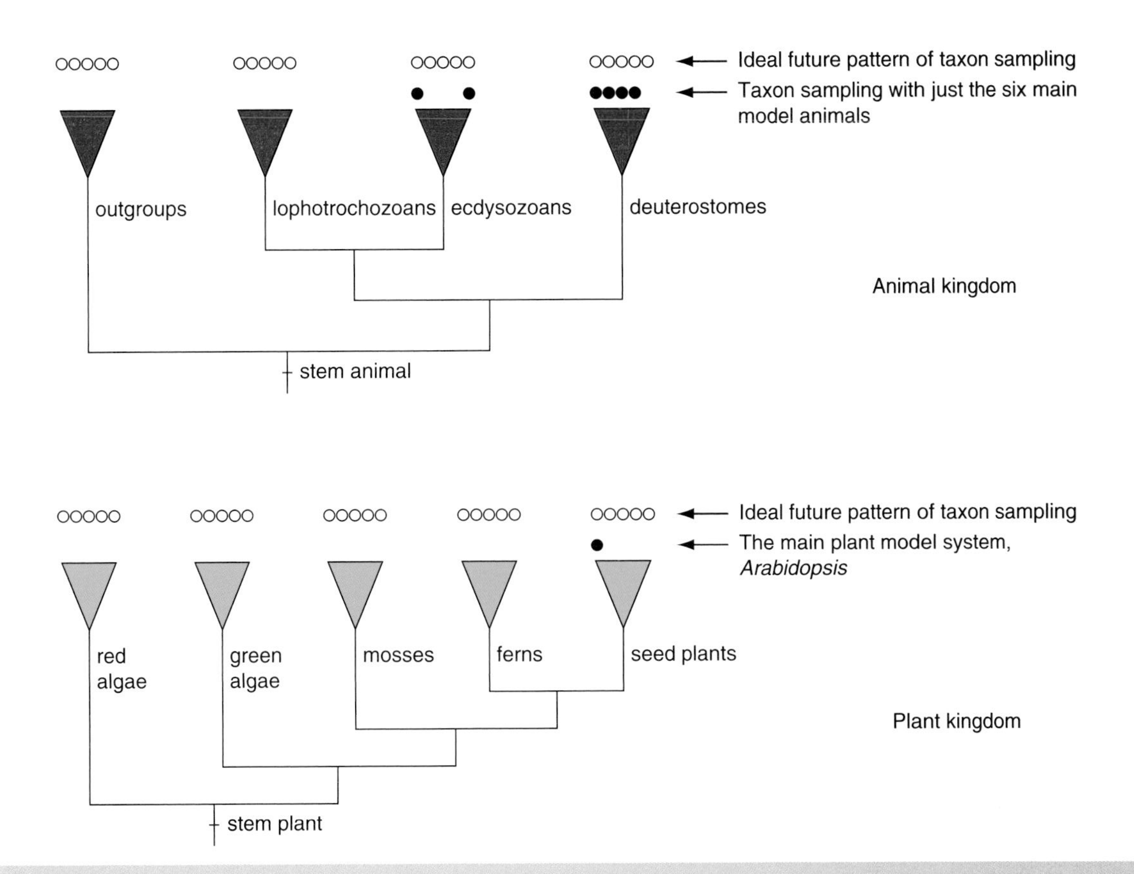

Figure 20.2 Two patterns of taxon sampling on the overall trees of animal and plant life. The pattern that is possible with the seven main 'model organisms' of developmental genetics is shown by filled circles; the desirable pattern is shown by open circles.

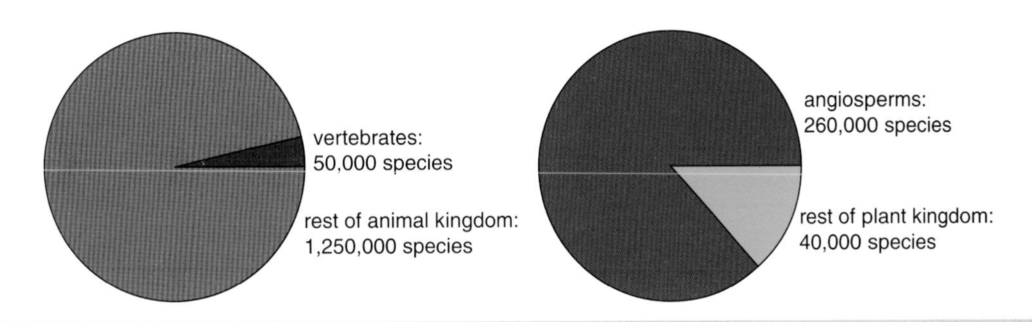

Figure 20.3 The different structures of the animal and plant kingdoms in relation to the proportion of each represented by the group characterised by the most complex body form: vertebrates and angiosperms, respectively.

no doubt there are some instances where it is indeed the case. Perhaps this result, like the comparison of vertebrate and angiosperm species numbers, is just an indicator of the arbitrariness of evolution. But alternatively, it may be telling us something more specific. For example, there is much to the making of a segment that is downstream of the determination of the antero-posterior polarity of the segment **primordium**. The molecular basis of the later stages is not yet well known. Perhaps when it becomes so, we will see that the high degree of conservation of the **segment polarity** genes is a molecular example of the morphological 'egg-timer' pattern noted in the previous chapter.

The next issue we need to address is that of exactly which details of the molecular basis of development should be focused on at any one taxon sampling point; that is, in any particular species. Earlier studies were in a sense descriptive. The DNA sequences of developmental genes were determined, with some interesting findings, notably the **homeobox** sequence that became one of the main foundations of molecular evo-devo. And the spatio-temporal **expression patterns** of the genes concerned were described. This basic descriptive information is still lacking for most developmental genes in most species, and there is much yet to be done using tried-and-tested techniques – for example, *in situ* hybridisation.

Notwithstanding the continued usefulness of 'older' techniques, especially when combined with good taxon sampling strategies, it is important to recognise that newer techniques have been developed, enabling a more experimental approach. Often, these involve gene knock-down, using, for example, various types of RNA interference (RNAi) systems[8]. The importance of these is that knowing where and when a gene is expressed, which is what *in situ* hybridisation tells us, is not the same as knowing what its product does, though it may allow some tentative inferences about its function to be made. But the actual function of the product is revealed more certainly by investigating what happens when it is absent. Given the rate of proliferation of new molecular techniques over the last two decades, we can look forward to the appearance of many more in the next two.

It has been argued that, despite the many advances of comparative developmental genetics, its accumulating body of data is qualitative rather than quantitative. In other words, even if the function of a gene product is understood, this understanding is restricted to the proteins it interacts with and whether the interaction is positive or negative – that is, activating or inhibiting. But what of the quantitative dynamics of a molecular developmental network, and the evolutionary changes that must take place in this as **lineages** diverge? There have been a few pioneering studies in this area[9] (Fig. 20.4), but it is still largely unexplored territory.

With regard to the general concepts that have arisen directly from molecular evo-devo studies, one of the most important is that of gene **co-option**. As we have seen in earlier chapters, the expectation that the independent (i.e. convergent) evolutionary 'invention' of morphological entities such as segments, limbs and eyes would be associated with entirely different molecular genetic bases has proved to be wrong. Perhaps the use of an ever-increasing array of techniques on an ever-expanding sample of taxa will lead to clarification of the reasons for the high frequency of gene co-option as a form of developmental repatterning at the molecular level. This in turn will help to clarify the nature of the difference between **homology** and **homoplasy**, which once seemed relatively simple but is now known to be rather complex[10].

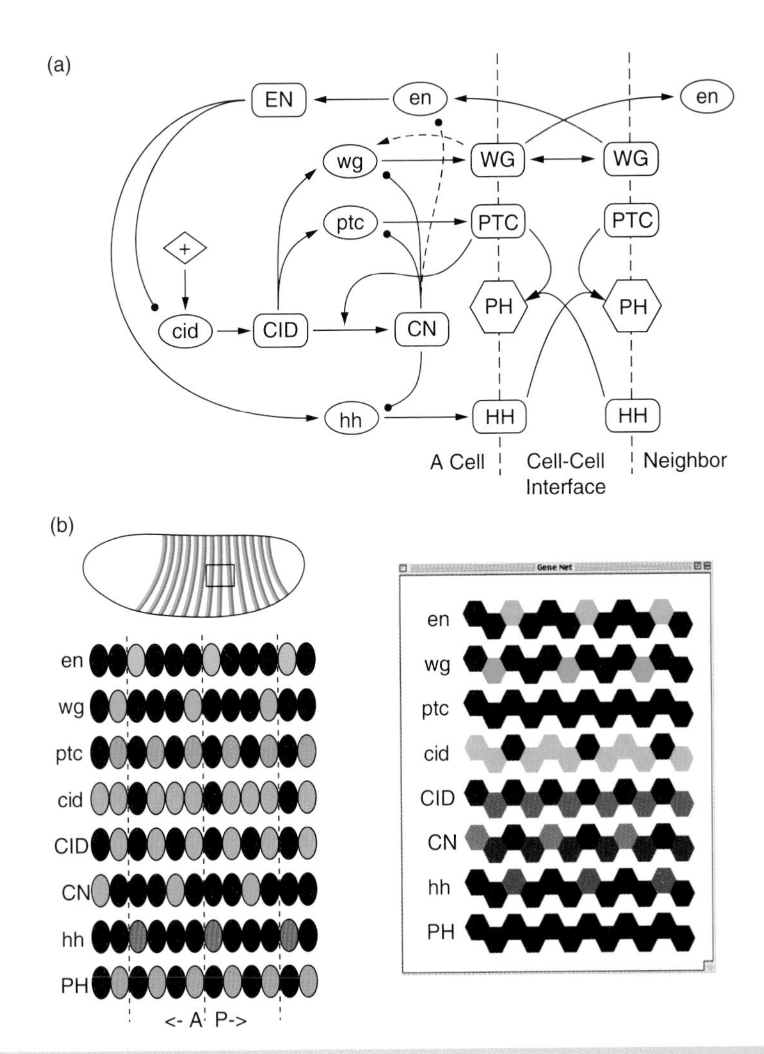

Figure 20.4 Quantification of the robustness (versus sensitivity) of a particular developmental system – the segment polarity module in *Drosophila* – to changes in the linkages between genes/proteins. (a): Interaction network. Solid arrows – linkages included in all computer simulations; dashed arrows – linkages only included in some. EN – Engrailed; WG – Wingless; HH – Hedgehog; PTC – Patched; PH – Patched-Hedgehog complex; CID – Cubitus interruptus full-length protein; CN – Cubitus interruptus repressor fragment. Upper case – proteins; lower case – mRNAs. (b): The actual pattern of expression of several segment polarity genes (left) was relatively easy to reproduce in most of its details (right), through simulations that included the solid and dashed lines; but not by simulations based on the solid lines only. Each colour represents a different segment polarity gene/protein. Lines ending in a black circle have the same meaning as arrows ending in a bar in other figures – i.e. they are inhibitory signals. (Redrawn with permission from von Dassow *et al*. 2000 *Nature*, **406**.)

20.3 Integrative Evo-Devo and General Evolutionary Theory

There are three kinds of integration to be considered here. These are: (i) integration of evo-devo concepts with each other; (ii) integration of evo-devo concepts with relevant facts; and (iii) integration between evo-devo and other areas of evolutionary theory.

With regard to the first of these, it is clear that many concepts can already be inter-related. We saw in the previous chapter that there is probably a relationship between the concept of developmental **modularity** and that of the **phylotypic stage**. Specifically, the least modular phase of a **developmental trajectory** may be the least **evolvable**, and so the most highly conserved. A major goal for the future is to improve the overall connectedness of evo-devo concepts, so that an equivalent of this book's Chapter 19 written, say, in the year 2030, will reveal more connections than we can see at present, and an improved understanding of those that we can already see. In other words, the discipline will have advanced significantly in its conceptual structure.

With regard to the second form of integration, between concepts and data, this might at first seem an odd link to highlight because ultimately these things surely *must* be connected: after all, concepts arise out of scrutiny of data of various kinds. While this is true in general, it is easy to find examples, in evo-devo, of poor connections between concepts and data. This was certainly true in the early days of the concept of **developmental constraint** (or bias). The famous treatment of this in terms of **spandrels** by Richard Lewontin and Stephen Jay Gould[11] was inspiring but very abstract. Even now, the number of cases in which the concept of **developmental bias** can be linked with a specific data-set in a meaningful way is rather small. Those that we have considered in earlier chapters include mammalian neck vertebrae, centipede segments and butterfly wing-spots. It is much to be hoped that the number of cases increases significantly in the future.

It is important to emphasise that 'data' should include not just the development of *extant* organisms and its molecular basis but also *palaeontological* data, especially where it includes information on the development of the organisms concerned. Both embryonic and post-embryonic development should be included here. Palaeontological information on the former is still very limited and open to different interpretations, as we noted in Chapter 17; but future fossil finds will inevitably include, albeit not often, more information on the **embryogenesis** of extinct forms. Regarding the latter (post-embryonic development), there are already some impressive palaeontological studies, such as those on the patterns of post-embryonic development of trilobites[12] (Fig. 20.5). These, like their neontological equivalents, can inform general concepts about the ways in which development is repatterned in evolution.

The Gould/Lewontin approach to developmental constraint, referred to above, can also be used to exemplify the third type of integration: that between evo-devo concepts and classical evolutionary theory. Their criticism of the latter, which we examined in Chapter 13, was that it was too adaptationist, and that it ignored the importance of developmental constraint in determining the direction of evolution. One of the many responses their criticism received from supporters of the classical theory was from a quantitative geneticist, Jim Cheverud[13], who said: 'the genetic variance/covariance

Figure 20.5 A rare case of being able to study development in some detail in fossils. This case-study is of the post-embryonic development of trilobites, and in particular their development of segments – note increasing segment number as development proceeds from an early stage (left) to later ones (centre and right). (Photographs courtesy of Nigel Hughes.)

matrix of quantitative genetic theory measures developmental constraints.' In other words – existing theory already takes the concept of developmental constraint on board, so what's all the fuss about?

This is a good example of people who probably agree with each other appearing not to because of the use of different 'languages' – i.e. the different sets of terms that are built up within individual disciplines such as quantitative genetics and evo-devo. Quantitative genetics is the name given to population-based studies of continuously-varying characters, often in relation to their response to selection. Most morphological characters, and hence the underlying developmental processes, are of this kind. So, there should be a particularly close relationship between quantitative genetics on the one hand and evo-devo on the other. Clearly there was not such a relationship at the time of the debate between Gould/Lewontin and Cheverud (the late 1970s and early 1980s); but the situation has improved since then and will hopefully improve further in the near future.

We can now see that quantitative geneticists are interested in understanding the structure of character correlations and the ways in which this structure influences the nature of evolutionary changes in the characters concerned. Proponents of evo-devo are interested in understanding the developmental basis of those correlations, which are often measured only in adults by quantitative geneticists. These endeavours are complementary; this is part of the overall complementarity of evo-devo and classical evolutionary theory, which was stressed in the previous chapter.

An important challenge for the future, for proponents of *all* disciplines involved in the study of evolution, is to forge sufficient links to reveal other cases that constitute only apparent, rather than real, disagreements. This will help, in turn, to reveal the few remaining real areas of tension, such as the scale dependence or independence of evolution, and in particular the question of whether the origins of **novelties** and **body plans** require an explanation that is more than just accumulated **micro-evolutionary** change over long periods of time.

Many scientists working in evo-devo today, whether in its molecular, organismic or palaeontological 'wings', believe that novelties and body plans *do* require some special explanation. However, not all of them feel this way. Also, it is not yet clear what the special explanation, if it is called for, should be. It should almost certainly *not* be **macromutation**, as proposed by an early student of the evolution of development, Richard Goldschmidt[14], among others. Perhaps it should lie instead in developmental bias, and specifically in the fleeting appearance of 'unusual variants' that are not normally present in the **standing variation** of a population. But, as we saw in Chapter 19, this is very much still an open question. It would be great if the next two decades see it being answered, but perhaps this is expecting too much.

Figure 20.6 Representatives of multicellular creatures (thus ones characterised by a developmental process) that belong to kingdoms other than those of the animals and plants, which have been the focus of attention in this book, and in the majority of evo-devo studies carried out to date. They are: a 'toadstool' (kingdom Fungi); and a 'seaweed' (kingdom Heterokontophyta; including the brown algae and diatoms).

20.4 Wider Challenges

There is an undefined number of these; a useful exercise for both individuals and seminar groups is to think of challenges for the future that are not included below. Here, we will look at just three very different ones:

First, in addition to improving the pattern of taxon sampling across the animal and plant kingdoms, it will be important for evo-devo studies to extend, much more than they have done so far, to other taxa that include significant multicellular forms. The most obvious examples are the fungi and the brown algae (Fig. 20.6). These provide an opportunity to see if the evolution of their development is similar to that of animals and plants, despite their independent origins of multicellularity.

Second, as well as extending phylogenetically, evo-devo studies should extend in terms of the range of topics included. For example, much of what has been done so far has involved a focus on the development of form rather than of function. However, all developmental stages must function physiologically, and some of them (e.g. **larvae**) must also have a behavioural repertoire. Evolutionary biology includes comparative physiological and behavioural studies, but the links between these and evo-devo are still in their infancy. They need to be strengthened in the near future. There are many interesting questions lurking within these links. For example, taking the case study of evolutionary switches between feeding and non-feeding echinoid larvae (introduced in Chapter 1), not only is larval structure affected, in terms of the gain or loss of feeding 'arms', but so too must be larval behaviour and probably also some aspects of larval physiology.

Finally, we should all strive to become philosophers of science as well as scientists, so that we more fully understand the nature of our shared endeavour. One important aspect of this is to cultivate a recognition of science's ideological and aesthetic sides. The unresolved issue of the mechanisms at work in the origin of evolutionary novelties and body plans is a good example here. Those who take the 'accumulated micro-evolution' view have a vision of a scale-independent process whose beauty lies in the operation of a single mechanism over all time-scales. Equally, those who take the 'unusual variants' view envisage a process whose beauty lies in its unique ability to achieve what can be considered to be the most important evolutionary transitions. In both cases, aesthetic considerations are important.

In the end, though, despite the fact that scientists, like mathematicians, set much store by the beauty of a theory, its ultimate test lies not in its beauty but in its ability to explain the way things are. Early theories of the evolution of development, which dressed themselves up as universal laws, such as the 'biogenetic law' of **recapitulation,** may have had a beauty, but it was a false beauty that was not robust in the face of the facts. There may yet be laws to be discovered that are real, and it is science's duty to discover them if they exist. But for now we must get on with the groundwork of extending our studies in all directions, in terms of both techniques and taxa. And we must hope that some beautiful generalisations, even if not universal laws – which are few and far between in biology anyhow – will emerge from this continuing endeavour.

SUGGESTIONS FOR FURTHER READING

It is hard to suggest literature sources that relate to the *future* of evo-devo and its relationship with more general evolutionary theory, as opposed to its past. Two useful types of material to read in this context are: (i) 'perspective' papers that argue for what the future focus should be; and (ii) publications by philosophers of evo-devo.
A good, provocative, perspective paper arguing that the central thrust of evo-devo should be 'evolvability' is:

Hendrikse, J.L., Parsons, P.A. and Hallgrimsson, B. 2007 Evolvability as the proper focus of evolutionary developmental biology. *Evol. & Dev.* **9**: 393–401.

Two recent books looking at evo-devo and related areas from a philosophical angle, both in the 'Cambridge Studies in Philosophy and Biology' series, are:

Amundson, R. 2005 *The Changing Role of the Embryo in Evolutionary Thought: Roots of Evo-Devo*. Cambridge University Press, Cambridge.
Robert, J.S. 2004 *Embryology, Epigenesis and Evolution: Taking Development Seriously*. Cambridge University Press, Cambridge.

And of course, the other thing that those interested in the future of a developmental approach to evolution should do is to keep abreast of the latest findings in the field, as they continue to appear in the appropriate journals.

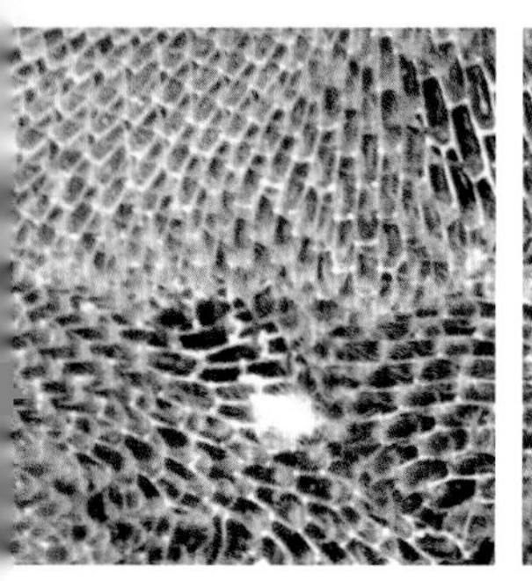
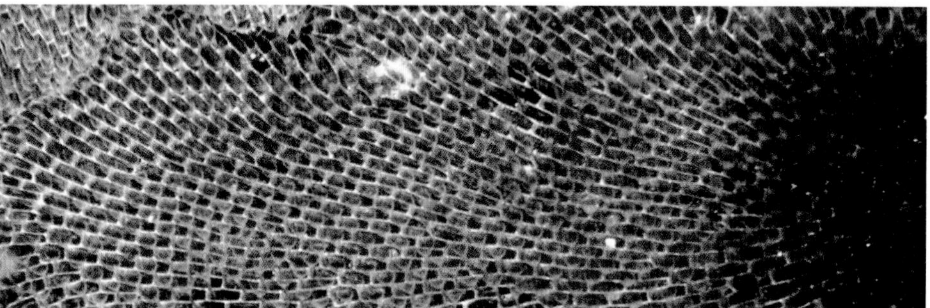

Glossary

A

Adaptive hypothesis: Hypothesis about the possible reason why a characteristic of an organism might be advantageous, that is, lead to enhanced fitness, either in general or, more usually, under particular environmental conditions.

Adaptive landscape: A three-dimensional model for the way natural selection works on combinations of the values of two characters. The character values form the base of the landscape, while variation in fitness values is represented by the topography – with fitness peaks representing local maxima in fitness.

Adaptive scenario: Used in two ways: first, as a synonym for adaptive hypothesis; second, in a negative sense to indicate the difficulty in testing such hypotheses – those who wish to be especially negative in this way often use 'adaptive story-telling'.

Allele: One form of a gene. When a population is polymorphic at a particular gene locus, the alternative forms of the gene are called alleles. Each individual in the population can have two copies of the same allele (homozygous) or one of each (heterozygous). However, many polymorphisms involve three or more alleles.

Allometric growth: Growth in which one part of the organism grows faster than another, so that increased size is accompanied by altered shape. Also referred to as allometry. The alternative is isometric growth or isometry.

Amborellaceae: A family of angiosperms (flowering plants). This is thought to be one of the most basal of angiosperm families. Another very basal family is the Hydatellaceae.

Amniota: The group of tetrapod vertebrates characterised by possession of the complex amniote egg that allows a fully terrestrial existence. This type of egg has a series of membranes including the outer chorion and the inner amnion, after which it is named and within which the embryo is found. Amniota includes the mammals, birds and 'reptiles'; but excludes the amphibians.

Angiosperm: Flowering plant. The angiosperms constitute a monophyletic group of more than a quarter of a million species. Angiosperms dominate the current land flora now; but in earlier times it was dominated by non-flowering plants (e.g. mosses, ferns, conifers).

Annelida: The phylum of segmented worms. This includes the familiar earthworms; also leeches and marine polychaete worms.

Anura: The taxon (usually ranked as an order) that includes all species of frogs and toads. The name means 'tail-less'. This contrasts with the other main amphibian group, the urodeles (newts, salamanders), which do have tails.

Apomorph: Term used in cladistics to refer to characters that are unique to particular species or clades of species and are therefore helpful in reconstructing the phylogeny of the group concerned.

Apoptosis: Programmed cell death. This is an integral part of development. It can help to achieve a pattern of morphogenesis, such as the apoptosis of inter-digital cells in the feet of most vertebrates causing separation of the digits as opposed to webbing.

B

Basal: Used to describe those members of a taxon that have departed least from the form of the stem lineage of that taxon. Now used instead of 'primitive', as this latter word can be taken to imply that evolution is necessarily progressive, which it is not.

Bilateria: The clade of all animals that have evolved from the original bilaterally-symmetrical animal, which may have lived in the early Cambrian, but probably originated earlier still, for example in the Vendian. Although most of today's bilaterians have retained their bilateral symmetry, some (e.g. echinoderms such as starfish) have lost it.

Blastoderm: Name given to the early, blastula-equivalent stage, in some animals, notably insects. The blastoderm of *Drosophila* and most other insects is initially syncytial; then later it becomes cellular.

Blastula: The earliest stage in the embryonic development of many animals. It is typically a hollow ball of cells. It is produced by the repeated cleavage of the fertilised egg or zygote.

Body plan: The general plan or layout of the body of animals belonging to a particular higher taxon, such as the vertebrates with their internal skeleton, or the bivalve molluscs with their paired shells.

Brassicaceae: Family of angiosperms that includes the model plant *Arabidopsis thaliana*. Also included are many common vegetables such as turnip, cabbage and radish. Contains more than 3000 species altogether.

Bryozoa: Phylum of animals also known as 'moss animals' or ectoprocts. Bryozoans are typically colonial animals; each colony is made up of many individual zooids that are genetically identical. These colonies grow on suitable intertidal and subtidal surfaces, such as rocks and seaweeds. Characterised by having feeding structures called lophophores, supporting a ring of tentacles around the mouth.

Burgess Shale: Name given to a Cambrian Period fossil-rich area of British Columbia, Canada, and to the fossils (collectively) found there.

C

Cambrian explosion: Phrase referring to the sudden (in geological terms) appearance of many animal taxa in a relatively short period of time shortly after the start of the Cambrian Period. It is still not clear whether this reflects a burst of evolutionary radiation at that time or is due to a sudden increase in 'fossilisability'.

Canalisation: A developmental system is said to be canalised if it is able to continue following certain pathways ('canals'), despite perturbations either from the environment or from a mutation in one of the genes affecting it.

Canonical: Used to describe a signalling pathway, such as the Wnt pathway. When more than one kind of signalling involving the agent concerned is discovered, the originally-discovered form of signalling is described as canonical.

Case study: An in-depth study of some process, normally based on a particular species or small group of species, which can be used to illustrate the way the process concerned works in general. For example, we talk about the Darwin's finches case study of evolutionary radiation of species.

Cell differentiation: The process by which the initially similar cells of the early embryo become different in type, with some becoming, for example, nerve cells, others blood cells, and so on. Also used for the process of generating specific cell types later in life from undifferentiated stem cells.

Centromere: A point on a eukaryote chromosome that usually appears as a constriction when the chromosome is viewed with a light microscope. This is the position where spindle fibres attach during cell division prior to pulling the duplicated chromosomes apart into what will be the two daughter cells.

Cephalochordata: Also known as Acraniata, this is the subphylum of Chordata into which the basal forms known as lancelets, or 'amphioxus', fall.

Chirality: Refers to the 'handedness' of a structure, at both morphological and molecular levels. For example, a snail's shell is a chiral structure. If it is coiled to the right, it is dextral; if to the left, then it is sinistral.

Choanoflagellata: A group of single-celled organisms characterised by possession of a flagellum surrounded by a collar of finger-like projections (or microvilli). The choanoflagellates are thought to be the sister-group of the animals.

Chordata: The phylum of the animal kingdom into which the vertebrates fall. The other sub-phyla within Chordata, in addition to Vertebrata, are the Cephalochordata and the Urochordata.

Chromatin: The material of which chromosomes are made. It consists of a mixture of DNA and proteins (the latter mostly belonging to the group called histones), which are wrapped around each other and then coiled and 'supercoiled' with the effect that a very long DNA molecule can fit into a relatively short chromosome.

***Cis*-acting:** Refers to regulatory regions on the same chromosome, often in the vicinity of the gene concerned. The opposite is trans-acting. In reality these usually act jointly by interaction between both kinds of factors.

Clade: Equivalent to the term 'monophyletic group'; that is, all the descendant lineages from a particular stem lineage. Gives its name both to the discipline of cladistics (reconstruction of evolutionary pattern through analysis of clades) and to cladograms (branching diagrams showing patterns of lineage splitting).

Cladistics: The approach to analysing evolutionary pattern pioneered by the German entomologist Willi Hennig. Based on the use of derived ('apomorph') characters rather than primitive ('plesiomorph') ones, it was instrumental in the explicit recognition of, and rejection of the use of, paraphyletic groups.

Cladogenesis: The splitting of one lineage into two (or more) descendant lineages. Often used synonymously with speciation (the origin of new species), though speciation can also occur without lineage splitting (in which case the process is called anagenesis).

Cladogram: A simplified version of an evolutionary tree, which retains only information on the sequence of lineage-splitting that occurred within the overall taxonomic group under study.

Cleavage: Process of repeated division of the fertilised egg or zygote, leading to the earliest developmental stage of an animal, which is usually called a blastula.

Cnidaria: The phylum of basal, radially symmetrical animals that includes the sea anemones, the corals, the jellyfish and the hydroids.

Coadaptation: Used in two ways. Internally to an organism, it refers to the way in which interacting parts are adapted to function jointly, whether proteins in a signalling pathway or bones in a limb. Externally, it refers to the way in which interacting organisms, such as flowering plants and pollinating insects, are adapted to live together.

Cohort: Subgroup within a population, defined in terms of age. For example, a species of animal or plant that breeds annually and has a life-span of about 5 years will typically have 5 annual cohorts in each population.

Colinearity: The phenomenon whereby the order of Hox genes along the chromosome is the same as their order of expression along the antero-posterior axis of the body. Strictly, this is spatial colinearity. In some animals there is also temporal colinearity because anterior Hox genes are expressed first, posterior ones last.

Congeneric: Belonging to the same genus.

Conspecific: Belonging to the same species.

Convergence: The independent evolution of different lineages towards a similar solution to an adaptive 'challenge'. For example, the evolution of a tail in the first fishes and also in the early cetaceans (whales and dolphins). The similar structures produced by convergent evolution are homoplastic rather than homologous.

Co-option: The use of a gene and its product to perform a new function, usually as well as, rather than instead of, the original one. Sometimes this is also called the recruitment of the gene/protein to the new function. Co-option is thought to be a key process in the evolution of development at the molecular level.

Cryptic: Refers to features that make an animal camouflaged against its background and hence difficult for predators to see. Normally used in relation to pigmentation (as in the case of the peppered moth), but can also be used to refer to shape (as in the case of stick insects).

Ctenophora: A phylum of radially symmetrical basal animals superficially similar in appearance to jellyfish (which in fact belong to the phylum Cnidaria). The common name for ctenophores is comb jellies.

D

Daughter species: The two (or more) species produced from a single parental one following the splitting of a lineage by a speciation event (cladogenesis).

Denticle: Literally, 'little tooth'. Used to describe small, hard projections, often on the cuticle of an insect larva, but in other contexts too.

Derived: The opposite of basal. Refers to members of a clade that have departed the most from the stem lineage of that clade. For example, *Drosophila* is often said to be a very derived insect in that many of its features, including the simultaneous formation of all its segments, are atypical for the group, especially its most basal orders.

Deuterostome: One of the two primary subdivisions of the bilaterian animals. Means 'mouth second' (cf. protostome, 'mouth first'), referring to the fact that the anus rather than the mouth derives from the earliest opening in the embryo, the blastopore.

Developmental bias: The tendency of developmental systems to be more readily altered in some ways than others. This applies at both molecular and morphological levels. It is very common and *may* have a major directing role in evolution, though this possibility is still debated.

Developmental constraint: Sometimes used synonymously with 'developmental bias' (see above); but better used only for negative biases. Developmental changes that are 'hard' to produce represent relative constraint; those that appear to be impossible to produce represent absolute constraint.

Developmental drive: The opposite of developmental constraint. Refers to variations in development that are 'easy' to produce, which may result in the 'driving' of evolution in those directions.

Developmental gene: Any gene controlling or affecting any developmental process. Thus a very general category, including many sub-categories (e.g. Hox gene).

Developmental patterning: The production of any type of pattern during the course of development. Examples include production of the main body axes in bilaterian animals and the production of flowers in angiosperms.

Developmental repatterning: Evolutionary changes in developmental patterning, such that descendant species are characterised by patterning that is different from their ancestral species. There are four kinds of repatterning: heterochrony, heterotopy, heterometry and heterotypy.

Developmental trajectory: The route taken by a developing organism from a starting point of (usually) a fertilised egg, through a series of more-or-less well defined stages, and culminating in the reproductively mature adult.

Diploid: Possessing two copies (a maternal and a paternal one) of each gene, as opposed to just one (haploid). Most eukaryotes are diploid, though there are exceptions of various kinds, including plants that alternate between haploid and diploid growth forms, and animals in which the male is haploid, the female diploid.

Diptera: Order of insects generally referred to as flies. These insects have just a single pair of wings as opposed to the more typical two pairs, because the hind-wings have been transformed into flight balancing organs called halteres. The order, which contains more than one hundred thousand species, includes the model animal *Drosophila melanogaster*, together with such familiar creatures as houseflies, midges and mosquitoes.

Domain: The highest-level division of the living world, higher even than kingdom. Life is divided into the domains Bacteria, Archaea (also prokaryotic) and Eukaryota. The animal, plant and fungal kingdoms are all contained within the Eukaryota.

Dominance (genetic): When a natural population is polymorphic for a particular gene, one form of the gene (an allele) is often dominant over the alternative form in that it determines the phenotype of the heterozygote. For example, in the peppered moth the main melanic gene is dominant over the original form of the gene; thus the heterozygote is melanic. It is important not to confuse genetic dominance with evolutionary fitness: sometimes, a dominant gene produces a less fit phenotype than its recessive counterpart.

Downstream: Used in two senses, developmental and genetic. A developmental pathway can be thought of as a series of stages arranged in temporal sequence with an earlier stage being described as upstream of a later one. The genetic usage refers to the direction of transcription of a gene: a regulatory region can be described as upstream or downstream of the coding sequence, with the former meaning before the transcription start-site and the latter meaning after the stop-site.

Duplication and divergence: The phenomenon whereby, following the duplication (or more generally replication) of a structure at either the molecular level (e.g. a gene) or the morphological level (e.g. a limb-pair), the different copies diverge in function.

E

Echinodermata: The phylum of animals that includes starfish (sea stars), sea urchins, sea cucumbers, brittle stars and the superficially plant-like crinoids or 'sea-lilies'. These animals typically have pentaradial (five-fold) symmetry, but their ancestors were bilaterally symmetrical. The name means 'spiny skin'.

Ectopic expression: Refers to expression of a gene in the wrong place (in the same sense that pregnancies can be ectopic when the embryo implants outside of the uterus). Caused by mutation in the short term. In the long term, it may be important in evolution – in which case 'the wrong place' becomes, eventually, the right place (or one of them).

Embryogenesis: The development of the embryo. Contrasts with other developmental processes, such as post-embryonic growth, metamorphosis and regeneration. Developmental biology involves the study of all of these and is thus a more inclusive phrase than embryology.

Endosperm: The embryo-nourishing tissue of most plants. In angiosperms it is often a triploid tissue produced by a fusion of the second sperm nucleus of the pollen with the two polar nuclei associated with the egg.

Endo-symbiont: One organism living inside another, generally to their mutual benefit (as opposed to parasitism, in which one organism suffers). Eukaryote cells are thought to have arisen from such an arrangement in which smaller prokaryotic cells began to live inside larger ones. The smaller ones began to lose their own individuality and are seen today as mitochondria and (in plants) chloroplasts.

Epigenesis: Sometimes used as a synonym for development. However, more often used to describe those aspects of development that are 'outside of the genes'. This can refer to things that are literally on the outside of the DNA sequence, such as methyl groups; or more broadly to the control of developmental genes by other agents, as opposed to the reverse process. Development is best seen as involving control by genes and control of the genes.

Epigenetic landscape: Term introduced by C.H. Waddington in an attempt to picture developmental trajectories in terms of a system of valleys (possible developmental routes) in which a ball rolls, taking one particular route, which becomes the actual course of development in the organism concerned.

Epigenetics: A word to beware of, as it has been used in too many ways! (and the same might be said of epigenesis). It can mean simply the study of development. Or it can mean the study of development from an approach that emphasises the importance of agents other than the genes. And recently it often refers specifically to the study of patterns of DNA methylation and their developmental effects.

Epithelium: A sheet-like type of tissue. The name means being on the outside of something. Thus epithelial cells are found in the skin. But they are also found in other places such as the lining of the gut.

Eukaryote: The type of cell found in animals, plants and fungi, but not in bacteria. It is characterised by membrane-bound nucleus and organelles. The group of all organisms that evolved from the eukaryote stem lineage is known as the Eukaryota.

Evolvability: Used to describe the relative 'ease' of evolving in one way rather than another. Exact usage is not consistent between authors. Can be thought of as being inversely related to the amount of developmental constraint – the more developmentally constrained a character is the less evolvable it is and vice versa. However, with some usages of evolvability, it is inversely related to selective, rather than developmental, constraint.

Exaptation: A feature of an organism that was initially moulded by selection for one function but is later further elaborated by selection for another function. For example, the avian feather may be an exaptation, as it may have initially been selected for its thermal insulation properties and then, later, for its use in flight. The second kind of selection can also be referred to as exaptation, so the word can be used as a verb as well as a noun.

Exon: The coding regions of the genes of eukaryotes are divided into two types of section: introns and exons. All of the coding sequence is transcribed, producing an RNA molecule of the same length. But then some parts of the RNA, corresponding to the introns, are removed, while the other parts, corresponding to the exons, are retained and joined together into a shorter, mature mRNA molecule that leaves the nucleus and is translated into protein.

Expression pattern: The pattern in which a gene is expressed in space and time. Some developmental genes are only expressed very transiently and in small regions of the embryo. Others are expressed for a longer period and over greater areas.

Extant: The opposite of extinct. Humans, starlings and bracken (a fern species) are all extant. *Tyrannosaurus rex* is extinct. Sometimes used of higher-level taxa than species: for example, insects are extant, trilobites are extinct.

F

Fitness: A combination of the probability of surviving to adulthood, and the mean number of offspring produced, normally used in a within-species comparative sense in which the fitness of one variant is compared with that of another. When fitness differences occur in a population (as they almost always do), the frequencies of the variants concerned alter over time (by definition) due to natural selection.

Fluctuating asymmetry: Departure from perfect bilateral symmetry, measurable, for example, by subtracting the length of the right-hand one of a pair of structures, such as the wings of an insect, from the length of its left-hand counterpart. Thought to be increased by 'stress', such as extreme temperature during development. The name is misleading: the asymmetry is not fluctuating in time – rather, it just varies among the individuals of a population.

G

Gastropoda: One of the major classes within the phylum Mollusca, comprising tens of thousands of species of snails and slugs.

Gastrulation: The developmental process that leads from the earliest stage of animal development, the blastula, to the next stage, called the gastrula. In contrast to the earlier process of cleavage, in which cell division is the key process, gastrulation is characterised by considerable cell movement. A key part of this is the inward movement of cells in the blastopore region. In most animals, gastrulation produces the three main germ layers of the embryo – endoderm, mesoderm and ectoderm.

Gene duplication: The accidental duplication of a gene, for example as a result of DNA 'slippage', resulting in the possession of two copies, usually side-by-side on the chromosome. Included within 'mutation' in its broad sense.

Genet: In some kinds of plant, such as most grass species, reproduction can occur both sexually and vegetatively. In such cases, a 'genet' is a genetic individual, which may consist of an array of interconnected 'ramets', or physically separate (at least above the ground) plants.

Genetic assimilation: The process whereby a feature of an organism that initially only arose as a plastic response to an environmental cue becomes constitutive – that is, it arises without the cue that was previously necessary. This occurs due to selection for individuals with an enhanced propensity to produce the feature concerned.

Genetic drift: The 'random walk' of gene frequency that can occur in a population simply due to chance events, as opposed to the systematic change in gene frequency that occurs as a result of selection.

Genome: The totality of the genetic material in an individual or, more generally, a species. For example, it is said that the human genome is composed of 25,000 or so genes. This is true of all humans. But equally the exact genome of any one human is different from that of any other (except in the case of monozygotic twins) because of variation within and between populations.

Genotype: The combination of alleles present at a particular gene locus in a particular individual. If the alleles are the same, the individual is homozygous for that gene; if they are different, the individual is heterozygous.

Genotype-phenotype map: The way in which genotypes, at multiple gene loci, map to either two-dimensional (e.g. adult) or three-dimensional (developmental) phenotypes. A short and apparently simple phrase for a plethora of different processes, including interactions between alleles at a locus (dominance) and interactions between loci (epistasis).

Geographical range: The total area of the world where a species is found. At one extreme, some species are endemic (= restricted) to a single place, for example, a single lake in the case of freshwater species. At the other extreme, some species (e.g. humans) are found across all continents (a 'cosmopolitan' distribution).

Germ band: An early stage in the development of insects and other arthropods in which the embryo has become stretched along the antero-posterior axis and so has a head and tail end but few obvious features.

Germ layer: One of the three layers of tissue that characterise most animals from the gastrulation stage of development: endoderm, mesoderm and ectoderm. Each of these goes on to give rise to particular tissues/structures. For example, the gut is endodermal, the muscles mesodermal and the skin ectodermal. The most basal animals, sponges, lack these layers. Cnidarians, such as jellyfish, may only have two layers, but this is currently debated.

Glyptodont: A kind of extinct mammal with what appears to be an integral bony shell, thus making glyptodonts superficially similar to the (distantly related) turtles. Glyptodonts are closely related to the extant armadillos, which are armoured rather than shelled.

Gradualism: The view that evolution always occurs gradually, and can thus be characterised by a slowly shifting mean character value at the population level. The opposing view is saltationism, wherein most important evolutionary changes are thought to occur suddenly, through large-effect mutations. The latter view is now known to be wrong *in general*, but it applies in some specific instances, such as the switch from a dextral to a sinistral gastropod.

Gymnosperm: The group of plants that includes the conifers and the (less well-known) cycads. Although this is now thought to be a paraphyletic group (with respect to the angiosperms), it continues to be widely used.

H

Haploid: Having only a single copy of each gene. Typically, prokaryotes, such as the bacterium *Escherichia coli*, are haploid; while eukaryotes are diploid, with two copies of each gene; but there are exceptions to this 'rule'.

Hemichordata: A phylum of deuterostome animals. They were originally thought to be the sister-group of the chordates but are now known to be the sister-group of the echinoderms. They include the solitary 'acorn worms' and another group, called pterobranchs, which are colonial.

Hemimetabolous: Used to describe the development of insects in which the adult stage is reached through a series of nymphal stages, each of which typically resembles the adult more than the one before it. Example – the insect order Orthoptera, which includes grasshoppers, crickets and locusts. Cf holometabolous.

Heritability: The proportion of the phenotypic variation in a particular character (e.g. adult body length) in a population that is due to genetic rather than environmental effects. This is sometimes referred to as 'broad-sense' heritability, because there is also a narrow-sense, more technical meaning of heritability within the discipline of quantitative genetics.

Heterochrony: Evolutionary change in the relative timing of different developmental processes/events.

Heterometry: Evolutionary change in the relative amounts or concentrations of developmental substances, such as the amount of the product of one developmental gene relative to others.

Heterotopy: Evolutionary changes in the relative spatial distribution of developmental processes/events.

Heterotypy: Evolutionary change in the type of some developmental agent, such as the protein product of a developmental gene, as opposed to changes in its amount, or its spatio-temporal distribution. This involves mutation in the coding region rather than the regulatory region of the gene concerned.

Heterozygous advantage: Refers to cases of selection acting on a polymorphism at a particular gene locus in which the heterozygote is the fittest of the three genotypes – i.e. it is fitter than both homozygotes. This constitutes a form of balancing selection.

Heterozygous: Having different alleles present at a single gene locus. Refers to individual organisms, whereas 'polymorphic' refers to populations.

Histone: A type of protein that is found closely associated with DNA in all eukaryote chromosomes. The combination of the two is referred to as chromatin, which just means 'chromosomal material'.

Holometabolous: Used to describe the development of insects in which the adult stage is reached from a very different larval stage by a series of major changes collectively called metamorphosis, which occurs in the (immobile) pupa. Example – the insect order Lepidoptera, which includes the butterflies and moths; a lepidopteran pupa is often referred to as a chrysalis. Cf hemimetabolous.

Homeobox: A sequence of 180 base-pairs in a gene, which, when translated, will produce a 60-amino acid sequence in the corresponding protein (the homeodomain). More highly conserved in evolution than most other DNA sequences.

Homeodomain: A sequence of 60 amino acids in a protein that forms a DNA-binding region of the protein and thus shows that it is a transcription factor.

Homeosis: The phenomenon of having 'the right structure in the wrong place', such as, in a fly, having legs growing out of the head where antennae should be. Homeotic transformations can occur in response to environmental effects; but they are more commonly caused genetically, in which case the mutations involved are called homeotic mutations.

Homology: Structures at both molecular and morphological levels whose similarity reflects shared ancestry rather than convergent evolution.

Homoplasy: The 'opposite' of homology. Structures at both molecular and morphological levels whose similarity reflects convergent evolution rather than shared ancestry.

Homozygous: Having both paternal and maternal copies of a particular gene the same. Opposite of heterozygous.

Hormone: A developmental and/or physiological agent that travels round the body via a transport system (such as blood) and thus is able to have long-range effects. An example is the insect moulting hormone, ecdysone. Biochemically, hormones are heterogeneous: some are peptides (small proteins), while others are steroids.

Housekeeping gene: A gene whose product is involved in the routine metabolic processes, such as glycolysis, which occur in most or all cells.

Hox genes: A subgroup of homeobox-containing genes characterised by expression in a graded antero-posterior pattern; and occurrence (usually) in discrete chromosomal clusters.

Hydatellaceae: Family of flowering plants. Currently thought to be one of the most basal families of extant angiosperms. Another very basal angiosperm family is the Amborellaceae.

I

Imaginal discs: Approximately disc-shaped pieces of tissue found in pairs in the head and thoracic segments of the larvae of holometabolous insects. It is from these that much of the adult will be built during metamorphosis. Their abdominal counterparts are referred to as histoblast nests and have a similar role.

Instar: Insect larval stage, separated from other such stages by moults. The number of instars varies between taxa. For example, *Drosophila melanogaster* has three, whereas some butterflies have five or six.

Integrin: Type of protein involved in connecting the cell to its neighbours and/or to the extracellular matrix. Dynamic patterns of integrin expression/function can lead to cell movement.

Intron: Part of a gene that is transcribed but then removed from the RNA product. See also exon.

Isometric growth: The kind of growth that occurs when different body parts grow in proportion to one another, with the result that shape does not change with size. This is much rarer than the alternative, allometric growth, which is the norm.

K

Karyotype: The visible appearance of the set of chromosomes of a cell from a particular species, as seen through a light microscope. A key feature of the karyotype is the number of chromosome pairs (e.g. 23 in humans, 4 in *Drosophila melanogaster*). Other features include chromosome sizes and the positions of the centromeres.

L

Lamarckian: Refers to theories of evolutionary mechanism that involve the inheritance of characteristics acquired during the lifetime of a parent; or, put another way, the effect of use or disuse of a character in one generation on its degree of development in the next. After the 18th/19th-century French biologist Lamarck.

Larva: A post-embryonic developmental stage that, rather than looking like a miniature adult, takes a completely different form. The forms of larvae are very different in different groups of animals. Examples that are referred to in this book include molluscan trochophore larvae, crustacean nauplius larvae, echinoderm pluteus larvae and the well-known tadpole larvae that most species of frogs go through on their way to adulthood. Developmental trajectories that include one or more larval stage are called indirect; those without larvae are called direct.

Ligand: The name given to a molecule that binds to a receptor on the outside of a cell and causes a signal to pass into the cell. Many such signals have important developmental consequences.

Lineage: An evolutionary line of descent leading from an ancestor in the distant past to either a present-day (extant) descendant or, in some cases, to an extinct species – the last descendant of that lineage. Since the overall 'shape' of evolution is a tree rather than a line, a lineage is always part of something bigger. It is a way of focusing in on one particular series of ancestors and descendants – for example, from an australopithecine species to *Homo sapiens*.

Locus (plural loci): Any point along a chromosome that is the site of a particular gene. Different DNA sequences at a locus are known as alleles.

Long-germ: A descriptor of early insect development, indicating that the whole animal (including all of its segments) is represented in the initial germ-band stage. The 'model animal' *Drosophila* (along with other flies) provides an example of this kind of development. Cf short-germ.

M

Macro-evolution: Used in two ways. Some authors use it to describe all evolutionary changes that are trans-specific; i.e. all except micro-evolutionary changes. Other authors use it to describe the kind of evolution that produces a range of species within a genus or family; in which case higher-level changes are called mega-evolution. In this book the latter usage is adopted.

Macromutation: A mutation that causes a major developmental/phenotypic effect. Examples include that causing a switch from a dextral to a sinistral shell, and the one causing the growth of legs where the antenna should be in a fly. Most are highly detrimental in fitness terms and do not contribute to evolution, but there are some exceptions to that rule, including the dextral/sinistral mutation.

MADS-box genes: Group of transcription-factor-producing genes that contain a conserved region rather like that of the homeobox genes in its length and function. Although found in animals, plants and fungi, this group of genes is particularly important in the plant kingdom, where it has undergone much diversification, producing many genes that are important in plant development.

Mega-evolution: The biggest type of evolutionary change, such as that which produces a novelty or new body plan. Examples include the evolutionary origin of the turtle shell and of the vertebrate skeleton. There is still much debate as to whether mega-evolution is explicable in terms of many micro-/macro-evolutionary changes compounded over long periods of time; or whether it also includes evolutionary processes that are rare or non-existent in the micro-evolutionary realm.

Melanism: The occurrence and spread of darkly-pigmented (melanic) phenotypes in populations. Most frequently encountered in the form of industrial melanism, where many species of moths (and other insects) became melanic in areas affected by the soot pollution of the industrial revolution.

Mesoderm: One of the three germ layers that are found in early embryos of most animal species (from the gastrulation stage). The mesoderm lies between the ectoderm and endoderm; as development proceeds, it gives rise to particular tissue types (e.g. muscle) and particular organs (e.g. the heart).

Mesomutation: Refers to mutations that are intermediate in magnitude of developmental and phenotypic effect between micro- and macro-mutations.

Metamorphosis: Used to describe any dramatic developmental change that occurs after initial development but before adulthood. Can occur early or late in development. Examples include frog-to-tadpole, caterpillar-to-butterfly and trochophore larva-to-juvenile snail.

Metapopulation: Regional population, consisting of many local populations connected with each other by migration, sometimes diffuse but sometimes directed along corridors of appropriate habitat that connect the local population patches.

Micro-evolution: Within-species evolutionary changes, including those that lead to the formation of races, varieties and sub-species. The divergent products of micro-evolution can thus by definition still interbreed. Most micro-evolution at the morphological level is based on genetic variation that is ultimately derived from micromutations.

Micromutation: Mutation whose effect is so small that it is not individually discernible. Many continuously variable characters (e.g. body length) are underlain by micromutational effects at many gene loci; sometimes in conjunction with several mesomutations.

Mimicry: There are two forms of this. In Batesian mimicry, edible species (e.g. of butterfly) come to resemble poisonous ones through selection. In this case the poisonous species is the 'model' and the edible one resembling it is the mimic. In Muellerian mimicry, many poisonous species come to resemble each other, again through a process of selection.

Model animal/plant/organism: Used especially in relation to developmental genetic studies to refer to an organism that has become a 'workhorse' about which much is known, representing a higher taxon about many of whose other member species very little is known – for example, *Drosophila* as a model arthropod, the mouse as a model mammal and *Arabidopsis* as a model flowering plant.

Modern synthesis: Name given to the synthesis of different approaches to evolution that took place in the mid-20th century. These approaches included population genetics, ecology, systematics and palaeontology, but notably excluded developmental biology.

Modularity: The phenomenon whereby a developing organism is composed of several different parts, or modules, whose development is quasi-autonomous, that is, semi-independent of the others. Examples include limb buds in animals and leaf primordia in plants. Also used to describe the existence of quasi-autonomous individuals within a colony, in cases where there is a colonial growth form.

Monophyletic: The type of taxonomic group where all the descendants from a particular ancestral species are included. Now regarded as the only type of taxonomic grouping to be 'natural'.

Morphogen: Developmentally-active substance whose concentration varies across a region of an embryo, eliciting different responses in different places, partly due to the concentration-dependence of the response. An example is the Sonic hedgehog protein in the mid-dorsal region of vertebrate embryos.

Morphospace: Used to describe the realm of possible morphologies (in the sense of an n-dimensional space, where n represents the number of morphological characters). It is generally agreed that morphospace is very unevenly populated with the morphologies of actual creatures.

Mutation: A change in the DNA base sequence of a gene (or, in the broader sense, of any part of the genome, including non-genic regions). Somatic mutations occur in the body (or soma); these are not inherited and so do not contribute to evolution. Germ-line mutations occur in eggs and sperm or the cells that lead to these; consequently they are inherited and constitute the genetic basis for evolutionary change.

N

Nauplius larva: The type of larva found in most crustaceans and which has been used to identify, as crustaceans, some groups with unusual adults (e.g. barnacles). It is a dispersive larva, with three pairs of appendages.

Nematoda: The phylum of roundworms that includes the model animal *Caenorhabditis elegans*. Roundworms have a relatively simple body form and are unsegmented (in contrast to annelid worms).

Neoteny: A type of heterochrony involving altered relative timing in the development of the reproductive system relative to the rest of the body (or soma). Specifically, in neoteny, the development of one or more somatic characters is slowed so that what were juvenile features appear in the adult.

Neural tube: Stage in the development of the central nervous system (CNS) in vertebrate embryos when the future brain/spinal cord take the form of a simple tube. The tube, which is later much elaborated, arises from the earlier neural plate stage by the upturning and eventual fusion of the left and right edges of the plate.

Neutral: Refers to mutations that have no effect on fitness and are thus neutral in selective terms. The fate of such mutations in populations is determined by the random process of genetic drift. Many supposedly-neutral mutations may in fact have tiny effects on fitness – in this case the phrase 'nearly neutral' is often used.

Notochord: Stiff rod of tissue extending from anterior to posterior ends of the body of a chordate. In basal chordates such as amphioxus, this rod persists into adulthood and forms the main skeletal support for the body. In vertebrates this function is taken over by the vertebral column or spine; the notochord is present only in the embryo and subsequently disappears, though parts of it contribute to the inter-vertebral discs of the spine.

Novelty: An evolutionary novelty (or innovation) is something, usually a morphological structure, that seems fundamentally new compared with what went before. An example is the turtle shell. However, all apparently new structures in evolution arise through the modification of previous ones; in novelties the modification is more radical than usual.

O

Oncogene: Gene that, when mutated, can cause tumour formation. Since a tumour is the result of uncontrolled cell proliferation, the un-mutated product of the oncogene must have a role, direct or indirect, in regulating cell division, restricting it to a level consistent with the replacement of dying cells.

Ontogeny: Another word for development. It normally refers to the complete developmental trajectory from egg to adult. Hence embryogenesis is only a part of the ontogeny of an individual. Post-embryonic development, including metamorphosis where applicable, is the other part.

Outgroup: A taxonomic group that is close or closest to, but outside of, the one being studied. For example, the closest outgroup to the vertebrates is thought to be the urochordates; the closest outgroup to the Bilateria is thought to be the Cnidaria.

P

Paedomorphosis: A form of heterochrony involving the accelerated development of reproductive features relative to other (somatic) ones. The result of this is that what were juvenile forms become reproductively mature, and hence adults. Examples include paedomorphic salamanders.

Pancrustacea: The name given to the clade including all groups of crustaceans and also the insects. It has only been used in recent years after the realisation that insects and crustaceans, rather than insects and myriapods, are sister-groups. Previously, the term Atelocerata was used to include insects and myriapods, but this has now been rejected as an invalid (polyphyletic) grouping.

Paraphyletic: The kind of taxonomic group that includes some but not all of the descendants of the LCA of that group. Now generally avoided by evolutionary biologists but still commonly used by lay-people. Examples include Reptilia which, since birds evolved from one reptilian lineage, is paraphyletic unless we call birds reptiles.

Parasegment: One of the developmental units along the antero-posterior axis of an insect embryo. These constitute early-stage developmental modules that later give way to the visible segments. Parasegments and segments are out-of-phase with each other by approximately half a segment, so that the posterior part of parasegment n corresponds with the anterior part of segment $n + 1$.

Pattern formation: The formation of one-, two- and three-dimensional patterns involving not single cells but whole populations of cells. Used by many authors as synonymous with 'morphogenesis'. Can be contrasted with the within-cell changes leading to the creation of a particular cell type (e.g. an epithelial cell) – see 'cell differentiation'.

Phenotype: The result, sometimes visible on the outside of an organism's body but often not, of its having a particular genotype. Examples include banding patterns on snail shells, the heights of pea plants (as in Mendel's pioneering work) and the shapes of red blood cells in humans suffering from the inherited disease sickle cell anaemia. Often thought of as a three-dimensional feature of the adult; but should be thought of from a developmental perspective, so it becomes a four-dimensional feature of the life-cycle. It can also be influenced by the environment.

Phylogeny: The pattern of relatedness of a group of organisms, usually represented diagrammatically in the form of an evolutionary (or phylogenetic) tree, showing a particular pattern of lineage branching within the overall taxon concerned. The study of phylogenetic relationships is called phylogenetics or systematics.

Phylotypic stage: In many animal phyla, the various member species are often most similar in their ontogeny, not right at the start (the fertilised egg) but a short time later. This point of maximum similarity is called the phylotypic stage. Examples include the pharyngula in vertebrates and the germ-band stage in insects.

Phylum: A high-level taxon representing a major group of animals, examples being arthropods, chordates and molluscs. Groupings above the phylum level are generally referred to as superphyletic – for example, Deuterostomia.

Placozoa: An animal phylum represented by just a single genus (*Trichoplax*). It is characterised by a very simple body plan and is a basal group within the animal kingdom.

Plasticity: The phenomenon whereby the developmental trajectory of an animal or plant is influenced by one or more environmental factors. Plastic effects on development can have continuous or discrete effects. The former are referred to as developmental reaction norms. The latter include seasonal effects producing polyphenisms, such as the alternation between wet and dry season forms in a population.

Pleiotropy: The production of many different phenotypic effects by one gene. This should be regarded as the rule rather than the exception.

Plesiomorph: Term used in cladistics to refer to characters that are found throughout a taxon and in related taxa. They are therefore not useful in helping to reconstruct the phylogeny of the group concerned. (cf. apomorph).

Polygeny: The contribution of many genes ('polygenes'), each normally having a smallish effect, to a continuously-variable character such as body size.

Polymorphism: The occurrence of two or more genetically-determined 'morphs' or phenotypes in a population at the same time. Examples include widespread polymorphism at the molecular level and diverse cases of polymorphism at the morphological level, such as those involving different pigmentation patterns on the shells of snails or the wings of butterflies.

Polyphenism: The occurrence of two or more forms that are *not* genetically determined in a population. Polyphenisms can occur in the same characters (e.g. butterfly wing pigmentation patterns) as polymorphisms.

Polyphyletic: Invalid taxonomic group that is un-natural in the sense that it includes species from different clades with different stem lineages. Groups previously thought to be monophyletic, and then realised to be polyphyletic, are abandoned. An example is the group 'Articulata', including the annelids and arthropods, which are now known not to be sister-groups.

Polyploidy: Refers to tissues, organisms or species that have more than two copies (= diploid, 2n) of each gene and each chromosome. Includes triploidy (3n), tetraplody (4n) and higher levels of ploidy. The endosperm of many plants is a triploid tissue. Some species arise through increases in ploidy.

Polytene chromosome: Name given to the 'giant' chromosomes found in some tissues of some organisms, most famously the salivary glands of flies, including the model organism *Drosophila melanogaster*. They arise from repeated division of the chromosomes without accompanying cell division.

Positional information: Information that cells receive as to where they are in an embryo or other developmental stage of an organism. Often, this information results from exposure to a gradient in the concentration of a morphogen.

Primordium: The beginning of something in development. For example, in a tetrapod, the very earliest stages of the limb buds can be referred to as the limb primordia (plural). Each one of them is a limb primordium (singular).

Progenesis: A form of heterochrony in which the development of reproductive maturity is speeded up relative to the development of the soma (or body).

Prokaryote: Type of cell, or organism composed of one or more such cells, where there is no nucleus and no membrane-bound organelles such as mitochondria. Bacteria are prokaryotes; animals, plants and fungi, in contrast, are eukaryotes. Eukaryotic cells arose from prokaryotic cells in early evolution.

Promoter: An important and ubiquitous regulatory region of a gene, typically found immediately upstream of the coding region. This is in contrast to other regulatory regions (enhancers) that may lie a considerable distance away.

Protostome: Animal in which the mouth appears 'first' in development, in the sense that it derives from the earliest embryonic opening, the blastopore. It thus appears before the other end of the alimentary canal, the anus. (cf. deuterostome.) Most phyla of bilaterian animals are protostomes (including arthropods, nematodes, annelids and molluscs), while only a few (including chordates) are deuterostomes.

Punctuated equilibrium: A pattern of evolution in which most or all morphological change is concentrated into speciation events, which thus 'punctuate' long periods of time in which there is stasis of form in the non-splitting lineages. The heated debate on this topic in the 1970s has given way to the acceptance, by most evolutionists, that evolution is sometimes gradual, sometimes punctuational, and sometimes in between.

Q

Quantitative trait loci (QTLs): The gene loci that underlie the observed variation in continuously-variable characters, e.g. body length. Typically, many loci (20+) contribute to the variation in a quantitative trait. The mutations that are involved are usually a mixture of micromutations and mesomutations.

R

Ramet: An 'individual' plant in the sense of one that appears to be separate from other neighbouring plants of the same species. However, these are often connected below ground and have arisen from each other by vegetative reproduction, in which case they are genetically identical and together comprise a genet. A familiar example of this duality in the nature of an 'individual' is the buttercup.

Reaction norm: The pattern of variation in development or adult phenotype that is caused by variation in one or more environmental factors. Often plotted graphically, with character value on the y-axis and value of environmental variable (e.g. temperature) on the x-axis. Sometimes referred to as the developmental reaction norm, which in turn is sometimes abbreviated to DRN.

Recapitulation: In many cases of comparisons between the developmental trajectories of two different animals, the more 'advanced' (or complex) one will go through developmental stages that resemble ancestral forms in some way; such as the appearance of gill clefts in human embryos. So the ontogeny of the more complex animal recapitulates some of the features of its less complex 'cousin'; and thus, presumably, of their last common ancestor. However, there is no recapitulation of the adult forms (of cousins or ancestors) as some have claimed.

Recessive: In genetics, the opposite of dominant. In a heterozygote, a version (allele) of a gene that, in combination with an alternative allele, does not have its developmental/phenotypic effects manifested.

Regulatory gene: A gene that regulates the activity of one or more others. Developmental genes are necessarily regulatory. However, the reverse is not necessarily true, because a gene that regulates the activity of a housekeeping gene may not have developmental effects.

RNA polymerase: Enzyme that attaches to DNA in the process of transcription of a gene and facilitates the production of an RNA strand whose base sequence is complementary to that of its DNA template.

Rotifera: Phylum of animals individually called rotifers. There are several thousand species of these. They are typically very small (up to a few millimetres). The name means 'wheel-bearing', and refers to roundish structures that can be seen at the anterior end of these animals.

S

Saltationism: Opposite view of evolution to gradualism. Saltationists take the view that many (or, in more extreme versions of the theory, most) important changes in evolution take the form of sudden large jumps (saltations) caused by macromutations in developmental genes. Now generally thought to be incorrect in relation to most evolutionary changes, but a few such changes (a small but not yet quantifiable proportion of them) probably *are* saltational.

Segment identity: In a segmented animal such as an arthropod, the body can be largely characterised by the number and the identity of its segments. The former is self-explanatory. The latter refers to the different kinds of segment, each with their characteristic form or identity. For example, some segments will be leg-bearing, others not. The Hox genes are crucial in determining segment identity.

Segment polarity: Used to refer both to the antero-posterior polarity of segments and to the group of genes, including the well-known *engrailed* gene, that are responsible for conveying this polarity on the segments.

Short-germ: A descriptor of early insect development, indicating that only the anterior end of the animal (including its head) is represented in the initial germ-band stage. Beetles provide an example of this kind of development. Cf long-germ.

Signalling pathway: A series of interactions between genes and their protein products that can be thought of as starting with the receipt of a signal at a cell's periphery and the transfer of this signal to the nucleus via transmembrane receptors, cytoplasmic intermediates and nuclear transcription factors.

Sister-group: Used of a group (at any level in the taxonomic hierarchy) that is thought to be the most closely related one to the same-level group that is under consideration. More specific versions applicable to particular levels include 'sister-species'.

Soma: Name given to the body of a multicellular organism as opposed to its germ line (the gametes and their precursor cells). Some forms of heterochrony involve a change in the relative rates of development of somatic and reproductive structures.

Somitogenesis: The name given to the development of segments (or somites) in the body of a vertebrate. This process of segmentation proceeds from anterior to posterior, and is primarily a mesodermal process in contrast to arthropod segmentation, which is primarily ectodermal.

Spandrel: In architecture, a spandrel is a triangular surface where two round arches meet at a right-angle to each other under a dome. The term was brought into evolutionary biology to illustrate the point that not all morphological features of an organism need be adaptive. Rather, some may simply be inevitable consequences of a particular overall structure, in the same way that spandrels are inevitable consequences of the joining of arches.

Speciation: The origin of new species. Often occurs (and has been most studied) in relation to the splitting of one parent species into two or more daughter species.

Speciose: Used to refer to a taxon, at any level, which has an unusually high number of species. For example, *Drosophila* is a very speciose genus because it contains more than 2000 species, which is much more than in a typical genus.

Standing variation: The pool of variation that exists in a natural population at a given moment of time. This can refer to the genes themselves (the 'gene pool') or to the developmental and phenotypic variation that results (partly) from the underlying variation in genes.

Stem cell: A cell that is still capable of generating a range of differentiated cell types. One which can generate several but not all cell types is described as pluripotent, while one that can generate all the cell types of the organism concerned is described as totipotent. Stem cells exist not just in embryos and other developmental stages but also in adults as a source for new differentiated cells to replace those that die.

Stochastic: Refers to a process that is effectively random. The opposite of systematic or deterministic. In population genetics, genetic drift is a stochastic process, while natural selection is a deterministic one.

Strepsiptera: An insect order with larviform females and two-winged males characterised by reduction of the forewing, in contrast to reduction of the hindwing, which characterises the better-known two-winged insect group, the Diptera ('flies').

Syncytium: A tissue or organism that is not composed of physically separate cells. Rather, there is a large area of cytoplasm with many nuclei but no dividing cell membranes. Some developmental stages or tissues in otherwise cellular creatures are syncytial – for example, the insect blastoderm. Also, some organisms are completely (or nearly so) syncytial; one of the best-known such groups is that containing the syncytial slime moulds.

T

Taxon (plural taxa): A group of related organisms at any level, from the lower levels of species and genera to the high levels of phylum and kingdom. Taxa are hierarchically arranged, as noted by Darwin in his 'groups within groups' approach to evolutionary pattern. Each taxon should correspond to a clade, because taxa that are not monophyletic are considered to be invalid groupings.

Teleost: Most of today's ray-finned fishes belong to a group called the Teleostei, which is distinguished by paired bracing bones in the tail. The few exceptions include the basal ray-finned fishes, the sturgeon and the paddle-fish.

Telomeres: The terminal regions of a chromosome. These are often characterised by large amounts of highly repetitive DNA. Changes in telomeres are associated with the ageing process.

Terminal target gene: Gene that is switched on at the end of a cascade of interactions among developmental genes. Often cell-type specific. The genes for the haemoglobin proteins are good examples.

Tetrapod: A vertebrate that belongs to the group (Tetrapoda) that derived from an ancient stem lineage that invaded the land. Includes both amphibians and amniotes. Although most do indeed have four legs, as their name suggests, some have lost them, most obviously the snakes. These are still tetrapods in a taxonomic sense, even though they are not in a functional sense.

Trans-acting: Refers to the control of a gene by mobile factors that act by coming towards the regulatory regions of a gene from elsewhere and binding to one or more of them, thereby helping to switch the gene on or off. Thus these factors are not stretches of DNA close to the target gene on the same chromosome. In fact, many trans-acting agents are 'transcription factors' – typically the protein products of regulatory genes that can reside anywhere in the genome. Trans-acting factors interact with cis-acting ones.

Transcription factor: A protein that binds to DNA and is involved in switching one or more genes on or off; or in regulating the rate of transcription. There are different classes of transcription factors, defined by the presence of particular motifs such as helices and 'zinc fingers'. Homeobox-containing genes make transcription factors.

Trans-environment fitness profile: A graphical way of picturing the degree to which a fitness change is specific to one environment (ecological adaptation), general across all environments (internal coadaptation), or somewhere in between these two extremes (which is likely to be the norm).

Transformation: The name given by D'Arcy Thompson to a systematic change of some sort in the grid on which the outline of one kind of animal has been drawn, with the result that the outline of another animal appears. An example would be the transformation of a rectangle into a parallelogram.

Transposon: Also known as a transposable element or a 'jumping gene'. A stretch of DNA that can transpose from one site in the genome to others. Generally thought of as a type of 'selfish DNA'. Transposition of one of these elements causes mutation in a gene into which it transposes.

Trochophore larva: A kind of larva shared by several phyla of animals, including annelids (segmented worms) and molluscs. It is translucent and has characteristic bands of cilia. It is normally lost from the life-cycle in species that leave the marine habitat and become freshwater or terrestrial.

Tunicate: An animal belonging to the chordate subphylum Urochordata. Commonly referred to as sea-squirts, they have a tadpole larva and a sessile adult. Recent work suggests that these animals are the closest living relatives of the vertebrates.

U

Upstream: Like 'downstream', used in two distinct ways: one developmental, the other genetic. In the context of developmental pathways, thought of as a temporal series of stages that are linked causally, any particular stage is downstream of an earlier one in the series. In gene structure, a regulatory region that is downstream of the coding region is one that can be found after the site where transcription stops.

Urbilaterian: The original bilaterian animal, from which the whole of the clade Bilateria, comprising most of the animal kingdom, derived. There is still much debate on what the urbilaterian looked like – perhaps a simple flatworm body plan; perhaps something much more complex.

Urochordata: One of the three subphyla of the phylum Chordata, the other two being the Cephalochordata (or Acraniata) and the Vertebrata. In terms of species numbers, the vertebrates are by far the biggest of the three; but the other two are interesting because of what they can tell us about chordate and vertebrate origins.

V

Vendian: Name of the last pre-Cambrian geological Period. It thus ends when the Cambrian begins – at about 543 MYA. The start of the Vendian is harder to define, but is generally taken to be about 650 MYA. The Vendian Period, or part of it, is sometimes referred to as the Ediacaran, after the Ediacaran Hills in Australia where the first Vendian fossils were discovered. These are called by many names, including the Vendobionta. There is still much debate about exactly what kinds of organisms they were.

Vestigial: Refers to a structure or organ that has become reduced in a particular lineage over evolutionary time to the point where it is tiny and may well be functionless. Examples include the human appendix and the tiny shells of some slugs.

Von Baerian divergence: Pattern often but not always seen in comparing the embryos of different animals within a higher taxon such as an order, class or phylum. The pattern is one of early similarity giving way to later difference. Numerous variations on this pattern are known, including differences in very early stages giving way to similarity, which later gives way to differences again. This pattern is referred to as the developmental egg-timer or hour-glass.

W

Wild-type: Term deriving from *Drosophila* genetics. In early laboratory experiments it was common to breed 'normal' flies (such as those that could be found in natural populations) with specific mutant flies (e.g. white eye, vestigial wing). The normal flies were referred to as wild-types. However, it is now clear that most natural populations are massively polymorphic, so the idea of a definite wild-type is illusory.

Z

Zooid: One of the units that make up a colony in the case of a colonial animal such as a bryozoan or some species of tunicate. Broadly equivalent to a ramet in the case of a vegetatively-reproducing plant.

APPENDIX 1

A Little Bit of History

The Incorporation of Genetics into Evolutionary Theory; and its Effects on the Ongoing Incorporation of Developmental Biology

The incorporation of genetics into evolutionary theory can be thought of as happening in five stages, as follows:

1. *Stage 1*: In 1866, Gregor Mendel[1] published his theory of particulate (rather than blending) inheritance. This solved the problem of maintaining the variation that natural selection acts upon and has thus become a major contribution to evolutionary theory, though it was not recognised as such at the time. Under a system of blending inheritance, in which each offspring is a 'blend' of its parents, **heritable** variation would rapidly decrease to zero; but if each offspring is merely a temporary association between the contributions of its parents, as we now know, variation can be maintained indefinitely. Mendel's work, which languished unrecognised for what it was for 34 years, was 're-discovered' in 1900 and brought into the limelight.
2. *Stage 2*: Mendel's work, which initially applied just to families of parents and their offspring, was extended to whole populations by Hardy and Weinberg. Also, the apparent distinction between discrete variation in populations (e.g. eye colours), which was clearly explicable in Mendelian terms, and continuous variation (e.g. in body height or length), which was apparently not, was dissolved by R.A. Fisher in 1918. He showed that the latter could be explained in terms of many genes, each of small effect, producing a normal distribution of **phenotypic** values rather than a small number of phenotypic categories.
3. *Stage 3*: This stage could be named after the title of Fisher's 1930 book[2]: *The Genetical Theory of Natural Selection*. Around this time, the 'great triumvirate' of pioneering

population geneticists – Fisher, Haldane and Wright – effectively merged Darwin's theory of natural selection with Mendel's theory of particulate inheritance to produce a combined theory. This gave a firm mathematical basis to natural selection, to which practical case studies could subsequently be related.

4 *Stage 4*: In the 'modern synthesis' of evolutionary theory, which began in the 1930s and continued into the 1940s and 1950s, palaeontology, systematics and field-based ecological genetics lined up with the mathematical theory of natural selection to produce what many then saw (wrongly) as the ultimate consolidation of evolutionary biology. In particular, the work of Dobzhansky[3] and Mayr[4] in the USA, and later that of Ford in Britain, continued the incorporation of genetics into our view of evolution.

   There is a major difficulty in working with continuous variation in natural populations: such variation is only partially heritable, and the separation of heritable and non-heritable components in nature is often impossible. Because of this difficulty, Dobzhansky, Ford and others worked on discrete variation that took the form of **polymorphisms**: respectively, of chromosomal inversion types in dipterans and pigmentation patterns in lepidopterans. The first of these involved looking at **karyotypes** rather than phenotypes; the second involved looking at a rather special and unrepresentative suite of phenotypic traits.

5 *Stage 5*: Many genes make enzymes. But variation in these does not usually show up in the 'visible phenotype'. However, the pioneering use of gel electrophoresis at the population level in the 1960s by Lewontin and Hubby (on dipterans) and Harris (on humans) revealed massive levels of polymorphism in these genes. This revelation gave rise to the controversy between those who thought the variation was acted upon by natural selection and those who thought it was selectively **neutral**. Work on polymorphism in enzyme-producing genes spread out to cover many **taxa** in the 1970s. One clear conclusion was that a high level of polymorphism characterised almost all sexually-reproducing species[5]. A less clear conclusion was that some of the variation was acted upon by selection while some was not – to this day we are unable to identify with certainty the relative proportions.

The incorporation of modern developmental biology into evolutionary theory began about 1980. We need to look at how the earlier incorporation of genetics affected (and is still affecting) this second expansion of evolutionary theory.

We will begin by considering the positive effects, and will concentrate on the three main ones, as follows. First, as already noted, the ability of Mendel's particulate theory of inheritance to explain the indefinite persistence of variation in a population was important for evolutionary theory. This variation is developmental as well as genetic. Second, a theory that has been put on a mathematical footing has a stronger basis than one that is merely verbal. The studies of Fisher, Haldane and Wright, and indeed of subsequent generations of theoretical population geneticists, have quantified the ways in which natural selection works. Since variation in developmental pathways ultimately interacts with natural selection, this is again beneficial to the overall endeavour of attempting to incorporate development into evolutionary theory. Third, the demonstration that the typical enzyme-producing gene is polymorphic rather than fixed was helpful in moulding our expectations about development-controlling genes.

This is especially true given the unfortunate tendency of some earlier non-evolutionary-minded developmental biologists to refer to the development of, for example, 'the frog', notwithstanding the fact that no two frogs are identical, and that the reason for this is partly genetic.

Now for the negative effects of the formulation of a genetic theory of evolution for the subsequent and ongoing attempts to produce a theory of evolution that includes development as well. Whether these 'effects' are indeed plural is a moot point. Perhaps there has just been one negative effect that has appeared in different guises in different decades. It takes the form of what has often been called 'genetic imperialism' – the tendency to see the gene as the supreme evolutionary player, thus relegating other players, such as developmental pathways, to a minor role at best.

One manifestation of this problem can be seen in the decade from 1930 to 1940. In contrast to the pluralistic approach that Darwin[6] employed in 1859, Fisher's 1930 book is very narrow in its approach. Thus, although it is recognised as one of the foundations of population genetics, its downplaying of development (to the point of virtual omission) caused an adverse reaction among those who thought that development should be given a central role. An example of this is Richard Goldschmidt's anguished comment[7]: 'I have tried for a long time to convince evolutionists that evolution is not only a statistical genetical problem, but also one of the developmental potentialities of the organism.' The sympathy which some present-day advocates of evo-devo have for Goldschmidt in this respect, despite his holding extreme **saltationist** views on evolution that most of them would reject, is reflected in the many recent references to his work, and indeed in dedications to him, including one at the front of the pioneering evo-devo book[8] *Embryos, Genes and Evolution*.

As ever, though, no school of scientific thought is homogeneous. So while Goldschmidt's exasperation with the 'evolutionists' of the time was justified in the case of Fisher, it was much less so in the case of Haldane, whose 1932 book[9] *The Causes of Evolution* (note the pluralism in the title) contains some interesting statements from our current perspective. Notably, at the end of his chapter 'What is Fitness?' Haldane says:

"To sum up, it would seem that natural selection is the main cause of evolutionary change in species as a whole. But the actual steps by which individuals come to differ from their parents are due to causes other than selection, and in consequence evolution can only follow certain paths. These paths are determined by factors which we can only very dimly conjecture. Only a thorough-going study of variation will lighten our darkness."

Thus, in a few sentences, Haldane presaged both the theoretical and practical evo-devo of the future, including the concept[10] of **developmental bias/constraint** and the variation in the spatio-temporal **expression patterns** of developmental genes that we now know occurs within, as well as between, species.

A later manifestation of the 'genetic imperialism' problem can be seen in the 1970s. Much of the primary literature of the time saw evolution at the population level (i.e. micro-evolution) in terms of an interplay between **mutation** and selection. Development simply did not enter the picture. There was often a Fisherian flavour to this work. Again, though, we must remember that no school of scientific thought is homogeneous, and hence that this criticism is more applicable to some authors of the primary papers of the 1970s than to others. However, the problem was not so much the primary literature but rather the more popular books based on it. In particular, Wilson and Bossert, in a book aimed at students, actually *defined* evolution as a change in the gene frequency of a population (what loss of Darwin's 'grandeur'!). And Richard Dawkins' very successful popular science book[11] *The Selfish Gene* managed to promulgate the idea that genes were the really important players in evolution, on the back of one rather specific topic (the role of genes and genetic relatedness in kin selection).

The reappearance of genetic imperialism in the 1970s caused a strong reaction, just as its earlier appearance in the 1930s did. Perhaps the most famous paper criticising it was what is now usually called the **spandrels** paper[12], written by Stephen J. Gould and Richard Lewontin. These authors stated that organisms are:

"so constrained by phyletic heritage, pathways of development and general architecture that the constraints themselves become more interesting and more important in delimiting pathways of change than the selective force that may mediate change when it occurs."

Notice the similarity to the quote from Haldane earlier; it is interesting to note that Gould and Lewontin did not refer to Haldane. The importance of Gould and Lewontin's critique of genetic imperialism, or what they called the adaptationist programme, is attested to by the title of a chapter in a recent book[13] by Pigliucci and Kaplan: 'A quarter century of spandrels'.

The threat of genetic imperialism is not dead, but it is fading. Current evolution texts aimed at a general student readership often have a little bit of evo-devo, but presented as a sort of afterthought – perhaps a single chapter close to the end. This is true, for example, of Ridley's *Evolution*, even in its most recent edition. But a newer text by Barton and co-authors, also entitled *Evolution*, does better. The time is approaching when no reputable text on evolution will be able to get away with downplaying the importance of development in our overall picture of the evolutionary process.

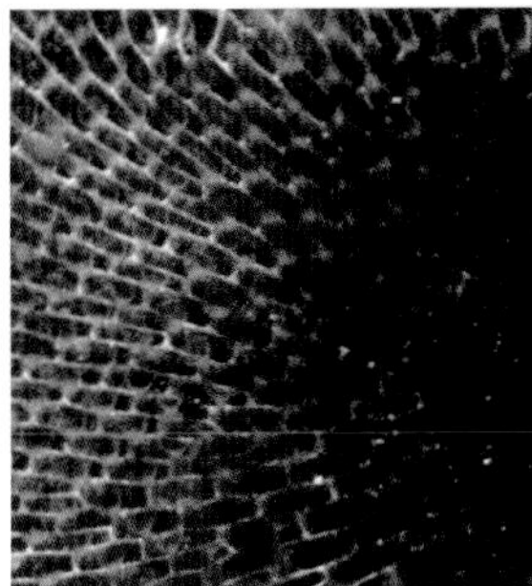

APPENDIX 2

Naming of Genes and Proteins

Most animal and plant **genomes** have more than 10,000 genes, and many of them have more than 20,000. Although not all of the genes in any species are involved in the developmental process, a good many of them are. So even if we restrict our attention to **developmental genes**, we should expect to find hundreds and perhaps thousands of them in any one type of creature.

Taking a comparative approach, which is essential in evo-devo, the number of developmental genes multiplies up by about 2 million – the number of known species of multicellular organisms. Of course, most developmental genes in one type of organism have **homologues** in other types, so this helps to keep the overall number of gene names down; but it is still a very large number. Confronted by this problem, we should have a single unified system for naming developmental genes that makes both remembering and understanding them possible, and even, ideally, easy.

The achievement of such a unified system will not be simple. What has happened to date is that workers on the main **model organisms** have each developed a system for their own organism. These differ from each other in ways that will shortly be described. In many cases, workers on non-model systems have used what is phylogenetically the nearest system – for example, those working on centipedes tend to use the *Drosophila* system, while those working on rats tend to use the mouse system, and those working on some species of **angiosperms** tend to use the *Arabidopsis* system.

However, sometimes workers on non-model organisms make a deliberate attempt to build their species name into their gene names. One way to do this is to use an abbreviated form of the genus/species name, often three letters long. An example is *Stm-cad* for the homologue of the *Drosophila* gene *caudal* (*cad*) in the centipede *Strigamia maritima*. Another is to build a shortened version of the common name of the species into the gene name. An example is *Amphi-En* for the homologue of the *Drosophila* gene *engrailed* (*en*) in the **cephalochordate** animal called the amphioxus. This is probably not a good idea because the widespread use of such a system would produce a plethora of shortened Latin or common names, most of which would only mean something to a small group of people.

Notice that italics are creeping in. One convention that is in common among most systems is the use of italics for the *gene*, whereas non-italics (i.e. Roman font) are used for the corresponding protein. Another common convention is to use a short, alphabetic or alphanumeric abbreviation for both gene and protein. Less consistent is the use of upper case letters. Below is an example to show the different usages of upper-case letters among the systems for human, mouse, zebrafish, *Drosophila* and *Arabidopsis* – i.e. our own species plus four of the seven main model organisms.

We will consider, below, a hypothetical gene whose abbreviated name is *ABC1*. We will assume that this gene, or a homologue of it, is found in each of the four species listed.

Human:

Human gene: *ABC1*
Human protein: ABC1

Mouse:

Mouse gene: *Abc1*
Mouse protein: Abc1

Zebrafish:

Zebrafish gene: *abc1*
Zebrafish protein: Abc1

Drosophila (fly):

Fly gene: *Abc1* OR *abc1*
Fly protein: Abc1

Arabidopsis (flowering plant):

Plant gene: *ABC1*
Plant Protein: ABC1

It can be seen from the above comparisons that there is more commonality among the various systems regarding proteins than genes. Usually, proteins have a capital initial letter, lower-case other letters, and are not in italics. So we will follow that convention throughout this book.

Genes present more of a problem, apart from the use of italics, which is common to all four systems. The greatest inconsistency is in relation to the use of upper or lower case for the initial letter of a gene name (or its abbreviation). One system applies to humans, mice and *Arabidopsis* (upper case always). A second system applies to zebrafish (lower case always). And a third system applies to *Drosophila* (upper OR lower case).

The rationale for the at-first-sight bizarre *Drosophila* system is this: if the mutation through which the gene was first discovered was dominant, the gene (and not just this mutation) gets an initial capital; but if that mutation was recessive, the gene gets an initial lower-case letter. This may have made sense once; but it does not do so now. It leads to some very strange results. For example, take the 'Hox genes', which incidentally is in Roman because this is the name of a group of genes, not an individual gene. In *Drosophila*, two of these genes are *abdominal A* (*abdA*) and *Abdominal B* (*AbdB*). Notice the difference, in terms of the presence or absence of initial capitalisation, both in the full and abbreviated versions of the gene names.

Because it makes sense to use a single system throughout this book, a choice needs to be made regarding the initial letter of a gene being upper or lower case. Since the *Drosophila* system has little to recommend it, the choice is between the other systems. The zebrafish system seems the best because: (i) the use of italics is sufficient in itself to identify something as a gene rather than an ordinary word – for example, *dorsal* in *Drosophila* is a gene involved in patterning the dorsal-ventral axis; and (ii) the lack of an initial capital in the gene makes the difference between it and its corresponding protein doubly clear, given the initial capital used for protein names (e.g. Dorsal).

With regard to the choice of names for individual genes, this is often derived from the developmental/phenotypic effects of mutations in them. Examples of this that come up in the text are:

- *bicoid*: 'bi-caudal' or two-tailed. This is the form taken by an embryo in which the anterior-determining gene product is absent or defective.
- *distal-less*: Lacking the distal (furthest from the body) parts of the limbs because of the absence or defectiveness of a gene product normally found there.
- *hedgehog*: 'Prickly' fly larva resulting from a mutation in a gene that causes a continuous 'lawn' of projecting denticles where normally there would just be stripes.
- *stamenoid petals*: Here the mutant plant has petals that have been transformed into stamen-like structures.

In other cases, gene names derive from different sources. Sometimes they are combinations of two separate things – see Chapter 3 for an example of this (*wnt*). In other cases, a sense of humour is involved – as in *sonic hedgehog*, whose hedgehog-based name, together with other related ones, was necessary when it was discovered that in vertebrates there were several hedgehog-type genes rather than just one.

Finally, a word on groups of genes. These can be defined in various ways. At the broad end of the spectrum there are categories such as 'developmental genes'. Within this category there are narrower ones, such as 'segmentation genes'. Within the latter are yet narrower categories, such as 'pair-rule genes'. Note that no italicisation or capitalisation is used for these. However, in cases where gene groups are named in a way that reflects the names of their constituent genes, it is conventional to use an initial capital. So, for example, there are the Hox genes (e.g. *hoxb4*) and the Pax genes (e.g. *pax6*). Although it could be argued that this is opening the door to confusion with protein names, the convention is well established; and it is almost always clear from the context that proteins are not being referred to – simply because the word Hox (or Pax, etc.) is immediately followed by 'genes', as above.

The use of a consistent system should be helpful to readers of this book, especially students who have not yet become familiar with one system at the expense of others. But it is important to bear in mind that readers using the reference list provided at the end of the book to access primary papers in the area of developmental genetics will have to adapt to the diversity of systems 'out there' in the literature more generally. (Some readers may also notice that, while I have used a consistent system in the text of this book, Figures reproduced here from elsewhere often do not conform to this system.)

APPENDIX 3

Geological Time

The Earth formed about 4.5 billion years ago (BYA). As far as we are aware, there was no life of any kind prior to about 4 billion years ago. The earliest known fossils date from between 4 and 3 BYA. From that span of time until the base of the Cambrian period, approximately 0.5 BYA, or, to be more exact, 543 million years ago (MYA), fossils are few and far between, and many of them are hard to allocate to particular taxa. After that point, fossils are plentiful, albeit patchy in their occurrence in both time and space.

A commonly accepted scheme for naming the eons (or aeons) in the history of the Earth is as follows (Fig. A3.1):

- 4.5 to 4.0 BYA: The Azoic eon ('no animals'; but really meaning 'no life').
- 4.0 to 2.5 BYA: The Archaean eon ('old')
- 2.5 to 0.5 BYA: The Proterozoic eon ('first animals'; but not clear when)
- 0.5 to 0.0 BYA: The Phanerozoic eon ('manifest animals'; but really 'life')

From an 'evolution of development' perspective, there is little need to subdivide the first three of these four great eons, because not much seems to have happened during them in terms of multicellular life-forms. However, we should make an exception for the latest part of the Proterozoic (from ~650–543 MYA) because there clearly *are* fossil multicells from this time-span. This is called the **Vendian** period.

This period and the many later ones dividing up the Phanerozoic are shown in Fig. A3.2. Also shown are the three eras of the Phanerozoic; eras being intermediate in scale between eons and periods. The use of three eras and about ten periods to divide up the Phanerozoic eon is helpful, because so much happened during this eon. Its rich fossil record compared to the earlier three eons means that we need to have a scheme for splitting it into manageable 'chunks' of time (Fig. A3.2).

Notice that both the eras and the periods within the Phanerozoic eon begin and end with non-round numbers; also that they are not of equal length (periods range from ~25 to ~80 MY in duration). This is because the dividing lines between them are not arbitrary, but rather are based on particular geological or palaeontological events.

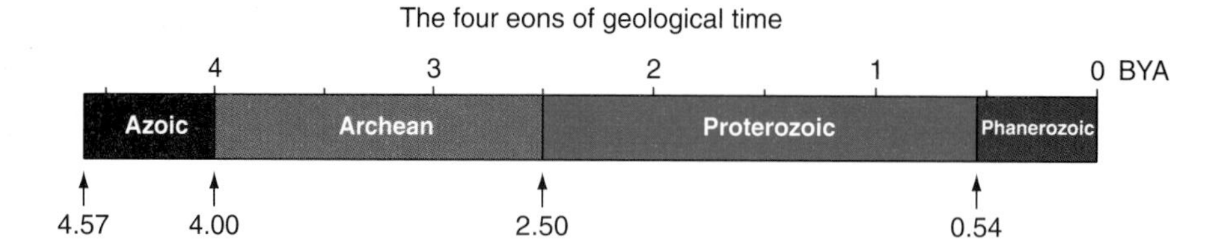

Figure A3.1 The four eons of Earth history. The names of all but the first are well established. However, the first one has been called, in addition to the Azoic, the Priscoan and the Hadean.

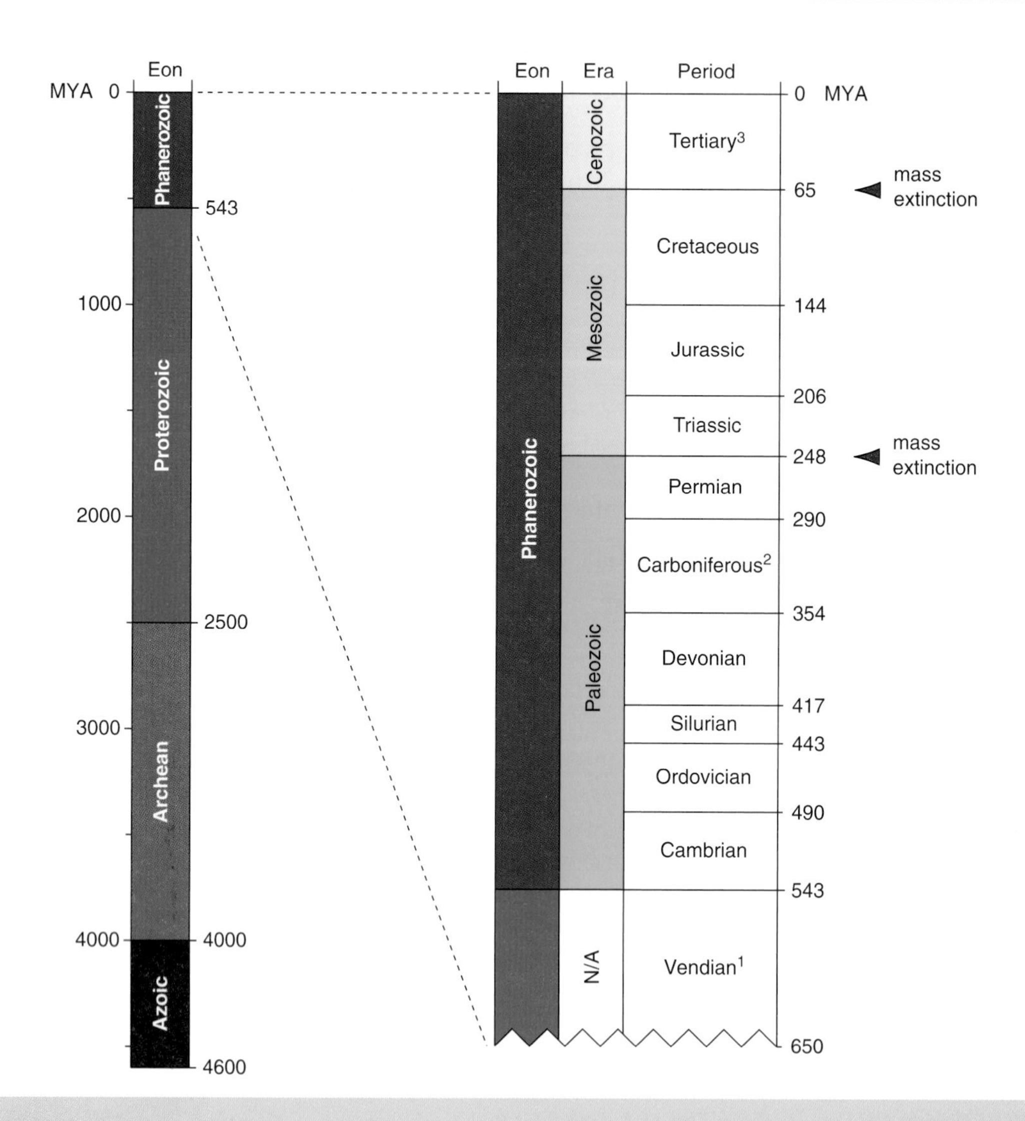

Figure A3.2 The eras and periods of the Phanerozoic eon, together with the Vendian period, which is the latest part of the Proterozoic eon. Some additional information, corresponding to the superscripts at three points on the figure, is as follows: (1) The date of the base of the Vendian is not agreed on – different publications use varied dates from 650 to 600 MYA. (2) The Carboniferous is divided in two by American authors, with the Mississippian (earlier) and Pennsylvanian (later) adjoining each other at about 318 MYA. (3) The most recent geological era, the Cenozoic, is short. In the figure it is shown as having just one period, the Tertiary. However, it is conventional to recognise the most recent 2.5 MY as a separate period, the Quaternary. Also, there is an alternative system, which is to use Palaeogene for the time from 65 to 28 MYA and the Neogene for the period from 28 MYA to the present.

For example, as can be seen from Fig. A3.2, the ends of the Palaeozoic and Mesozoic eras are defined by mass extinction events. The second of these, at 65 MYA, is particularly well known, even outside biology and geology, because this is when the charismatic dinosaurs became extinct (along with many other **taxa**). The earlier mass extinction (at ~250 MYA) is less well known, but was even more drastic in terms of the number of taxa affected. It was in this mass extinction event that the trilobites (one of the best-known extinct invertebrate groups) finally disappeared.

There are generally reckoned to have been at least four other mass extinction events. However, since the background level of extinction (number or percentage of species or other taxa that end per million years) is itself very variable, exactly what constitutes enough of a rise above the background rate to be described as a mass extinction event is not clear.

When we wish to refer to a narrower stretch of geological time than those represented by the various periods, we can do this in three ways. First, we can make use of the division of a period into Early, Middle and Late. Second, we can use the finer-scale 'stages', which are too numerous to name here, but to give an example there is a stage in the Early Cambrian called the Tommotian; stages are usually just a few million years in duration. Third, we can put an approximate date to an event in terms of MYA. Herein, this third method is used. This is because it is hard enough for students of evolution to get used to the names of the eons, eras and periods (almost 20 in all), without having to become familiar with the huge number of names of all the subdivisions of the periods.

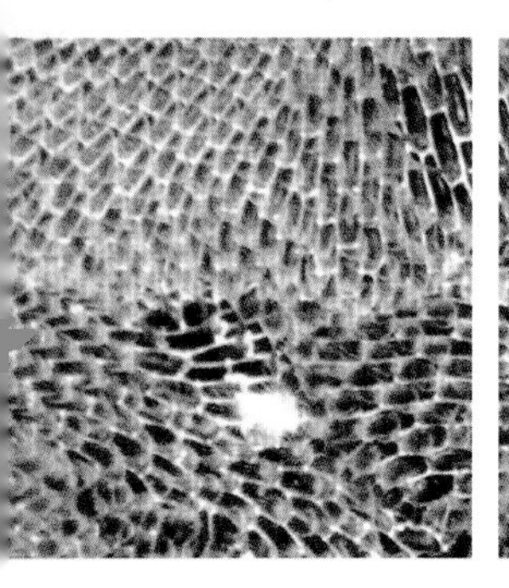
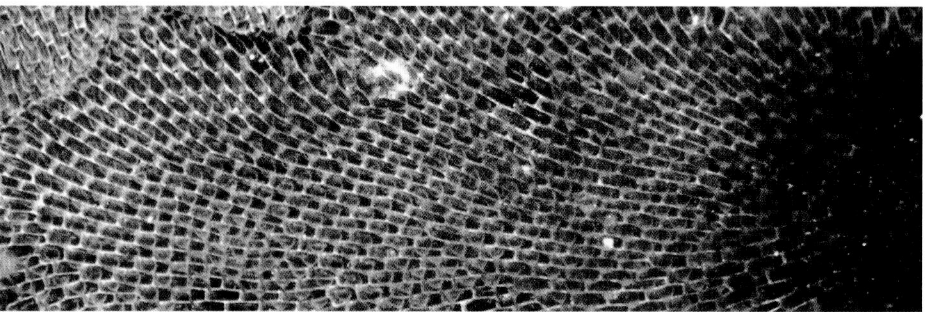

APPENDIX 4

Inferring Evolutionary Trees from Comparative Data

If evolution was a neat, orderly process, then inferring the pattern of relationship among a group of species, and thus the pattern of **lineage** splitting (i.e. evolutionary tree) through which they were produced, would be easy. However, since evolution is in reality a messy, unpredictable process, inferring patterns of relationship and the trees that produced them is not easy at all.

We will examine this issue in a very simple situation: three species (called S1, S2 and S3) in each of which we can 'measure' two characters (C1 and C2), each of which has two possible values or 'states'. Let's say that C1 has possible states A and a, while C2 has possible states B and b. For those who like less abstract examples, consider all of S1 to S3 to be species of snail. The character C1 is **chirality** (direction of coiling; sinistral or dextral), while C2 is the colour of the shell (plain pigmentation with no patterning; yellow or brown). We will further suppose that each character is fixed for a particular state in each species – that is, there is no intraspecific variation.

In the 'simple, orderly evolution' scenario, we might find a pattern of character states as follows:

- S1: C1-a; C2-b
- S2: C1-A; C2-B
- S3: C1-A; C2-B

This would argue for one of the two evolutionary trees (ET) shown on the left-hand side in Fig. A4.1. However, this pattern of character states cannot distinguish between them – that is, it cannot help us to infer the character states in the stem species. To do that, we would need data on an **outgroup** beyond our initial group of three species. Suppose this outgroup is another species, S4, which is thought to be quite closely

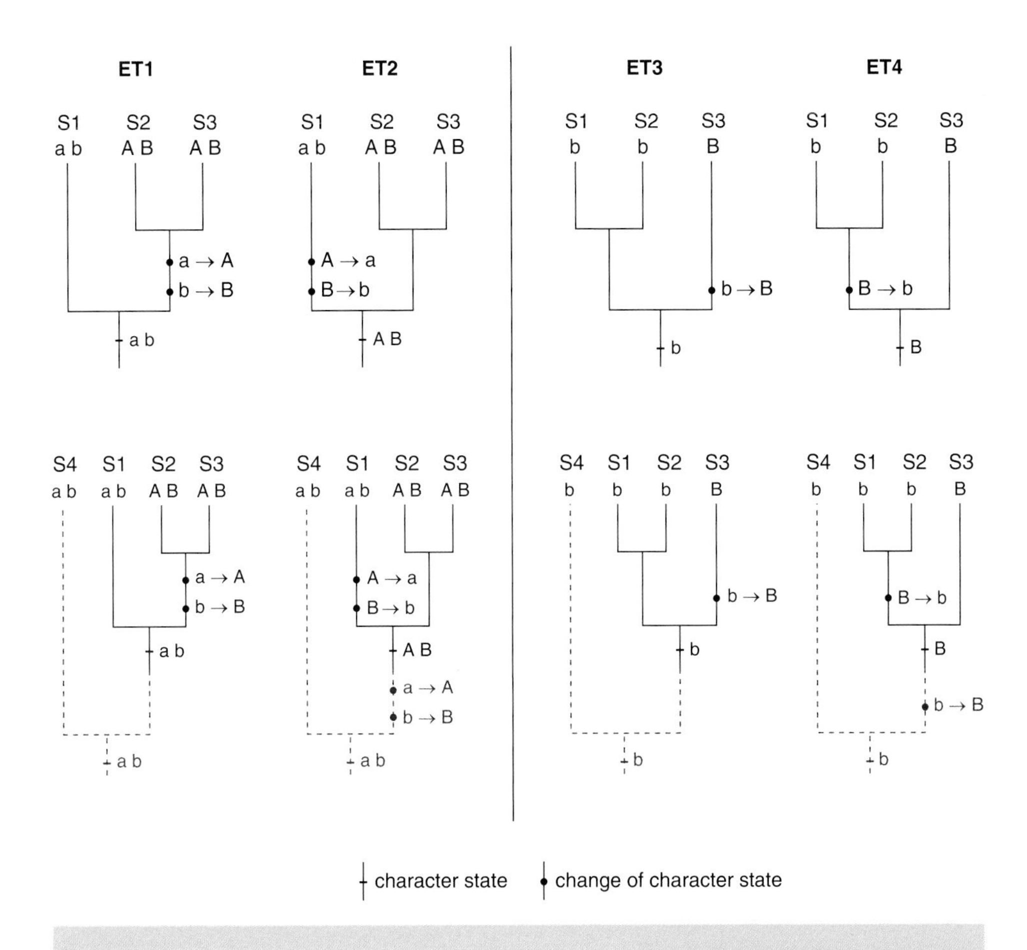

Figure A4.1 Possible evolutionary trees (ET) linking up a series of three species (S1–S3), and sometimes also (bottom) a fourth, outgroup, species, S4, using information on two characters (C1, C2), each of which can take two possible states (*A*, *a* for character 1; *B*, *b* for character 2). For further information, see text.

related to, but outside of, the group S1, S2, S3. If S4 has character states a, b, it seems more likely that ET1 is correct than ET2, because ET1 is simpler. This kind of argument is referred to as parsimony. Where two evolutionary trees can both produce the same result, we decide that the simpler one is more likely to be correct.

Now, here are three issues that arise in reality:

First, the issue of different characters showing different patterns, or 'character conflict'. An example of this issue can be seen below:

- S1: C1-a; C2-b
- S2: C1-A; C2-b
- S3: C1-A; C2-B

Here, if we use the pattern of character states in C1, we end up with the same inferred trees (LHS of Fig. A4.1) as before. However, if we use the pattern of states in C2, we end up with different inferred trees (RHS of Fig. A4.1). If we choose to believe that C2 is likely to be more reliable for some reason (see below) then, as previously, an outgroup S4 would help to infer the character state of the ancestor; or to distinguish between ET3 and ET4, which is the same thing put in a different way.

Second, the issue of character weighting. In the above approach, it is assumed that neither character is any more 'important', in the sense of being more useful in inferring the correct evolutionary tree, than the other. In reality, this may not be true. It is certainly not true of the examples given above – chirality and colour in snail shells. Chirality is known to be very resistant to change, whereas colour changes very readily. Knowing this means that in the case of 'character conflict' described above, we would rely more on the data for C1 (chirality) than that for C2 (colour).

This example, using two greatly contrasting characters in snails, involves character weighting being a help rather than a hindrance. But more often, when we are dealing with many rather similar characters, their relative weightings are unclear, and the question of whether we should attempt to weight them differently or not is very difficult to answer. This leads nicely into the third issue...

Third, the issue of multiple characters (with multiple states) in multiple species. Here, instead of having $2^{2\times3} = 64$ possible combinations of character states across the 3 species, we have a potentially enormous number. If there are 5 species, 5 characters and 3 possible states for each character (still much smaller than most actual datasets), then there are $3^{5\times5}$, which is more than 847 billion, possible combinations. The solution to this problem is to replace visual inspection of the data with computer-based analysis. There are many well-established software packages available, some based (or partly based) on the principle of parsimony that we noted above – for example, PAUP (Phylogenetic Analysis Using Parsimony – see http://paup.csit.fsu.edu/); others based on different approaches – such as that of Bayesian statistics (which is beyond the scope of this book).

It is important to realise that 'a character' can be virtually anything – well, anything that varies and whose 'state' can be determined. That means that 'characters' can be molecular as well as morphological. They can be as small as a single base in a stretch of DNA (which will automatically have four states – A, T, C and G). DNA (and protein) sequence information has been much used in the last three decades to infer evolutionary trees. When a tree derived from molecular data is congruent with one based on morphological data, we can be reasonable sure that it is correct (e.g. the tree of relatedness of the main groups of centipedes). When the trees derived from these two sources are very different, we have to admit that we do not really know which (if either) is correct (e.g. in relation to the main groups of sponges). It may be true that molecular data are better because the problem of **convergent** evolution is less of an issue in the molecular realm than in its morphological equivalent; but no-one should assume that convergence does not occur at the level of DNA and proteins – it does.

Finally, a word of caution on the use of embryological 'characters' (and other developmental, including **larval** characters) to infer evolutionary trees. Prior to the advent of molecular data, embryological characters were often thought to be more reliable than adult ones, and this may indeed be true. The two main subgroups of **bilaterian** animals – **protostomes** and **deuterostomes** – were established on the basis of such characters and have stood the test of time, since molecular data have subsequently confirmed them. However, since evo-devo involves, among other things, the mapping of **developmental repatterning** to trees (Chapter 11), we have to be careful not to become involved in circular reasoning.

The purpose of this brief appendix has been to demonstrate that a kind of diagram (a tree) that *looks* simple may have some very complex analysis underlying its production. It is always worth bearing this in mind.

FURTHER READING

A good, short, account of the use of parsimony in phylogenetic inference, and the methods of cladistics in general (from the man who formulated it) can be found in chapter 1 of:

Hennig, W. 1981 *Insect Phylogeny*. Wiley, Chichester.

For a more recent account, aimed at students, and going into more detail than I have done here, see:

Futuyma, D. J. 2005 *Evolution*. Sinauer, Sunderland, MA.

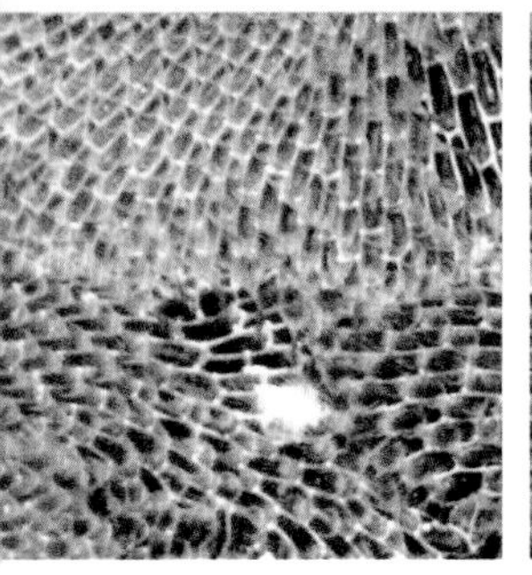
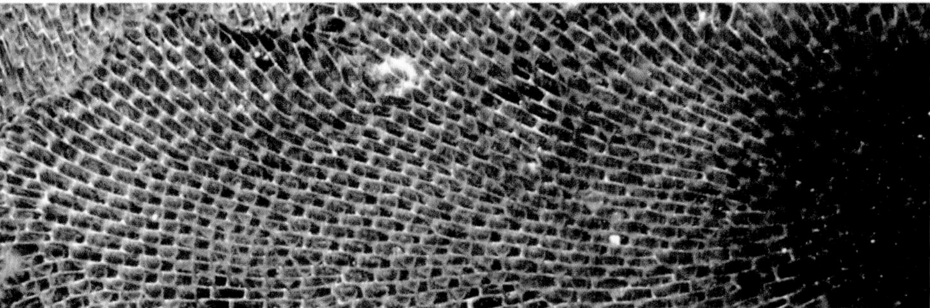

References

References for Part I

1 Darwin, C. (1859) *On the Origin of Species by Means of Natural Selection, or the Preservation of Favoured Races in the Struggle for Life*, J. Murray, London.

2 Minelli, A. (2003) *The Development of Animal Form: Ontogeny, Morphology and Evolution*, Cambridge University Press, Cambridge.

3 Wray, G.A. and Bely, A.E. (1994) *The Evolution of Developmental Mechanisms* (eds M. Akam, P. Holland, P. Ingham and G. Wray), Development Supplement, Company of Biologists, Cambridge.

4 Gilbert, S.F. (2007) *From Embryology to Evo-Devo: A History of Developmental Evolution* (eds M.D. Laubichler and J. Maienschein), MIT Press, Cambridge, MA.

5 McKinney, M.L. and McNamara, K.J. (1991) *Heterochrony: The Evolution of Ontogeny*, Plenum Press, New York.

6 Raff, R.A. (1996) *The Shape of Life: Genes, Development and the Evolution of Animal Form*, Chicago University Press, Chicago.

7 Buchholtz, E.A. and Stepien, C.C. (2009) Anatomical transformation in mammals: developmental origin of aberrant cervical anatomy in tree sloths. *Evolution & Development*, **11**, 69–79.

8 Roth, G. and Wake, D.B. (1985) Trends in the functional morphology and sensorimotor control of feeding behaviour in salamanders: an example of the role of internal dynamics in evolution. *Acta Biotheoretica*, **34**, 175–191.

9 Arthur, W. (2000) The concept of developmental reprogramming and the quest for an inclusive theory of evolutionary mechanisms. *Evol. & Dev.*, **2**, 49–57.

10 Park, I-H. *et al.* (2008) Reprogramming of human somatic cells to pluripotency with defined factors. *Nature*, **451**, 141–146; Hajkova, P. *et al.* (2008) Chromatin dynamics during epigenetic reprogramming in the mouse germ line. *Nature*, **452**, 877–881.

11 King, N. *et al.* (2008) The genome of the choanoflagellate *Monsiga brevicollis* and the origin of the metazoans. *Nature*, **451**, 783–788.

12 Dunn, C.W. *et al.* (2008) Broad phylogenetic sampling improves resolution of the animal tree of life. *Nature*, **452**, 745–749.

13 Haeckel, E. (1896) *The Evolution of Man: A Popular Exposition of the Principal Points of Human Ontogeny and Phylogeny*, Appleton, New York.
14 Von Baer, K.E. (1828) *Uber Entwicklungsgeschichte der Tiere: Beobachtung und Reflexion*, Borntrager, Konigsberg.
15 Panchen, A.L. (1992) *Classification, Evolution and the Nature of Biology*, Cambridge University Press, Cambridge.
16 De Beer, G.R. (1940) *Embryos and Ancestors*, Clarendon Press, Oxford.
17 Mendel, G. (1866) Versuche über Pflanzenhybriden. *Verhandlungen des naturforschenden Vereines in Brünn, Bd. IV für das Jahr 1865, Abhandlungen*, **4**, 3–47.
18 Haldane, J.B.S. (1932) *The Causes of Evolution*, Longman, London.
19 Thompson, D.A.W. (1917) *On Growth and Form*, Cambridge University Press, Cambridge.
20 Goldschmidt, R.(1940) *The Material Basis of Evolution*, Yale University Press, New Haven.
21 Goldschmidt, R. (1952) Homeotic mutants and evolution. *Acta Biotheoretica*, **10**, 87–104.
22 Waddington, C.H. (1957) *The Strategy of the Genes*, Allen & Unwin, London.
23 Waddington, C.H. (1956) Genetic assimilation of the bithorax phenotype, *Evolution*, **10**, 1–13.
24 Whyte, L.L. (1965) *Internal Factors in Evolution*, Tavistock Publications, London.
25 Gould, S.J. (1977) *Ontogeny and Phylogeny*, Harvard University Press, Cambridge, MA.
26 Watson, J.D. and Crick, F.H.C. (1953) Molecular structure of nucleic acids: a structure for deoxyribose nucleic acid. *Nature*, **171**, 737–738.
27 Berrill, N.J. (1961) *Growth, Development and Pattern*, Freeman, San Francisco, CA.
28 Pigliucci, M. and Kaplan, J. (2006) *Making Sense of Evolution: The Conceptual Foundations of Evolutionary Biology*, Chicago University Press, Chicago.
29 Duboule, D. (ed.) (1994) *Guidebook to the Homeobox Genes*, Sambrook & Tooze, Oxford.
30 Holland, P.W.H., Booth, H.A. and Bruford, E.A. (2007) Classification and nomenclature of all human homeobox genes. *BMC Biology*, **5**, Rec 47.
31 Lewis, E.B. (1963) Genes and developmental pathways. *Am. Zool.*, **3**, 33–56.
32 Nüsslein-Volhard, C. and Wieschaus, E. (1980) Mutations affecting segment number and polarity in *Drosophila*. *Nature*, **287**, 795–801.
33 Alber, F. *et al.* (2007) The molecular architecture of the nuclear pore complex. *Nature*, **450**, 695–701.
34 Wolpert, L. (1971) Positional information and pattern formation. *Current Topics in Developmental Biology*, **6**, 183–222.
35 Lawrence, P.A. (1992) *The Making of a Fly: The Genetics of Animal Design*, Blackwell, Oxford.
36 Arthur, W. (1988) *A Theory of the Evolution of Development*, John Wiley & Sons, Ltd, Chichester & New York.
37 Robert, J.S. (2003) *Keywords and Concepts in Evolutionary Developmental Biology* (eds B.K. Hall and W.M. Olson), Harvard University Press, Cambridge, MA.
38 Hutchinson, G.E. (1965) *The Ecological Theater and the Evolutionary Play*, Yale University Press, New Haven, CT.
39 Dobzhansky, T. (1943) Genetics of natural populations. IX. Temporal changes in the composition of populations of *Drosophila pseudoobscura*. *Genetics*, **28**, 162–186.
40 Turner, J.R.G. (1977) Butterfly mimicry: the genetical evolution of an adaptation. *Evol. Biol.*, **10**, 163–206.
41 Cain, A.J. and Sheppard, P.M. (1950) Selection in the polymorphic land snail *Cepaea nemoralis* (L.). *Heredity*, **4**, 275–294.
42 Lack, D. (1947) *Darwin's Finches: An Essay on the General Biological Theory of Evolution*, Cambridge University Press, Cambridge, UK.

43 MacArthur, R.H. (1972) *Geographical Ecology: Patterns in the Distribution of Species*, Harper & Row, New York.

44 Brakefield, P.M. *et al.* (1996) Development, plasticity and evolution of butterfly eyespot patterns. *Nature*, **384**, 236–242.

45 Karataglis, S.S. (1982) Combined tolerance to copper, zinc and lead by populations of *Agrostis tenuis*. *Oikos*, **38**, 234–241.

46 Birky, C.W. (2004) Bdelloid rotifers revisited. *Proc. Natl. Acad. Sci. USA*, **101**, 2651–2652.

47 Virtanen, R., Edwards, G.R. and Crawley, M.J. (2002) Red deer management and vegetation on the Isle of Rum. J. *Appl. Ecol.*, **39**, 572–583.

48 Hanski, I. (1998) Metapopulation dynamics. *Nature*, **396**, 41–48.

49 Clarke, B. and Murray, J. (1969) Ecological genetics and speciation in land snails of the genus *Partula*. *Biol. J. Linn. Soc.*, **1**, 31–42.

50 Charlesworth, B. (1994) *Evolution in Age-structured Populations*, 2nd edn, Cambridge University Press, Cambridge.

51 Lewontin, R.C. (1974) *The Genetic Basis of Evolutionary Change*, Columbia University Press, New York).

52 Kimura, M. (1979) The neutral theory of molecular evolution. *Sci. Am.*, **241** (5), 98–126.

53 Zhong, Y-F., Butts, T. and Holland, P.W.H. (2008) HomeoDB: a database of homeobox gene diversity. *Evolution & Development*, **10**, 516–518.

54 Saccheri, I.J. *et al.* (2008) Selection and gene flow on a diminishing cline of melanic peppered moths. *Proc. Natl. Acad. Sci. USA*, **105**, 16212–16217.

55 True, J.R. (2003) Insect melanism: the molecules matter. *Trends in Ecology and Evolution*, **18**, 640–647.

56 Hooper, J. (2002) *Of Moths and Men: Intrigue, Tragedy and the Peppered Moth*, Fourth Estate, London.

References for Part II

1 Jeffreys, A.J., Wilson, V. and Thein, S.L. (1985) Hypervariable 'minisatellite' regions in human DNA. *Nature*, **314**, 67–73.
2 Burke, T. and Bruford, M.W. (1987) DNA fingerprinting in birds. *Nature*, **327**, 149–152.
3 Butlin, R.K., Collins, P.M. and Day, T.H. (1984) The effect of larval density on an inversion polymorphism in the seaweed fly *Coelopa frigida*. *Heredity*, **52**, 415–423.
4 Goldschmidt, R. (1952) Homeotic mutants and evolution. *Acta Biotheoretica*, **10**, 87–104.
5 Averof, M. and Akam, M. (1995) *Hox* genes and the diversification of insect and crustacean body plans. *Nature*, **376**, 420–423.
6 Hintz, M. *et al.* (2006) Catching a 'hopeful monster': shepherd's purse (*Capsella bursa-pastoris*) as a model system to study the evolution of flower development. *J. Exp. Botany*, **57**, 3531–3542.
7 Mendel, G. (1866) Versuche über Pflanzenhybriden. *Verhandlungen des naturforschenden Vereines in Brünn, Bd. IV für das Jahr 1865, Abhandlungen*, **3**, 3–47
8 Horton, W., Hall, J. and Hecht, J. (2007) Achondroplasia. *The Lancet*. **370**, 162–172.
9 Mackay, T.F.C. (2001) Quantitative trait loci in *Drosophila*. *Nature Reviews Genetics*, **2**, 11–20.
10 Huxley, J.S. (1932) *Problems of Relative Growth*, Methuen, London.
11 Thompson, D.A.W. (1917) *On Growth and Form*, Cambridge University Press, Cambridge.
12 Mezey, J.G., Cheverud, J.M. and Wagner, G.P. (2000) Is the genotype-phenotype map modular? A statistical approach using mouse quantitative trait loci data. *Genetics*, **156**, 305–311.
13 Fisher, R.A. (1930) *The Genetical Theory of Natural Selection*, Clarendon Press, Oxford.
14 Whiting, M.F. and Wheeler, W.C. (1994) Insect homeotic transformation. *Nature*, **368**, 696.
15 Minelli, M. (2009) Phylo-evo-devo: combining phylogenetics with evolutionary developmental biology. *BMC Biology*, **7**, 36.
16 De Beer, G.R. (1940) *Embryos and Ancestors*, Clarendon Press, Oxford.
17 Gould, S.J. (1977) *Ontogeny and Phylogeny*, Harvard University Press, Cambridge, MA.
18 McKinney, M.L. and McNamara, K.J. (1991) *Heterochrony: The Evolution of Ontogeny*, Plenum Press, New York.
19 Andersson, G. (1990) *Proc. 7th Internat. Cong. Myriapodology* (ed. A. Minelli).
20 Ingham, P.W. (1988) The molecular genetics of embryonic pattern formation in *Drosophila*. *Nature*, **335**, 25–34.
21 Hughes, C.L. and Kaufman, T.C. (2002) Exploring the myriapod body plan: expression of the ten Hox genes in a centipede. *Development*, **129**, 1225–1238.
22 Kettle, C. *et al.* (2003) The pattern of segment formation, as revealed by *engrailed* expression, in a centipede with a variable number of segments. *Evol. & Dev.*, **5**, 198–207.
23 Stollewerk, A., Schoppmeier, M. and Damen, W.G.M. (2003) Involvement of *Notch* and *Delta* genes in spider segmentation. *Nature*, **423**, 863–865.
24 Chipman, A.D. and Akam, M. (2008) The segmentation cascade in the centipede *Strigamia maritima*: involvement of the Notch pathway and pair-rule homologues. *Developmental Biology*, **319**, 160–169.
25 Kim, J., Kerr, J.Q. and Min, G-S. (2000) Molecular heterochrony in the early development of *Drosophila*. *Proc. Natl. Acad. Sci. USA*, **97**, 212–216.
26 Carleton, K.L. *et al.* (2008) Visual sensitivities tuned by heterochronic shifts in opsin gene expression. *BMC Biology*, **6**, 22.
27 Kimura, M. (1983) *The Neutral Theory of Molecular Evolution*, Cambridge University Press, Cambridge.

28 Nei, M. (2007) The new mutation theory of phenotypic evolution. *Proc. Natl. Acad. Sci. USA*, **104**, 12235–12242.
29 Friedman, W.E. (2008) Hydatellaceae are water lilies with gymnospermous tendencies. *Nature*, **453**, 94–97.
30 Boorman, C.J. and Shimeld, S.M. (2002) The evolution of left-right asymmetry in chordates. *BioEssays*, **24**, 1004–1011.
31 Tagawa, M. and Aritaki, M. (2005) Production of symmetrical flatfish by controlling the timing of thyroid hormone treatment in spotted halibut *Verasper variegatus*. *Gen. Comp. Endocrinol.*, **141**, 184–189.
32 Saele, O., Silva, N. and Pittman, K. (2006) Post-embryonic remodelling of neurocranial elements: a comparative study of normal versus abnormal eye migration in a flatfish, the Atlantic halibut. *Journal of Anatomy*, **209**, 31–41.
33 Friedman, M. (2008) The evolutionary origin of flatfish asymmetry. *Nature*, **454**, 209–212.
34 Boycott, A.E. and Diver, C. (1923) On the inheritance of sinistrality in *Limnaea peregra*. *Proc. R. Soc. Lond., B* **95**, 207–213.
35 Sturtevant, A.H. (1923) Inheritance of direction of coiling in *Limnaea*. *Science*, **68**, 269–270.
36 Grande, C. and Patel, N.H. (2009) Nodal signalling is involved in left-right asymmetry in snails. *Nature*, **457**, 1007–1011.
37 Vonk, F.J. *et al.* (2008) Evolutionary origin and development of snake fangs. *Nature*, **454**, 630–633.
38 Sommer, R.J. and Sternberg, P.W. (1994) Changes of induction and competence during the evolution of vulva development in nematodes. *Science*, **265**, 114–118.
39 Kiontke, K. *et al.* (2007) Trends, stasis, and drift in the evolution of nematode vulva development. *Current Biology*, **17**, 1925–1937.
40 Geoffroy Saint-Hilaire, E. (1822) Considérations générales sur la vertèbre. *Mem.Mus.Hist. Nat.*, **9**, 89–119.
41 Holley, S.A. *et al.* (1995) A conserved system for dorsal-ventral patterning in insects and vertebrates involving *sog* and *chordin*. *Nature*, **376**, 249–253.
42 Lowe, C.J. *et al.* (2006), Dorsoventral patterning in hemichordates: insights into early chordate evolution. *PLoS Biology*, **4**, e291.
43 Gerhart, J. (2000) Inversion of the chordate body axis: are there alternatives? *Proc. Natl. Acad. Sci. USA*, **97**, 4445–4448.
44 He, C. and Saedler, H. (2005) Heterotopic expression of *MPF2* is the key to the evolution of the Chinese lantern of *Physalis*, a morphological novelty in Solanaceae. *Proc. Natl. Acad. Sci. USA*, **102**, 5779–5784.
45 Jernvall, J., Keranen, S.V.E. and Thesleff, I. (2000) Evolutionary modification of development in mammalian teeth: quantifying gene expression patterns and topography. *Proc. Natl. Acad. Sci. USA*, **97**, 14444–14448.
46 Arthur, W. (2000) The concept of developmental reprogramming and the quest for an inclusive theory of evolutionary mechanisms. *Evol. & Dev.*, **2**, 49–57.
47 Wang, Y-Q. and Su, B. (2004) Molecular evolution of *microcephalin*, a gene determining human brain size. *Human Molecular Genetics*, **13**, 1131–1137.
48 Ali, F. and Meier, R. (2008) Positive selection in *ASPM* is correlated with cerebral cortex evolution across primates but not with whole brain size. *Molecular Biology and Evolution*, **25**, 2247–2250.
49 Ponce de Leon, M.S. *et al.* (2008) Neanderthal brain size at birth provides insights into the evolution of human life history. *Proc. Natl. Acad. Sci. USA*, **105**, 13764–13768).

50 Gibert, J-M., Mouchel-Vielh, E., Queinnec, E. and Deutsch, J.S. (2000) Barnacle duplicate *engrailed* genes: divergent expression patterns and evidence for a vestigial abdomen. *Evolution & Development*, **2**, 194–202.

51 Briggs, D.E.G., Sutton, M.D., Siveter, David J. and Siveter, Derek J. (2005) Metamorphosis in a Silurian barnacle. *Proc. Roy. Soc. Lond. B*, **272**, 2365–2369.

52 Gould, S.J. and Vrba, E.S. (1982) Exaptation - a missing term in the science of form. *Paleobiology*, **8**, 4–15.

53 Brown, P. *et al.* (2004) A new small-bodied hominin from the Late Pleistocene of Flores, Indonesia. *Nature*, **431**, 1055–1061.

54 Beldade, P., Koops, K. and Brakefield, P.M. (2002) Developmental constraints versus flexibility in morphological evolution. *Nature*, **416**, 844–847.

55 Beldade, P., Brakefield, P.M. and Long, A.D. (2002) Contribution of *Distal-less* to quantitative variation in butterfly eyespots. *Nature*, **415**, 315–318.

56 Stevens, M. (2005) The role of eyespots as anti-predator mechanisms, principally demonstrated in the Lepidoptera. *Biological Reviews*, **80**, 573–588.

57 Jeffery, W.R. (2008) Emerging model systems in evo-devo: cavefish and microevolution of development. *Evolution & Development*, **10**, 265–272.

58 Jeffery, W.R. (2005) Adaptive evolution of eye degeneration in the Mexican blind cavefish. *Journal of Heredity*, **96**, 185–196.

59 Yang, T. *et al.* (2005) Transcriptional regulation network of cold-responsive genes in higher plants. *Plant Science*, **169**, 987–995.

60 Chinnusamy, V., Zhu, J. and Zhu, J-K. (2007) Cold stress regulation of gene expression in plants. *Trends in Plant Science*, **12**, 444–451.

61 Xiong, L., Lee, H., Huang, R. and Zhu, J-K. (2004) A single amino acid substitution in the *Arabidopsis* FIERY1/HOS2 protein confers cold signaling specificity and lithium tolerance. *The Plant Journal*, **40**, 536–545.

62 Cain, A.J. and Sheppard, P.M. (1950) Selection in the polymorphic land snail *Cepaea nemoralis* (L.). *Heredity*, **4**, 275–294.

63 Jones, J.S., Leith, B.H. and Rawlings, P. (1977) Polymorphism in *Cepaea*: a problem with too many solutions? *Annual Review of Ecology and Systematics*, **8**, 109–143.

64 Almond, J.E. (1985) The Silurian-Devonian fossil record of the Myriapoda. *Phil. Trans. Roy. Soc. Lond. B*, **309**, 227–237.

65 Shear, W.A. and Bonamo, P.M. (1988) Devonobiomorpha, a new order of centipeds (Chilopoda) from the Middle Devonian of Gilboa, New York State, USA, and the phylogeny of centiped orders. *Amer. Mus. Novit.*, **2927**, 1–30.

66 Munoz-Chapuli, R. *et al.* (2005) The origin of the endothelial cell: an evo-devo approach to the invertebrate/vertebrate transition of the circulatory system. *Evolution & Development*, **7**, 351–358.

67 Yoshida, M., Shigeno, S., Tsuneki, K. and Furuya, H. (2010) Squid vascular growth factor receptor: a shared molecular signature in the convergent evolution of closed circulatory systems. *Evol. & Dev.*, **12**, 25–33.

68 Grosskurth, S.E., Bhattacharya, D., Wang, Q. and Lin, J. J-K. (2008) Emergence of Xin demarcates a key innovation in heart evolution. *PloS ONE*, **3**, e2857.

69 Hamada, H., Meno, C., Watanabe, D. and Saijoh, Y. (2002) Establishment of vertebrate left-right asymmetry. *Nature Reviews Genetics*, **3**, 103–113.

70 Parkin, C.A. and Ingham, P.W. (2008) The adventures of Sonic Hedgehog in development and repair. I. Hedgehog signaling in gastrointestinal development and repair. *AJP: Gastrointest. Liver Physiol.*, **296**, G363–G367.

71 Dessaud, E. *et al.* (2007) Interpretation of the sonic hedgehog morphogen gradient by a temporal adaptation mechanism. *Nature*, **450**, 717–720.
72 Luo, Z-X. (2007) Transformation and diversification in early mammal evolution. *Nature*, **450**, 1011–1019.
73 McGregor, A.P. *et al.* (2007) Morphological evolution through multiple cis-regulatory mutations at a single gene. *Nature*, **448**, 587–590.
74 Payre, F. (2004) Genetic control of epidermis differentiation in *Drosophila*. *Int. J. Dev. Biol.*, **48**, 207–215.
75 Hennig, W. (1966) *Phylogenetic Systematics*, University of Illinois Press, Urbana, IL.
76 Medawar, P.B. and Medawar, J.S. (1977) *The Life Science: Current Ideas of Biology*, Wildwood House, London.
77 Rosen, D. (1984) *Evolutionary Theory: Paths into the Future* (ed. J.W. Pollard), John Wiley & Sons, Ltd, Chichester.
78 Owen, R. (1848) *On the Archetype and Homologies of the Vertebrate Skeleton*, John van Voorst, London; Panchen, A.L. (1994) *Homology: The Hierarchical Basis of Comparative Biology* (ed. B.K. Hall), Academic Press, London & New York.
79 Dunn, C.W.. *et al.*, (2008) Broad phylogenomic sampling improves resolution of the animal tree of life. *Nature*, **452**, 745–749.
80 Clark, R.B. (1964) *Dynamics in Metazoan Evolution: The Origin of the Coelom and Segments*, Clarendon Press, Oxford.
81 Aguinaldo, A.M.A. *et al.* (1997) Evidence for a clade of nematodes, arthropods and other moulting animals. *Nature*, **387**, 489–493.
82 Niedzwiedzki, G. *et al.* (2010) Tetrapod trackways from the early Middle Devonian period of Poland. *Nature*, **463**, 43–48.

References for Part III

1 Sato, A. *et al.* (1999) Phylogeny of Darwin's finches as revealed by mtDNA sequences. *PNAS*, **96**, 5101–5106.
2 Abzhanov, A. *et al.* (2004) *Bmp4* and morpological variation of beaks in Darwin's finches. *Science*, **305**, 1462–1465.
3 Abzhanov, A. *et al.* (2006) The calmodulin pathway and evolution of elongated beak morphology in Darwin's finches. *Nature*, **442**, 563–567.
4 Wright, S. (1931) Evolution in Mendelian populations. *Genetics*, **16**, 97–159.
5 Whyte, L.L. (1960) Developmental selection of mutations. *Science*, **132**, 954.
6 Whyte, L.L. (1965) *Internal Factors in Evolution*, Tavistock Publications, London.
7 Klingenberg, C.P. and Leamy, L.J. (2001) Quantitative genetics of geometric shape in the mouse mandible. *Evolution*, **55**, 2342–2352.
8 Simpson, G.G. (1953) *The Major Features of Evolution*, Columbia University Press, New York.
9 Gould, S.J. and Vrba, E.S. (1982) Exaptation – a missing term in the science of form. *Paleobiology*, **8**, 4–15.
10 Hu, D., Hou, L., Zhang, L. and Xu, X. (2009) A pre-*Archaeopteryx* troodontid theropod from China with long feathers on the metatarsus. *Nature*, **461**, 640–643.
11 Maynard Smith, J. *et al.* (1985) Developmental constraints and evolution. *Q. Rev. Biol.*, **60**, 265–287.
12 Galis, F. *et al.* (2006) Extreme selection in humans against homeotic transformations of cervical vertebrae. *Evolution*, **60**, 2643–2654.
13 Gould, S. J. (1977) *Patterns of Evolution* (ed. A. Hallam), Elsevier, Amsterdam.
14 Antonovics, J. and van Tienderen, P.H. (1991) Ontoecogenophyloconstraints? The chaos of constraint terminology. *Trends Ecol. Evol.*, **6**, 166–168.
15 Gould, S.J. and Lewontin, R.C. (1979) The spandrels of San Marco and the Panglossian paradigm: a critique of the adaptationist programme. *Proc. R. Soc. Lond. B.*, **205**, 581–598.
16 Gould, S J. (1989) A developmental constraint in *Cerion*, with comments on the definition and interpretation of constraint in evolution. *Evolution*, **43**, 516–539.
17 Arthur, W. (2001) Developmental drive: an important determinant of the direction of phenotypic evolution. *Evol. & Dev.*, **3**, 271–278.
18 Gould, S.J. (1983) *Hen's Teeth and Horse's Toes: Further Reflections in Natural History*, Norton, New York.
19 Darwin, C. (1859) *On the Origin of Species by Means of Natural Selection, or the Preservation of Favoured Races in the Struggle for Life*, J. Murray, London.
20 Wallace, A.R. (1897) *Darwinism: An Exposition of the Theory of Natural Selection, with some of its Applications*, Macmillan, London.
21 Cheverud, J.M. (1984) Quantitative genetics and developmental constraints on evolution by selection. *J. Theor. Biol.*, **110**, 155–171.
22 Beldade, P., Koops, K. and Brakefield, P.M. (2002) Developmental constraints versus flexibility in morphological evolution. *Nature*, **416**, 844–847.
23 Allen, C.E., Beldade, P., Zwaan, B.J. and Brakefield, P.M. (2008) Differences in the selection response of serially repeated color pattern characters: Standing variation, development, and evolution. *BMC Evolutionary Biology*, **8**, doi:10.1186/1471.
24 Minelli, A. and Bortoletto, S. (1988) Myriapod metamerism and arthropod segmentation. *Biol. J. Linn. Soc.*, **33**, 323–343.

25 Chipman, A.D., Arthur, W. and Akam, M. (2004) A double segment periodicity underlies segment generation in centipede development. *Current Biology*, **14**, 1250–1255.
26 Goodwin, B. (1994) *How the Leopard Changed its Spots: The Evolution of Complexity*, Weidenfeld & Nicolson, London.
27 Li, C. *et al.* (2008) An ancestral turtle from the late Triassic of southwestern China. *Nature*, **456**, 497–501.
28 Nagashima, H. *et al.* (2009) Evolution of the turtle body plan by the folding and creation of new muscle connections. *Science*, **325**, 193–196.
29 Kirschner, M. and Gerhart, J. (1998) Evolvability. *Proc. Natl. Acad. Sci. USA*, **95**, 8420–8427.
30 Hendrikse, J.L., Parsons, T.E. and Hallgrimsson, B. (2007) Evolvability as the proper focus of evolutionary developmental biology. *Evol. & Dev.*, **9**, 393–401.
31 Roth, S., Stein, D. and Nusslein-Volhard, C. (1989) A gradient of nuclear localization of the Dorsal protein determines dorso-ventral polarity in th *Drosophila* embryo. *Cell*, **59**, 1189–1202.
32 Holley, S.A. *et al.* (1995) A conserved system for dorsal-ventral patterning in insects and vertebrates involving *sog* and *chordin*. *Nature*, **376**, 249–253.
33 Geoffroy Saint-Hilaire, E. (1822) Considérations générales sur la vertèbre. *Mem.Mus.Hist. Nat.*, **9**, 89–119.
34 Chow, R.L., Altmann, C.R., Lang, R.A. and Hemmati-Brivanlou, A. (1999) Pax6 induces ectopic eyes in a vertebrate. *Development*, **126**, 4213–4222.
35 Bateson, W. (1894) *Materials for the Study of Variation, Treated with Especial Regard to Discontinuity in the Origin of Species*, Macmillan, London.
36 Lewis, E.B. (1963) Genes and developmental pathways. *Am. Zool.*, **3**, 33–56.
37 Duboule, D. (1994) *The Evolution of Developmental Mechanisms (Development 1994 Supplement)* eds M. Akam et al. Company of Biologists, Cambridge, UK.
38 Duboule, D. (2007) The rise and fall of Hox gene clusters. *Development*, **134**, 2549–2560.
39 Mendel, G. (1866) Versuche über Pflanzenhybriden. *Verhandlungen des naturforschenden Vereines in Brünn, Bd. IV für das Jahr 1865, Abhandlungen*, **4**, 3–47.
40 De Vries, H. (1910) *The Mutation Theory: Experiments and Observations on the Origin of Species in the Vegetable Kingdom*, Kegan Paul, Trench, Trübner & Co., London.
41 Wray, G. (2007) The evolutionary significance of *cis*-regulatory mutations. *Nature Reviews Genetics*, **8**, 206–216.
42 Shubin, N., Tabin, C. and Carroll, S. (2009) Deep homology and the origins of evolutionary novelty. *Nature*, **457**, 818–823.
43 Jacobs, D.K. *et al.* (2000) Molluscan *engrailed* expression, serial organization, and shell evolution. *Evol. & Dev.*, **2**, 340–347.
44 Catmull, J. *et al.* (1998) *Pax-6* origins - implications from the structure of two coral *Pax* genes. *Dev. Genes. Evol.*, **208**, 352–356.
45 Holland, L.Z. and Holland, N.D. (2001) Evolution of neural crest and placodes: amphioxus as a model for the ancestral vertebrate? *J. Anat.*, **199**, 85–98.
46 Donoghue, P.C.J., Graham, A. and Kelsh, R.N. (2008) The origin and evolution of the neural crest. *BioEssays*, **30**, 530–541.
47 Costa, M.M.R. *et al.* (2005) Evolution of regulatory interactions controlling floral asymmetry. *Development*, **132**, 5093–5101.
48 King, N. *et al.* (2008) The genome of the choanoflagellate *Monosiga brevicollis* and the origin of metazoans. *Nature*, **451**, 783–788.
49 Bakker, K. (1961) An analysis of factors which determine success in competition for food among larvae of *Drosophila melanogaster*. *Arch. Neerl. Zool.*, **14**, 200–281.

50 Braendle, C., Davis, G.K., Brisson, J.A. and Stern, D.L. (2006) Wing dimorphism in aphids. *Heredity*, **97**, 192–199.
51 Lewis, J.G.E. (1961) The life history and ecology of the littoral centipede *Strigamia (Scolioplanes) maritima* (Leach). *Proc. Zool. Soc. Lond.*, **137**, 221–248.
52 Kettle, C. and Arthur, W. (2000) Latitudinal cline in segment number in an arthropod species, *Strigamia maritima*. *Proc. Roy. Soc. Lond. B*, **267**, 1393–1397.
53 Vedel, V., Chipman, A.D., Akam, M. and Arthur, W. (2009) Temperature-dependent plasticity of segment number in an arthropod species: the centipede *Strigamia maritima*. *Evol. & Dev.*, **10**, 487–492.
54 Bell, D.L. and Galloway, L.F. (2008) Population differentiation for plasticity to light in an annual herb: adaptation and cost. *Am. J. Bot.*, **95**, 59–65.
55 Bouvet, J-M., Vigneron, P. and Saya, A. (2005) Phenotypic plasticity of growth trajectory and ontogenetic allometry in response to density for *Eucalyptus* hybrid clones and families. *Annals of Botany*, **96**, 811–821.
56 Sumner, S., Pereboom, J.J.M. and Jordan, W.C. (2006) Differential gene expression and phenotypic plasticity in behavioural castes of the primitively eusocial wasp, *Polistes canadensis*. *Proc. R. Soc. Lond. B*, **273**, 19–26.
57 Gibert, J-M., Peronnet, F. and Schlotterer, C. (2007) Phenotypic plasticity in *Drosophila* pigmentation caused by temperature sensitivity of a chromatin regulator network. *PLoS Genetics*, **3**, doi:10.1371/journal.pgen.0030030.
58 Waddington, C.H. (1956) Genetic assimilation of the bithorax phenotype. *Evolution*, **10**, 1–13.
59 Waddington, C.H. (1953) The genetic assimilation of an acquired character. *Evolution*, **7**, 118–126.
60 Gould, S.J., Young, N.D. and Kasson, B. (1985) The consequences of being different: sinistral coiling in *Cerion*. *Evolution*, **39**, 1364–1379.
61 Suzuki, Y. and Nijhout, H.F. (2006) Evolution of a polyphenism by genetic accommodation. *Science*, **311**, 650–652.
62 Stappenbeck, T.S., Hooper, L.V. and Gordon, J.I. (2002) Developmental regulation of intestinal angiogenesis by indigenous microbes via Paneth cells. *PNAS*, **99**, 15451–15455.
63 Leroi, A.M. (2000) The scale independence of evolution. *Evol. & Dev.* **2**, 67–77.
64 Erwin, D.H. (2000) Macroevolution is more than repeated rounds of microevolution. *Evol. & Dev.*, **2**, 78–84.
65 Prakash, S. (1972) Origin of reproductive isolation in the absence of apparent genic differentiation in a geographic isolate of *Drosophila pseudoobscura*. *Genetics*, **72**, 143–155.
66 Stanley, S.M. (1975) A theory of evolution above the species level. *Proc. Nat. Acad. Sci. USA*, **72**, 646–650.
67 Eldredge, N. and Gould, S.J. (1972) *Models in Paleobiology* (ed. T.J.M. Schopf), Freeman, San Francisco.
68 Wray, G.A., Levinton, J.S. and Shapiro, L.H. (1996) Molecular evidence for deep pre-Cambrian divergences among metazoan phyla. *Science*, **274**, 568–573.
69 Seilacher, A. (1989) Vendozoa: organismic construction in the Proterozoic biosphere. *Lethaia*, **22**, 229–239.
70 Yin, L. *et al.* (2007) Doushantuo embryos preserved inside diapause egg cysts. *Nature*, **446**, 661–663.
71 Bailey, J.V. *et al.* (2007) Evidence of giant sulphur bacteria in Neoproterozoic phosphorites. *Nature*, **445**, 198–201.

72 Haeckel, E. (1896) *The Evolution of Man : A Popular Exposition of the Principal Points of Human Ontogeny and Phylogeny*, Appleton, New York.

73 Dawkins, R. (1986) *The Blind Watchmaker*, Longman, London.

74 Furuya, H. and Tsuneki, K. (2003) Biology of dicyemid mesozoans. *Zool. Science*, **20**, 519–532.

75 Hanelt, B. *et al.* (1996) The phylogenetic position of *Rhopalura ophiocomae* (Orthonectida) based on 18S ribosomal DNA sequence analysis. *Mol. Biol. Evol.*, **13**, 1187–1191.

76 Poulin, R. (1995) Evolution of parasite life history traits: myths and reality. *Parasitology Today*, **11**, 342–345.

77 Valentine, J.W., Collins, A.G. and Meyer, C.P. (1994) Morphological complexity increase in metazoans. *Paleobiology*, **20**, 131–142.

78 Kristensen, N.P. (1981) Phylogeny of insect orders. *Ann. Rev. Entomol.*, **26**, 135–157.

79 Sempere, L.F., Cole, C.N., McPeek, M.A. and Peterson, K.J. (2006) The phylogenetic distribution of metazoan microRNAs: insights into evolutionary complexity and constraint. *Mol. Dev. Evol.*, **306B**, 575–588.

80 Ohno, S. (1970) *Evolution by Gene Duplication*, Springer-Verlag, New York.

References for Part IV

1 Watson, J.D. and Crick, F.H.C. (1953) Molecular structure of nucleic acids: a structure for deoxyribose nucleic acid. *Nature*, **171**, 737–738.

2 Duboule, D.(1994) *The Evolution of Developmental Mechanisms (Development 1994 Supplement)*, eds M. Akam et al. Company of Biologists, Cambridge, UK.

3 Sander, K. (1983) *Development and Evolution* (eds B.C. Goodwin, H. Holder and C.C. Wyllie), Cambridge University Press, Cambridge, UK, pp. 137–159.

4 Raff, R.A. (1996) *The Shape of Life: Genes, Development and the Evolution of Animal Form*, Chicago University Press, Chicago, IL.

5 Panchen, A.L. (1994) *Homology: The Hierarchical Basis of Comparative Biology* (ed. B.K. Hall), Academic Press, London & New York.

6 Arthur, W. (2003) *Keywords and Concepts in Evolutionary Developmental Biology* (eds B. Hall and W. Olson), Harvard University Press, Cambridge, MA.

7 Allen, C.E., Beldade, P., Zwaan, B.J. and Brakefield, P.M. (2008) Differences in the selection response of serially repeated color pattern characters: Standing variation, development, and evolution. *BMC Evolutionary Biology*, **8**, doi:10.1186/1471.

8 Boutros, M. and Ahringer, J. (2008) The art and design of genetic screens: RNA interference. *Nat. Rev. Gen.*, **9**, 554–566.

9 Von Dassow, G., Meir, E., Munro, E.M. and Odell, G.M. (2000)The segment polarity network is a robust developmental module. *Nature*, **406**, 188–192.

10 Abouheif, E. *et al.* (1997) Homology and developmental genes. *Trends in Genetics*, **13**, 432–433.

11 Gould, S.J. and Lewontin, R.C. (1979) The spandrels of San Marco and the Panglossian paradigm: a critique of the adaptationist programme. *Proceedings of the Royal Society of London B.*, **205**, 581–598.

12 Hughes, N.C., Minelli, A. and Fusco, G. (2006) The ontogeny of trilobite segmentation: a comparative approach. *Paleobiology*, **32**, 602–627.

13 Cheverud, J.M. (1984) Quantitative genetics and developmental constraints on evolution by selection. *Journal of Theoretical Biology*, **110**, 155–171.

14 Goldschmidt, R. (1940) *The Material Basis of Evolution*, Yale University Press, New Haven, CT.

References for Appendix 1

1 Mendel, G. (1866) Versuche über Pflanzenhybriden. *Verhandlungen des naturforschenden Vereines in Brünn, Bd. IV für das Jahr 1865, Abhandlungen*, **4**, 3–47.
2 Fisher, R.A. (1930) *The Genetical Theory of Natural Selection*, Clarendon Press, Oxford.
3 Dobzhansky, T. (1943) Genetics of natural populations. IX. Temporal changes in the composition of populations of *Drosophila pseudoobscura*. *Genetics*, **28**, 162–186.
4 Mayr, E. (1942) *Systematics and the Origin of Species*, Columbia University Press, New York.
5 Lewontin, R.C. (1974) *The Genetic Basis of Evolutionary Change*, Columbia University Press, New York.
6 Darwin, C. (1859) *On the Origin of Species by Means of Natural Selection, or the Preservation of Favoured Races in the Struggle for Life*, J. Murray, London.
7 Goldschmidt, R. (1940) *The Material Basis of Evolution*, Yale University Press, New Haven.
8 Raff, R.A. and Kaufman, T.C. (1983) *Embryos, Genes and Evolution: The Developmental Genetic Basis of Evolutionary Change*, Macmillan, New York.
9 Haldane, J.B.S. (1932) *The Causes of Evolution*. (Longman, London).
10 Maynard Smith, J. *et al.* (1985) Developmental constraints and evolution. *Q. Rev. Biol.*, **60**, 265–287.
11 Dawkins, R. (1976) *The Selfish Gene*, Oxford University Press, Oxford.
12 Gould, S.J. and Lewontin, R.C. (1979) The spandrels of San Marco and the Panglossian paradigm: a critique of the adaptationist programme. *Proc. R. Soc. Lond. B.*, **205**, 581–598.
13 Pigliucci, M. and Kaplan, J. (2006) *Making Sense of Evolution: The Conceptual Foundations of Evolutionary Biology*, Chicago University Press, Chicago, IL

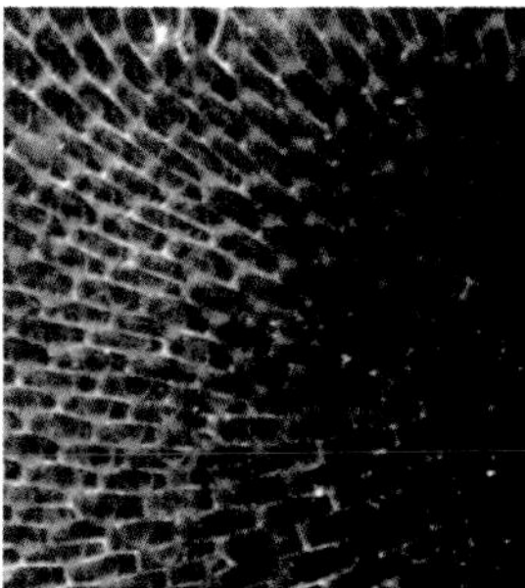

Index